Texturen in Forschung und Praxis

Textures in Research and Practice

Proceedings
of the International Symposium
Clausthal-Zellerfeld, October 2-5, 1968

Editors
J. Grewen · G. Wassermann

Springer-Verlag Berlin Heidelberg GmbH 1969

JOHANNA GREWEN
Institut für Metallkunde und Metallphysik der
Technischen Universität Clausthal

GÜNTER WASSERMANN
Institut für Metallkunde und Metallphysik der
Technischen Universität Clausthal

ISBN 978-3-662-13127-5   ISBN 978-3-662-13125-1 (eBook)
DOI 10.1007/978-3-662-13125-1

© by Springer-Verlag Berlin Heidelberg 1969
Ursprünglich erschienen bei Springer-Verlag Berlin Heidelberg 1969
Softcover reprint of the hardcover 1st edition

Library of Congress Catalog Card Number: 70-82425
Title-No. 1597

VORWORT

Die Bedeutung der Texturen hat in den letzten Jahren erheblich zugenommen. Die mit der Verformung verbundenen Änderungen der Kristallorientierungen lassen sich heute auch bei Vielkristallen weitgehend deuten, bieten aber gegenüber der Verformung der Einkristalle eine Reihe zusätzlicher Probleme. Noch mehr gilt dies für die Vorgänge der Erholung und Rekristallisation als Grundlage der Rekristallisationstexturen. Heterogene Bestandteile beeinflussen die Anisotropie der Eigenschaften, da sie Texturbildung hervorrufen oder vermindern können, doch sind gerade hier noch viele Fragen offen.

Auch in der Praxis der Metallverarbeitung und der Fertigung von Bauteilen zeigt sich mehr und mehr, daß eine wirklich werkstoffgerechte Herstellung von Halbzeug sowie die Weiterverarbeitung, insbesondere durch Umformung, ohne Kenntnis der Texturen und ihrer Wirkungen nicht mehr auskommen kann. Die bewußte Erzeugung bestimmter Texturen gewinnt bei althergebrachten Materialien (Stahlbleche) und neuen Metallen (Titan, Zirkon) zunehmend an Interesse.

Es schien daher wünschenswert, das Thema Texturen in einem besonderen Symposium zu behandeln, das vom Institut für Metallkunde und Metallphysik der Technischen Universität Clausthal vom 2.-5. Oktober 1968 in Clausthal-Zellerfeld veranstaltet wurde. Die Vorträge umfassen ein weites Gebiet der Wissenschaft und Technik, in dem Texturen eine Rolle spielen. Lediglich die magnetischen Eigenschaften wurden nicht berücksichtigt, um das Programm nicht zu umfangreich werden zu lassen.

Um den Preis dieses Buches möglichst niedrig zu halten, war es erforderlich, die meisten Vorträge zu kürzen. Wir sind den Autoren dankbar, daß sie unseren Kürzungsvorschlägen zugestimmt haben. Zur Platzersparnis mußte auch auf eine

IV

Wiedergabe der umfangreichen und lebhaften Diskussionen, auf deren Berücksichtigung schon bei der Programmgestaltung besonderer Wert gelegt worden war, sowie auf schriftliche Diskussionsbemerkungen verzichtet werden.

Der Druck des Buches wurde vor allem durch die Unterstützung des Vereins von Freunden der Technischen Universität Clausthal ermöglicht, dem wir für seine Hilfe herzlich danken möchten. Ferner danken wir auch an dieser Stelle allen, die zum Gelingen des Symposiums beigetragen haben.

Clausthal-Zellerfeld im März 1969   Die Herausgeber

V

PREFACE

The importance of textures has grown considerably in recent years. The changes in crystal orientation associated with deformation can now largely be explained, although poly-crystalline materials offer many additional problems compared with deformation of single crystals. This is even more true of recovery and recrystallisation which form the basic processes of annealing textures. Heterogeneous components influence the anisotropy of the properties by promoting or inhibiting texture formation, and here there are still many unanswered questions.

Furthermore it is becoming increasingly clear that in metal forming operations and in the design of engineering structures knowledges of textures and their effects become more and more essential. The disgned development of textures, i.e. texture tailoring, looks increasingly interesting both for traditional materials, such as steel sheet, and for new metals as titanium or zirconium.

For these reasons, it was thought desirable that there should be a symposium devoted to textures, this was organized by the Institut für Metallkunde und Metallphysik der Technischen Universität Clausthal, and took place on 2-5 October, 1968, in Clausthal-Zellerfeld. The papers cover a wide range of scientific and practical aspects of textures. Magnetic properties were however, excluded in order to keep the program within bounds.

To keep the price of this book reasonable, it was necessary to shorten most of the papers. We are grateful to the authors who readily accepted our proposals. To save space, both oral and written discussion had to be omitted.

We heartily thank the Verein von Freunden der Technischen Universität Clausthal. Its support made it possible to publish this book. We also thank all those who contributed to the symposium.

Clausthal-Zellerfeld, March 1969   The Editors

| Die Planung des Symposiums und Auswahl der Vorträge besorgte ein wissenschaftliches Komitee, dem folgende Herren angehörten: | The Symposium was organized and the papers selected by a Scientific Committee composed of: |

P.A.Beck, University of Illinois, Urbana, USA
W.G.Burgers, Technische Hogeschool Delft, Delft, Holland
E.Schmid, Universität Wien und Österreichische Akademie der
          Wissenschaften, Wien, Österreich
G.Wassermann, Technische Universität Clausthal, Clausthal-
          Zellerfeld, Deutschland

Das örtliche Komitee bildeten: | Local Committee:

H.Ahlborn, Technische Universität Clausthal, (Secretary)
J.Grewen, Technische Universität Clausthal
E.Hornbogen, Universität Göttingen, Göttingen, Deutschland
H.Hu, at the time of the Symposium visiting scientist of the
      Institut für Metallkunde und Metallphysik der Tech-
      nischen Universität Clausthal

| Die Durchführung des Symposiums wurde ermöglicht durch Zuwendungen von: | The Symposium was supported by grants from: |

Bundesministerium für wissenschaftliche Forschung, Bonn
Kultusministerium des Landes Niedersachsen, Hannover
Verein von Freunden der Technischen Universität Clausthal
Verein Deutscher Eisenhüttenleute, Düsseldorf
Wirtschaftsvereinigung Nichteisen-Metalle, Düsseldorf

INHALTSVERZEICHNIS

CONTENTS

1 TEXTURBESTIMMUNG UND -BESCHREIBUNG [+)]
TEXTURE DETERMINATION AND DESCRIPTION

1.1 Verfahren zur Texturbestimmung
Texture Determination

von F.Haeßner [++)]

Abstract

At first a definition of the term "texture" is given.
Methods for texture determination will be classified into
those giving pole densities and those giving orientation
distributions. Furthermore, these methods will be discussed
in regard to the information they give. Methods arranged in
this manner will be examined in relation to their practical
availability and will be critically compared. The most
important method, the determination of textures with x-rays
and counters, is surveyed in detail. Possible future
developments will be discussed.

1.1.1 Einführung

Über die Verfahren zur Texturbestimmung ist wiederholt zusammenfassend berichtet worden [1-4]. Die anläßlich eines Symposiums über "Methoden der Texturbestimmung" 1967 in Berlin
gehaltenen Vorträge wurden in der Zeitschrift "Kristall und
Technik" veröffentlicht [5-7].

Angesichts dieses Sachverhaltes wird in dem vorliegenden Beitrag darauf verzichtet, im Detail erneut auf die vielen möglichen Verfahren einzugehen. Im Vordergrund der Betrachtungen
soll vielmehr die Frage stehen:"Welche Verfahren sind für gegebene Problemstellungen am zweckmäßigsten anzuwenden".

[+)]Der Symposiums-Beitrag von J. Grewen, D. Sauer und H.-P.
Wahl: Quantitative Texture Determination without a Random
Sample erscheint in Scripta Met.3(1969). Er behandelt die
quantitative Texturbestimmung mit Hilfe des Rechenautomaten.

[++)]Max-Planck-Institut für Metallforschung, Institut für
Sondermetalle, Stuttgart, Bundesrepublik Deutschland

Die Verfahren selbst sollen danach unterteilt werden, welchen
Informationsgehalt sie zu liefern in der Lage sind.

Der Kreis derer, die an Texturuntersuchungen interessiert
sind, kann auf Grund ganz verschiedener Aufgabenstellungen
grob in drei Gruppen unterteilt werden. Dementsprechend unterschiedlich sind die Anforderungen an das Untersuchungsverfahren: So benötigt man beispielsweise in der Produktion, bei
der Qualitätskontrolle und in Prüfstellen oft ein rasches und
einfaches Verfahren zum Feststellen, ob eine bestimmte, er-
wünschte bzw. unerwünschte Textur vorliegt oder nicht (Gruppe "A"). In Entwicklungs- und Forschungslaboratorien möchte
man dagegen häufig Texturen bzw. Texturdetails in einer an-
gemessenen Zeitspanne sehr exakt bestimmen (Gruppe "B").
Schließlich kann es, meist bei ausgesprochen wissenschaftli-
chen Problemen, darauf ankommen, möglichst viel Information
über eine Textur zu erhalten. Der zeitliche und apparative
Aufwand spielt dabei eine untergeordnete Rolle (Gruppe "C").

1.1.2 Allgemeine Gesichtspunkte

Um eine sinnvolle Einteilung der einzelnen Meßverfahren vor-
nehmen zu können, ist es zweckmäßig, zuvor auf drei Fragen-
komplexe einzugehen: Was will man messen? (Problemstellung)
– Wie genau will man messen? (Auflösungsvermögen) – Wie will
man das Resultat mitteilen? (Darstellung).

1.1.2.1 Problemstellung, Bedeutung des Begriffes "Texturen"

Unter der Textur eines vielkristallinen Werkstoffes versteht
man die Gesamtheit der Orientierungen seiner Kristalle. Diese
Definition beinhaltet drei in ihren Konsequenzen für die Texturbestimmung wichtige Aussagen:

1. Die Textur ist eine statistische Größe.

Das heißt, die mit dem Problem einer repräsentativen Stichpro-
benentnahme zusammenhängenden Fragen müssen bei einer Messung
berücksichtigt werden. Die Einhaltung der durch die Statistik
geforderten Grenzen ist dann besonders wichtig (und meist
schwierig), wenn man Aussagen über Texturinhomogenitäten machen will oder wenn aus kleinen Abweichungen in der Feinstruk-
tur von Texturen Folgerungen gezogen werden sollen.

2. Die Textur ist eine Häufigkeitsverteilung, die von drei
   Orientierungsparametern abhängt.

Wenn man die Orientierung eines Kristalliten in einer Probe
vollständig beschreiben will, benötigt man drei Orientierungsparameter, beispielsweise die drei Eulerschen Winkel
φ_1, ϕ und φ_2 zwischen dem Koordinatensystem des Kristalliten
und dem Bezugssystem der Probe. In einem vielkristallinen
Werkstoff besitzt jeder Kristallit seinen eigenen Satz von
Orientierungsparametern. Es ist im allgemeinen nicht möglich
und auch nicht zweckmäßig, diese Werte für alle Kristalle des
Werkstoffes einzeln anzugeben. Es genügt, wenn man die relative Häufigkeit $\frac{dV}{V}$ (V: Gesamtvolumen der Probe) angeben kann,
mit der Kristalle der verschiedenen Orientierungen vorkommen.
Definiert man die dreidimensionale Orientierungsverteilungsfunktion $f(\varphi_1, \phi, \varphi_2)$ folgendermaßen

$$\frac{dV}{V} = f(\varphi_1, \phi, \varphi_2) \frac{\sin \phi}{8 \cdot \pi^2} \, d\varphi_1 \, d\varphi_2 = f(g)dg , \qquad (1)$$

so charakteresiert die Funktion f eindeutig die relative Häufigkeit aller in der Probe vorkommenden Kristallorientierungen. Aus der Orientierungsverteilungsfunktion lassen sich alle anderen, weniger umfassenderen Beschreibungsarten für die
Textur, wie zum Beispiel Flächenpolfiguren, gewinnen [8].

Bislang gibt es keine einfache Methode, um die Orientierungsverteilungsfunktion zu ermitteln. Bei allen praktisch bedeut-
samen Meßverfahren begnügt man sich daher mit der Bestimmung
von Größen, die sich aus der Orientierungsverteilungsfunktion
durch Mittelwertbildung ergeben.

3. Die Gestalt sowie die gegenseitige Anordnung der Kristallite werden durch den Texturbegriff nicht erfaßt.

Zwei Proben identischer Textur können somit, zumindest theoretisch, ganz verschiedene Kristallitmorphologien aufweisen.
Eine Texturbeschreibung sollte daher ergänzt werden durch
eine Angabe von Kristallitgestalt und Anordnung. Zur Erfassung dieses Komplexes dürften statistische Funktionen, die sogenannten Korrelationsfunktionen, besonders geeignet sein [9].
Eine gewisse Kenntnis über die Kristallitgestalt und Anordnung
wird im übrigen vor einer Texturmessung sowieso benötigt, um
die Probe zweckmäßig auswählen zu können.

1.1.2.2 Auflösungsvermögen von Meßverfahren

Die bei einer Texturmessung interessierende Größe, die Orientierungsverteilungsfunktion, läßt sich mathematisch in eine Reihe entwickeln nach Funktionen, die der Kristallsymmetrie und der Probensymmetrie genügen [10-12]. Je mehr Koeffizienten dieser Reihenentwicklung mit einem Meßverfahren bestimmt werden können, umso größer ist offenbar das Auflösungsvermögen des Verfahrens. Als quantitatives Maß für das Auflösungsvermögen kann man den Grad der möglichen Reihenentwicklung benutzen [13].

Die Verfahren zur Texturbestimmung lassen sich in zwei Gruppen einteilen. Bei den Methoden der ersten Gruppe, den sogenannten d i r e k t e n V e r f a h r e n, bestimmt man für eine möglichst große Anzahl von Kristalliten die Orientierungsparameter und ermittelt daraus die Orientierungsverteilungsfunktion. Diese Verfahren sind sehr mühsam. Sie setzen eine Meßsonde voraus, die kleiner ist als die Kristallitgröße.

Bei den Verfahren der zweiten Gruppe, den i n d i r e k t e n V e r f a h r e n, mißt man die Richtungsabhängigkeit des Mittelwertes einer Eigenschaft des texturbehafteten Werkstoffes. Falls die entsprechende Eigenschaft für den Einkristall bekannt ist und die Wechselwirkung zwischen den einzelnen Kristalliten sowie der Korngrenzeneinfluß erfaßt werden können, läßt sich die Textur daraus ableiten [8]. In der Praxis werden wegen ihrer Einfachheit fast ausschließlich die Verfahren der zweiten Gruppe benutzt.

Nach dem oben angegebenen Kriterium für das Auflösungsvermögen haben die Verfahren der ersten Gruppe ein unendlich großes Auflösungsvermögen, sie liefern also die höchstmögliche Information über die Textur. Das Auflösungsvermögen der Verfahren der zweiten Gruppe wird demgegenüber begrenzt durch einen von zwei Faktoren [13]: Der eine Faktor ist gegeben durch die Stufe des Tensors, durch den die zur Messung herangezogene Eigenschaft dargestellt werden kann. Der zweite Faktor ist gegeben durch die Anzahl der unabhängig gemessenen Mittelwerte und die Gittersymmetrie des Werkstoffes. Danach haben beispielsweise alle Meßverfahren, die auf elastischen Anisotropiemessungen bzw. magnetischen Anisotropiemessungen beruhen, bei Werkstoffen kubischer Symmetrie ein Auflösungs-

Tab.1. Maximal erreichbares Auflösungsvermögen bei Verwendung von 3 verschiedenen Flächenpolfiguren für Gitter unterschiedlicher Symmetrie (nach BUNGE)

Gittersymmetrie	Auflösungsvermögen
triklin S_2	1
monoklin C_{2h}	2
orthorhombisch D_{2h}	4
tetragonal D_{4h}	10
hexagonal D_{6h}	16
kubisch O_h	34

vermögen von 4 bzw. 6. Das heißt, die Orientierungsverteilungsfunktion kann bestenfalls bis zum 4ten bzw. 6ten Koeffizienten, also nur sehr unvollkommen, bestimmt werden.

Von den indirekten Meßverfahren haben diejenigen das größte

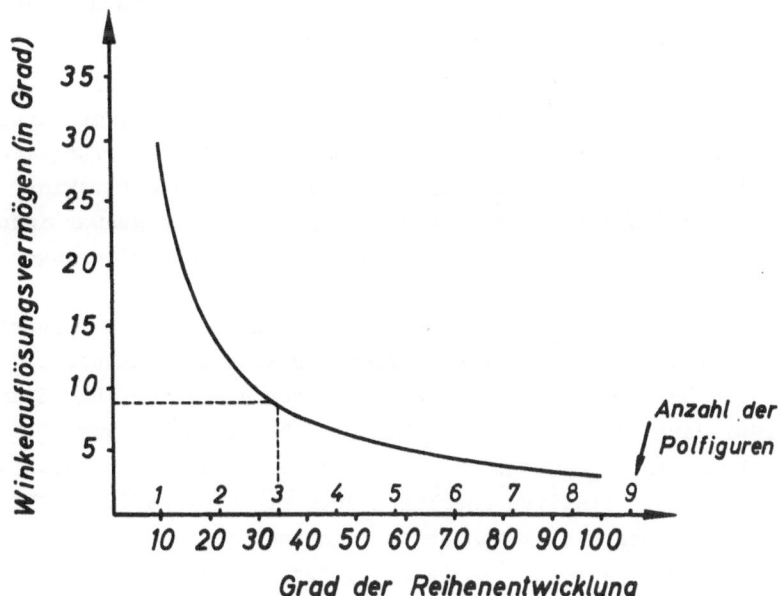

Abb.1. Maximal erreichbares Winkelauflösungsvermögen für Werkstoffe kubischer Symmetrie in Abhängigkeit von der Anzahl der gemessenen Flächenpolfiguren (nach BUNGE [8])

Auflösungsvermögen, bei denen mit Hilfe von Beugungsphänome-
nen Flächenpolfiguren ermittelt werden. In Tab.1 ist angegeben, welches Auflösungsvermögen durch Messung von drei verschiedenen Flächenpolfiguren für Werkstoffe unterschiedlicher
Symmetrie erreicht wird. Für die bei metallischen Werkstoffen
wichtigen Gitter kubischer und hexagonaler Symmetrie ist das
Auflösungsvermögen besonders hoch. Im Falle der kubischen
Symmetrie steigt das Auflösungsvermögen außerdem sehr stark
mit wachsender Zahl n der gemessenen Flächenpolfiguren, nämlich gemäß (12n - 2). Diese Zunahme des Auflösungsvermögens
bedeutet, daß immer mehr Texturdetails unterschieden werden
können. Ein anschauliches Maß hierfür ist das Winkelauflösungsvermögen. In Abb.1 ist der Verlauf dieser Größe für den
Fall kubischer Symmetrie in Abhängigkeit von der Anzahl der
gemessenen Flächenpolfiguren wiedergegeben.

1.1.2.3 Darstellung von Texturen

Die Art der zweckmäßigsten Darstellung von Texturen hängt
einerseits davon ab, welche Information über die Textur vorliegt, und andererseits davon, welche Aussagen hauptsächlich
benötigt werden. Aus Gründen der Anschaulichkeit wird man immer bestrebt sein, die Textur graphisch wiederzugeben.

Am schwierigsten ist die übersichtliche Darstellung der dreidimensionalen Orientierungsverteilungsfunktion. Hierfür liegen bislang drei verschiedene Vorschläge vor [14-16]. Eine
kritische Beurteilung dieser Vorschläge ist im gegenwärtigen
Zeitpunkt verfrüht, da noch zu wenig praktische Erfahrungen
vorliegen.
Sind von der Orientierungsverteilungsfunktion nur Mittelwerte
bekannt, die von zwei Orientierungsparametern abhängen, so
ist die Darstellung wesentlich einfacher als zuvor. Die Stan-
dardwiedergabe ist in diesem Fall die "Flächenpolfigur" [17].
Sie kann direkt gemessen werden und liefert einen anschaulichen und quantitativen Überblick über die Häufigkeit der Flächenpole einer bestimmten Netzebenenschar im Bezugssystem der
Probe (z.B. {111}-Pole im Bezugssystem von Walzrichtung (WR)
und Querrichtung (QR) eines Bleches). Eine zweite Darstell-
lungsart sind die "inversen Polfiguren" [18]. In ihnen wird
die Häufigkeit einer bestimmten Probenrichtung im Koordina-

tensystem des Kristalls wiedergegeben (z.B. die Richtung der
Blechnormale (NR) im Orientierungsdreieck). Obwohl die inversen Polfiguren rein mathematisch weniger Information enthalten als die Flächenpolfiguren, lassen sich in ihnen charakteristische Merkmale einer Textur oft besonders deutlich erkennen. Der Hauptnachteil der inversen Polfiguren besteht darin,
daß sie aus den üblichen Meßdaten nur mit Hilfe umständlicher
Auswerteverfahren erhalten werden können [11,12,14].

Weitere Darstellungsarbeiten, wie die Angabe von "idealen Lagen" [19], von "Faserachsen" [19] oder eines Kristallausrich-
tungsgrades bzw. Orientierungsindex sind recht unvollkommen.
Sie haben oft nur den Charakter prägnanter Kurzbeschreibungen
der Textur.
Selbstverständlich kann man die Textur auch völlig unanschaulich durch Angabe der Koeffizienten der Reihenentwicklung der
Orientierungsverteilungsfunktion wiedergeben [15]. Für alle
die Probleme, die mit dem Einfluß der Textur auf Materialei-
genschaften zusammenhängen, ist diese mathematische Art der
Texturdarstellung sicher die adäquateste.

1.1.3 Untersuchungsverfahren

Grundsätzlich kann man zur Ermittlung von Texturen alle Verfahren verwenden, die die innere Anisotropie des Werkstückes
wiederzugeben vermögen. Hinsichtlich ihres Informationsgehal-
tes und ihrer praktischen Durchführbarkeit unterscheiden sich
die Verfahren stark. Im folgenden sollen die wesentlichen
Merkmale der einzelnen Verfahren bzw. Verfahrensgruppen und
ihre Anwendungsbereiche aufgeführt werden.

1.1.3.1 Direkte Untersuchungsverfahren

Zur Ermittlung der Orientierung einzelner Kristallite hat man
lichtoptische, röntgenographische und elektronenmikroskopische Verfahren herangezogen. Um über die Textur genügend ge-
naue Angaben machen zu können, steht man vor der Notwendig-
keit, eine große Zahl einzelner Orientierungsmessungen durchführen zu müssen. Dementsprechend zeitraubend und mühsam sind
diese Verfahren. Sie werden daher auch nur dann angewandt,
wenn es keine andere Untersuchungsmöglichkeit gibt oder wenn
es auf den hohen Informationsgehalt ankommt, den die direkten

Verfahren liefern (Gruppe "C"). Die Untersuchungstechnik erfordert im Falle der lichtoptischen und röntgenographischen
Methoden relativ große, einheitlich orientierte Materialbereiche von mindestens 10 /um Durchmesser. Für einen sehr
feinkörnigen oder stärker verformten Werkstoff eignet sich
nur die elektronenmikroskopische Methode.

1.1.3.1.1 Lichtoptische Verfahren. Bei diesen Verfahren nützt
man zur Orientierungsbestimmung solche Erscheinungen auf der
Probenfläche aus, bei denen es sich um rein orientierungsbe-
dingte Phänomene handelt. Lichtoptisch sichtbare[+] Spuren auf
der Oberfläche, deren kristallographische Natur bekannt sein
muß, lassen sich sowohl durch physikalische als auch durch
chemische Maßnahmen erzeugen. Zur ersten Art von Erscheinungen gehören beispielsweise die Spuren von kohärenten Zwil-
lingsgrenzen mit der Oberfläche oder die Gleitlinien auf Oberflächen. Zur zweiten Art gehören die Ätzgruben auf Oberflächen. Die höchste Genauigkeit in der Orientierungsbestimmung
läßt sich bei diesen Verfahren entweder durch Ausmessen einzelner Ätzgruben in einem Zweikreisgoniometer [22] oder durch
Reflexion von konvergentem Licht an vielen Ätzgruben, sogenannte Lichtfiguren [23], erzielen. Die Genauigkeit hängt im
übrigen entscheidend von der Güte der Oberflächenspuren, d.h.
von der mitunter zeitraubenden Präparationstechnik ab. Der
apparative Aufwand für die lichtoptischen Verfahren ist klein.
Da sich andererseits Oberflächenspuren nicht immer erzeugen
lassen, ist die Untersuchungstechnik nicht allgemein anwendbar. Sie ist bisher auf grobkörnige, rekristallisierte Werkstoffe beschränkt geblieben [22,24-27].
Von dieser Verfahrensgruppe gibt es viele Varianten [28].

1.1.3.1.2 Röntgenographische Verfahren. Die Orientierung der
einzelnen Kristallite wird hier mit Hilfe von Lauaufnahmen
ermittelt. Je nach Probenbeschaffenheit kann in Transmissions-
oder Reflexionstechnik gearbeitet werden. Verglichen mit den
lichtoptischen Verfahren ist für diese Untersuchungstechnik
der apparative Aufwand groß, insbesondere wenn man vorgegebe-

---

[+] Prinzipiell kann zum Ausmessen auch das Elektronenmikroskop
herangezogen werden [21]. Der Nutzen steht jedoch in keinem
Verhältnis zum Aufwand.

ne Kristallite erfassen will. Demgegenüber ist die Probenpräparation viel einfacher. Im Prinzip könen mit diesem Verfahren a l l e Werkstoffe untersucht werden, sofern die Korngröße nicht zu klein ist. Zur Texturbestimmung ist die Methode nur selten herangezogen worden [29].

1.1.3.1.3 Elektronenmikroskopische Verfahren. Mit Hilfe der
elektronenmikroskopischen Feinbereichsbeugung ist es möglich,
Orientierungen zu bestimmen. Da der Bündelquerschnitt der
Elektronen sehr klein gemacht und der Strahl genau lokalisiert
werden kann, lassen sich auf diese Weise Materialbereiche von
1 /um Ausdehnung und darunter gezielt untersuchen. Mit dieser
Technik ist es somit möglich, auch in verformten Werkstoffen
Aussagen über Orientierungen zu machen.

Der apparative und präparative Aufwand - es müssen elektronenmikroskopisch gut durchstrahlbare Folien hergestellt werden -
ist bei dieser Untersuchungsmethode erheblich. Dafür liefert
das Verfahren aber auch Aussagen, die auf andere Weise nicht
gewonnen werden können [30]. Die Methode läßt sich grundsätzlich auf alle Werkstoffe anwenden. Sie ist in jüngster Zeit
zur Texturbestimmung an kalt gewalztem Kupfer, Niob und 70-30
Messing herangezogen worden [30-35]. Im Falle des Kupfers wurden die Daten von 512 einzelnen Orientierungsbestimmungen be-
nützt, um zum ersten Mal die oben erwähnte Orientierungsverteilungsfunktion d i r e k t zu bestimmen [15].

Die wichtigsten Merkmale der direkten Verfahren sind in Tab.2
schematisch gegenübergestellt.

Tab.2. Die wichtigsten Merkmale der direkten Untersuchungsverfahren

Ausge-nutztes Phänomen	Typ. Vertre-ter	Beurteilung			Begrenzg. durch	An-wen-dung
		präpar. appar. Aufwand		prakt.Durch-führung		
Lichtre-flexion	Ätzgru-ben	mäßig	klein	langwierig, viel Übung, unsicher	Werkstoff-zustand	spezielle Probleme
Röntgen-beugung	Laue-aufnah-me	klein	mäßig	langwierig	feines Korn	
Elektro-nenbeu-gung	Feinbe-reichs-beugung	groß	groß	sehr lang-wierig, viel Übung		

1.1.3.2 Indirekte Untersuchungsverfahren

Es gibt eine ganze Reihe richtungsabhängiger Eigenschaften,
die für eine Texturbestimmung herangezogen werden können.
Bei diesen Verfahren werden im allgemeinen immer mehrere
Kristallite gleichzeitig erfaßt. Dadurch geht die Zuordnung
des Meßresultates zu einem bestimmten Korn verloren. Wenn
die Untersuchungssonde sehr groß ist im Vergleich zum Kri-
stallitdurchmesser, so handelt es sich um ideal integrieren-
de Meßverfahren. Die Schlüsse, die man aus Mittelwertsmes-
sungen anisotoper Eigenschaften ziehen kann, reichen von
rein qualitativen Angaben bis zur quantitativen Texturbestimmung.

1.1.3.2.1 Qualitative und halbquantitative Verfahren. Diese
Verfahren sind im wesentlichen nur als Test geeignet auf bestimmte, meist bekannte Texturen. Ihre Bedeutung liegt darin,
daß man mit ihrer Hilfe recht oft in einfacher und rascher
Weise feststellen kann, ob sich ein Material für einen gegebenen Verwendungszweck eignet. Sie entsprechen somit den Anforderungen der Gruppe "A".
Ein häufig benutzes Verfahren besteht darin, durch Anätzen
des Werkstoffes bestimmte Kristallflächen freizulegen und
die Reflexion eines gerichteten Lichtstrahles zu verfolgen.
"Diese Methode des "maximalen Schimmers" ist sehr geeignet,
um sich rasch ein Urteil zu bilden, ob überhaupt eine gere-
gelte Kristallitorientierung vorhanden ist. Ihre Anwendung
ist auch dann zu empfehlen, wenn man die Textur àn sich kennt
und weiß, in welchen Lagen ungefähr Reflexion eintritt, und
nur prüfen will, ob z.B. eine bestimmte Glühbehandlung die
gewünschte Textur erzeugt hat" [36]. Das Verfahren ist oft
benutzt worden. Für stärker verformte Werkstoffe eignet es
sich allerdings nicht. Der apparative Aufwand ist minimal.
Man benötigt lediglich eine Lichtquelle, die paralleles Licht
gibt, und einen Drehtisch. Durch verfeinerte Ätzverfahren wie
die Kornfärbungsätzung und die Schraffurätzung lassen sich
für manche Werkstoffe die Aussagemöglichkeiten über die Textur erweitern [37]. Zur Auswertung wird dann ein Mikroskop
benötigt.
Die plastischen Eigenschaften von Metallen sind besonders
stark anisotrop. Dennoch kann man aus der Messung ihrer Rich-

tungsabhängigkeit im allgemeinen nur wenig Rückschlüsse auf
die Textur selbst ziehen. Das liegt in erster Linie daran,
daß die plastischen Eigenschaften extrem störungsempfindlich
sind. Infolgedessen hängt der Mittelwert einer plastischen
Eigenschaft in einem polykristallinen Werkstoff außer von der
Textur auch erheblich von den Korngrenzen und von der Morpho-
logie der Gesamtheit seiner Körner ab. Die beiden wichtigsten
Untersuchungsverfahren für Bleche, die auf der Messung plastischer Eigenschaften beruhen, der Näpfchentiefziehversuch [38]
und die Ermittlung von Breiten- und Dickenänderung im Zugversuch (R-Faktor-Bestimmung), sind deshalb für eine Texturbestimmung nur insoweit geeignet als man schon weiß, wie bestimmte Texturen diese Größen beeinflussen. Ihr Wert für die
Praxis besteht darin, daß sie Aussagen darüber liefern können,
wie sich ein texturbehaftetes Blech bei Umformprozessen verhalten wird [39]. Der Informationsgehalt der R-Faktor-Bestimmung über das anisotrope Werkstoffverhalten ist größer als
der des Näpfchentiefziehversuches, weil bei der R-Faktormes-
sung die Veränderungen in drei aufeinander senkrecht stehenden
Materialrichtungen ausgewertet werden, beim Tiefziehversuch
dagegen nur in zwei Richtungen.

Weitere auf der Anisotropie plastischer Eigenschaften beruhen-
de Untersuchungsverfahren sind die Reißlängenprüfung [40], das
Biegeverfahren [40] und die Mikrohärteprüfung. Während die ersten beiden Verfahren reine Prüfverfahren auf die Werkstoff-
anisotropie sind, ist es mit Hilfe von Mikrohärtemessungen gelungen, im Falle von Zircaloy-2 qualitative Polfiguren zu konstruieren [41]. Dieses Verfahren ist bei einfachen Texturen im
Prinzip auf alle hexagonalen Metalle übertragbar. Zur Festlegung einer noch unbekannten Textur ist diese Methode jedoch
nicht geeignet. In Anbetracht der erheblichen Vorarbeit, die
geleistet werden muß, bis man aus Mikrohärtemessungen Rückschlüsse auf die Textur ziehen kann, ist der Einsatz des Verfahrens zur Qualitätskontrolle nur bei großen Prüfzahlen und
dann anzuraten, wenn andere Untersuchungsmöglichkeiten, beispielsweise wegen der Probenform, ausscheiden.

Elastische Eigenschaften können stark anisotrop sein, sie hängen aber weniger empfindlich von Gitterfehlern ab als die
plastischen Eigenschaften [42,43]. Sie sollten sich daher eher

für eine Texturuntersuchung eignen. Das Auflösungsvermögen
und damit der Informationsgehalt, den die elastischen Verfahren über die Textur liefern können, ist jedoch niedrig, da
der Elastizitätstensor von niedriger Stufe ist. Ferner ist
die Messung der Richtungsabhängigkeit elastischer Eigenschaften recht umständlich. Die elastischen Verfahren haben
infolgedessen für die Texturbestimmung, außer zur Klärung
spezieller, wissenschaftlicher Probleme [44], keine praktische Bedeutung erlangt.

Verfahren, die auf der Messung magnetischer Eigenschaften beruhen, sind oft zur Überprüfung von Texturen ferromagnetischer Werkstoffe herangezogen worden [45]. Die Methode stellt
jedoch kein Verfahren dar zur Bestimmung unbekannter Texturen. Dazu ist einerseits das Auflösungsvermögen zu gering,
andererseits kann das Auftreten magnetischer Vorzugsrichtun-
gen auch von Faktoren abhängen, die nicht ausschließlich kristallographischer Natur sind. Die Bedeutung der magnetischen
Methoden beruht in erster Linie darauf, daß es mit Hilfe einfacher Meßverfahren (Drehscheibenversuch [46]) möglich ist,
die für die Praxis wichtigen Größen von Lage und Betrag der
magnetischen Anisotropie rasch festzustellen. Für die Produk-
tions- und Qualitätskontrolle magnetischer Werkstoffe ist
der Drehscheibenversuch sehr geeignet.

Die wichtigsten Merkmale dieser Verfahrensgruppe sind in
Tab.3 schematisch gegenübergestellt.

1.1.3.2.2 Quantitative Verfahren. Diese Verfahren beruhen auf
der Beugung von monochromatischen Röntgen-, Neutronen- oder
Elektronenstrahlen an vielen Kristalliten. Der für die Texturbestimmung folgenreichste Unterschied in den drei Strahlungsarten ist ihre Eindringtiefe. Sie ist in den meisten Metallen für die Elektronenstrahlen am kleinsten und für die
Neutronenstrahlen am größten. Mit Neutronenstrahlen lassen
sich somit auch große Probenvolumina untersuchen. Man erhält,
selbst in Fällen grobkörniger Werkstoffe, repräsentative Mittelwerte. Demgegenüber können mit Elektronenstrahlen nur ganz
dünne Schichten erfaßt werden. Elektronenstrahlen eignen sich
daher besonders zur Untersuchung von Texturinhomogenitäten,
beispielsweise an der Oberfläche von Werkstoffen.

Tab.3. Die wichtigsten Merkmale der indirekten, qualitativen
und halbquantitativen Untersuchungsverfahren

Ausgenutztes Phänomen	Typ. Vertreter	Beurteilung			Begrenzung durch	Anwendung
		präpar. appar. Aufwand		prakt. Durchführung		
Lichtreflexion	Dislozierte Reflexion	sehr klein	sehr klein	einfach, rasch	Werkstoffzustand	Überprüfg. auf Textur
Anisotropie plastischer Eigenschaften	R-Faktor Bestimmung	mäßig	mäßig	routinemäßig	Probenform	Test auf Umformverhalten, Qualitätskontrolle
	Härteeindruck	klein	klein	einfach, viel Vorarbeit	feines Korn, Gittertyp	Überprüfg. auf Textur, Qualitätskontrolle
Anisotropie magnetischer Eigenschaften	Drehscheibenversuch	klein	klein	einfach, rasch	Werkstoffart	Test auf magnetisches Verhalten, Qualitätskontrolle

Für die Texturbestimmung haben die röntgenographischen Diffraktometerverfahren mit Abstand die größte Bedeutung erlangt. Wegen ihrer großen Anpassungsfähigkeit sind sie fast immer einsetzbar. Sie entsprechen etwa den Anforderungen der Gruppe "B". Neutronentrahlen sind bisher nur selten für Texturmessungen verwandt worden. Da der apparative Aufwand erheblich und eine geeignete Strahlenquelle selten verfügbar ist, dürfte die Neutronenbeugung auch in Zukunft nur in Sonderfällen benutzt werden (Gruppe "C"). Elektronenstrahlen schließlich sind wiederholt zur Texturbestimmung eingesetzt worden [47]. Sie haben jedoch wegen ihrer geringen Eindringtiefe und wegen des erforderlichen experimentellen Aufwandes keine größere Bedeutung erlangt (Gruppe "C").

Das Untersuchungsprinzip der modernen Diffraktometerverfahren ist für alle drei Strahlungsarten gleich. Ausgangspunkt ist die Braggsche Reflexion eines monochromatischen Strahles SS' an einer vorgegebenen Netzebenenschar (hkl) der zu untersu-

chenden Probe (P) (Abb.2). Eine Beugung ist in all den Fällen
möglich, in denen die Netzebenennormale *w* auf dem Reflexions-
kegel R Liegt. Bei den Diffraktometerverfahren werden von dem
Beugungskegel B nur die in Richtung PL verlaufenden Strahlen
ausgeblendet und mit einem Detektor (D) gemessen. Diese Intensität kann denjenigen Netzebenen der Schar (hkl) zugeordnet werden, deren Normalen *w* (h,k,l) in Richtung PO zeigen.
Die Verfahren liefern also Aussagen über die Lage von Netz-
ebenennormalen (Flächenpole) im probeneigenen Koordinatensystem (Polfiguren)[+].

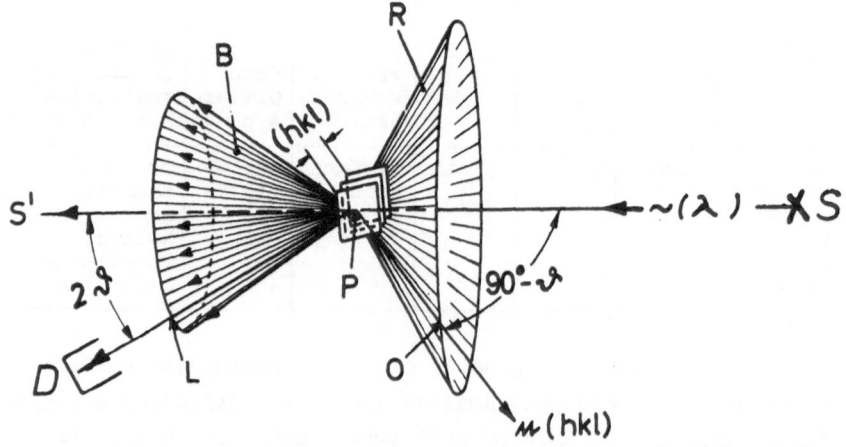

Abb.2. Zum Prinzip der Texturbestimmung bei Diffraktometerverfahren.

(Braggsche Reflexion an der Netzebenenschar (hkl); es gibt
n·λ = 2 d·sin θ; der Primärstrahl SS', der gebeugte Strahl
PL sowie die Netzebenennormale *w*(hkl) in Richtung PO liegen
in einer Ebene, der Meßebene des Goniometers)

Damit bei feststehender Anordnung von Primärstrahl und Detektor der reflektierte Strahl von allen Netzebenen der Schar
(hkl) in den Detektor gelangen kann, muss die Probe in viele

[+] Inverse Polfiguren lassen sich aus diesen Daten nur auf um-
ständliche Weise berechnen [11,12,14]. Ihre direkte Ermittlung ist, allerdings recht unvollkommen, bei Werkstoffen
geringer Gittersymmetrie (z.B. Uran [48]) oder bei Verwendung kurzwelliger Strahlung (z.B. MoK_α Strahlung [49],
AgK_α Strahlung [50]) möglich.

Positionen gebracht werden (Abtasten der gesamten Lagenkugel). Um dies zu bewerkstelligen, sind eine Reihe von Verfahren entwickelt worden, die sich durch die Gestalt der Unter-
suchungsproben und die Art der Probenbewegung unterscheiden [51].

Röntgenographische Goniometerverfahren. Ein brauchbares Untersuchungsverfahren sollte es erlauben, einen möglichst großen Teil der Lagenkugel so abzutasten, daß die vom Detektor
gemessene Intensität ohne Korrektur (Probenabsorption, variable Untergrundstreuung, Defokussierungseffekte) der Poldich-
te auf der Lagenkugel proportional ist. Die zu untersuchende
Probe sollte möglichst einfach geformt und leicht herstellbar
sein. Da die aufgezählten Anforderungen von keinem Verfahren
gleichzeitig erfüllbar sind, muß man stets einen Kompromiß
eingehen.

Wenn die gesamte Lagenkugel ohne Intensitätskorrektur mit
einem einzigen Verfahren abgetastet werden soll, dann kann
man kugelförmige Proben benützen, die in Reflexion untersucht
werden [52,53]. Dabei wird für jede Stelle in der Polfigur
über ein anderes, wegen der Probenkrümmung meist recht klei-
nes, Materialvolumen gemittelt. Die auf diese Weise erhaltenen Polfiguren sind verhältnismäßig genau. Dennoch muß man
mit Fehlern in der Größenordnung bis zu 10 % rechnen [54].

Streng betrachtet, benötigt man auch beim Kugelprobenverfah-
ren aus zwei Gründen Intensitätskorrekturen: Zunächst ist
die Höhe des Streuuntergrundes eine Funktion der Textur. Er
muß also für jeden Meßwert eigens berücksichtigt werden.
Durch Verwendung von Detektoren, welche die Streuintensität
in der Nähe der Bragg-Reflexion messen, lassen sich diese
Daten experimentell ohne weiteres gewinnen [55]. Ferner hängt
die reflektierte Intensität außer von der Textur auch von
der Anordnung und Art der Gitterfehler im Werkstoff ab (Ex-
tinktionseinflüsse). Eine exakte Erfassung dieser Störein-
flüsse ist praktisch nicht möglich. Da beide Effekte im allgemeinen klein sind, werden sie bei der Korrektur der Meßwerte meist nicht berücksichtigt. Es ist jedoch fraglich, ob
ein solches Vorgehen auch im Falle sehr scharfer Rekristallisationstexturen (hohe Intensitäten, wenig Gitterfehler)

gerechtfertigt ist.

Wenn man eine Korrektur der Meßwerte in Kauf nehmen will, so
läßt sich die gesamte Lagenkugel an einer planparallelen Probe geeigneter Dicke durch eine Kombination der SCHULZ'schen
Reflexionstechnik [56] mit der Durchstrahltechnik nach
DECKER, ASP und HAKKER [57] abtasten. Die Polverteilung kann
mit der Reflexionstechnik bis zu einem Winkel von etwa 70°
von der Probenebenennormale und mit der Durchstrahltechnik
bis zu etwa 40° von der Probenebene erfaßt werden. Durch eine
geeignete Anpassung der korrigierten Meßwerte im Überlappungs-
bereich der beiden Methoden läßt sich somit die gesamte Polfigur gewinnen. Wenn die gesamte Polfigur vorliegt, kann die
Übertragung von Intensitätsdaten in statistische Einheiten
rein rechnerisch vorgenommen werden [58]. Der kritische Punkt
dieses Verfahrens ist die Korrektur der Meßwerte und die Anpassung. Polfiguren, die auf diese Weise gewonnen wurden,
weisen daher meistens Intensitätsverfälschungen in der Grös-
senordnung von 10 bis 30 % auf [54].

Die kombinierte Rückstrahl-Durchstrahltechnik ist die Stan-
dardmethode, um Flächenpolfiguren von Blechen aufzunehmen[+).
Ihre Anwendung ist immer dann zu empfehlen, wenn von einem
Werkstück planparallele Proben hergestellt werden können. Die
Methode ist für alle Materialien geeignet, insofern sie nicht
zu grobkörnig sind. Da die Größe des Untersuchungsbereiches,
über den bei diesem Verfahren gemittelt wird, etwa um den
Faktor 100 variiert werden kann, lassen sich mit der gleichen
Anordnung auch Texturinhomogenitäten studieren. Geeignete Goniometer und Probenhalter, mit denen die notwendige Änderung
der Probenstellung automatisch vorgenommen werden kann, sind
kommerziell erhältlich. Beim Übergang von der Rückstrahl-
zur Durchstrahltechnik muß die Probe allerdings von Hand neu
eingespannt werden. Die vom Detektor[++) gemessene Intensität

---

[+) Andere Kombinationsmöglichkeiten [59] haben wegen kompli-
zierterer Intensitätskorrekturen keine weite Verbreitung
gefunden. Auch die Benutzung der Reflexion 1. und 2. Ordnung oder der Einsatz von zwei verschiedenen Wellenlängen
lohnt sich nicht.

[++) Als Detektor verwendet man am besten einen Szintillations-

(Fortsetzung s.S. 17)

wird im allgemeinen als Kurve, Zahlenkolonne oder Lochstrei-
fen registriert. Die weitere Auswertung erfolgt meist manu-
ell. Eine bedeutende Einsparung an Zeit und Arbeitskräften
läßt sich bei großen Probenserien durch den Einsatz vollau-
tomatischer Polfigurenschreiber erreichen. In diesen Geräten
werden die gemessenen Intensitäten korrigiert und von einer
automatischen Vorrichtung direkt als Polfigur gezeichnet. Die
Zeit für die Herstellung einer vollständigen Polfigur wird
damit auf etwa eine Stunde reduziert [60]. Von derartigen Geräten existieren bislang nur Prototypen. In jüngster Zeit
sind auch Rechenautomaten zum vollautomatischen Zeichnen von
Polfiguren eingesetzt worden [50]. Die reine Rechen- und Zei-
chenzeit der Computeranlage betrug 30 Minuten.

Wenn die Textur eines Bleches zu den Ebenen WR-NR und QR-NR
symmetrisch ist, kann bei Verwendung geeignet angeschnittener
Proben, die aus zusammengeklebten Einzelblechen bestehen, auf
eine Eichprobe ohne Vorzugsorientierungen verzichtet werden [53].

Häufig begnügt man sich damit, nur den Teil der Polfigur zu
bestimmen, der durch die Reflexionstechnik oder durch die
Durchstrahltechnik allein erfaßbar ist. Das gilt insbesondere für die Reflexionstechnik, da diese geringere Anforderungen an die Probenpräparation stellt[++) und nur verhältnismäßig kleine Intensitätskorrekturen erfordert. Zur Umrechnung
der Intensitätsdaten in statistische Einheiten benötigt man
in diesen Fällen allerdings eine Probe ohne Vorzugsorientierungen. Eine solche ist mitunter schwer zu beschaffen. Dieses abgekürzte Bestimmungsverfahren ist immer dann zu empfehlen, wenn die Polfigur als solche bekannt ist und nur gewisse
Details festgestellt werden sollen.

Für Texturbestimmungen an Drähten und Stangen sind die erwähnten Verfahren oft unzweckmäßig. Andere Untersuchungstech-
niken [61,62], auf die hier nicht näher eingegangen werden

(Fortsetzung Fußnote von S.16)
zähler. Dann läßt sich Strahlung unerwünschter Wellenlänge
(Streustrahlung; Eigenstrahlung einer radioaktiven Probe)
mit Hilfe eines Diskriminators wegfiltern.
[++)Die Proben müssen nur e i n e ebene Fläche aufweisen und
genügend dick sein.

soll, sind für diese Fälle meist besser geeignet.

Tab.4. Für die Texturbestimmung bemerkenswerte Unterschiede
bei der Benutzung von Neutronen- und Röntgenstrahlen (nach
KAJAMAA [63]

Eigenschaft	Neutronenbeugung	Röntgenbeugung
Strahlintensität	niedrig	hoch
Strahldurchmesser (relativ)	etwa 10^2	etwa 1
Strahlhomogenität	gut	ungewiß
Auflösungsvermögen	mäßig	gut
Absorption	niedrig (Ausnahmen für einige Elemente, wie beispielsweise Cd, seltene Erden)	hoch
Probenvolumen (relativ)	etwa 10^5	etwa 1
Probenpräparation	leicht	Folienherstellung
Messung schwerer Elemente	verhältnismäßig leicht	schwer
Begrenzung durch Korngrenze	kaum	ja
Extinktion	hoch	niedrig
Zusätzliche magnetische Streuung	ja	nein
Möglichkeit der Probenbehandlung (z.B. Glühen Verformen) zwischen den Messungen	ja	begrenzt

Neutronographische Goniometerverfahren. Die Benutzung von Neutronenstrahlen hat für die Texturbestimmung den Vorteil, daß infolge der geringen Absorption alle Bereiche der Lagenkugel am gleichen Probenvolumen gemessen werden können. Aus dem gleichen Grund ist die Zahl der erfaßten Kristallite um mehrere Größenordnungen höher als im Falle der Röntgenbeugung. Ferner kann die Probe mit einer größeren Genauigkeit hergestellt werden. In Tab.4 sind eine Reihe von Unterschieden

aufgeführt, die bei der Benutzung von Neutronenstrahlen im
Vergleich zu Röntgenstrahlen bemerkenswert sind.

Die große Eindringtiefe der Neutronenstrahlen hat zur Folge,
daß die Intensitätskorrekturen infolge Absorption verhältnismäßig genau berücksichtigt werden können. Inwieweit dieser
Vorteil durch Extinktionseffekte, die bei Neutronenstrahlen
hoch sind, wieder zunichte gemacht wird, ist im gegenwärtigen
Stadium schwer zu entscheiden. Polfiguren an gewalztem Kupfer, die mittels Neutronenbeugung im Rückstrahl- und Durch-
strahlverfahren bestimmt wurden, zeigen jedenfalls eine besonders hohe Genauigkeit [64].

Tab.5. Die wichtigsten Merkmale der indirekten, quantitativen
Untersuchungsverfahren

Ausge-nutztes Phänomen	Typ. Vertre-ter	Beurteilung			Begren-zung durch	Anwendung
		präpar. appar. Aufwand		prakt. Durch-rührung		
Röntgen-beugung	Gonio-meter-aufnah-me	klein	mäßig	einfach, routine-mäßig	grobes Korn	Quantitati-ve Textur-bestimmung, Standard-verfahren
Neutro-nenbeu-gung	Gonio-meter aufnah-me	klein	sehr groß	lang-wierig, viel Erfah-rung		Quantitati-ve Textur-bestimmung, spezielle Probleme
Elektro-nenbeu-gung	Gonio-meter-aufnah-me	groß	groß	schwie-rig, viel Er-fahrung	Proben-form, grobes Korn	Quantitati-ve Textur-bestimmung, spezielle Probleme

Neutronenstrahlen sind verhältnismäßig früh für Texturuntersuchungen an Aluminium [65], Nickeldrähten [66], einigen ku-
bisch-raumzentrierten Metallen [67] und Uranstäben [68] eingesetzt worden. Quantitative Texturbestimmungen an Blechen
sind jedoch erst in allerjüngster Zeit vorgenommen worden
[63,64]. Es ist kaum anzunehmen, daß die Neutronenbeugung
eine größere Bedeutung für die Texturbestimmung in der Praxis erlangen wird.
Die wichtigsten Merkmale dieser Verfahrensgruppe sind in
Tab.5 schematisch zusammengestellt.

1.1.4 Entwicklungstendenzen

Zwei Entwicklungstendenzen, die für zukünftige Texturuntersuchungen von erheblicher Bedeutung sein werden, zeichnen
sich gegenwärtig ab.

Die eine, mehr praktisch ausgerichtete, ist das Bestreben,
Texturen in immer kürzerer Zeit immer genauer zu messen. Der
Einsatz von automatischen Polfigurenschreibern und Rechenan-
lagen weist in diese Richtung. Dadurch liegt der zeitbestim-
mende Faktor heute oft nicht mehr in der Auswertung und Darstellung der Meßunterlagen sondern im Gewinnen der Meßwerte.
Eine wesentliche Reduzierung der Aufnahmezeit für eine Polfigur dürfte daher nur durch raschere Probenbewegung bei Verwendung höherer Primärstrahlintensitäten zu erreichen sein.
Da durch die Verwendung von Computern die gesamte Auswertung
sehr verfeinert werden kann, ist es ferner möglich, die einzelnen Meßwerte individuell zu korrigieren und die Übertragung der Daten in statistische Einheiten rein rechnerisch –
also ohne Eichproben – mit größerer Genauigkeit als bisher
vorzunehmen [55, 69].

Die zweite Entwicklungsrichtung ist mehr grundsätzlicher Art.
Sie betrifft die Berechnung der dreidimensionalen Orientierungsverteilungsfunktion aus den meßtechnisch leicht zugäng-
lichen Flächenpolfiguren. Die theoretischen und praktischen
Voraussetzungen dafür liegen heute vor [8,14,15,64]. Sofern
genügend genau gemessene Flächenpolfiguren von möglichst vielen Netzebenen vorliegen, läßt sich diese Berechnung mit Hilfe einer Rechenanlage ohne Schwierigkeit durchführen. Hiervon wird sicher in der Forschung in der nächsten Zeit immer
stärker Gebrauch gemacht werden.

Literatur

1. K.v. Gehlen: Beiträge Miner. u. Petrogr. 7 (1960) 340.

2. G. Wassermann und J. Grewen: Texturen metallischer Werkstoffe, Springer, Berlin 1962, S. 14.

3. I.L. Dillamore und W.T. Roberts: Met. Rev. 10 (1965) 271.

4. C.S. Barrett und T.B. Massalski: Structure of Metals. McGraw Hill, New York 1966, S. 193.

5. G. Wassermann: Kristall u. Technik 3 (1968) 255.

6. H.J. Bunge und J. Ehlert: Kristall u. Technik 3 (1968) 313.

7. F. Lihl: Kristall u. Technik 3 (1968) 271.

8. Siehe H.J. Bunge: Dieses Symposium, Kap. 1.2.

9. Siehe E. Kröner: Dieses Symposium, Kap. 5.2.

10. A.S. Wiglin: Fiz. Twordowo Tela 2 (1960) 2463.

11. H.J. Bunge: Mber. D. Akad. Wiss. Berlin 3(1961)97.

12. R.J. Roe: J. Appl. Phys. 36(1965)2024.

13. H.J. Bunge: Fiz. Metal. Metallov. 13 (1962) 512.

14. R.O. Williams: Trans. Met. Soc. AIME 242 (1968) 105 − s.a. Zitat 8.

15. H.J. Bunge und F. Haeßner: J. Appl. Phys. 39 (1968) − s.a. Zitat a.

16. G. Dorn, K. Lücke und G. Ibe: Verhandl. DPG (VI) 3 (1968) 102 − s.a. K. Lücke, dieses Symposium, Kap. 1.3.

17. Zitat 2) S. 18.

18. Zitat 2) S. 92.

19. Zitat 2) S. 86.

20. ASTM-Designation E 81-54 T (1954).

21. D.I. Lajner, E.I. Kruppnikova-Perrina und A.S. Baj: Trud. Gos. nauchno.-issled. i proekt inst. obrab. tsvet metal 20 (1961) 143.

22. C.S. Barrett und L.H. Levenson: Trans. AIME 137 (1940) 76.

23. H. Gengnagel und W. Schwab: Z. Metallkde. 57 (1966) 281.

24. C.S. Barrett und F.W. Steadman: Trans. AIME 147 (1942) 57.

25. J.S. Bowles und W. Boas: J. Inst. Metals 74 (1948) 501.

26. G.E.G. Tucker und P.C. Murphy: J. Inst. Metals 81 (1952/53) 235.

27. F. Assmuss, K. Detert und G. Ibe: Z. Metallkde. 48 (1957) 344.

28. Zitat 2) S. 104/112.

29. C.G. Dunn: J. Appl. Phys. 25 (1954) 233, 30 (1959) 850.

30. F. Haeßner, U. Jakubowski und M. Wilkens: phys. stat. sol. 7 (1964) 701.

31. K. Lücke, H. Perlwitz und W. Pitsch: phys. stat. sol. 7 (1964) 733.

32. S. Horiuchi und I. Gokyu: J. Jap. Inst. Metals 28 (1964) 555.

33. F. Haeßner, U. Jakubowski und M. Wilkens: Met. Sci. Eng. 1 (1966)30.

34. F. Haeßner und W. Hemminger: Z. Metallkde. 58 (1967) 104.

35. F. Haeßner und D. Keil: Z. Metallkde. 58 (1967) 220.

36. Zitat 2) S. 102/103.

37. Zitat 2) S. 104.

38. Zitat 2) S. 599.

39. Siehe Kap. 4.8 u. die entsprechenden Beiträge in Kap. 5 dieses Symposiums.

40. Zitat 2) S. 114.

41. P.L. Rittenhouse und M.L. Picklesimer: Trans. Met. Soc. AIME 236 (1966) 496.

42. Zitat 2) S. 559.

43. Siehe E. Schmid: Dieses Symposium, Kap. 5.1.

44. Y.C. Liu und G.A. Alers: Trans. Met. Soc. AIME 236 (1966) 489.

45. Zitat 2) S. 115.

46. O. Dahl und J. Pfaffenberger: Z. Phys. 71 (1931) 93.

47. Zitat 2) S. 99.

48. G.B. Harris: Phil. Mag. 43 (1952) 113.

49. H. Takechi, H. Kato und S. Nagashima: Trans. Met. Soc. AIME 242 (1968) 56.

50. A.J. Heckler, J.A. Elias und A.P. Woods: Trans. Met. Soc. AIME 239 (1967) 1241.

51. Zitat 2) S. 58.

52. L.K. Jetter und B.S. Borie: J. Appl.Phys. 24 (1953) 532.

53. F. Wever und H. Bötticher: Arch. Eisenhüttenw. 34 (1963) 205.

54. D. Schläfer: Kristall u. Technik 3 (1968).

55. G. R. Love: Trans. Met. Soc. AIME 242 (1968) 746.

56. L.G. Schulz: J. Appl. Phys. 20 (1949) 1030.

57. B.F. Decker, E.T. Asp und D. Harker: J. Appl. Phys. 19 (1948) 388.

58. J.Grewen, A Segmüller und G. Wassermann: Arch.Eisenhüttenw. 29 (1958) 115.

59. Zitat 2) S. 67.

60. K. Lücke, H.D. Mengelberg, R. Alan und G. Burmeister: J. Sci. Instr.: Veröffentlichung demnächst.

61. Zitat 2) S. 37.

62. H.J. Bunge: Exper. Technik der Physik X (1962) 338.

63. J. Kajamaa: Trans. Met. Soc. AIME 242 (1968) 973.

64. H.J. Bunge und J. Tobisch: Z. Metallkde. 59 (1968) 471.

65. A.H. Weber: Phys. Rev. 73 (1948) 1285.

66. B.N. Brockhouse: Can. J. Phys. 31 (1953) 353.

67. R.A. Swalin und A.H. Geisler: Trans. AIME 206 (1956) 1259.

68. J. Laniesse, M. Englander und P. Meriel: C.R. Acad. Sci., Paris 249 (1959) 2576.

69. J. Grewen, D. Sauer und H.-P. Wahl: Scripta Met. 3 (1969).

1.2 Texturbeschreibung durch dreidimensionale Polfiguren
Textures in Three-Dimensional Pole Figures

von H.J. Bunge [+)]

Abstract

In order to describe uniquely the orientation of a crystal
in a polycrystalline material, three parameters are required,
for example the three Euler angles. Therefore, the
orientation distribution or the texture is characterzised by
a function of three variables. Pole figures and inverse pole
figures are different two-dimensional projections of these
three-dimensional function. The three-dimensional
distribution function is not directly measurable, but it can
be calculated from the pole figure data. For this calculation
we know two methods. One of these is a successive
approximation, and the other a series expansion method. Some
examples will be given.
The series expansion method has the advantage that its
coefficients show a simple correlation with the mean values
of physical properties, which are orientation and texture
dependent. Therefore, these mean values can be calculated
in a simple manner. This will be shown for the E-modul and
the R-value of plastic anisotropy.

Um die Orientierung eines Kristalliten in einer vielkristallinen Probe vollständig beschreiben zu können, benötigt man
drei Parameter. Die Stellung einer Kristallrichtung im Raum
kann zunächst durch zwei Parameter - etwa sphärische Polarkoordinaten - festgelegt werden. Das Kriställchen besitzt dann
jedoch noch einen Freiheitsgrad einer Drehung um diese Richtung herum, der durch einen dritten Parameter festgelegt werden muß. Es gibt nun sehr viele verschiedene Möglichkeiten,
diese Orientierungsparameter zu wählen. Man kann auch mehr
als drei Parameter verwenden, die jedoch dann nicht alle von-

[+)] Deutsche Akademie der Wissenschaften zu Berlin, Institut
für Metallische Spezialwerkstoffe, Dresden, Deutsche
Demokratische Republik

einander unabhängig sein können (z.B. die Millerschen In-
dizes der mit Walz- bzw. Normalrichtung zusammenfallenden
Kristallrichtungen (hkl) [uvw]).

Im folgenden werden wir die Orientierung häufig durch die
drei Eulerschen Winkel charakterisieren. Ausgehend von einer
Nullage (der Würfellage) wird der Kristallit zunächst um
seine z-Achse um φ_1, dann um seine x-Achse um ϕ und noch einmal um seine z-Achse um φ_2 gedreht. Die so erreichte Orientierung ist dann durch die drei Eulerschen Winkel φ_1, ϕ, φ_2
charakterisiert (Abb. 1).

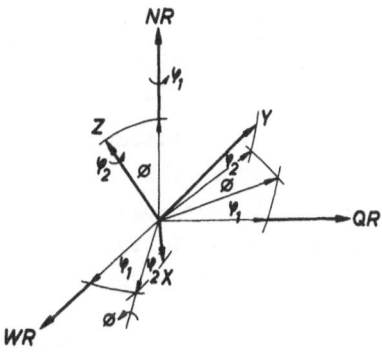

Abb. 1. Zur Definition der Eulerschen Winkel, WR =
Walzrichtung; NR = Normalrichtung; QR = Querrichtung

In einer vielkristallinen Probe hat nun jeder Kristallit
seine eigenen Orientierungsparameter. Wir interessieren uns
jedoch im allgemeinen nicht für den einzelnen Kristall, sondern nur für die relative Häufigkeit, mit der Kristalle der
verschiedenen Orientierungen auftreten. Sie wird durch eine
Funktion von drei Variablen, die dreidimensionale Orientierungsverteilungsfunktion oder auch dreidimensionale Polfigur beschrieben. Es sei dV das Volumen aller Kristalle der
Orientierung $(\varphi_1, \phi, \varphi_2)$ innerhalb des Orientierungsinter-
valles $(d\varphi_1, d\phi, d\varphi_2)$ und V das Gesamtvolumen der Probe.
Dann ist die Orientierungsverteilungsfunktion f $(\varphi_1 \phi \varphi_2)$
gegeben durch

$$\frac{dV}{V} = f\,(\varphi_1 \phi \varphi_2) \cdot \frac{\sin \phi}{8\pi^2}\, d\phi \; d\varphi_1 \; d\varphi_2 \qquad (1)$$

Sie ist in Vielfachen der regellosen Verteilung ausgedrückt.
Die Funktion $f(\varphi_1 \ \phi \ \varphi_2)$ gibt Antwort auf die Frage, wie häufig Kristalle einer beliebig vorgebbaren Orientierung in der
vielkristallinen Probe vorkommen. Sie ist somit das ange-
strebte Ziel einer jeden Texturuntersuchung.

Die dreidimensionale Funktion $f \ (\varphi_1 \ \phi \ \varphi_2)$ ist nicht direkt
meßbar. Eine Methode, bei der man die drei Parameter $\varphi_1 \ \phi \ \varphi_2$
vorgeben und dann sofort das Volumen dV aller Kristallite
dieser Orientierung messen könnte, gibt es bisher nicht. Man
kann natürlich die Orientierungsparameter jedes einzelnen
Kriställchens einer Probe bestimmen und daraus die Funktion
konstruieren. Das ist auch in einigen Fällen gemacht worden [1, 2, 3]. Das Verfahren ist jedoch so mühsam und lang-
wierig, daß es nur in Ausnahmefällen anwendbar ist.

Im allgemeinen wird man zur Ermittlung der dreidimensionalen Verteilungsfunktion von experimentell bestimmten Polfiguren ausgehen müssen. In der Polfigur sind nur jeweils
zwei Orientierungsparameter - nämlich die Orientierung einer bestimmten Kristallrichtung - vorgegeben. Über den dritten Orientierungsparameter - eine Drehung um diese Richtung
herum - ist gemittelt. Zwischen den Polfiguren und der dreidimensionalen Verteilungsfunktion bestehen also Beziehungen
der Form

$$P_{hkl} \ (\theta \ \beta) = \frac{1}{2\pi} \int_0^{2\pi} f \ (\varphi_1 \ \phi \ \varphi_2) \ d\gamma \qquad (2)$$

Dabei ist γ der Drehwinkel um die festgehaltene Kristallrichtung herum. Polfiguren sind so zu sagen zweidimensionale "Projektionen" der dreidimensionalen Verteilungsfunktion. Aus einer gewissen Anzahl verschieden indizierter
Polfiguren muß dann die dreidimensionale Verteilungsfunktion - zumindest näherungsweise - bestimmt werden. Hierfür
sind im wesentlichen zwei verschiedene Methoden angegeben
und benutzt worden, ein iteratives Näherungsverfahren, das
von einer gewissen ersten Näherung ausgeht und diese dann
schrittweise verbessert bis die theoretisch berechneten Polfiguren mit den experimentellen übereinstimmen [4] und ein
Reihenentwicklungsverfahren [5, 6].

Zur Lösung der Gleichung (2) kann man – wie es erstmals von WIGLIN [7] getan wurde – die Orientierungsverteilungsfunktion f ($\varphi_1 \not\varphi \varphi_2$) in eine Reihe nach verallgemeinerten Kugelfunktionen entwickeln.

$$f\,(\varphi_1 \not\varphi \varphi_2) = \sum_{\lambda=0}^{\infty} \sum_{\mu=1}^{M(\lambda)} \sum_{\nu=1}^{N(\lambda)} C_\lambda^{\mu\nu}\, \ddot{T}_\lambda^{\mu\nu}(\varphi_1 \not\varphi \varphi_2) \qquad (3)$$

Die verallgemeinerten Kugelfunktionen $\ddot{T}_\lambda^{\mu\nu}$ können dabei so gewählt werden, daß sie invariant gegenüber der Kristallsymmetrie sowie der statistischen Probensymmetrie sind [8]. Die letztere ist bei Blechen beispielsweise die orthorhombische Symmetrie, und sie ist von der Kristallsymmetrie vollständig unabhängig. Analog entwickelt man die Polfiguren in Reihen nach Kugelflächenfunktionen (die der Probensymmetrie genügen).

$$P_{hkl}\,(\theta\ \beta) = \sum_{\lambda=0}^{\infty} \sum_{\nu=1}^{N(\lambda)} F_\lambda^\nu\,(hkl)\, \dot{k}_\lambda^\nu\,(\theta\ \beta) \qquad (4)$$

Die Koeffizienten dieser Reihen können aus den gemessenen Polfiguren durch Integration berechnet werden

$$F_\lambda^\nu\,(hkl) = \int_0^\pi \int_0^{2\pi} P_{hkl}\,(\theta\ \beta)\, \dot{k}_\lambda^\nu\,(\theta\ \beta)\, \sin\theta\ d\theta\ d\beta \qquad (5)$$

Setzt man nun die Reihen Gl. (3) und Gl. (4) in die Grundgleichung Gl.(2) ein, so erhält man eine Beziehung zwischen den Koeffizienten

$$F_\lambda^\nu\,(hkl) = \frac{4\pi}{2\pi + 1} \sum_{\mu=1}^{M(\lambda)} C_\lambda^{\mu\nu}\, \dot{k}_\lambda^\mu\,(hkl) \qquad (6)$$

Darin sind die $\dot{k}_\lambda^\mu$ (hkl) Kugelflächenfunktionen, die invariant gegenüber der Kristallsymmetrie sind.

Sind die Koeffizienten F_λ^ν (hkl) für eine Reihe verschiedener Polfiguren bekannt, so können durch Auflösung des linearen Gleichungssystems Gl. (6) die Koeffizienten $C_\lambda^{\mu\nu}$ der gesuchten Funktion f ($\varphi_1 \not\varphi \varphi_2$) ermittelt werden. Setzt man sie in Gl. (3) ein, so kann man die Funktion f ($\varphi_1 \not\, \varphi_2$) für beliebige Orientierungen $\varphi_1 \not\, \varphi_2$ berechnen.

Die Anzahl der Unbekannten M (λ) in Gl. (6) wächst mit dem Grad λ. Für eine vorgegebene Anzahl von Polfiguren kann Gl. (6) daher nur bis zu einem Maximalwert von λ aufgelöst werden. Die Koeffizienten $C_\lambda^{\mu\nu}$ höherer Ordnung können nicht eindeutig bestimmt werden. Dementsprechend gibt es mehrere verschiedene Funktionen f ($\varphi_1 \not\, \varphi_2$), die die Gleichungen Gl. (2) exakt erfüllen. Sie unterscheiden sich durch Koeffizienten höherer Ordnung ihrer Reihenentwicklung. Diese Mehrdeutigkeit der Lösung von Gl.(2) bei einer endlichen Anzahl von Polfiguren ist prinzipieller Natur, unabhängig davon ob die Lösung durch Reihenentwicklung erhalten wurde oder nicht [9]. Bei Verwendung von drei Polfiguren und kubischer Kristallsymmetrie beträgt z.B. λ max = 34, das entspricht einem Winkelauflösungsvermögen von etwa 9° (Abb. 1, Kap. 1.1, S. 5).

Es ist vielleicht nicht überflüssig, sich diese Mehrdeutigkeit an einem anschaulichen Beispiel zu verdeutlichen. Abb.2 zeigt eine zweidimensionale Verteilungsfunktion, von der zwei eindimensionale Projektionen (Polfiguren) bekannt sein sollen. Sie ist von der darunter gezeichneten verschmierten Verteilungsfunktion nicht zu unterscheiden. Wie Abb. 3 zeigt, kann auch das Umgekehrte der Fall sein. Es können Maxima der Verteilungsdichte, also "Ideallagen" vorgetäuscht werden, die in Wirklichkeit gar nicht vorhanden sind. Diese Mehrdeutigkeit kann nur durch eine größere Anzahl von Polfiguren reduziert werden.

Nach der oben geschilderten Methode wurde die Orientierungsverteilung in einem 90 % kaltgewalzten Kupferblech aus vier mit Neutronenbeugung gemessenen Polfiguren bestimmt [10]. Das Ergebnis ist in Abb. 4a in Schnitten für konstantes φ_2 dargestellt. Man sieht, daß eine ganze kontinuierliche Reihe von Orientierungen maximaler Orientierungsdichte auftritt. Noch besser erkennt man das, wenn man die zweidimensionalen Schnitte zu einem dreidimensionalen Modell zusammenfügt.

In Abb. 4b ist die Orientierungsverteilung für einen kalt

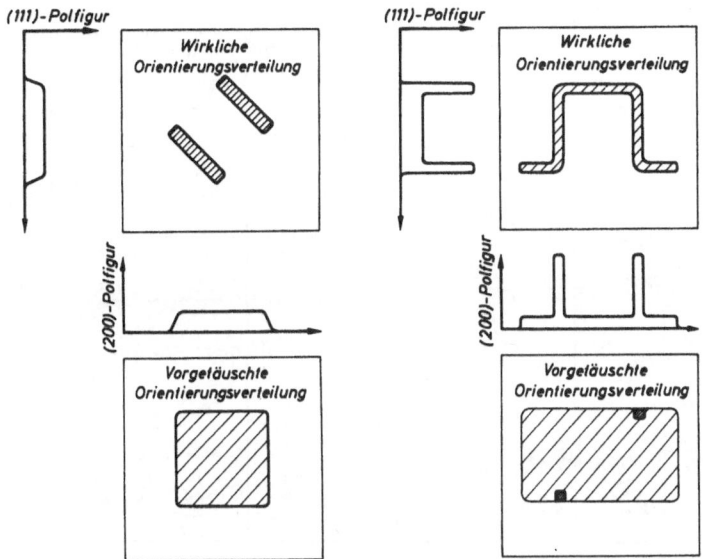

Abb.2 (links). Zweidimensional schematisches Beispiel für das begrenzte Auflösungsvermögen bei
der Bestimmung der Orientierungsverteilungsfunk-
tionen aus zwei gegebenen Polfiguren

Abb.3 (rechts). Vortäuschung von Ideallagen infolge der Mehrdeutigkeit der Orientierungsverteilungsfunktion (zweidimensional schematisch)

gewalzten Stahl dargestellt [11] und zwar in Schnitten für
konstantes φ_1 (also senkrecht zur Schnittebene von Abb.4a).
Sie wurde aus drei röntgenographisch gemessenen Polfiguren
[12] berechnet. Man sieht, daß auch in diesem Falle eine
kontinuierliche Verteilung bevorzugter Orientierungen vorliegt, allerdings mit komplizierterem Querschnitt als im
Falle der Walztextur des Kupfers.

Sieht man von der Form des Querschnittes ab und betrachtet
nur die Punkte maximaler Dichte, so bilden sie eine Linie
im Orientierungsraum, die man die "Skelettlinie" nennen
könnte. Überträgt man die Punkte der Skelettlinie in ste-
reographische Projektion, so erhält man die Abb. 5. Im Falle des Kupfers liegt also die Walzrichtung auf einer Kurve

30

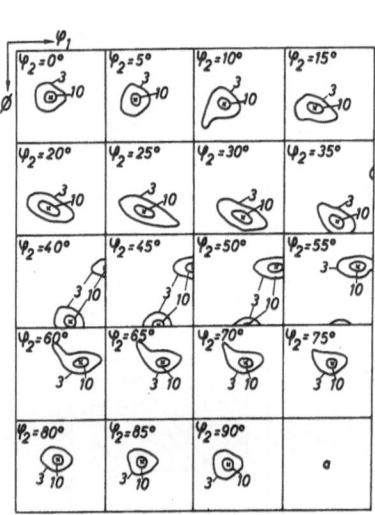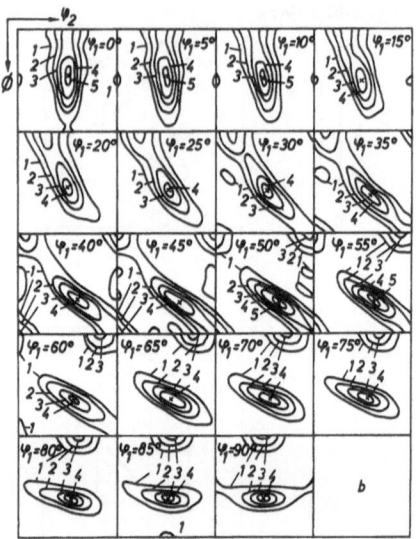

Abb.4a und b. Die Orientierungsverteilungsfunktion f (φ_1 ∅ φ_2)

a) Kupferblech, 90% kaltgewalzt, bestimmt aus vier neutronographisch bestimmten Polfiguren

b) Kaltgewalztes Stahlblech, bestimmt aus drei röntgenographisch bestimmten Polfiguren

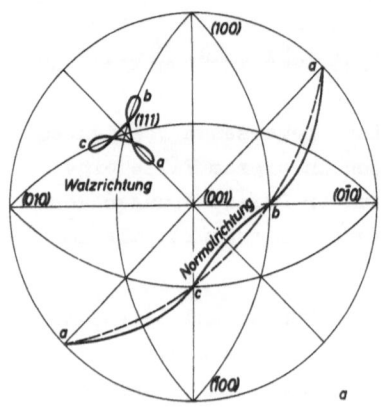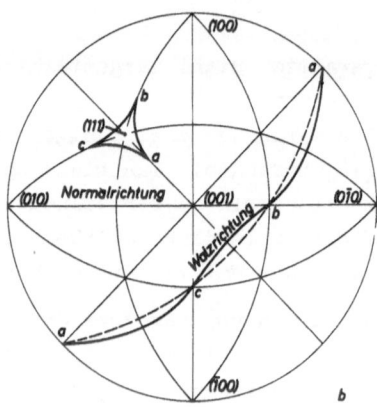

Abb.5a u.b. "Skelette" der Walztextur des Kupfers (a) und des Stahles (b) in stereografischer Projektion

in der Nähe von [111], während die Normalrichtung in der
Nähe des dazu senkrechten Großkreises liegt. Beim Stahl ist
es im wesentlichen umgekehrt. Hier liegt die Normalrichtung
nahe bei [111], während die Walzrichtung alle Orientierungen des dazu senkrechten Großkreises einnimmt.

Die beiden Texturen gehen also in erster Näherung durch
Vertauschung von Walz- und Normalrichtung auseinander hervor. Sie sind so zu sagen komplementär zueinander.

In Einzelheiten der Verteilung unterscheiden sie sich allerdings, so z.B. in der Form der Streuung um die Skelettlinie herum, der Intensität längs der Skelettlinie und durch
einige in der Eisentextur zusätzlich vorhandene Ideallagen.
Darauf soll jedoch hier nicht eingegangen werden.

Die von GREWEN und WASSERMANN [13, 14] eingeführten be-
schränkten Fasertexturkomponenten sind im dreidimensionalen
Orientierungsraum ebenfalls durch eindimensional ausge-
dehnte Verteilungen charakterisiert im Gegensatz zu den Ideallagen, die durch punkt- bzw. kugelförmige Verteilungen
repräsentiert werden. In diesem Sinne entsprechen also beide Texturen am besten dem Grewen-Wassermann Modell.

Auch mit Hilfe der zweiten, oben erwähnten Methode, der
iterativen Methode, wurden ganz ähnliche Ergebnisse erhalten. Von WILLIAMS [4] wurde auf diese Weise die Walztextur
von Kupfer, Aluminium und Messing untersucht. Allerdings
wurden dabei etwas andere Orientierungsparameter und eine
andere Art der Darstellung, sogenannte biaxiale Polfiguren,
verwendet. In Abb. 6 ist die so erhaltene Orientierungsverteilung für Kupfer wiedergegeben. Mit der durch Reihenentwicklung erhaltenen Funktion ist sie allerdings erst nach
einer Koordinatentransformation vergleichbar. Ein Vorteil
dieser Methode besteht darin, daß das Auftreten negativer,
also physikalisch sinnloser Werte, von vornherein vermieden
wird. Andererseits gibt das Verfahren jedoch keine Möglichkeit abzuschätzen, inwieweit die gefundene Lösung eindeutig ist. Dadurch wird eine höhere Genauigkeit des Ergeb-
nisses vorgetäuscht. Die Unsicherheit der Lösung ist auch
in diesem Falle natürlich durch die Anzahl der verwendeten
Polfiguren bedingt.

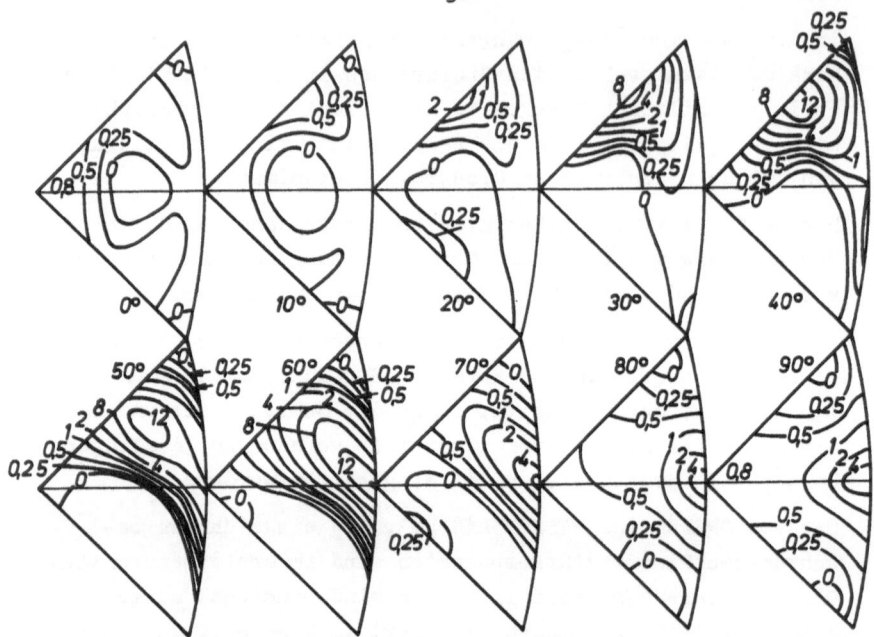

Abb.6. Biaxiale Polfiguren für ein kaltgewalztes Kupferblech
(nach WILLIMAS)

In vielen Fällen wird die Orientierungsverteilungsfunktion
f (g) als Gewichtsfunktion bei der Berechnung von Mittel-
werten $\bar{E}$ orientierungsabhängigerEigenschaften E (g) der
Kristallite in vielkristallinen Materialien benötigt [9].

$$\bar{E} = \oint E \ (g) \ f \ (g) \ dg \qquad (7)$$

Drückt man nun f (g) durch die Reihenentwicklung Gl.(3) aus

$$\bar{E} = \sum_{\lambda=0}^{\infty} \sum_{\mu=1}^{M(\lambda)} \sum_{\nu=1}^{N(\lambda)} C_\lambda^{\mu\nu} \oint E \ (g) \ \overset{\cdot\cdot}{T}_\lambda^{\mu\nu} \ (g) \ dg, \qquad (8)$$

so verschwinden in vielen Fällen die auftretenden Integrale
von einem bestimmten Grad λ an identisch. Der zu berechnen-
de Mittelwert hängt dann nur von sehr wenigen Texturkoeffi-
zienten$C_\lambda^{\mu\nu}$ ab, er ist also bei Kenntnis dieser Koeffizienten sehr leicht zu berechnen. Bedeutet E (g) z. B. die Komponenten des Elastizitätstensors, so gehen im Falle kubischer Kristalle nur drei Koeffizienten C_4^{11}, C_4^{12}, C_4^{13} in
Gl.(8) ein [15]. Von allen anderen Texturkoeffizienten ist

der Mittelwert der elastischen Eigenschaften unabhängig.

Von besonderem Interesse ist gegenwärtig die plastische Anisotropie [16]. Soll ein Kristall in x-Richtung um A % ge-
dehnt, in y-Richtung um q · A % und in z-Richtung um (1–q)·
A % verkürzt werden, so ist dazu im allgemeinen die Betätigung von 5 Gleitsystemen erforderlich [17]. Nach den Annahmen von TAYLOR [18] sind das diejenigen Systeme, die die
geringste Summe M der Gleitbeträge erfordern. Dieser Taylorfaktor M hängt von der Orientierung g des Kristalles und von
dem Querkontraktionsverhältnis q ab [19]. Für ein vielkristallines Material erhält man dann einen mittleren Taylorfaktor.

$$\bar{M}\ (q) = \oint M\ (q,g)\ f\ (g)\ dg \qquad (9)$$

Wird ein Probestab aus dem betreffenden Material frei ge-
dehnt, so ist das Querkontraktionsverhältnis q zunächst
nicht bekannt. Es wird sich jedoch so einstellen [20], daß
M ein Minimum wird.

$$\bar{M}\ (q) = \text{Min} \qquad (10)$$

In Abb. 7 ist $\bar{M}$ (q) für ein warm gewalztes Stahlblech dargestellt und zwar für einen in Walzrichtung und einen unter

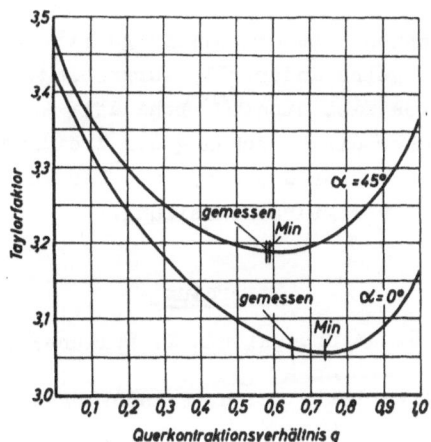

Abb.7. Der Taylorfaktor M als Funktion des Querkon-
traktionsverhältnisses q für ein warmgewalztes Stahlblech in Walzrichtung und unter 45°

45° geschnittenen Probestab [21]. Dazu ist jeweils das gemessene Querkontraktionsverhältnis angegeben. Die Übereinstimmung ist befriedigend jedoch nicht besonders gut. Da es
sich nur um Näherungsrechnungen handelt und der Kurvenver-
lauf in der Umgebung des Minimums sehr flach ist, ist das
auch kaum zu erwarten. In Abb. 8 schließlich ist der Verlauf

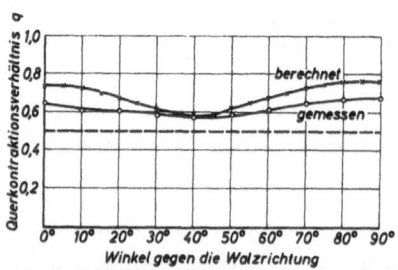

Abb.8. Verlauf des berechneten und gemessenen Quer-
kontraktionsverhältnisses in der Walzebene eines
warmgewalzten Stahlbleches

von q_{min} in der Walzebene des Bleches dargestellt, zusammen mit den experimentell bestimmten q–Werten. Für diese
Berechnungen wurden die Koeffizienten $C_{\lambda}^{\mu\nu}$ bis $\lambda = 10$ verwendet. Die Orientierungsverteilungsfunktion selbst wurde
dazu gar nicht benötigt.

Neben der Berechnung von Orientierungsmittelwerten können
auch eine ganze Reihe anderer Texturprobleme sehr bequem
mit Hilfe der Koeffizienten $C_{\lambda}^{\mu\nu}$ behandelt werden. Das ist
ein weiterer, über die Berechnung der dreidimensionalen
Polfigur hinausgehender Grund für die Verwendung der Reihen-
darstellung in der Texturbeschreibung.

Literatur

1. F. Haeßner, U. Jakubowski und M. Wilkens: phys.stat.sol.
   7(1964)701.
2. K. Lücke, H. Perlwitz und W. Pitsch: phys.stat.sol. 7
   (1964)733.
3. H.J. Bunge und F. Haeßner: Appl.Phys., im Druck.
4. R.O. Williams: Trans.Met.Soc.AIME 242(1968)104.
5. H.J. Bunge: Z. Metallkde. 56(1965)872.
6. R.J. Roe: J.Appl.Phys. 36(1965)2024.

7. A.S.Wiglin: Fiz.Twordowo Tela 1 (1959) 261.

8. H.J.Bunge: Mber. Dt.Akad.Wiss.Berlin 7 (1965) 351.

9. H.J.Bunge: Mathematische Methoden der Texturanalyse, Akademieverlag, Berlin 1969.

10. H.J. Bunge und J. Tobisch: Z.Metallkde 59 (1968) 471.

11. H.J. Bunge: phys.stat.sol. 26 (1968) 167.

12. H. Takechi, H. Kato und S. Nagashima: Trans Met. Soc. AIME 242 (1968) 56.

13. J. Grewen und G. Wassermann: Acta Met. 3 (1955) 354.

14. G. Wassermann und J. Grewen: Texturen metallischer Werkstoffe,Springer, Berlin 1962.

15. H.J. Bunge: Kristall u. Technik 3 (1968) 431.

16. J. Grewen und G. Wassermann in: O. Kienzle, Hersgb., Mechanische Umformtechnik, Springer, Berlin 1968,S.93.

17. R.v. Mises: Z.angew.Math.u.Mech. 8 (1928) 161.

18. G.J. Taylor: J.Inst.Metals 62(1938)307.

19. H.J. Bunge: Veröffentlichung demnächst.

20. W.F.Hosford und W.A.Backofen: 9th Sagamore Conf. AMRA, Fundamental of Deformation Processing, Univ.Press, Syracuse 1964, S. 259.

21. H.J. Bunge und W.T. Roberts: Veröffentlichung demnächst.

1.3 Beispiele für die Anwendung eines dreidimensionalen Orientierungsraumes zur Ermittlung und Darstellung von Orientierungsverteilungen
Examples for the Use of a Three-Dimensional Orientation Space
for the Investigation and Representation of Orientation
Distribution

von K. Lücke[+)]

Abstract

A crystal orientation is defined by three parameters.
Therefore, a three-dimensional orientation space is
necessary for the representation of orientation distribution.
It is usefull to fit the three orientation coordinates to
the respective problem. The utility of such a three-
dimensional representation and the selection of the
orientation coordinates will be discussed with two examples,
namely the orientation distribution of Fe-Si single crystals
as a result of deformation and growth selection, and the
recrystallization textures of copper and brass after
rolling at different temperatures.

1.3.1 Einleitung

Wie schon des öfteren ausgeführt worden ist[1,2], ist die
Angabe einer zweidimensionalen Polfigur für eine vollständige und eindeutige Beschreibung von Orientierungsverteilungen nicht ausreichend. Es bedarf dazu vielmehr der Angabe der Verteilung der Orientierungsdichte in einem dreidimensionalen Orientierungsraum.

Auf der anderen Seite hat sich jedoch gezeigt, daß die bis-
herige Beschreibung der Orientierungsverteilung mittels Polfiguren in vielen Fällen ausreicht, ein gewisses Verständnis der Orientierungsmannigfaltigkeit zu übermitteln [3],
so daß es nahe liegt, die dreidimensionale Beschreibung
als eine unnütze Komplizierung anzusehen. Im folgenden soll

[+)] Institut für Allgemeine Metallkunde und Metallphysik der
Technischen Hochschule Aachen, Aachen, Bundesrepublik
Deutschland

daher anhand zweier praktischer Beispiele gezeigt werden,
daß in manchen Fällen die dreidimensionale Beschreibungsform nicht nur unumgänglich notwendig ist, sondern sogar
zu einer einfacheren Wiedergabe der Orientierungsverteilung führt als die bisher allein angewandte Polfigurenmethode.

Zur Durchführung der dreidimensionalen Analyse mußten in
beiden Beispielen die Orientierungen der zu untersuchenden
Kristalle einzeln vermessen werden. Die Koordinaten des
Orientierungsraumes wurden hier jeweils so gewählt, daß eine
möglichst einfache Darstellung der Ergebnisse erhalten wurde. Es soll im folgenden besonders auf die bei der praktischen Durchführung solcher Analysen auftretenden Fragen
eingegangen werden.

1.3.2 Orientierungsverteilung nach der Rekristallisation
gewalzter Fe-Si-Kristalle

Im ersten Beispiel, das einer Arbeit von IBE und LÜCKE [4]
entnommen ist, wurde die Orientierung der bei der Rekristallisation verformter Fe-3% Si-Einkristalle auftretenden
neuen Kristalle relativ zum Gitter des verformten Einkri-
stalles vermessen. Abb. 1 gibt die {100}-Pole der auf diese
Weise erhaltenen 270 Kristalle wieder, wobei die Matrixorientierung jeweils in die Standardprojektion gelegt ist.
Man sieht deutliche Abweichungen von einer regellosen Verteilung, ist jedoch nicht in der Lage, daraus klar eine
Vorzugsorientierung zu erraten. (Wenn überhaupt, würde man
hier eine Orientierung ähnlich der der Würfellage angeben,
was jedoch - wie sich zeigen wird - nicht den Tatsachen
entspricht.) Hier ist es daher nötig, eine dreidimensionale
Orientierungsanalyse vorzunehmen.

Dabei erhebt sich die Frage nach der Wahl der dafür zu be-
nutzenden Koordinaten. Da man eine Kristallorientierung in
eine vorgegebene andere stets durch eine Drehung überfüh-
ren kann, soll der Orientierungsraum hier durch die räum-
lichen Drehkoordinaten φ, θ und ψ aufgespannt werden [5,6].
Dabei gibt ψ den Drehwinkel an und φ und θ sind die Polarkoordinaten der Drehachse, mit deren Hilfe die Kristall-

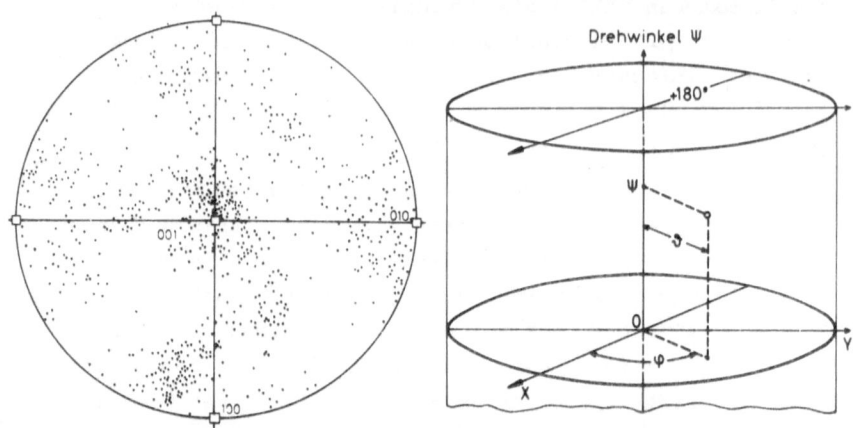

Abb.1. {100}-Polfigur für
270 durch Wachstumsausle-
se entstandene Kristalle

Abb.2. Dreidimensionaler
Orientierungsraum bei der
Verwendung der räumlichen
Drehkoordinaten φ, θ und ψ

orientierung in die Matrixorientierung überführt werden
kann. In dem man etwa θ und φ in einer stereographischen
Projektion und senkrecht darüber jeweils den Drehwinkel ψ
aufträgt, erhält man den in Abb. 2 wiedergegebenen Orientierungsraum. Aus Darstellungsgründen werden im folgenden
allerdings nur Projektionen dieses Raumes betrachtet.

Im vorliegenden Fall wurden für 27o Kristalle die Koordinaten θ , φ und ψ jeweils mit Hilfe eines Rechenprogrammes
durch einen Großrechner ermittelt. Allerdings entsteht hier
die Schwierigkeit, daß wegen der kubischen Symmetrie insgesamt 24 solcher Drehungsmöglichkeiten für die Überfüh-
rung einer vorgegebenen Orientierung in eine andere vorgegebene Orientierung existieren, die alle gleichberechtigt
sind. Man erhält so zur Beschreibung einer Orientierung in
Wirklichkeit 24 Koordinatentripel, d. h. 72 Zahlen an Stelle von 3 Zahlen und somit hier insgesamt 270 x 14 $\approx$ 6500
Zahlenangaben. Das eigentliche Problem bei der Darstellung
einer Orientierungsverteilung liegt in der Bewältigung

dieses umfangreichen Zahlenmaterials, z. B. darin, für jeden
Kristall nur eine der 24 möglichen Drehungen möglichst ge-
schickt auszuwählen. Dabei soll das einzige Auswahlprinzip
darin bestehen, eine möglichst übersichtliche Beschreibung
zu erhalten.

Es wurde daher zunächst die anschaulichste der 24 Drehungen, nämlich jeweils die mit dem kleinsten Drehwinkel, die
als Desorientierung bezeichnet werden soll, ausgewählt. Die
in Abb. 3 wiedergegebene Verteilung dieser Drehwinkel zeigt

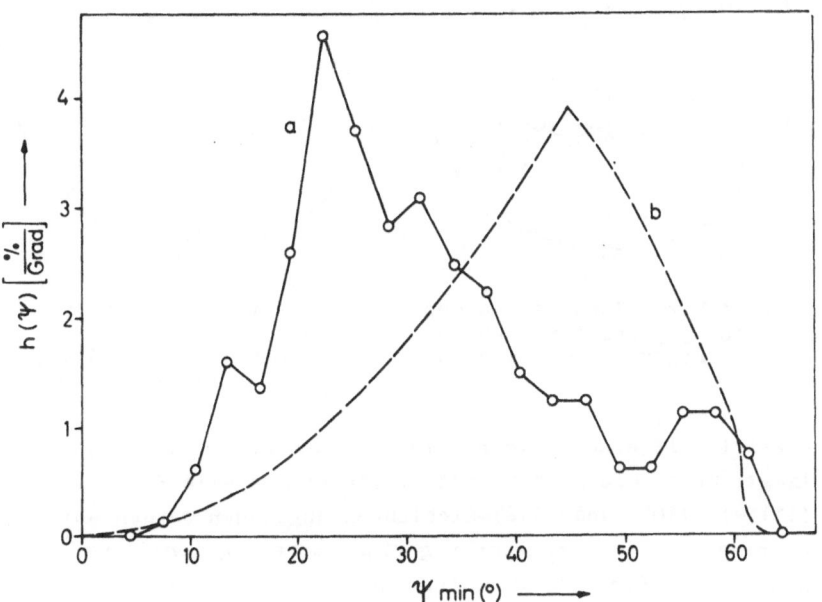

Abb.3. Verteilung der jeweils kleinsten Drehwinkel für
die betrachteten 270 Kristalle (gestrichelt gezeichnet: Verteilung der kleinsten Drehwinkel bei regelloser Orientierungsverteilung)

ein Maximum bei etwa 27°. Man könnte daher geneigt sein,
diesen Winkel als Idealdrehwinkel zu betrachten, jedoch
zeigt die in Abb.4 wiedergegebene Verteilung der den Winkeln der Abb. 3 entsprechenden Drehachsen keinerlei Vor-
zugslagen. Man könnte daraus zu schließen versuchen, daß
hier eine Verteilung vorliegt, bei der zwar der Drehwinkel
aber nicht die Drehachse definiert ist, aber auch das ist

- wie sich zeigen wird - nicht korrekt.

Nachdem hier also die Auswahl aus den 24 Rotationen zunächst auf Grund der Größe des Drehwinkels getroffen worden
ist, soll diese nunmehr nach der Lage der Drehachsen erfolgen. So ist in Abb.5 jeweils diejenige der 24 Drehachsen

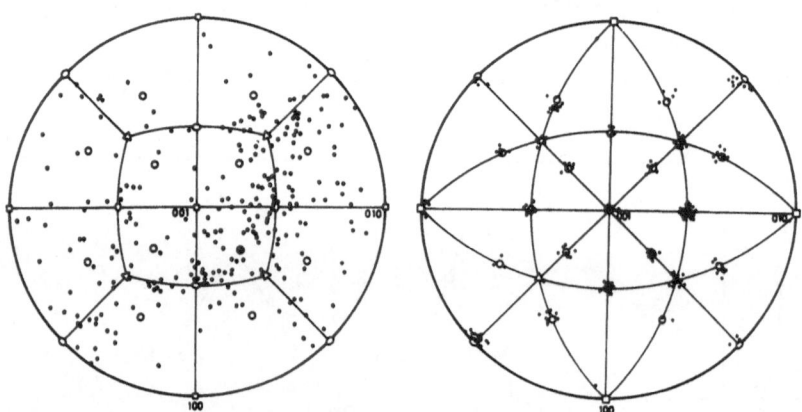

Abb.4.Verteilung der zu den   Abb.5.Verteilung derjeni-
in Abb.3 gezeigten Drehwin-   gen der 24 Drehachsen,
keln gehörigen Drehachsen   die einem {100}-, {111}-,
                                   {110}-oder {112}-Pol am
                                   nächsten liegen

gezeigt, die einer "niedrig indizierten Achse" am nächsten
liegt, wobei als niedrig indizierte Achsen alle ⟨100⟩-,
⟨111⟩-, ⟨110⟩- und ⟨112⟩-Richtungen angesehen werden sollen.
Hier hat man zwar eine sehr geringe Streuung der Drehachsen,
aber eine gleichzeitige Zulassung dieser verschiedenen
Kristallachsentypen als Idealachsen würde zu einer sehr un-
einheitlichen Orientierungsbeschreibung führen.

Da in Abb.5 die ⟨110⟩-Achsen offensichtlich bevorzugt be-
legt sind, ist in Abb.6 stets diejenige der 24 Drehachsen
aufgetragen, die einer ⟨110⟩-Achse am nächsten liegt. Auch
dann erhält man nur eine Streuung von wenigen Grad. Das
zeigt auch Abb.7, wo die aus Abb.6 entnehmbare Verteilungsdichte dieser Drehachsen als Funktion des Abstandes λ von
der zugehörigen ⟨110⟩-Achse wiedergegeben ist. Wenn man jeweils die Drehachse auswählen würde, die statt einer ⟨110⟩-

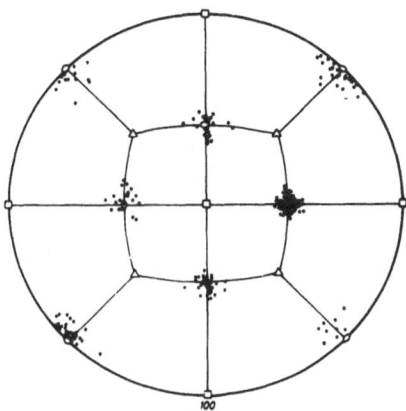

Abb.6. Verteilung derjenigen der 24 Drehachsen, die
jeweils einem {110}-Pol am nächsten liegen

Achse einer ⟨100⟩-, ⟨111⟩- bzw. ⟨112⟩-Achse am nächsten
liegt, erhielte man die übrigen in Abb. 7 wiedergegebenen
Kurven. Man sieht, daß die Verteilung um die ⟨11o⟩-Achsen
am schärfsten ist, daß diese Achsen also am besten als Ide-
aldrehachsen geeignet sind.

Betrachtet man schließlich die Verteilung der zu den in
Abb.6 dargestellten Drehachsen gehörigen Drehwinkel, so findet man zwei klare Maxima (Abb.8), die ihren Schwerpunkt
bei 95,9° und 153,3° besitzen. Das bedeutet, daß die Orientierungsverteilung aus zwei Komponenten besteht, deren Ideallagen beschrieben werden können durch Drehungen um einen
⟨110⟩-Pol der Matrix um Winkel von 153,3° bzw. 95,9°, oder
- was wegen der Zweizähligkeit der ⟨110⟩-Achsen damit identisch ist - durch Drehungen

$$26,7° \ ⟨111⟩ \ \text{(Hauptkomponenten) und}$$
$$84,1° \ ⟨111⟩ \ \text{(Nebenkomponenten) [7].} \qquad (1)$$

Diese Ideallagen und deren Streuungen lassen sich nun nach-
träglich wieder in einer Polfigur veranschaulichen. So erhält man die {100}-Polfigur der Abb. 9, indem man alle Orientierungen durch Symmetrieoperationen so transformiert,
daß symmetrische Lagen zusammenfallen.

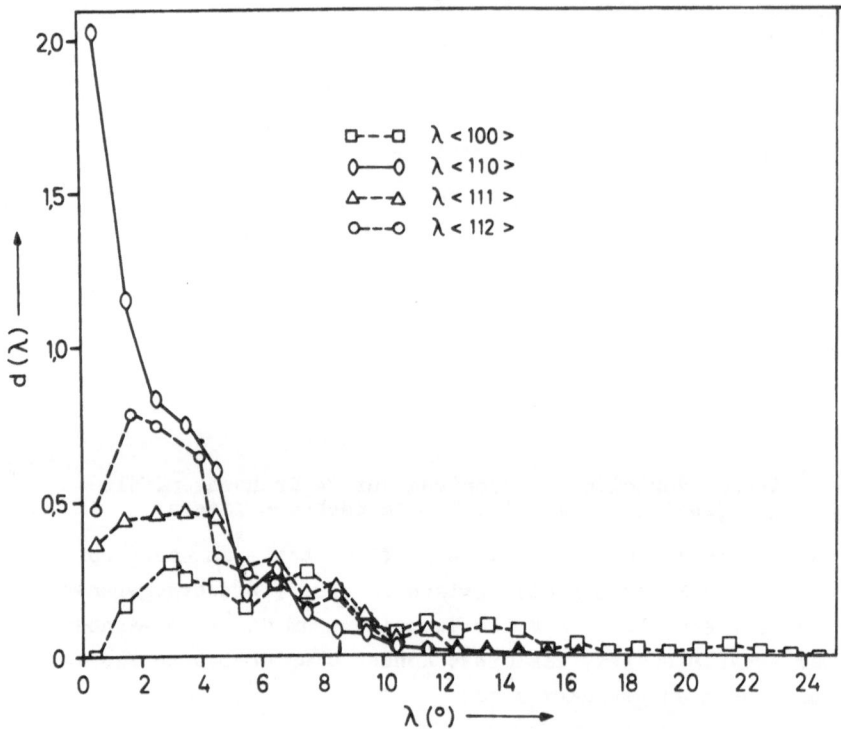

Abb.7. Ausgezogene Kurve: Verteilung der Dichte der
in Abb. 6 wiedergegebenen Drehachsen als Funktion
des Abstandes λ von den jeweiligen {110}-Achsen.
Die übrigen Kurven stellen die entsprechenden Verteilungen dar, wenn man die Drehachsen ausgewählt
hätte, die entweder einem {100}-, einem {111}- oder
einem {112}-Pol am nächsten liegen

Es existiert also im vorliegenden Fall eine recht einfach
zu beschreibende Textur, von der es jedoch nicht möglich
war, sie direkt aus der Polfigur (Abb. 1) ohne die hier wiedergegebene dreidimensionale Analyse zu erkennen. Bei dieser
Analyse sind allerdings noch einige Fragen offengeblieben,
etwa warum man nicht direkt die Winkel ψ = 26,7° und 84,1°,
sondern zunächst deren Supplementwinkel erhält, oder warum
man als kleinsten Drehwinkel zwar sofort ψ ≈ 27° findet
(Abb. 3), jedoch die zugehörige Drehachse nicht gleich als
≈⟨110⟩ erkennbar ist (Abb. 4). Jedoch auch diese Fragen
lassen sich leicht durch Betrachtung der Struktur des hier
gewählten Orientierungsraumes verstehen [6].

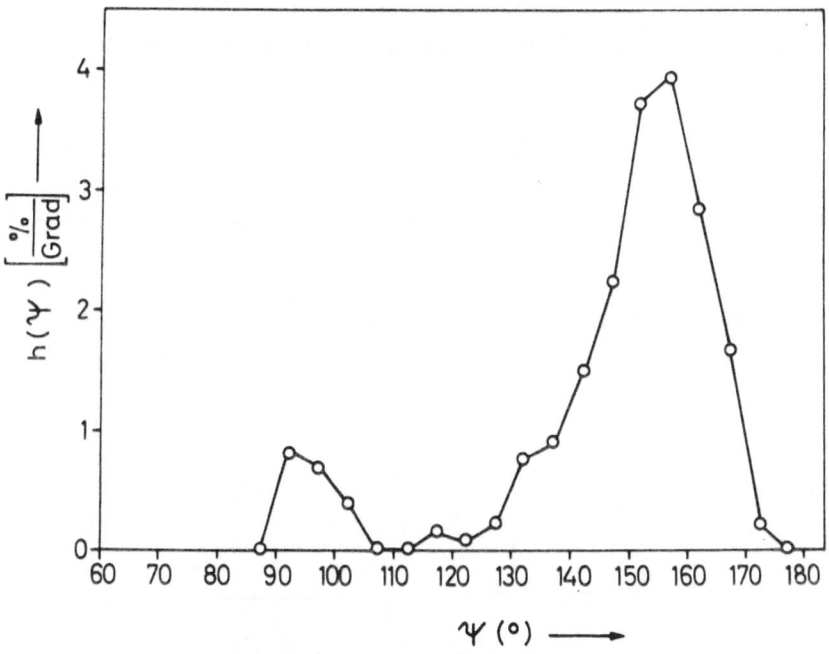

Abb.8. Verteilung der zu den in Abb.6 gezeigten
Drehachsen gehörigen Drehwinkel

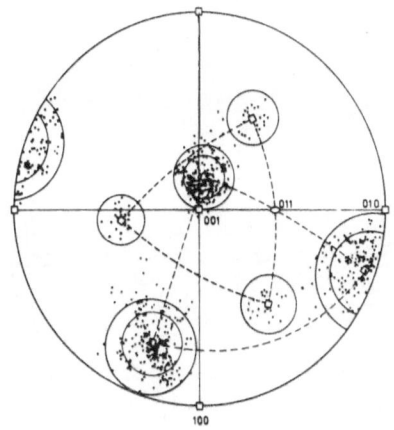

Abb.9. {100}-Polfiguren der 270 betrachteten Kristalle nach Durchführung geeigneter Symmetrieoperationen

1.3.3. Die Walztexturen kubisch-flächenzentrierter Metalle

Während die beim Walzen von Messing erhaltene Textur recht
einfach ist, ist die beim Walzen von Kupfer entstandene
ziemlich unübersichtlich und ein Beispiel dafür, daß man
aus einer Polfigur nicht die wahre Orientierungsverteilung
ableiten kann. Dies zeigt sich hier beispielsweise darin,
daß bei den Versuchen, eine solche durch Erraten von Ideal-

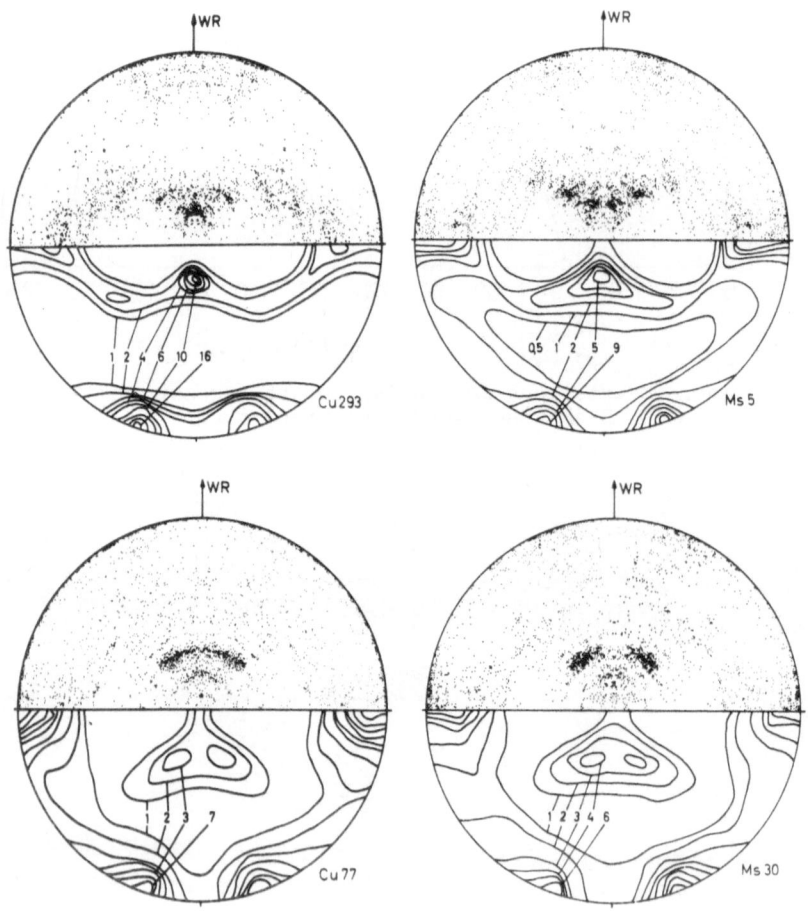

Abb.10. Röntgenographisch (jeweils untere Hälfte) und elektronenmikroskopisch (jeweils obere Hälfte) ermittelte {111}-
Polfiguren der vier untersuchten Materialien

45

lagen zu erhalten, verschiedene Autoren zu verschiedenen Ergebnissen kamen [3]. Daher wurde an Kupfer bei 77°K (Cu 77)
und Kupfer bei Raumtemperatur (Cu 293) gewalzt, Messing mit
5% Zn (Ms 5) und mit 30% Zn (Ms 30), beide bei Raumtemperatur gewalzt, die Orientierungsverteilung durch Einzelver-
messung der verschieden orientierten Bereiche mit Hilfe der
elektronenmikroskopischen Feinbereichsbeugung bestimmt [8 -
10]. Im folgenden sollen die dabei von PERLWITZ, LÜCKE und
PITSCH [10] erhaltenen Ergebnisse diskutiert werden.

Abb.10 zeigt zunächst, daß die röntgenographischen und elektronenmikroskopisch gewonnenen Polfiguren gut übereinstimmen. Zur dreidimensionalen Darstellung der Ergebnisse wurden
die drei Orientierungskoordinaten so gewählt, daß zwei von
ihnen, nämlich φ_3 und φ_2, die Lage der Walzebenennormale angeben und die dritte φ_1, die Lage der Walzrichtung beschreibt. Wie Abb. 11 zeigt, ist φ_3 = 0 durch den Großkreis
[001] - [$\bar{1}$11] gegeben. Da die Walzrichtung auf einem Großkreis in 90° Abstand von der Walzebenennormale liegt und
sich die zu verschiedenen φ_2 aber gleichen φ_3 gehörigen 90°
Großkreise im gleichen Punkt schneiden (s. Abb. 11), wird
dieser Schnittpunkt als Nullpunkt für die Zählung von φ_1 gewählt.

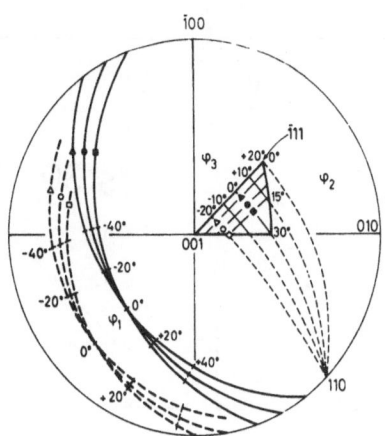

Abb.11. Zur Definition der Orientierungsparameter
φ_1, φ_2 und φ_3

46

Da, wie Abb. 12 zeigt, für etwa 75% aller gefundenen Orientierungen die Walzebenennormale in das φ_3-Intervall von
-5° bis $+ 5^{\circ}$ fällt, sollen der Einfachheit halber hier nur
diese Orientierungen betrachtet werden. Die Abb. 13 ergibt
so als Schnitt bei $\varphi_3 = 0^{\circ}$ ($\pm$ 5) durch den 3-dimensionalen
Orientierungsraum die Höhenliniendarstellungen der Häufig-
keitsverteilungen als Funktion von φ_1 und φ_2. Für Messing
mit 30% Zn erhält man, wie bereits auf Grund der Polfigur
zu erwarten war, ein ausgeprägtes Maximum. (In Abb. 13 ergeben sich zwar 2 Maxima, die jedoch als Folge der kubischen Symmetrie kristallographisch identisch sind.) Ähnlich
ist die Textur des bei $-196^{\circ}C$ gewalzten Kupfers, während
bei der von Messing mit 5% Zn, die in der Polfigur als

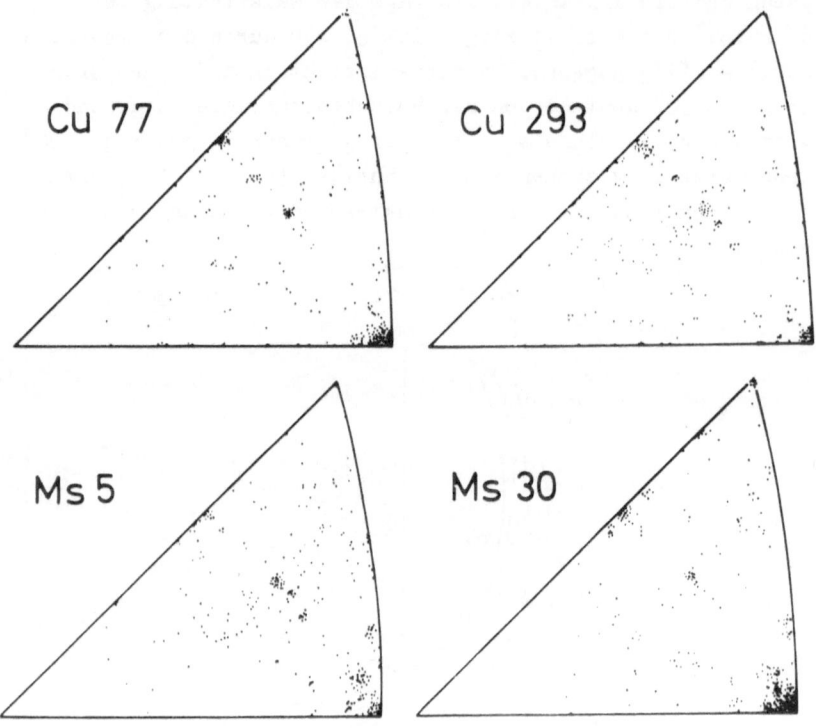

Abb.12. Lage der Walzebenennormalen bei den elektronenmikros
kopisch bestimmten Orientierungsverteilungen für die
vier untersuchten Materialien

Übergangstextur erscheint, in der Orientierungsverteilung
ein zweites Maximum auftritt. Für bei Raumtemperatur gewalztes Kupfer erhält man schließlich drei Maxima, d. h.
die Kupfer-Textur setzt sich aus drei Ideallagen zusammen,
die näherungsweise durch die Indizierungen

$$I: (011) [2\bar{1}1]; \quad II: (\bar{1}23) [7\bar{4}5]; \quad III: (\bar{1}12) [1\bar{1}1] \qquad (2)$$

gekennzeichnet sind.

Die Lage der Maxima ist für die verschiedenen hier untersuchten Texturtypen, die das gesamte Spektrum der bei Kfz.-
Metallen gefundenen Walztexturen umfassen, annähernd die
gleiche. Dasselbe gilt für die Streuung, die um die Ideallagen in Richtung φ_1 etwa um den Faktor 3 größer ist als
in Richtung φ_2 (und auch – wie man zeigen kann – φ_3), so

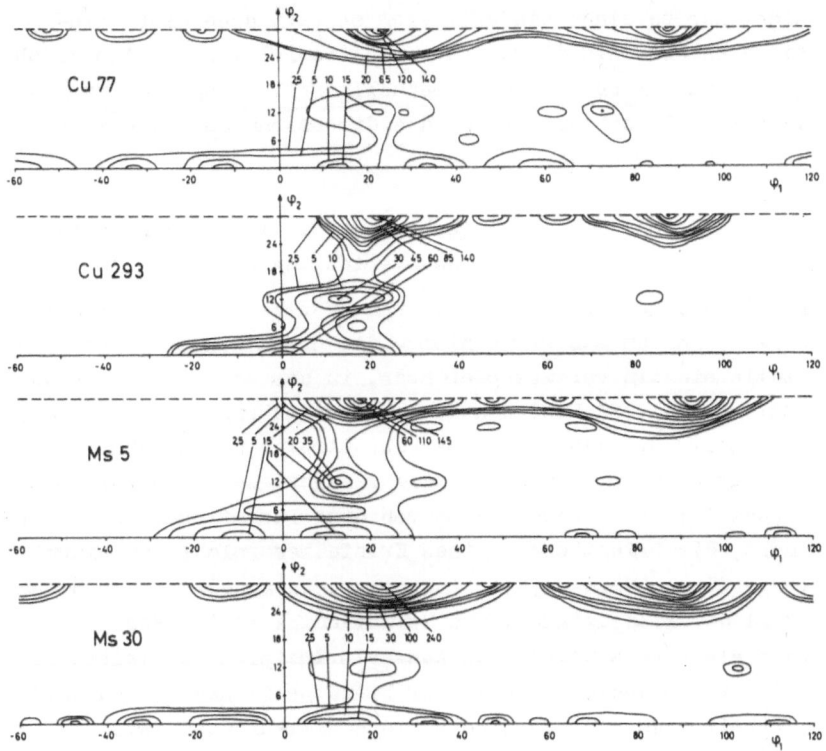

Abb.13. Höhenliniendarstellung der Orientierungsverteilung
in der Ebene $\varphi_3 = 0$ als Funktion von φ_1 und φ_2 für die vier
untersuchten Materialien

daß diese Texturen also unvollständige Fasertexturen [7] (mit
der Blechebenennormale als Faserachse) darstellen. Diese
Befunde legen die Annahme nahe, daß die Walztexturen ku-
bisch-flächenzentrierter Metalle grundsätzlich aus den drei
durch Gl. (2) gegebenen Maxima aufgebaut sind. Das würde bedeuten, daß man zur Beschreibung dieser Texturen lediglich
drei Zahlen benötigt, die die Besetzungshäufigkeit der drei
Ideallagen angeben. (Die bisher bisweilen versuchte Anwendung nur eines Texturparameters wäre hiernach unzulässig).
Durch die Hinzunahme der hier vernachlässigten Orientierungen mit größeren φ_3-Werten ändert sich daran nichts wesent-
liches.

Auch diese Untersuchung ist ein Beispiel dafür, daß die
drei-dimensionale Wiedergabe von Orientierungsverteilungen
nicht nur zu einem eindeutigeren sondern auch einem einfacheren Ergebnis führen kann als die Polfigurenmethode, obgleich natürlich die experimentelle Ermittlung und die ge-
dankliche Durchdringung der dreidimensionalen Verteilung
ungleich schwieriger ist.

1.3.4 Allgemeine Bemerkungen zur Darstellung 3-dimensionaler Orientierungsverteilungen

Eine dreidimensionale Orientierungsverteilung kann erhalten
werden, in dem man - wie hier - die Orientierungen der Kristalle einzeln vermißt oder aber, in dem man eine große Anzahl von zu verschiedenen Netzebenen gehörigen Polfiguren
ermittelt. Die Hauptfrage bei der Darstellung solcher Verteilungen ist die nach der Wahl der geeigneten Koordinaten.
Grundsätzlich ist jeder Satz von drei Parametern, der es erlaubt, die Orientierung eines Kristalles relativ zu einem
vorgegebenen Bezugssystem anzugeben, geeignet und vom Standpunkt der Eindeutigkeit und Vollständigkeit der Beschreibung gleichberechtigt. Als Auswahlprinzipien verbleiben somit nur die Gesichtspunkte der Einfachheit und der Anschaulichkeit, denen stets eine gewisse Willkür anhaftet.

Besonders gründlich sind die oben erwähnten räumlichen Drehkoordinaten [5,6,11] und von BUNGE die Euler-Koordinaten [2,
12] untersucht. Von diesen erscheinen uns die Drehkoordinaten

anschaulicher, da eine Drehung leichter vorstellbar ist als
eine Summe von 3 Drehungen. Sie sollten auch physikalisch
bedeutsamer sein, wenn - wie hier im ersten Beispiel - die
Idealdrehachse einer kristallographischen Achse entspricht.
Die unbequeme 24-fache Multiplizität der Beschreibung erhält man natürlich für beide Koordinatensysteme und es ist
aus Gründen der praktischen Erkennbarkeit einfacher Orien-
tierungszusammenhänge häufig nicht ausreichend, irgend eine
dieser 24 Darstellungsmöglichkeiten zu wählen. Dieses geht
aus dem ersten Beispiel dieser Arbeit hervor, wo die naheliegende Wahl der Desorientierung zu keinem anschaulichen
Ergebnis geführt hat (Abb. 4).

Eine ganz andere Art der Darstellung erhält man, wenn man
statt direkt die Dichte in 3-dimensionalem Orientierungsraum als Funktion der drei gewählten Orientierungsparameter
anzugeben ("direkte Darstellung"), diese Dichte nach einem
geeigneten System von Funktionen entwickelt und dann die
Orientierungsverteilung durch den (wiederum dreidimensional
unendlichen) Satz der Entwicklungskoeffizienten beschreibt
("transformierte Darstellung"). Dies wurde erstmals von BUN-
GE [2] auf der Basis der Eulerkoordinaten und inzwischen von
DORN, IBE und LÜCKE [11] für die räumlichen Drehkoordinaten
durchgeführt. Solch transformierte Darstellungen haben die
Nachteile, sehr unanschaulich zu sein und in dem häufig vorkommenden Fall ausgeprägter Vorzugsorientierung, wo man bei
der direkten Darstellung mit wenigen in ihrer Bedeutung
leicht erkennbaren Zahlenangaben die Verteilungen beschreiben kann (s. z. B. Gl. (1)) und (2)), gerade eine sehr große
Anzahl von Entwicklungskoeffizienten zu benötigen. Von Vorteil ist hingegen, daß man das Funktionssystem so wählen
kann, daß es bereits die kubische Symmetrie enthält und somit der Koeffizientensatz nicht mehr die 24-fache Multiplizität besitzt. Es ist daher zweckmäßig, die transformierte
Darstellung dann anzuwenden, wenn es weniger auf das an-
schauliche Verständnis der Orientierungsverteilung sondern
mehr auf die rechnerische Ermittlung daraus folgender makroskopischen Größen ankommt, wie etwa bei der Berechnung
des mittleren Elastizitätsmoduls einer texturbehafteten
Probe.

Schließlich sei noch darauf hingewiesen, daß man auch bei
einer "inversen" Darstellung von Orientierungsverteilungen,
wie sie im zweiten Beispiel dieser Arbeit angewandt worden
ist, eine 3-dimensionale Beschreibungsform benötigt, und daß
auch dort die erwähnten grundsätzlichen Probleme, wie etwa
die 24-fache Multiplizität, auftreten.

Literatur

1. K. Lücke: Z. Metallkde. 52(1961)1.

2. H.J. Bunge: Z. Metallkde. 56(1965)872.

3. Vgl. hierzu: G. Wassermann und J. Grewen: Texturen metallischer Werkstoffe, Springer, Berlin 1962.

4. G. Ibe und K. Lücke: Arch. Eisenhüttenw. 39(1968)693.

5. J.K. Mackenzie: Acta Met. 12(1964)223.

6. G. Ibe und K. Lücke: Veröffentlichung im Druck.

7. W. Heye: Dieses Symposium, Kap. 2.2.

8. K. Lücke, H. Perlwitz und W. Pitsch: phys. stat. sol. 7(1964)733.

9. F. Haeßner, U. Jakobowski und H. Wilkens: Mat. Sci. Eng. 1(1966)30.

10. H. Perlwitz, K. Lücke und W. Pitsch: Veröffentlichung im Druck.

11. G. Dorn, G. Ibe und K. Lücke: Veröffentlichung demnächst.

12. H.J. Bunge: Dieses Symposium, Kap. 1.2.

2 VERFORMUNGSTEXTUREN
DEFORMATION TEXTURES

2.1 Tension and Compression Textures
Zug- und Drucktexturen
by G. Y. Chin [+)]

Zusammenfassung

Ausgehend von der Kristallplastizität wird die Entstehung von
Fasertexturen in gezogenen Drähten und druckbeanspruchten
Blechen untersucht. Das allgemeine Problem enthält zwei Teil-
bereiche: Ein Verformungsmodell, das die Auswahl der aktiven
Gleitelemente ermöglicht und einen Verformungsmechanismus,
der die potentiellen Verformungselemente für diese Auswahl
liefert. Es wird gezeigt, daß von allen bisher vorgeschlagenen Vorformungsmodellen das TAYLOR'sche vom Standpunkt der
Texturbildung aus das realistischste ist. Der Verformungsmechanismus in diesem Modell ist jedoch auf $\{111\}\langle 110\rangle$
Gleitung beschränkt und reicht deshalb für eine Deutung der
bei kubisch flächenzentrierten Metallen beobachteten Texturunterschiede nicht aus. Es werden deshalb im Rahmen eines
verallgemeinerten TAYLOR'schen Modells der Verformung andere
Verformungsmechanismen, die in letzter Zeit im Zusammenhang
mit Texturtheorien vorgeschlagen wurden, diskutiert. Es sind
dies: Quergleitung, latente Verfestigung, Stapelfehlerbildung
bei der Verformung (Gleitung auf $\{111\}\langle 112\rangle$) und mechanische Zwillingsbildung. Störungen durch die Entstehung von
Deformationsbändern werden ebenfalls berücksichtigt. Durch
die Verwendung eines Rechenautomaten wird eine derartige Untersuchung sehr erleichtert.

Die Ergebnisse zeigen, daß in Drähten kubisch flächenzentrierter Metalle mit hohen Werten von γ/Gb das Vorherrschen
von $\langle 111\rangle$ auf Kolineares Gleiten (Quergleiten) und die Bildung von Deformationsbändern zurückzuführen ist. Das gleiche

---

[+)] Bell Telephone Laboratories, Incorporated, Murray Hill,
New Jersey, USA

Vorherrschen bei sehr niedrigen Werten von γ/Gb ist hauptsächlich auf Koplanares Gleiten (latente Verfestigung) und
instrinsische Stapelfehlerbildung zurückzuführen. Das hohe
⟨100⟩-Maximum nahe Ag wird hervorgerufen durch passende
Zwillingsbildung, die hauptsächlich dann auftritt, wenn die
Faserachse durch normales Gleiten ⟨111⟩ erreicht hat. Die
Drahttexturen von kubisch raumzentrierten und die Drucktex-
turen von kubisch flächen- und raumzentrierten Metallen können in gleicher Weise gedeutet werden.

2.1.1 Introduction

Much progress has been made in recent years toward an
improved understanding of deformation textures. Since most
of the works have been summarized in several excellent
reviews [1-4], there will be no attempt to repeat them here.
Rather, the aim of the present paper is to view the problem
of crystal plasticity in perspective and to examine some of
the current theories of texture in terms of this perspective.
Attention will be focused on the explanation of the wire
textures of fcc metals for which data are most complete and
to which most of the theories have been directed. The
relevant works on the wire textures of bcc metals and the
compression textures of both cubic metals will be discussed
subsequently. The related topic on rolling textures is
reviewed by HEYE[5] in the next chapter (2.2).

2.1.2 Crystal Plasticity and the Selection of Active Slip
Systems

It is generally known that deformation textures are
developed as a result of lattice rotation brought on by
crystallographic shear (slip and/or twinning) during plastic
flow. The lattice rotation can be calculated if the active
shear systems are known at each stage of the deformation
process. In the usual unconstrained tension test, the shear
from one slip system is sufficient to accommodate the tensile
deformation. Since slip obeys a critical resolved shear
stress criterion[6], the active slip system is one for which
the Schmid factor is a maximum among all equivalent slip

systems. This amounts to saying that the inverse of the
Schmid factor, $M = \sigma/\tau = \gamma/\epsilon$, is a minimum, where σ and ϵ
are the tensile stress and strain and τ and γ are the
resolved shear stress and shear strain.

A grain embedded in a polycrystalline aggregate, on the other
hand, cannot change its shape freely because of constraints
from adjacent grains. In this case, slip from at least five
independent slip systems (to accommodate five independent
strains) is generally required [7]. TAYLOR [8] advances a
principle of minimum work and hypothesized that among all
combinations of (five) slip systems which are capable of
satisfying the imposed strain, the active combination is
that one for which the sum of the glide shears is a minimum.
Again, this is equivalent to saying that the value
$M = \Sigma\gamma_j/\epsilon$ is a minimum, in analogy to the single slip case.

Later work by BISHOP and HILL[9] strengthened Taylor's
theory by showing that is fulfills the yield criterion;
i.e. the shear stress reaches the critical level for slip
on the active slip systems chosen by Taylor without
exceeding it on the inactive slip systems. This point is
extremely important because other methods of selection, such
as those proposed by SACHS [10], COX and SOPWITH [11], WEVER
and SCHMID [12], BOAS and SCHMID [13], KOCHENDÖRFER [14],
CALNAN and CLEWS [15], and TUCKER [16] do not satisfy the
imposed strain or else the yield condition is violated.
There is a tendency to consider all crystals in an aggregate
deforming by the same stress state, such as a tensile stress
in tension or a biaxial stress in rolling, thus neglecting
the stresses arising from constraints necessary to satisfy
an imposed strain. As a result, continuity of stress and
strain across grain boundaries is violated[9]. It may
perhaps be argued that the grain interior can deform by single
or duplex slip, while the difference between the shape
change resulting from such slip and that demanded by the
macroscopic deformation is somehow made up by multiple slip
in the grain boundary region. The value of such an approach
has been questioned by TAYLOR[17]. Experimentally, FLEISCHER
and BACKOFEN [18] have shown that in bicrystals of square

cross-section,where a grain is constrained only on one
surface, the grain boundary multiple slip region extends to
0.5 - 0.75 of the crystal volume. For a crystal bounded on
two sides, the extend of penetration from each boundary is
even greater [19]. Hence for a completely embedded crystal,
single and duplex slip regions probably do not exist throughout much of the grain this way.

In addition to satisfying the yield condition, the Taylor
theory provides strain continuity across grain boundaries
by assuming that each grain deforms in the same way as the
aggregate. The condition of stress continuity, however is
not satisfied. Hence in actual crystals the strain is
probably nonuniform. Nevertheless, the Taylor model of deformation represents the most realistic model to date that
is capable of being analyzed from the standpoint of texture
development [+]. For this reason, until a more realistic model
comes along, any suggested mechanisms of deformation textures
would best be analyzed in terms of the Taylor framework.
Taylor's texture theory, based on {111} ⟨110⟩ slip, is considered first. Other mechanisms such as cross-slip, latent
hardening, deformation faulting and mechanical twinning which
have been suggested in recent theories, are treated subsequently. The question of deformation banding - a sign of nonuniform strain - will also be considered.

2.1.3 Taylor's Theory of Wire Textures ({111} ⟨110⟩ Slip)

Using his minimum work criterion of selecting active slip
systems, Taylor calculated the lattice rotations based on
{111}⟨110⟩ slip for fcc metals. His result for axis-
symmetric tension, such as occurs in wire-drawing, are shown
in Fig. 1. The tail and head of an arrow correspond to the
positions of the tensile axis before and after a 2.37 %
extension, respectively. Double arrows indicate

[+]Other deformation models [20-23] treated the early stages
of yielding where elastic strains are important. These
treatments apply to a randomly oriented aggregate only
and the results indicate that after a total strain of 6 -
10 times the elastic yield strain, the deformation becomes
essentially that described by Taylor.

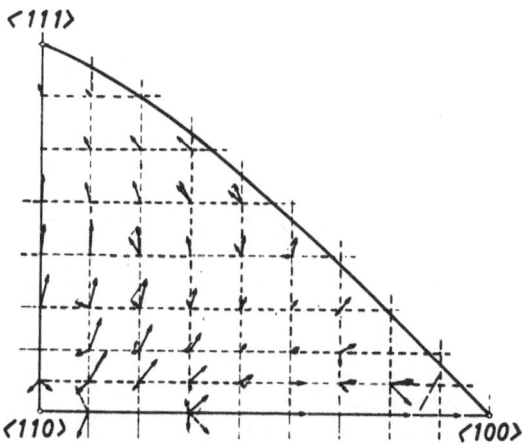

Fig.1. Lattice rotations for axisymmetric tension, determined by TAYLOR [8] for {111}⟨110⟩ slip. The sense of rotation is reversed for compression; 2.37% strain

equally favored combinations of five slips; any rotation in-between them could equally well occur. The predicted rotations for axisymmetric compression are exactly the reverse of those for tension. Unfortunately, many other equally favored combinations were left out of Fig. 1 due to an oversight on the signs of the equations relating the glide shears to the external strains[17]. Thus when BARRETT and LEVENSON[24] tested Taylor's theory by examining the lattice rotations of some 25 grains in a compressed aluminum block, only "about half of the grains rotate as predicted, about a third do not, and the rest are uncertain." Actually, when the complete Taylor theory was solved using techniques of linear programming and with the help of a computer[25], the results, shown in Fig. 2, indicate that nineteen of the grains (75%) examined by BARRETT and LEVENSON[24] rotate as predicted.

In addition, Fig. 2 shows that the complete Taylor theory predicts three regions of rotation: region A where the rotation is toward ⟨100⟩; region B, toward ⟨111⟩; and region C,

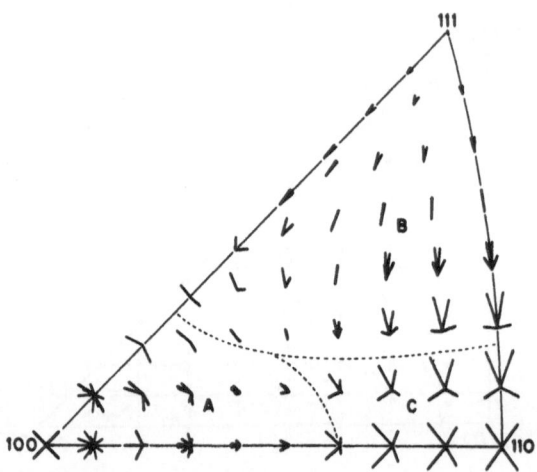

Fig.2. Computer-plotted complete results based on
TAYLOR's theory [8], as obtained by CHIN, MAMMEL a.
DOLAN [25], for {111}⟨110⟩ in axisymmetric
tension; 5% strain

either ⟨100⟩ or ⟨111⟩. This behaviour has been observed in Al,
Cu, Au, Ag and 85-15 brass single crystals studied by AHL-
BORN [26]. A typical result, for Cu, is shown in Fig. 3.

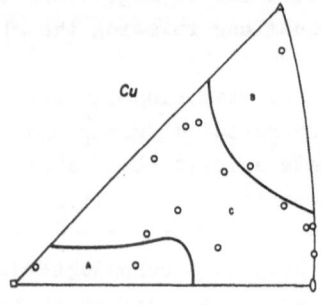

Fig.3. Experimental results
obtained by AHLBORN [26] on
wire-drawing of Cu single
crystals. The three regions
of rotation correspond to
those of Fig.2

The solid lines separate the three regions of rotation as
predicted from the Taylor theory, although the observed
region C is somewhat larger than predicted. In comparing
his single crystals data with the incomplete Taylor results

(Fig. 1), AHLBORN[27] had noted that crystals near [110] rotate away from [110], contrary to prediction. This criticism is no longer necessary in view of the complete results of Fig. 2.

CHIN, MAMMEL and DOLAN[28] also charted some theoretical path of rotation by assuming an initial combination of slips allowed by the Taylor theory and then, after rotating to a new position as a result of a small incremental strain (5%), calculating a new combination of slips which most nearly matches the initial combination. This was done by minimizing the difference between the two sets of shears. The process is then repeated, generating a complete path of lattice rotation. Three such paths issued from one initial orientation are shown in Fig. 4a; They are very similar to those observed by AHLBORN[26], see Fig. 4b.

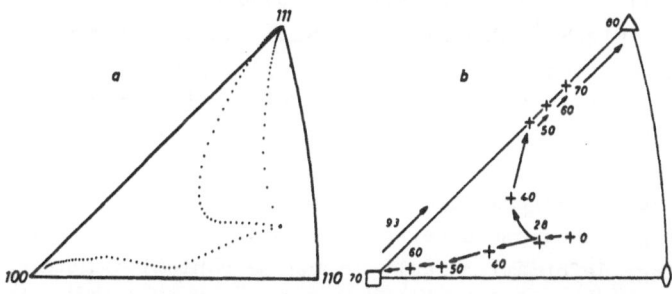

Fig.4a. Computer-plotted paths of rotation based on Taylor's theory, as obtained by CHIN, MAMMEL and DOLAN [28], for {111}⟨110⟩ in axisymmetric tension

Fig.4b. Experimentally determined paths of rotation obtained by AHLBORN [26] on a gold crystal for increasing degree of deformation between 0 and 93%

Other tests are in basic support of the Taylor theory. The value of the orientation factor M, mentioned in the Introduction, relates the applied stress-strain curves of different orientations to the basic shear stress – shear strain curve and can be tested. For a randomly oriented aggregate, the average value of M for axisymmetric flow is 3.067[8,9,29], in good agreement with observation. In recent years, single crystals of different orientations

have been tested in environments similar to those
encountered by an embedded grain. Direct stress—strain
measurements of crystals deformed in plane strain
compression (simulated rolling) have upheld the Taylor
theory fairly well[30-32]. The predicted active slip systems
have also been verified metallographically [30,32]. Less
direct measurements on crystals deformed in wire—drawing
are somewhat conflicting at present[33,34], with some of
the complications possibly due to "cocking" and to the onset
of mechanical twinning in crystals such as Cu—7% Al. The
question of twinning will be taken up later.

Despite such notable success, one major shortcoming of
Taylor's theory is that it treats {111}⟨110⟩ slip only and
treats all twelve slip systems alike. As a result, the
Taylor theory is unable to explain the variation of the
duplex ⟨111⟩ + ⟨100⟩ wire texture found among fcc metals
[25], see Fig.5[+). Consequently, with the aid of the linear
programming technique applied previously [25,28,29], we
decided to incorporate into the Taylor framework the
various deformation mechanisms proposed in recent theories,
which by themselves either lack the rigor of the minimum
work model of deformation or provide no model at all. The
following mechanisms have been treated: cross—slip, latent
hardening, deformation faulting and mechanical twinning.
Two other recent proposals have not yet been considered.
These include HAESSNER's {100}⟨110⟩ slip mechanism[37]

[+)]It has been suggested[36] that the decrease of ⟨100⟩ at
very low values of γ/Gb may not be as large as shown in
Fig. 5. A careful re-evaluation of the Co-10%Fe wire
texture using Co-Kα radiation and balanced Fe and Mn
filters to obtain true Bragg reflection showed that the
texture actually consists of ⟨111⟩ + ⟨411⟩ after 99.3%RA.
Evaluation of the {111} and {200}-pole densities gave
86% ⟨111⟩ + 14% ⟨411⟩ and 88% ⟨111⟩ + 12% ⟨411⟩ respectively.
Hence the results of Fig. 5 are not significantly altered
if ⟨411⟩ is considered belonging to ⟨100⟩. The value of
5% ⟨100⟩ in Fig. 5 was estimated on the basis of interpreting the {111} and {200} densities in terms of a ⟨111⟩
+ ⟨100⟩ duplex texture. The single crystal results of
AHLBORN [26] are also in basic agreement with the shape
of Fig. 5.

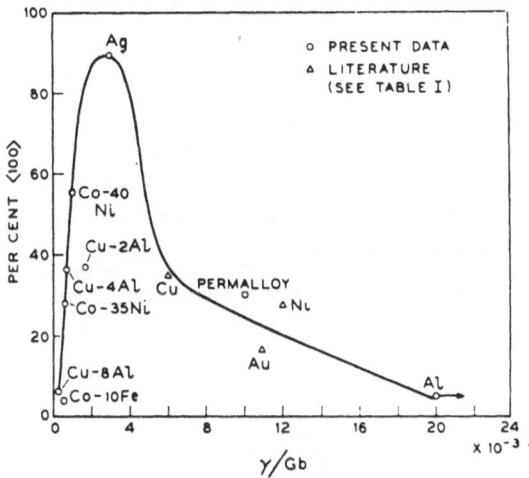

Fig.5. The variation of ⟨111⟩ + ⟨100⟩ duplex wire texture of fcc metals as a function of the stacking fault energy parameter γ/Gb, where γ is the stacking fault energy, G is the shear modulus and b the strength of the Burgers's vector (after ENGLISH and CHIN [35])

for high stacking fault energy metals; evidence for such slip is rather limited. The possibility of incorporating dislocation reaction mechanisms such as that proposed by LIU [38] has yet to be exploited.

2.1.4 Cross—Slip (Colinear Slip)

There is ample evidence from single slip experiments that cross—slip occurs more easily in high stacking fault energy metals and at high testing temperature, see Fig. 6 [39]. One is thus tempted to generalize such a cross—slip mechanism to the multiple slip situation in constrained deformation, so that all {111}⟨110⟩ slip system pairs sharing a common slip direction (colinear slip) are favored. Thus among the several equivalent slip system combinations selected by the minimum work criterion, one can seek out and maximize the activities of the colinear pairs. The notion of cross—slip has been suggested by BROWN[40] for wire textures, and by SMALLMAN and GREEN [41] and DILLAMORE and ROBERTS [42] for

sheet textures, but has been criticized [37] that at large
stresses all metals should be able to undergo cross-slip.
In the present view, even though stresses may be high
enough to cause cross-slip in all metals, the low stacking
fault energy alloys still would choose other slip
combinations (e.g. coplanar slip as discussed in next
section) whenever they are available. Similarly, the high
stacking fault energy metal would seek out the colinear
pairs among various allowed combinations.

Fig. 7 shows the predicted rotations based on maximized
colinear slip within the confines of the minimum work

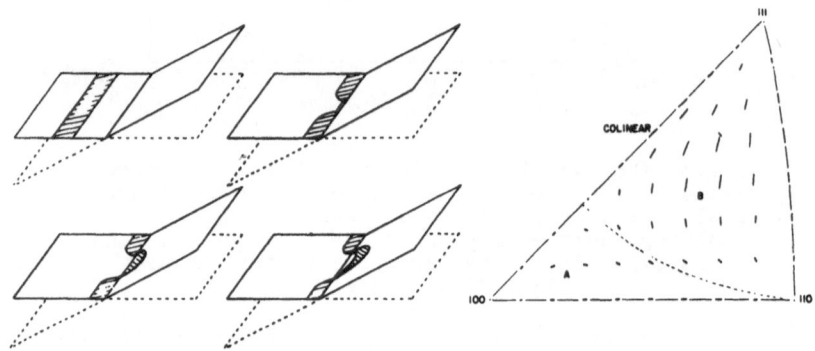

Fig.6   **Fig.7**

Fig.6. Model of the cross-slip process (after NABARRO and
BASINSKI [39]). The dissociated dislocation must combine
prior to movement into the cross-slip plane. Hence cross-
slip is easiest for high stacking fault energy metals

Fig.7. Computer-plotted lattice rotations obtained for {111}
⟨110⟩ slip in which the activities of colinear pairs have
been maximized within the confines of minimum work.
Axisymmetric tension; 5% strain

criterion. As compared with Fig. 2, it may be seen that
those crystals within the previous region C are biaxed
toward the ⟨111⟩ orientation, thus partially accounting for
the drop in the ⟨100⟩ component for high stacking fault
energy metals, see Fig. 5.

2.1.5 Latent Hardening (Coplanar Slip)

Another type of {111}⟨110⟩ slip modification involves the
concept of latent hardening as proposed by BISHOP [43] and
CALNAN [44]. The only latent hardening data available at the

time was tensile overshoot, which involves the activation
of the conjugate slip system after prolonged activity on
the primary system has rotated the tensile axis past the
[111] – [100] symmetry line. Based on this concept, Bishop
chose among the several equivalent combinations of slip
systems the one for which the largest glide shear is as
large as possible where latent hardening exceeds active
hardening. But if latent hardening equals active hardening,
then the combination is chosen such that the smallest glide
shear is as large as possible. His predicted wire textures
are shown in Fig. 8. Bishop claims that for a metal where
latent hardening is large, such as for brass, the ⟨100⟩
texture is unstable and rotates to ⟨111⟩, Fig. 8b; whereas
both components are stable if latent and active hardenings
are equal, Fig. 8a. Bishop's interpretation is somewhat
open to question. Since his results prior to the latent
hardening selection are identical to those obtained with
the Taylor theory, Fig. 2 shows that whatever the method of
selection among equivalent combinations, the ⟨100⟩
orientation cannot rotate more than about 15° away and is
hence "stable" as long as the minimum work criterion
remains in force. It may also be seen from Bishop's
results, Fig. 8, that the rotation of ⟨100⟩ cannot be very
far. Basic criticisms of the tensile overshoot concept
have also been offered [37,45].

On the other hand, our understanding of latent hardening
has improved in recent years. Experiments have been
performed by which pairs of slip systems other than primary-
conjugate relationship have been tested [46–48]. These test
results have revealed that prior slip on one system hardens
all other systems more than itself except a coplanar system,
where the hardening is about equal. There is a tendency for
hardening on non-coplanar systems to increase with
decreasing stacking fault energy. For Al, the increase is
about 20% [47]; for Cu [48] and Ag [46], about 40%. These
data were obtained by first activating one system and then
the other, and it is not obvious that they are applicable
to simultaneous activity of both systems. Generally,
intersections between extended dislocations are expected

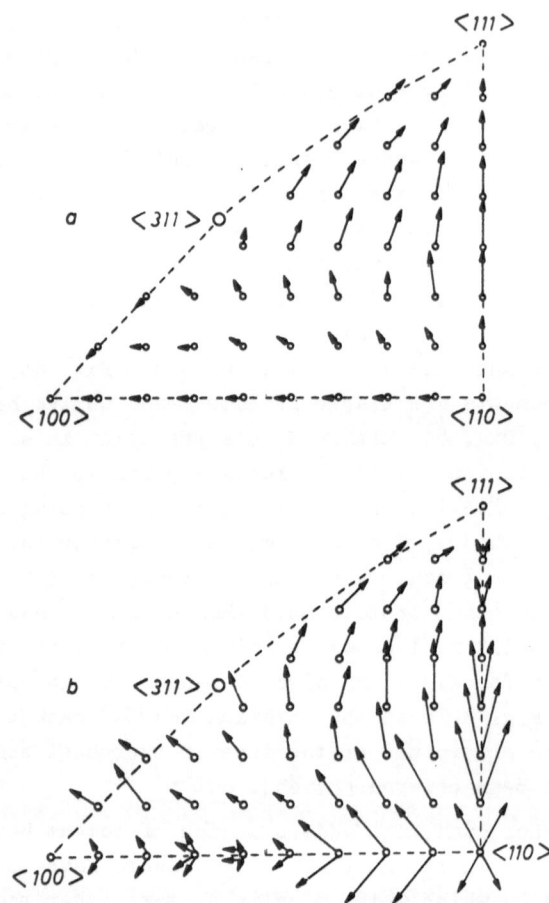

Fig.8. Lattice rotations obtained by BISHOP [43] for {111}
⟨110⟩ slip whereby (a) the smallest allowed shear (within
the confines of minimum work) is maximized and (b) the
largest allowed shear is maximized

to be more difficult than unextended ones, Fig. 9 [39].
Hence a true test of the latent hardening concept should
involve a selection among the equivalent slip system
combinations those which maximize coplanar slip. Fig.10
shows the predicted rotations when coplanar slip is thus
favored within the confines of the minimum work criterion.
Somewhat surprisingly, the results again indicate a bias

63

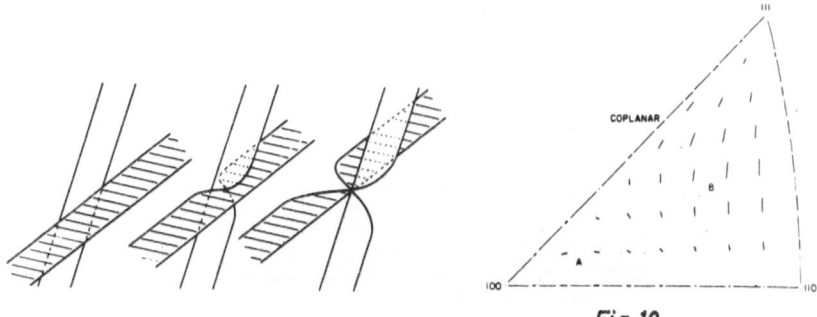

Fig.10

Fig.9. Model of dislocations from intersecting planes (after
NABARRO and BASINSKI [39]). Intersection is more difficult
with extended dislocations

Fig.10. Computer-plotted lattice rotations obtained for
{111}⟨110⟩ slip in which the activities of coplanar pairs
have been maximized within the confines of minimum work.
Axisymmetric tension; 5% strain

of crystals in previous region C (See Fig. 2) toward the
⟨111⟩orientation. Hence latent hardening (coplanar slip)
also favors an increase of the ⟨111⟩ texture as Bishop had
concluded, although the increase does not come from the
⟨100⟩ instability as Bishop stated.

2.1.6 Deformation Faulting ({111}⟨112⟩Slip)

For energetic reasons, a {111}⟨112⟩ dislocation is normally
split into two Shockley partial dislocations of the {111}
⟨112⟩ type (see Fig. 11) connected by a strip of stacking
fault whose width is inversely proportional to the value
of stacking fault energy of the material. Thus widely-
spaced stacking faults are rather common in cold-worked
low stacking fault energy metals. Usually the separated
partial dislocations "zig-zag" in correlated sequential
movement – first along [$\bar{1}2\bar{1}$] and then along [$11\bar{2}$], Fig.11.
The result would then be the same as for normal (111)[$01\bar{1}$]
slip (Fig. 11a) and there would be no alteration of the
pattern of texture development except the extra latent
hardening on non-coplanar systems as described previously.
However, if the partials become widely separated, the

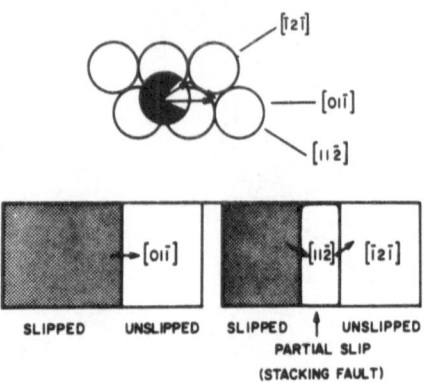

Fig.11. Hard-sphere model of [01T̄] slip via [T̄2T̄] (leading) and [112̄](trailing) partials. If the partials are unextended, (a) the motion is the same as for [01T̄]. If extended (b), a strip of intrinsic stacking fault exists between the two partials

texture development may be altered because the effective slip direction is switched from ⟨110⟩ to ⟨112⟩. This is the mechanism proposed by HU, CLINE and GOODMAN [3] to account for the appearance of the brass-type rolling texture and for the dominance of the ⟨111⟩ wire texture component of low stacking fault energy metals. It appears from their analysis of (110)[T̄12] and (112)[11T̄] rolling that the authors consider the independent motions of both partials equally likely. In effect, however, probably only the leading partial can do so. With reference to Fig. 11, there is a chance that the leading partial, [T̄2T̄], may move ahead creating a strip of instrinsic faulting [49], while the trailing partial, [112̄], is pinned behind. The slip movement would then be equivalent to (111)[T̄2T̄] in which case the lattice rotation and hence the texture development would be altered from the usual (111)[01T̄] slip pattern. If the leading partial is pinned, there is no way for the trailing partial to "unzip" the stacking fault and move past the pinning obstacle. Besides the repulsion between the partials, movement of the trailing partial after unzipping constitutes high energy faulting, a far less

likely event [50]. Hence if deformation faulting is to
provide independent slip, it probably does so through the
motion of the leading partial only. One difficulty with
this process is that only one partial per plane can move
this way, since otherwise the sequential motion of the
trailing partial would result in effective [01$\bar{1}$] slip. Other
considerations regarding stacking fault propagation have
been given by ESCAIG, FONTAINE and FRIEDEL [51].

By considering, then, {111}⟨112⟩ slip by intrinsic faulting
and applying the minimum work criterion for the selection
of the active slip systems, we have obtained (again via
computer methods) the predicted lattice rotations shown
in Fig. 12 for tension. As expected, the results are

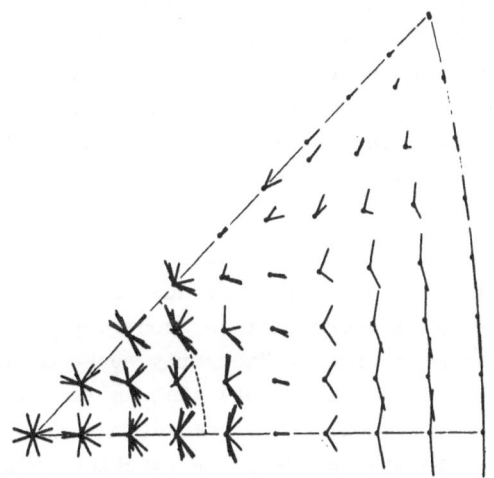

Fig.12. Computer-plotted lattice rotations for
{111}⟨112⟩ intrinsic faulting, according to minimum
work criterion. Axisymmetric tension; 5% strain

different from those obtained with {111}⟨110⟩ slip. In
particular, the [100] position is no longer stable as was
with {111}⟨110⟩ slip. Although it is possible for crystals
near [100] to rotate to [100], continued rotation would
lead them to the [111] stable orientation. Thus intrinsic
faulting favors a strong ⟨111⟩ wire texture, reinforcing
the effect of coplanar slip for very low stacking fault

energy alloys. This result is obtained without analogy to hcp wire textures [3].

2.1.7 Mechanical Twinning ({111}⟨112⟩Twin)

If the motion of the partial dislocation described above generates a stacking fault on successive (111) layers, the result is a mechanical twin. Since the orientation within a twin differs drastically from that of the matrix, mechanical twinning is expected to play an important role in the development of deformation textures. AHLBORN and WASSERMANN [52] have found that the ⟨100⟩ component of the wire texture of Ag is extremely intense (see Fig. 5). They reasoned that the elongation in wire drawing favors mechanical twinning of the ⟨111⟩ orientation, which is then converted to ⟨511⟩ whence a ⟨100⟩ position may be reached by additional slip. They also argued that the shape change does not favor twinning of the ⟨100⟩ orientation. The twinning mechanism was later applied to rolling [53]. A number of studies have established that fcc metals of low stacking fault energy do twin, especially when deformed at low temperatures and high strain rates [54].

If twinning favors the ⟨100⟩ orientation at the expense of ⟨111⟩ in Ag, then ⟨100⟩ should continue to dominate in still lower stacking fault energy metals, contrary to the texture reversal as shown in Fig. 5. The early works on tensile pulling of single crystals showed that as the stacking fault energy of the metal is decreased, the size of the twins decreases so that at very low stacking fault energies the twins practically become indistinguishable from stacking faults [49]. VENABLES [49] reasoned that with decreasing stacking fault energy, twin nucleation is made easier but propagation becomes more difficult. It was thus thought[35] that the decreased propagation might mean a corresponding decrease in the twinned volume, hence the decrease in the ⟨100⟩ component. However, AHLBORN [26] observed profuse twinning in wire drawing of 85-15 brass single crystals, and CHIN, HOSFORD and MENDORF [55] detected large scale twinning in plane strain compression of Co-8%Fe crystals

at very early stages of deformation. Both of these alloys
are expected to have low stacking fault energies. Hence the
argument based on decreasing twinned volume is inadequate.

Recently the mechanics of twinning has been incorporated
quantitatively into the minimum work criterion [55], with
interesting consequences on the nature of texture
development. In the new analysis, the incremental work done
by a stress σ_x causing a strain ϵ_{xx} is equal to

$$W = \sigma_x \epsilon_{xx} = \tau_s \sum_i s_i + \tau_t \sum_i t_i, \qquad (1)$$

where τ_s and τ_t are the critical resolved shear stresses
for slip and for twinning respectively, and s_i and t_i are
the corresponding shears. (Only the first term was
considered previously). The orientation factor M then
becomes

$$M = \frac{\sigma_x}{\tau_s} = \frac{1}{\epsilon_{xx}} \left[\sum_i s_i + \alpha \sum_i t_i \right] \qquad (2)$$

where $\alpha = \tau_t/\tau_s$. As in previous treatments, the operative
combination of slip and twin systems is found from one for
which M is a minimum. It was further demonstrated that for
{111}⟨110⟩slip and {111}⟨112⟩twinning applicable to fcc
crystals, mixed slip and twinning can only occur for values
of α between $1/\sqrt{3}$ and $2/\sqrt{3}$. If $\alpha > 2/\sqrt{3}$, only slip is
possible; and if $\alpha < 1/\sqrt{3}$, only twinning is allowed.

The modified Taylor theory was applied to selected
orientations of crystals deformed in plane strain
compression. The theory was tested with single crystals
of Co-8%Fe. Metallographic and x-ray pole-figure observations
verified the predicted active slip and twin systems.
Correlation of the stress-strain curves was also favorable.
Similar x-ray pole figure observations on rolled Ag
crystals studied by HEYE and WASSERMANN [56] were also
explained on this basis.

The theory has also been applied via linear programming
techniques to the case of axisymmetric flow [57]. It was
shown quantitatively that in axisymmetric tension (wire-

drawing),the $\langle 111 \rangle$ orientation is very favorable for twinning while $\langle 100 \rangle$ is not, in agreement with Ahlborn and Wassermann's earlier deductions. In addition, the new results cover a l l axial orientations, showing that as soon as the value of α drops below $2/\sqrt{3}$, some twinning is expected for all orientations with the exception of a 10° region surrounding $[100]$. The predicted fraction of the macroscopic strain taken up by twinning varies with orientation, as shown in Fig. 13 for the example of $\alpha=1$.

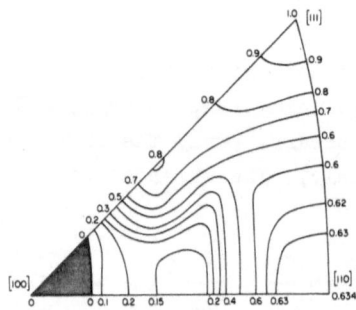

Fig.13. Fraction of strain in mixed $\{111\}\langle 110 \rangle$ slip and $\{111\}\langle 112 \rangle$ twinning that can be accommodated by twinning, according to minimum work criterion for axisymmetric tension. Calculations were made assuming the critical resolved shear stresses for twinning and slip are equal ($\alpha=1$)

It is zero in the $[100]$ vicinity (shaded region) and increases to 1 at $[111]$, in accord with AHLBORN's [27] observations on drawn crystals. The expected active twin systems and the twinned orientations were also determined[57]. Fig. 14 shows a standard (011) stereographic projection. Assuming the original wire axis lies within the triangle $[100]$ – $[110]$ – $[111]$, the twinned orientation will then lie within one of the four shaded triangles depending on which of four $\{111\}$ planes twinning has occurred. (The twinning direction does not enter into the reorientation). The different types of shading apply to the occurrence of twinning within certain values of α in accordance with the minimum work criterion. The black regions, for example, designate the allowable twinned orientations for

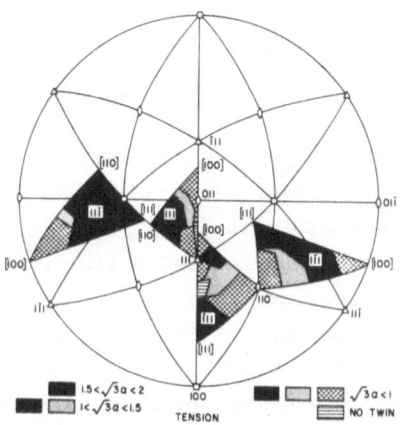

Fig.14. (110) standard stereographic projection
showing the reorientation of the [100] – [110] –
[111] triangle by twinning on each of four {111}
planes. The different shadings refer to twinning
allowed by the minimum work criterion for different
values of α. Axisymmetric tension. Note ease of
twinning near [111] but not [100]

$1.5/\sqrt{3} < α < 2/\sqrt{3}$. It may be seen that for a matrix wire
axis at [111], twinning occurs on (11$\bar{1}$), ($\bar{1}$11) and (1$\bar{1}$1)
(but not (111)), converting it to three different variants
of the ⟨511⟩ type. A [100] matrix, on the other hand, does
not twin until $α < 1/\sqrt{3}$, in which case it twins on all four
{111} planes and is converted to a ⟨221⟩ type orientation.

Both Figures 13 and 14 may shed new light on the wire
texture reversal problem. For metals such as Ag, twinning
probably comes with some measure of difficulty. Hence in
a polycrystalline material most crystals deform by slip
fairly extensively before the stress is built up to the
level at which twinning becomes favored. By this time, these
crystals would have rotated to the ⟨111⟩ vicinity. From
Fig. 13, all of the incremental strain can theoretically be
taken up by twinning, and all of the twinned orientation
would then be at ⟨511⟩, whence the rotation to ⟨100⟩ with
further slip. On the other hand, if the twinning becomes
so easy that it occurs b e f o r e the wire axis has

rotated to [111] by slip, only a portion of the strain is accommodated by the twins, Fig. 13. Moreover, of the four {111} twin planes, only (11$\bar{1}$) converts the wire axis to the ⟨100⟩ vicinity, Fig. 14. Hence the ⟨111⟩ texture may not be appreciably affected in these easily-twinned alloys.

One difficulty with the above rationale is that there is not apparent reason why the ⟨111⟩ texture, formulated after some initially complicated sequence of slip and twinning, would not t h e n twin to ⟨511⟩. This difficulty may be removed by assuming that initial twinning tends to suppress subsequent twinning. The assumption is certainly reasonable since the ends of twins are often observed to stop on other twins, presumably for strain compatibility reasons. In analogous ways, twinning is known to become more difficult as the grain size is decreased [58]. Thus the texture will depend in an important way on whether twinning occurs early or late in the deformation process.

2.1.8 Deformation Banding

All the topics treated thus far - colinear and coplanar slips, deformation faulting, mechanical twinning - are still within the confines of a broadened Taylor theory governed by the principle of minimum work and the assumption of homogeneous deformation. Yet deformation bands - a sign of inhomogeneous deformation - are a common feature in cold-worked metals as shown by the works of BARRETT [4], AHL-BORN [26,59], and REED and MC HARGUE [60] among others. It is thus important to examine the conditions under which deformation bands are formed without violating the basic principle of minimum work, and the role these bands play in texture formation.

One type of deformation banding is already allowed by the basic Taylor theory. For each orientation, there may be several equivalent combinations of slip systems which satisfy the minimum work criterion (see Fig. 2). Hence deformation banding (Type 1) can result from the operation of a different allowable combination in each of several portions of a grain. A large number of observed deformation bands may be explained this way.

On the other hand, there is another type of deformation banding (Type 2) within which the slip systems differ from those selected on the basis of homogeneous deformation. This appears to be the type studied by AHLBORN [59], who carefully examined the deformation banding behavior of $\langle 100 \rangle$ and $\langle 111 \rangle$ crystals deformed in wire drawing. He observed that the $\langle 100 \rangle$ crystal splits up into deformation bands bounded by $\{100\}$ boundaries. The rotations of two neighboring bands were directed along two orthogonal $\langle 100 \rangle$ – $\langle 111 \rangle$ symmetry lines toward the $\langle 111 \rangle$ symmetry lines toward the $\langle 111 \rangle$ position. Banding was not observed in the $\langle 111 \rangle$ crystal. Since the $\langle 100 \rangle$ orientation cannot rotate past about 15° as long as slip occurs only on $\{111\}\langle 110 \rangle$ systems and as long as the flow is homogeneous, Fig. 2, the instability of $\langle 100 \rangle$ must be ascribed to a different set of slips in the bands. CHIN and WONSIEWICZ [61] analyzed this problem and concluded that such bands can form if the work done (i.e. the M factor) by slip within the bands is less than that for homogeneous deformation, and if the bands can be arranged such that the net strain practically matches that for the overall deformation. The boundary separating neighboring deformation bands can be deduced from the necessity of strain continuity across it. On this basis, it was shown that for wire drawing of the [100] orientation, deformation bands within which duplex slip with primary-conjugate relationship are favored. The predicted band boundaries initially lie on $\{100\}$ planes as Ahlborn observed, and the predicted rotations are along $\langle 100 \rangle$ – $\langle 111 \rangle$, again as observed, see Fig. 15. A theoretical pole figure of five bands studied by Ahlborn matches the observed pole figure fairly well, see Fig. 16. It was further shown that for the [111] orientation, deformation bands with net strains approximating the circular cross-section are unlikely; hence homogeneous axisymmetric tension is favored at [111].

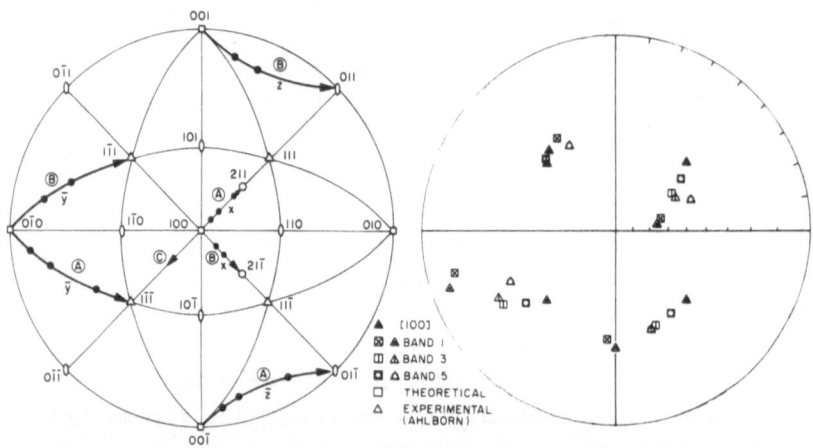

Fig.15. (100) standard stereographic projection showing the rotation of axes (x, y, z) of two groups of deformation bands (A und B) analyzed by CHIN and WONSIEWICZ [61]. The active slip systems within the bands are of primary-conjugate relationship; strain continuity is preserved across the band boundary x-y; and the net strain from the two types of bands is axisymmetric flow except for a ε_{xy} term. The dotted points refer to the positions of 5 deformation bands studied by AHLBORN [59].

Fig.16. Theoretical{111} pole figure of three of the bands (A-type in Fig.15) determined from Fig. 15. Comparison with Ahlborn's experimental pole figure data is good. The other two bands also compare well but have been omitted to avoid confusion

2.1.9 A Synthesis

Using the principle of minimum work and assuming homogeneous deformation, we have derived the predicted lattice rotations for wire-drawing of fcc metals based on various deformation mechanisms proposed in recent texture theories. The effect of inhomogeneous deformation (banding) as a perturbing factor was also considered. The trends of lattice rotations

for the various mechanisms are summarized in Table 1. It
must be emphasized that the predicted final texture for
each mechanism corresponds to the independent action of
that particular mechanism a l o n e , and that the actual
texture will depend on the resultant of the complex
competing mechanisms. The first four textures are derived
on the basis of an initial random distribution of crystals.
To the extent that this is not so, the results will of
course be altered but accountable. A question mark is
placed in the twinning column since the final texture
depends on the instantaneous distribution of crystal
orientations at the time twinning occurs. Hence the exact
influence of twinning must be correlated with the
simultaneous competing slip processes both before and after
twinning. A similar question mark is placed in the
deformation banding column. One type of banding (Type 1)
which is expected to occur for most orientations is
accounted for by the basic $\{111\}\langle110\rangle$ slip mechanism. The
second type (Type 2) appears to be associated only with
certain symmetrical orientations and hence cannot be
properly categorized with those mechanisms covering all
orientations. To the extent that it occurs prevalently at
[100], however, it is an important contributor to the
decrease of the $\langle100\rangle$ texture component. On the other hand,
the persistence of the $\langle100\rangle$ component in polycrystalline Ag
wire suggests that such banding may be suppressed if the
structure is disturbed by extensive prior twinning; hence
it may not be of general applicability to low stacking
fault energy metals.

Thus with reference to Fig. 5 and Table 1, it may be argued
that the predominance of the $\langle111\rangle$ component at very high
values of γ/Gb is due primarily to colinear slip and
deformation banding. The same predominance at very low
values of γ/Gb is probably due to coplanar slip and
intrinsic faulting. The large $\langle100\rangle$ peak near Ag is accounted
for by opportune twinning – chiefly after the axial
orientations of the crystals have reached the $\langle111\rangle$
position via a slip process.

TABLE 1. Summary of Results

Deformation Mechanism	Favored by	Predicted Final Wire Texture	Ref. Figs.
{111}⟨110⟩ slip	all γ/Gb	70%⟨111⟩+30%⟨100⟩	2
Colinear slip (cross slip)	high γ/Gb	77%⟨111⟩+23%⟨100⟩	7
Coplanar slip (latent hardening)	low γ/Gb	77%⟨111⟩+23%⟨100⟩	10
Intrinsic Faulting	low γ/Gb	100%⟨111⟩	12
Deformation twinning	low γ/Gb	?	14
Deformation banding	all γ/Gb?	100%⟨111⟩ ?	15

Having thus synthesized such a picture of the development
of wire textures of fcc metals, we are now in a position
to examine within the present framework the wire textures
of bcc metals as well as the compression textures of both
cubic metals.

2.1.10 Compression Textures of fcc Metals

BARRETT and LEVENSON [24] have reported that the compression
texture of Al consists of a strong ⟨110⟩ component with
considerable spread to ⟨311⟩ plus weak intensities
scattered to ⟨100⟩ and with ⟨111⟩ relatively empty, Fig.
17a. This is the type of texture expected on the basis of
{111}⟨110⟩ slip. When the sense of rotation in Fig. 2 is
reversed (for compression), the predicted texture consists
of an orientation spread in region C, plus a certain degree
of sluggishness of crystals of region A to rotate into C.
Since colinear (cross) slip is presumably favored with Al,
the predicted ⟨110⟩ texture is probably sharper, see Fig.7.
Unfortunately, quantitative experimental data of other
metals are not available for comparison.

BARRETT and LEVENSON [24] also examined the compression
texture of 70-30 brass, Fig. 17b. Here the texture still
consists of a strong ⟨110⟩ component with some spread to

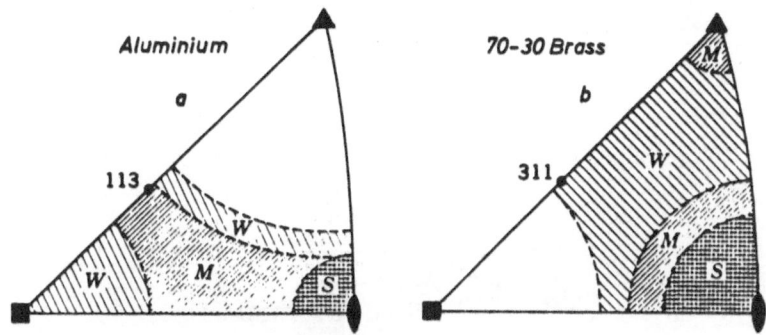

Fig.17.Compression texture (97% RA) of Al and
70-30 brass determined by BARRETT and LEVENSON
[24].W = Weak, M = Medium, S = Strong

⟨311⟩, but now the ⟨100⟩ orientation is empty and some
intensity appears near ⟨111⟩. The emptying of the ⟨100⟩
vicinity in favor of ⟨111⟩ most likely comes from mechanical
twinning. In compression, the ⟨100⟩ (and somewhat ⟨100⟩)
vicinity twins most easily while ⟨111⟩ is most reluctant
to twin, see the black shaded regions of Fig. 18. At [100],
all four twin planes can operate, converting the compression
axis to ⟨221⟩, which is only 15° from ⟨111⟩ on the ⟨111⟩ −
⟨110⟩ line. (Incidentally, while the ⟨110⟩ is also favored
to twin, it transforms into itself (its negative, to be
strict), see Fig. 18). If slip becomes favored after
twinning, Figure 2 ({111}⟨110⟩ slip) or 10 (coplanar slip)
suggests that the ⟨111⟩ intensity of brass should decrease
at still higher reductions. Similar conclusions are reached
if instrinsic faulting is important, see Fig. 19.

2.1.11 Tension and Compression Textures of bcc Metals

Slip in bcc metals has been variously described as occurring
on {110}⟨111⟩, {112}⟨111⟩, {123}⟨111⟩ systems as well as
mixed slip composed of all three. In axisymmetric flow the
sense of rotation for {110}⟨111⟩ slip in compression is the
same as that for {111}⟨110⟩ in tension and vice versa.
Thus the predicted rotations for {111}⟨110⟩ slip in
tension, Fig. 2, are directly applicable for {110}⟨111⟩
in compression. The predicted rotations for the other three

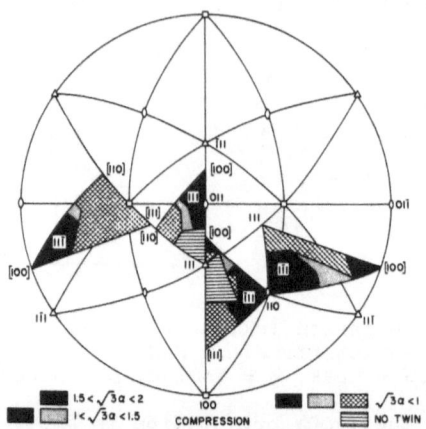

Fig.18. Same as for Fig. 14; case of axisymmetric
compression. Note ease of twinning near [100] but
not [111]

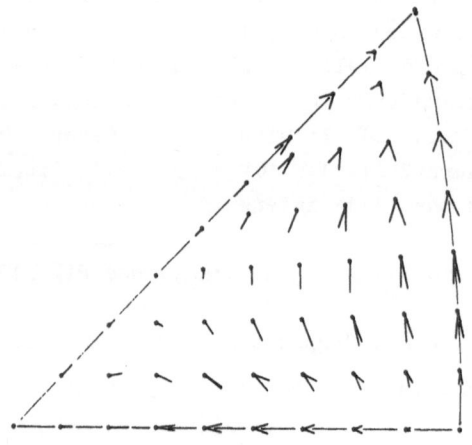

Fig.19. Same as for Fig. 12; case of axisymmetric
compression. Note the rotation is not the reverse
of that in Fig. 12 because of the unidirectional
character of intrinsic faulting similar to
twinning. (Cf. Figs. 14 and 18)

modes are similar [25]. They all show a major ⟨111⟩ texture plus a minor ⟨100⟩ texture, as observed by BARRETT [4]. Similarly, the reversed rotations of Figs. 2 are applicable to {110}⟨111⟩ in tension. A strong ⟨110⟩ texture with spread to ⟨311⟩ is then predicted. The spread is less marked if mixed slip is assumed [25]. The reported data all point to a simple ⟨110⟩ wire texture [1-4], but no inverse pole figures have been presented to show the actual orientation spread, if any. A related problem at present is that the slip systems for various bcc metals have not been clarified.

2.1.12 Concluding Remarks

It has been the attempt in this review to show, by utilizing the principle of minimum work, how the effects of the most prominent deformation mechanisms on the fiber textures of drawn wires and compressed sheets may be rationalized. Some of these mechanisms play a more significant role than others depending on alloy composition, deformation temperature and initial orientation distribution. It is obvious that no single mechanism can explain all of the observations, not even most of them. Only when all of the competing mechanisms are viewed in perspective is there hope for a reasonably simple understanding of a complex problem.

Looking beyond, it is certain that as our understanding of the interactions between deformation mechanisms is improved, so will our understanding of texture development. The interaction between slip and twinning, the merging of microtwins and stacking faults, and a more generalized and quantitative treatment of deformation banding, as well as slip system interactions under simultaneous multiple slip conditions are a few of the topics that look for clarification. Texture-wise, one must go beyond more pole figure work; correlation of stress-strain curves, metallographic verification of assumed deformation elements, and the degree of plastic constraint imparted to the deforming crystal are considerations of no less importance. Finally, the same type of treatment accorded to the fiber

texture problem should be extended to the case of rolling
and other types of deformation.

Acknowledgements

This paper might not have been written without the generous
efforts of W. L. Mammel, whose application of linear
programming techniques to the generalized minimum work
theory has made the unification of various deformation
mechanisms a practical reality. I am also deeply grateful
to M. T. Dolan for obtaining the computer plots. Many of
the issues were clarified through enlightening discussions
held with A. T. English, W. F. Hosford, T. Tisone and B. C.
Wonsiewicz. Finally, I am indebted to J. H. Wernick for
his valuable encouragements and discussions throughout the
course of this investigation.

References

1. G. Wassermann and J. Grewen: Texturen metallischer Werkstoffe, Springer, Berlin 1962.

2. I.L. Dillamore and W.T. Roberts: Met. Rev. 10(1965)271.

3. H. Hu, R.S. Cline and S.R. Goodman in: Recrystallization,
   Grain Growth and Textures, H.Margolin, Ed.,Amer.Soc.
   Metals, Metals Park, Ohio 1966, p. 295.

4. C.S. Barrett and T.B. Massalski: Structure of Metals,
   3rd ed., McGraw-Hill, New York 1966.

5. W. Heye: This volume,chapter 2.2.

6. E. Schmid: Proc.Int.Congr.Mech., Delft 1924,J.Waltman,
   Jr., Delft 1925, p. 342.

7. R. von Mises: ZAMM 8(1928)161.

8. G.I.Taylor: J.Inst.Metals 62(1938)307 - S.Timoshenko
   Anniversary Volume, MacMillan Co.,New York 1939,p.218.

9. J.F.W. Bishop and R. Hill: Phil.Mag. 42 (1951)414,1298.

10. G. Sachs: Z. VDI 72(1928)734.

11. H.L. Cox and D.G. Sopwith: Proc. Phys. Soc.(London)
    49(1937)134.

12. F. Wever and W.E. Schmid: Z.Metallkde. 22(1930)133.

13. W. Boas and E. Schmid: Z. techn. Phys. 12(1931)71.

14. A. Kochendörfer: Plastische Eigenschaften von Kristallen
    und metallischen Werkstoffen, Springer, Berlin 1941,
    p. 22.

15. E.A. Calnan and C.J.B. Clews: Phil.Mag. 41(1950)1085.

16. G.E.G. Tucker: J. Inst. Metals 82(1954)655 - Acta Met. 12(1964)1093.

17. G.I. Taylor: Proc. Colloqu. Deformation and Flow of Solids, Madrid 1955, Springer, Berlin 1956, p. 3.

18. R.L. Fleischer and W.A. Backofen: Trans.Met.Soc. AIME 218(1960)243.

19. C. Elbaum: Trans.Met.Soc. AIME 218(1960)444.

20. T.H. Lin: J. Mech. Phys. Solids 5(1957)143.

21. B. Budiansky and T.T. Wu: Proc. 4th Congr.Appl.Mech., 1962, p. 1175.

22. E. Kröner: Acta Met. 9(1961)155.

23. J.W. Hutchinson:J. Mech.Phys.Solids 12 (1964)11,25.

24. C.S. Barrett and L.H. Levenson: Trans. AIME 137(1940)112.

25. G.Y. Chin, W.L. Mammel and M.T. Dolan: Trans.Met.Soc. AIME 239(1967)1854.(See also:H.Siemes; Z.Metallkde.58 (1967)228.)

26. H. Ahlborn: Z.Metallkde. 56(1965)205, 411.

27. H. Ahlborn: Z.Metallkde. 57(1966)877.

28. G.Y. Chin, W.L. Mammel and M.T.Dolan: Trans.Met.Soc. AIME 239(1967)1111.

29. G.Y. Chin and W.L.Mammel: Trans.Met.Soc. AIME 239(1967) 1400.

30. G.Y. Chin, E.A. Nesbitt and A.J. Williams: Acta Met. 14(1966)467.

31. W.F. Hosford: Acta Met. 14(1966)1085.

32. B.C. Wonsiewicz: Private communication.

33. W.F. Hosford: Trans.Met.Soc. AIME 233(1965)329.

34. G. Mayer and W.A. Backofen: Trans.Met.Soc. AIME 242 (1968)1587.

35. A.T. English and G.Y. Chin: Acta Met. 13(1965)1013.

36. J. Grewen and G. Wassermann in : Recrystallization, Grain Growth and Textures, H. Margolin, Ed.,Amer.Soc.Metals, Metals Park, Ohio 1966, p. 382.

37. F. Haeßner: Z. Metallkde. 54(1963)98.

38. Y.C. Liu: Trans.Met.Soc. AIME 230(1964)656.

39. F.R.N. Nabarro; Z.S. Basinski and.D.B. Holt: Adv.Physics 13(1964)193.

40. N.B. Brown: Trans.Met.Soc. AIME 221(1961)236.

41. R.E. Smallman and D. Green: Acta Met. 12(1964)145.

42. I.L. Dillamore and W.T. Roberts: Acta Met. 12(1964)281.

43. J.F.W. Bishop: J.Mech. Phys. Solids 3(1954)130.

44. E.A. Calnan: Acta Met. 2(1954)865.

45. G.A.Alers and Y.C.Liu: Trans.Met.Soc. AIME 239(1967)210.

46. B. Ramaswami, U.F. Kocks and B. Chalmers: Trans.Met. Soc. AIME 233(1965)927.

47. U.F.Kocks and T.J. Brown: Acta Met. 14(1966)87.

48. P.J. Jackson and Z.S. Basinski: Can.J.Phys. 45(1967)707.

49. J.A. Venables in: Deformation Twinning, R.E. Reed-Hill et al Eds., Gordon and Breach, New York 1964, p.77.

50. J.P. Hirth and J. Lothe: Theory of Dislocations, McGraw-Hill, New York 1968, pp.296-300.

51. B. Escaig, G. Fontaine and J. Friedel: Can.J.Phys. 45(1967)481.

52. H.Ahlborn and G.Wassermann: Z.Metallkde. 54(1963)1.

53. G.Wassermann: Z. Metallkde. 54(1963)61.

54. See Deformation Twinning, R.E. Reed-Hill et al Eds., Gordon and Breach, New York 1964.

55. G.Y. Chin, W.F. Hosford and D.R.Mendorf: To be published in Proc. Roy. Soc..

56. W. Heye and G. Wassermann: phys.stat.sol. 18(1966). K107- Scripta Met. 2(1968)205.

57. G.Y. Chin, W.L. Mammel and M.T. Dolan: Submitted for publication in Trans.Met.Soc. AIME.

58. D. Hull: Acta Met. 9(1961)191.

59. H. Ahlborn in: Recrystallization, Grain Growth and Textures, H. Margolin, Ed., Amer.Soc.Metals, Metals Park, Ohio 1966, p. 374.

60. R.E. Reed and C.J. McHargue: Trans.Met.Soc. AIME 239(1967)1604.

61. G.Y. Chin and B.C. Wonsiewicz: Submitted for publication in Trans.Met.Soc. AIME.

2.2 Verformungstexturen von Blechen
Deformation Textures of Sheets

von W. Heye[+]

Abstract

The development of a deformation texture is the result of
the sum of orientation changes of all crystals in a
polycrystalline aggregate. Fundamental understanding of the
development of a texture in polycrystals, therefore,
requires the knowledge of orientation changes in single
crystals under the same conditions of deformation.

The orientation of a single crystal changes according to
the various modes of deformation, singly or in combination,
during the deformation process.

An assumption for the operation of a deformation mechanism
is that the critical stress is exceeded, and that the shape
change produced by the deformation process corresponds to
the shape change produced by the crystallographic
deformation mechanism. The critical stresses of these
mechanism are dependent on the stacking fault energy, which
is of predominant influence on mechanical twinning.

Questions concerning the application of results of single
crystals to the development of texture in polycrystals
will be discussed for fcc metals. It will be shown that
a single crystal is not preserved during deformation. It
forms deformation bands and changes into polycrystals.
Therefore, the orientation change of single crystals can
be implied to texture development in polycrystalline
aggregates.

Es ist möglich, die Diskussion über die Texturausbildung
beim Blechwalzen von zwei Standpunkten aus zu verfolgen.
Man kann einmal von den Endtexturen vielkristalliner Bleche

[+] Institut für Metallkunde und Metallphysik, Technische
Universität Clausthal, Clausthal-Zellerfeld, Bundesrepublik Deutschland

ausgehen, um analysierende Rückschlüsse auf ihre Entstehung
zu ziehen, oder man untersucht die Orientierungsänderungen
von Einkristallen unter gleichen Verformungsbedingungen.
Die Verformungstextur eines vielkristallinen Bleches ergibt
sich dann als die Summe der Orientierungsänderungen der
Einzelkristalle. Über diese zweite synthetische Methode soll
hier berichtet werden.

2.2.1 Diskussion des Spannungs- und des Formänderungszustandes beim Walzen und Auswahl der Verformungselemente

Einer aufgezwungenen Formänderung folgt ein Kristall durch
die Betätigung kristallographischer Verformungsmechanismen:
Gleiten und mechanische Zwillingsbildung. Sie können nur
dann betätigt werden, wenn eine für das Metall und den Verformungsmechanismus charakteristische kritische Spannung
überschritten wird. Diese kritische Spannung τ_0 und die
nach ihrer Überschreitung in den Verformungssystemen tatsächlich wirkenden Spannungen τ lassen sich nur unter einachsigen Spannungszuständen in einfacher Weise ermitteln.
Es gilt dann das Schmid'sche Schubspannungsgesetz:

$$\tau = \sigma \cos\chi \cos\lambda \qquad (1)$$

Bei mehrachsigen Spannungszuständen ist die Aufgabe wesentlich schwieriger und in der Regel nur unter vereinfachenden Annahmen möglich. Es ist deshalb versucht worden,
die Bestimmung der in den Verformungssystemen wirkenden
Spannungen zu umgehen.

Unter Berücksichtigung der v. MISES'schen Voraussetzung [1],
daß für eine allgemeine Formänderung stets fünf voneinander unabhängige Gleitsysteme betätigt werden müssen, können nach TAYLOR [2] nur diejenigen Kombinationen von Gleitsystemen betätigt werden, die zu einem Minimum der aufzu-
wendenden Formänderungsarbeit führen. Die daraus abgeleitete Analyse der Verformungsvorgänge blieb zunächst nur
auf den Zugversuch beschränkt. Erst in letzter Zeit wurde
sie auch für den Walzvorgang durchgeführt. Dabei brachte
indessen erst die Berücksichtigung der mechanischen Zwillingsbildung einen wesentlichen Fortschritt [3]. Die Be-

deutung der mechanischen Zwillingsbildung für die Entstehung von Walztexturen haben zuerst SCHMID und WASSER-
MANN [4] an hexagonalen Metallen als wesentlichen Einfluß-
faktor erkannt und diskutiert. Erst viel später wurde diese Betrachtungsweise auch auf die kubisch flächenzentrierten Metalle angewendet [5-9].

CHIN, HOSFORD und MENDORF [3] haben eine Taylor-Analyse des
Walzvorganges durchgeführt, bei der als vereinfachende Vor-
aussetzung ein ebener Formänderungszustand angenommen wird.
Der Spannungszustand muß dabei räumlich sein.

Es werden nur solche Kombinationen von Verformungssystemen
betätigt, für die die aufzuwendende Formänderungsarbeit
bzw. der mittlere Orientierungsfaktor M

$$M = \frac{\sigma_x}{\tau_s} = \frac{1}{\epsilon_{xx}} \left[\sum_i S_i + \alpha \sum_i t_i \right] \tag{2}$$

$$\alpha = \frac{\tau_t}{\tau_s}$$

zu einem Minimum wird. ϵ_{xx} ist darin die inkrementale Formänderung in x-Richtung, s_i die Abgleitung im Gleitsystem i,
$t_i = v_i s$ die durch Verzwillingung auf dem Zwillingssystem i
bewirkte makroskopische Formänderung, wenn v_i das verzwillingte Volumen und s der Betrag der Schiebung ist. τ_s und
τ_t sind die kritischen Schub- bzw. Zwillingsspannungen.

Für "plane strain compression", einem Druckversuch, bei dem
außer der in Druckrichtung x erzwungenen Formänderung ϵ_x
nur eine davon abhängige ϵ_z in z-Richtung zugelassen wird,
ergibt das von CHIN, HOSFORD und MENDORF [3] durchgeführte Experiment an Einkristallen aus Kobalt mit 8% Eisen eine
gute Übereinstimmung mit der Theorie. Es ist damit indessen nur der Nachweis erbracht, daß die Taylor-Analyse zu
richtigen Ergebnissen führt, wenn die getroffenen Verein-
barungen richtig oder eine gute Näherung sind. Es ist jedoch nicht sicher, daß die für "plane strain compression"
exakt gültige Voraussetzung einer ebenen Formänderung für
das Walzen eine gute Näherung ist.

Ein anderes Näherungsverfahren, das nachstehend besprochen
werden soll, ist die Annahme eines ebenen Spannungszustandes mit einer Druckspannung σ_x in der Normalenrichtung und
einer Zugspannung σ_z in der Walzrichtung, während die dritte der Hauptspannungen σ_y in der Querrichtung vernachlässigt wird. Der Formänderungszustand wird dann im allgemeinen räumlich sein.

Die in den Verformungssystemen tatsächlich wirkenden Spannungen τ lassen sich unter dieser Voraussetzung näherungsweise recht einfach berechnen

$$\tau = \sigma \underbrace{(\cos\chi_x \cos\lambda_x + \cos\chi_z \cos\lambda_z)}_{\mu}$$
$$\chi\lambda \ < \ 90^\circ$$

(3)

$\chi_{x,z}$ sind die Winkel zwischen den Richtungen x (Normalenrichtung) bzw. z (Walzrichtung) und der Gleitebenen-Normalen,
$\lambda_{x,z}$ die entsprechenden Winkel zur Gleitrichtung. Diese
Beziehung wurde zuerst von TUCKER [10] angegeben, jedoch
mit negativem Vorzeichen der von der Druckspannung herrührenden Komponente $\cos\chi_x \cos\lambda_x$. Die Druckspannung σ_x
verursacht allein aufgebracht jedoch die gleiche kristallographische Formänderung wie die allein aufgebrachte Zugspannung σ_z. Werden sie wie beim Walzen gleichzeitig aufgebracht, dann addieren sich beide in ihrer Wirkung.

Dieses zunächst nur für Gleiten abgeleitete Schubspannungsgesetz gilt auch für mechanische Zwillingsbildung, wenn
für $\lambda_{x,z}$ die Winkel der Schiebungsrichtung η_1 bzw. $\chi_{x,z}$
der Zwillingsebenen-Normalen zu den Richtungen x und
z eingesetzt werden.

Die äußeren Spannungen induzieren indessen nicht nur eine
Schubspannung in den Verformungssystemen. Sie haben ebenfalls unmittelbaren Einfluß auf die mit der Abgleitung
bzw. Schiebung verbundenen Gitterdrehungen. Diese erfolgen
immer so, daß die kristallographische und die aufgezwunge-
ne Formänderung übereinstimmen.

Wenn Aussagen über die Möglichkeit zur Betätigung eines
kristallographischen Verformungssystems gemacht werden sollen, dann müssen also auch die mit der Abgleitung und
Schiebung verbundenen makroskopischen Formänderungen bekannt sein. Es seien deshalb zunächst die dafür erforderlichen mathematischen bzw. kristallographischen Zusammenhänge abgeleitet:

Für Gleiten lassen sich die Formänderungen in den drei
Hauptrichtungen leicht berechnen. Die inkrementalen Formänderungen in den Richtungen x, y und z ergeben sich zu

$$\varepsilon_{xx} = -a \sin\alpha_x \cos\lambda_x$$
$$\varepsilon_{yy} = a \sin\alpha_y \cos\lambda_y \qquad (4)$$
$$\varepsilon_{zz} = a \sin\alpha_z \cos\lambda_z$$

a ist der Betrag der Abgleitung im aktiven Gleitsystem,
$\alpha_{x,y,z}$ sind die Winkel zwischen der Normalen auf die
Schiebungsebene, die aufgespannt wird durch die Gleitrichtung und die Gleitebenen-Normale, zu den Richtungen x, y
und z, $\lambda_{x,y,z}$ die entsprechenden Winkel zur Gleitrichtung.

Die integralen Formänderungen, hervorgerufen durch Gleiten
ergeben sich damit zu

$$\varepsilon_{x,y,z} = \int \varepsilon_{xx,yy,zz} \, d\varphi \qquad (5)$$

wobei dφ die Orientierungsänderung kennzeichnet.

Die für die Formänderung bei mechanischer Zwillingsbildung
von JOHNSON [11] angegebene Beziehung

$$\frac{l_1}{l_0} = \sqrt{1 + 2s \sin\chi \cos\lambda + s^2 \sin^2\chi} \qquad (6)$$

läßt sich nicht direkt zur Berechnung der makroskopischen
Formänderung verwenden, weil sie die Längenänderung einer
vorgegebenen Matrixrichtung angibt und unberücksichtigt
läßt, daß diese Richtung durch die Zwillingsbildung verändert wird.

Die tatsächlichen makroskopischen Formänderungen, die sich
als Summe der Längenänderungen der Matrixrichtungen und
ihrer Orientierungsänderungen ergeben, lassen sich errechnen

mit

$$\epsilon_{t_{x,y,z}} = v_t \, (\Delta l_{x,y,z} - 1)$$

$$\Delta l_{x,y,z} = \frac{1}{\sqrt{\phi_{x,y,z}}} \qquad\qquad (7)$$

v_t ist das verzwillingte Volumen, $\phi_{x,y,z}$ ein Orientierungsfaktor, der sich wie folgt aus dem Betrag der Schiebung s
und den Winkeln $\lambda_{x,y,z}$, der Schiebungsrichtung, $\chi_{x,y,z}$,
der Zwillingsebenen-Normalen und $\alpha_{x,y,z}$ der Schiebungs-
ebenen-Normalen zu den Richtungen x, y und z ergibt:

$$\phi_{x,y,z} = (\cos\lambda_{x,y,z} - s \cos\chi_{x,y,z})^2 + \cos^2\chi_{x,y,z} +$$

$$\cos^2\alpha_{x,y,z} \qquad\qquad (8)$$

Eine Taylor-Analyse wäre mittels der für die Spannungen
und Formänderungen abgeleiteten Beziehungen auch für den
betrachteten ebenen Spannungs- und damit verbundenen räumlichen Formänderungszustand möglich. Es darf dann aber für
eine vorgegebene Abgleitung oder Schiebung nur eine Formänderung $\epsilon_{x,y,z}$ bzw. $\epsilon t_{x,y,z}$ möglich sein.

Das ist zwar bei mechanischer Zwillingsbildung tatsächlich
der Fall, nicht aber bei einer Formänderung durch Gleiten.
Dort sind, wie an einem speziellen Beispiel noch gezeigt
wird, beim Abgleiten auf einem Gleitsystem mindestens zwei
unterschiedliche äußere Formänderungen möglich.

Deswegen wird zunächst nach einem mathematisch einfacheren
Weg gesucht. Statt einer Rechnung über die Formänderungsarbeit werden die in den Verformungssystemen tatsächlich
wirkenden Spannungen nach [3] errechnet, um zusammen mit
noch aufzustellenden Formänderungskriterien zu einer Aussage über die Möglichkeit zur Betätigung eines Verformungssystems zu gelangen.

Bei einer inkrementalen Formänderung durch Gleiten muß
erfüllt sein:

$$|\epsilon_{zz}| \quad , \quad |\epsilon_{xx}| \geq |\epsilon_{yy}| \tag{9}$$

bei einer integralen Formänderung ist die Vorschrift enger
und lautet:

$$|\epsilon_z| > |\epsilon_x| > |\epsilon_y| \tag{10}$$

Die Formänderung bei mechanischer Zwillingsbildung erfolgt
in einer vom Gleiten stark abweichenden Weise. Während die
Gleitung in inkrementalen Abgleitungen, verbunden mit einer
kontinuierlichen Gitterdrehung und Formänderung vor sich
geht, führt die mechanische Zwillingsbildung zu einer spontanen Umorientierung des Gitters, die mit einer ebenso
spontanen Formänderung verbunden ist. Deswegen muß die
Übereinstimmung dieser durch Zwillingsbildung bewirkten
Formänderung mit der durch das Walzen aufgezwungenen besser als bei der Gleitung sein. Es muß

$$|\epsilon_{t_z}| \geq |\epsilon_{t_x}| \ggg |\epsilon_{t_y}| \tag{11}$$

mit einer Breitung ϵ_{t_y} möglichst gleich null, erfüllt sein.
Das ist immer dann der Fall, wenn die Ebene der Schiebung
in der Ebene Walzrichtung-Blechnormale liegt. Der Orientierungsfaktor φ in Gleichung (8) vereinfacht sich dann
mit $\alpha = 90^\circ$ zu

$$\varphi_{x,z} = (\cos\lambda_{x,z} - s \cos\chi_{x,z})^2 + \cos^2\chi_{x,z} \tag{12}$$

Ein Verformungssystem wird immer dann betätigt, wenn der
Orientierungsfaktor μ nach (3) einen Maximalwert hat und
außerdem die Formänderungsbedingungen (9) bis (11) für
dieses System erfüllt sind.

Offen ist bis jetzt noch die Frage nach der besseren Näherung für das Walzen; ist es der zuerst diskutierte ebene
Formänderungszustand oder ein ebener Spannungszustand mit
einer räumlichen Formänderung?

Für die mechanische Zwillingsbildung stimmen beide Theorien
überein, da in beiden (11) erfüllt sein muß. Grundsätzliche

Unterschiede sind jedoch beim Gleiten vorhanden, die eine
Näherung läßt Breitung zu, die andere dagegen nicht.

Um hier eine Entscheidung zu treffen, wurde die Breitung
von Einkristallen aus Kupfer mit unterschiedlichen Ausgangsorientierungen und einer vielkristallinen Probe beim
Walzen gemessen. Das Ergebnis zeigt Abb.1, in der die Breitung in der Mittelschicht einiger besonders charakteri-
stischer Einkristalle neben der Breitung des Vielkristalls

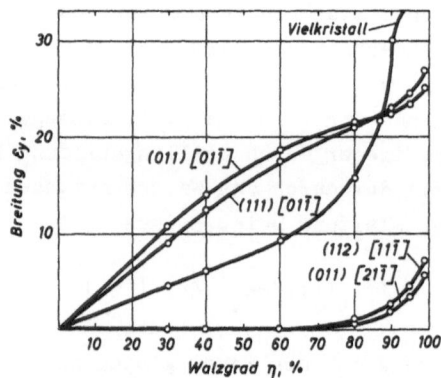

Abb.1. Breitung in der Mittelschicht von gewalztem
Kupfer- Einkristallen und einer vielkristallinen
Probe

in Abhängigkeit vom Walzgrad aufgetragen ist: Man erkennt
eine sehr starke Abhängigkeit der Breitung von der Kristallorientierung. Kristalle mit den beim Kupfer als stabil
bekannten Orientierungen $(011)[21\bar{1}]$ und $(112)[11\bar{1}]$ breiten
bis zu einem Walzgrad von über 70% fast nicht, während
Kristalle mit den Ausgangsorientierungen $(011)[01\bar{1}]$ und
$(111)[01\bar{1}]$ beim gleichen Walzgrad um ca. 20% gebreitet
haben. Außerdem zeigen die starken Breitungen bei Walzgraden über 70% deutlich, daß die Annahme eines ebenen Formänderungszustandes für das Walzen eine unzureichende Näherung ist.

Bekräftigt wird diese Tatsache durch die starke Breitung
einer vielkristallinen Probe. Bis zu einem Walzgrad von
ungefähr 70% ist ihre Breitung ein Mittelwert über die
Breitungen der Einzelkristalle. Darüber hinaus steigt die

Breitung noch stärker an als diesem Mittelwert entspräche.

Benachbarte Körner behindern also die Breitung der Einzelkristalle nicht, wie von CHIN, HOSFORD und MENDORF [3] ver-
mutet wurde. Die höhere Verfestigung des Vielkristalls
führt im Gegenteil zu einer Zunahme der Breitung. Es ist
darum richtiger, den Walzvorgang als einen ebenen Spannungszustand mit einer räumlichen Formänderung zu betrachten.

Ein weiterer Gesichtspunkt ist, daß die früher als selbstverständlich geltende Vorstellung eines sich als Ganzes ver-
formenden Kristalls unzutreffend ist. Bei einer Verformung
sowohl von Einkristallen wie auch polykristallinem Material
kubisch flächenzentrierter und kubisch raumzentrierter Metalle unter mehrachsigen Spannungszuständen wie z. B. beim Walzen, wird eine bandförmige Aufspaltung der Kristalle beobachtet. Diese Bänder werden nach BARRETT als "Deformationsbänder" bezeichnet.

Die Bildung von Deformationsbändern wurde bis in die jüngste
Zeit bei allen Theorien über die Ausbildung der Verformungstexturen vernachlässigt, obwohl von BARRETT und Mitarbeiter
[12-14] eingehende Untersuchungen über den Mechanismus der
Bildung von Deformationsbändern durchgeführt wurden und sie
bereits auf ihren möglichen Einfluß auf die Texturausbildung
hinwiesen. Dieser Gedanke wurde erst in neuerer Zeit wieder
aufgegriffen [8,9,15-20].

2.2.2 Stabilität und Änderung von Kristallorientierungen

Die für einen ebenen Spannungszustand abgeleiteten Kriterien
zur Betätigung eines kristallographischen Verformungssystems mit Berücksichtigung des Einflusses der Bildung von
Deformationsbändern sollen im folgenden am Beispiel kub.
flz. Metalle diskutiert werden.

Metalle dieser Struktur sind deswegen besonders interessant,
weil bei ihnen durch eine Variation der Stapelfehlerenergie,
z. B. durch Zulegieren einer zweiten löslichen Komponente
unterschiedliche Kombinationen von aktiven Verformungsmechanismen und damit Walztexturen auftreten können. So kann die
Textur des Kupfers durch Zulegieren von Zink in die Textur

vom Messing-bzw. Silbertyp überführt werden. Den gleichen
Effekt erhält man bei einer Verringerung der Verformungstem-
peratur. Da einerseits eine eindeutige Beziehung zwischen der
auf den Schubmodul und Burgers-Vektor des entsprechenden Me-
talles bezogenen Stapelfehlerenergie γ/G.b und dem Textur-
typ besteht, andererseits die Stapelfehlerenergie selbst tem-
peraturunabhängig ist, kann die Stapelfehlerenergie nur über
ihren Einfluß auf die "Aktivierungsenergie" der verschiedenen Verformungsmechanismen, wie Gleiten und mechanische Zwillingsbildung mittelbar die Orientierungsänderung der Kristalle und damit die Textur beeinflussen.

Da ein Einkristall in seinem kristallographischen Verfor-
mungsverhalten wesentlich einfacher als ein Vielkristall zu
beobachten ist, wurden Einkristalle aus Kupfer und Silber gewalzt, um die daran gewonnenen Erkenntnisse dann auf die Texturausbildung vielkristalliner Bleche kubisch flächenzentrierter Metalle zu übertragen.

2.2.2.1 Kubisch flächenzentrierte Metalle mit hoher Stapelfehlerenergie

Bei Metallen mit hoher Stapelfehlerenergie, wie z. B. Kupfer,
erfolgt die Verformung nur durch Gleiten nach {111}⟨110⟩.
Bereits von PICKUS und MATHEWSON [21] wurden hierfür eine
Reihe von Kristallorientierungen angegeben, die ihre Orientierung beim Walzen nicht verändern sollen. HIBBARD und
YEN [22] haben ihre Anzahl eingeschränkt. Um eine Quer-
schnittsverminderung beim Walzen zu ermöglichen, sollen nur
solche Gleitrichtungen betätigt werden können, die mit der
Walzrichtung einen kleineren Winkel als 45° einschließen. Als
stabile Orientierungen verbleiben dann nur (011)[100], (011)
[21$\overline{1}$] und (112)[11$\overline{1}$]

Die eingangs abgeleiteten Formänderungskriterien erlauben
jedoch auch größere Winkel als 45° zwischen aktiver Gleitrichtung und Walzrichtung. Außerdem blieb die Aufspaltung in
Deformationsbänder unberücksichtigt.

Es ist also notwendig, sowohl die Änderungen als auch die
Stabilität von Kristallorientierungen unter den neuen Ge-
sichtspunkten zu diskutieren. Von den im Verlaufe dieser

Untersuchungen gewalzten Kupfereinkristalle werden hier als
Beispiele nur die mit den nach HIBBARD und YEN [22] stabilen Orientierungen (011)[100], (011)[21$\bar{1}$], (112)[11$\bar{1}$] und
ferner eine (111)[01$\bar{1}$] - Ausgangsorientierung besprochen.

2.2.2.1.1 (111)[01$\bar{1}$] - Ausgangsorientierung. Abb. 2 zeigt
ein Gefüge parallel zur Ebene WR-BN des um 60% gewalzten
Kupfereinkristalles mit (111)[01$\bar{1}$]-Ausgangsorientierung. Es
läßt deutlich erkennen, wie der Kristall in Deformationsbändern aufgespalten ist. Die Aufspaltung ist sehr ausgeprägt
in der Zwischenschicht zwischen der Oberfläche und der Mittelschicht des Bleches, aber nur schwach in der Mittelschicht selbst. Wie eine elektronenmikroskopische Durch-
strahlungsaufnahme dieser Schicht zeigt, sind auch in ihr
bereits Mikrobänder vorhanden. Solche Mikrobänder entstehen
in Übereinstimmung mit den Untersuchungen an kubisch raumzentrierten Blechen [17,15] und kubisch flächenzentrierten
Drähten [20] aus einer durch die Verformung entstandenen ver-
längerten Zellstruktur. Sie übertragen von Band zu Band eine
Orientierungsdifferenz von max. 4°, ihre Grenzen sind also
Kleinwinkelkorngrenzen. Da die wahre Verformung in der Zwischenschicht größer als in der Mittelschicht ist, sind die
Deformationsbänder in der ersteren schon stärker ausgebildet.

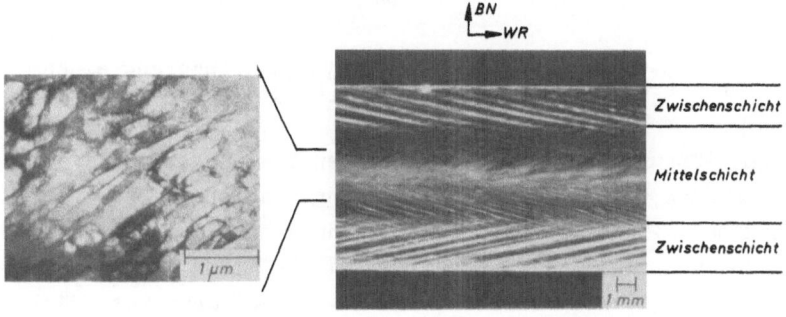

Abb.2. Deformationsbänder in einem Kupfer-Einkristall
mit (111)[01$\bar{1}$]-Ausgangsorientierung

92

Etwa ab einer wahren Verformung von 60% bilden sich Gruppen
von Mikrobändern, sog. Übergangsbänder [15], die die in der
höher verformten Zwischenschicht auftretenden größeren
Orientierungsdifferenzen übertragen. Bei den in der Zwischenschicht lichtmikroskopisch sichtbaren Bandgrenzen handelt es sich also nicht um Großwinkelkorngrenzen, sondern
um solche Übergangsbänder. Bei weiterer Verformung bewegen
sich diese Übergangsbänder dann immer weiter in die Blechmitte hinein, bis sie sich über den gesamten Blechquer-
schnitt erstrecken.

Abb. 3 zeigt in einer (111)-Standardprojektion die {111}
⟨110⟩-Gleitsysteme in diesem Kristall. An die Gleitsysteme

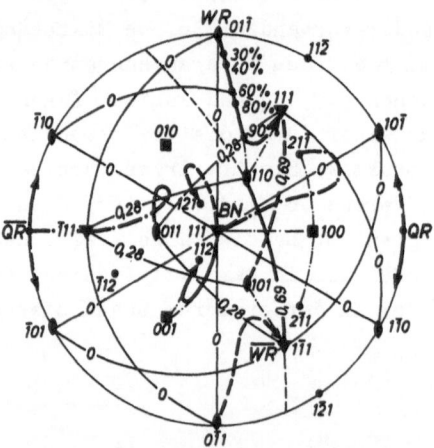

Abb.3. Gleitsysteme und Orientierungsänderung eines
Kupfer-Einkristalls mit (111)[01T]-Ausgangsorien-
tierung

sind ihre nach (3) für einen ebenen Spannungszustand errechneten Orientierungsfaktoren angeschrieben. Die Gleitsysteme mit der größten Schubspannung sind demnach (1T1)
[110] und (11T)[T1T], wobei für das letztere die Gegenrich-
tung [101] der [T0T] Gleitrichtung eingetragen ist. Würde
sich der Kristall als Ganzes verformen, ergäbe sich als resultierende Gleitrichtung [01T]. Da sie in der Walzrichtung
liegt, wäre die Orientierung also stabil.

Tatsächlich spaltet der Kristall jedoch in unterschiedlich

orientierte Deformationsbandgruppen auf. Dies wurde zusätz-
lich durch die Aufnahme von nicht über den gesamten Blech-
querschnitt integrierenden {220}-Polfiguren nachgewiesen.
In der einen Bandgruppe der Mittelschicht ist primär das
Gleitsystem $(1\bar{1}1)[110]$, in der anderen $(11\bar{1})[\bar{1}0\bar{1}]$ betätigt.
Eine Abgleitung auf diesen Gleitsystemen steht nicht im
Widerspruch zu der aufgezwungenen Formänderung.

Die Orientierungsänderungen der Walzrichtung, Blechnormalen
und Querrichtungen sind in Abb.3 für die eine Bandgruppe
in der Mittelschicht durch dick ausgezogene Pfeile, für
die andere Bandgruppe durch dick gestrichelte Pfeile,
dargestellt. Durch die Änderung der Winkel $\chi_{x,z}$ und $\lambda_{x,z}$
wird gemäß (3) in den primären Gleitsystemen, z.B. im
System $(1\bar{1}1)$ $[110]$ der einen Bandgruppe in der Mittelschicht, die Schubspannung immer kleiner, während sie in
den Gleitsystemen $(111)[01\bar{1}]$ und $(111)[10\bar{1}]$ ansteigt. Nach
einem Verformungsgrad von ca. 90% erfolgt die Gleitung
dann zusätzlich auf diesen sekundären Gleitsystemen mit
einer resultierenden $[11\bar{2}]$-Gleitrichtung. Beim Erreichen
der $(112)[11\bar{1}]$-Orientierung werden die sekundären Gleitsysteme gleichberechtigt mit dem primären Gleitsystem
$(1\bar{1}1)[110]$. Die aus den drei Gleitrichtungen resultierende $[11\bar{1}]$-Richtung liegt dann in der Walzrichtung und die
Orientierung wird nicht mehr verändert.

Die zweite Gruppe der Deformationsbänder in der Mittelschicht läuft in entsprechender Weise in die dazu symmetrische zweite Komponente der stabilen Endlage $(121)[\bar{1}1\bar{1}]$.

Die in Abb.1 bereits dargestellten Breitungen in der Blechmitte dieses Kristalls stimmen bis zu einem Verformungsgrad
von 90% sehr gut mit den nach (4) und (5) berechneten überein. Die Breitungen lassen sich hier besonders leicht berechnen, da bis zu hohen Verformungsgraden in den Bändern
Einfachgleitung erfolgt.

In der Zwischenschicht zwischen der Blechoberfläche und der
Mittelschicht werden die gleichen Gleitsysteme wie in der
bereits diskutierten Mittelschicht betätigt. Dabei entstehen ebenfalls zwei Gruppen von Deformationsbändern. Davon erfährt die eine mit der entsprechenden Mikrobandgruppe

der Mittelschicht identische Orientierungsänderungen bis in die stabile $(112)[11\bar{1}]$-Endorientierung. Da bei gleichem Walzgrad die wahre Verformung in der Zwischenschicht höher ist, sind dort die Orientierungsänderungen, d.h. auch die Orientierungsdifferenzen zwischen den Bändern, lediglich größer. Bereits nach einem Walzgrad von 80% ist in der Zwischenschicht die Orientierung erreicht, die die Mittelschicht erst nach einem Walzgrad von 90% angenommen hat.

Die mit der Abgleitung verbundene Gitterdrehung in der schon diskutierten Bandgruppe ist mit einer erheblichen Breitung verbunden. Die sich kontinuierlich verändernde Drehachse liegt dabei in der Nähe der Querrichtung. Bei Abgleitung auf dem gleichen Gleitsystem ist jedoch auch eine kompliziertere Gitterdrehung mit einer zusätzlichen Drehung um die sich kontinuierlich in ihrer Orientierung verändernde Walzrichtung möglich. Sie wäre mit einer wesentlich geringeren Breitung verbunden. Diese Gitterdrehung erfolgt nun tatsächlich in der zweiten Deformationsgruppe in der Zwischenschicht. Ihre Orientierungsänderung ist ebenfalls in Abb. 3 strich-punktiert eingezeichnet. Die Endorientierung dieser Bandgruppe ist $(21\bar{1})[\bar{1}1\bar{1}]$, die eine andere Komponente der stabilen $(112)[11\bar{1}]$-Orientierung ist.

2.2.2.1.2 $(011)[100]$ – Ausgangsorientierung. Eine der nach den älteren Vorstellungen stabilen Orientierungen ist (011) $[100]$. Einkristalle des Kupfers mit dieser Ausgangsorientierung spalten beim Walzen indessen ebenfalls in Deformationsbänder auf. Dadurch wird die Orientierung allerdings erst bei hohen Verformungsgraden instabil.

Unter den schon diskutierten Voraussetzungen der Orientierungsänderungen mit Berücksichtigung der Bildung von Deformationsbändern sind bei den Metallen der Kupfergruppe nur zwei Orientierungen stabil. Es sind dies die schon lange als ideale Lagen zur Beschreibung der Kupfertextur verwendeten Orientierungen $(011)[21\bar{1}]$ und $(112)[11\bar{1}]$.

2.2.2.1.3 $(011)[21\bar{1}]$ – Ausgangsorientierung. Abb. 4a zeigt für $(011)[21\bar{1}]$ – Kristalle in einer (100)-Standardprojektion die $\{111\}\langle110\rangle$-Gleitsysteme mit ihren Orientierungsfaktoren.

Die Gleitsysteme mit den größten Schubspannungen sind demnach
(1̄T̄T̄)[110] und (111)[10T̄], deren resultierende Gleitrichtung
[21T̄] in der Walzrichtung liegt. Die Orientierung wird deshalb durch Gleiten auf diesen Systemen nicht verändert. Die
Neigung zur Bildung von Deformationsbändern ist sehr gering,
durch geringere Bevorzugung entweder des Gleitsystems (1T̄T̄)
[110] oder (111)[10T̄] in unterschiedlichen Kristallbereichen
wird lediglich eine sehr schwache Streuung in den Polfiguren
eines gewalzten (011)[21T̄]-Kupfereinkristalls im Sinne einer
Drehung um die [1T̄1]-Querrichtung beobachtet. Bei sehr hohen
Walzgraden (>95%) werden zusätzlich, jedoch in untergeordnetem
Ausmaß, die Gleitsysteme (111)[01T̄] und (1T̄T̄)[01T̄] betätigt. Das führt, wenn man die Bildung von Deformationsbändern berücksichtigt, zu einer Streuung der Belegungen der
Polfigur eines um 99% gewalzten Kupfereinkristalls (Abb.4b)
mit einer (011)[21T̄]-Ausgangsorientierung im Sinne einer
Drehung um die nahe der Walzrichtung liegende [11T̄]-
Richtung.

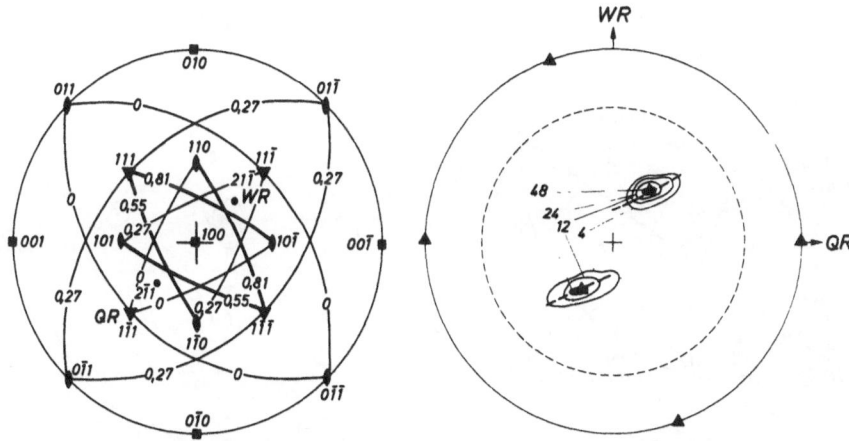

Abb.4. Kupfer-Einkristall mit (011)[21T̄]-Ausgangsorientierung
    a) Gleitsysteme   b) {111}-Polfigur nach einer
                           Verformung durch Walzen um 99%

2.2.2.1.4 (112)[11T̄]-Ausgangsorientierung. In Abb. 5a sind
in einer (110)-Standardprojektion die Gleitsysteme mit ihren
Orientierungsfaktoren in einem nach (112)[11T̄] orientierten

Kupfereinkristall wiedergegeben. Die Betätigung der Systeme
$(1\bar{1}\bar{1})[110]$, $(\bar{1}1\bar{1})[110]$, $(111)[10\bar{1}]$ und $(111)[01\bar{1}]$ mit der
größten Schubspannungen ($\mu = 0,57$) führt, da die resul-
tierende Gleitrichtung aller drei Gleitrichtungen $[11\bar{1}]$ in
der Walzrichtung liegt, zu einer Stabilität dieser Orientierung. Zu Beginn der Verformung entstehen schwach ausge-
bildete Deformationsbänder durch geringe Abgleitung entwe-
der in der $[110]$- bzw. in der aus $[01\bar{1}]$ und $[10\bar{1}]$ resultierenden $[11\bar{2}]$-Gleitrichtung in unterschiedlichen Kristallbereichen. Es tritt eine schwache Streuung der Belegungen
der Polfiguren im Sinne einer Drehung um die $[1\bar{1}0]$-Querrich-
tung auf. Diese Streuung wird bei sehr hohen Walzgraden ab-
gelöst durch Streuungen im Sinne einer Drehung um die nahe
der Blechnormalen liegenden $[111]$-Richtung. Die Ursache dafür ist das Auftreten von Mehrfachgleitungen in den primär
entstandenen Deformationsbändern. Zur primären Abgleitung,
z. B. in $[10\bar{1}]$-Richtung, kommt eine solche in aus $[110]$ und
$[10\bar{1}]$ resultierender $[21\bar{1}]$-Richtung hinzu. Das erzwingt die
Gitterdrehung um $[111]$.

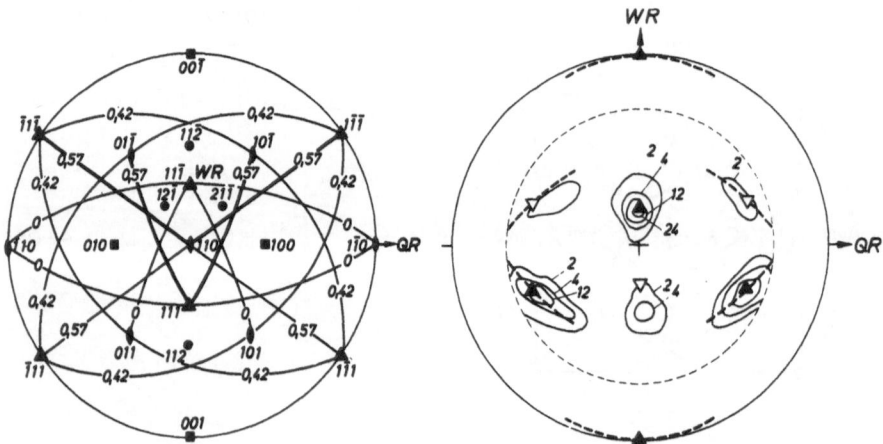

Abb.5. Kupfer-Einkristall mit $(112)[11\bar{1}]$-Ausgangsorientierung
a) Gleitsysteme   b) $\{111\}$-Polfigur nach einer
                         Verformung durch Walzen um 99%

2.2.2.1.5 Übertragung der Ergebnisse auf vielkristallines
Material. Da Kupfereinkristalle beim Walzen in Deformations-

bänder aufspalten, also "vielkristallin" werden, können die
an Einkristallen gewonnenen Ergebnisse direkt auf die Texturausbildung des vielkristallinen Kupfers übertragen werden.

In ihrer Orientierung keine Veränderung erfahren nur Kristalle mit den Ausgangsorientierungen (011)[21T] und (112)
[11T]. Alle anderen ändern ihre Orientierung bis in eine
dieser stabilen Lagen. Die Textur des vielkristallinen
Kupfers ist demzufolge eine Überlagerung dieser Orientierungen mit ihrem nach hohen Verformungsgraden auftretenden
Streulagen.

In Abb. 6a werden die durch Addition aus den Endlagen der
gewalzten (011)[21T]- und (112)[11T]-Kupfereinkristalle
konstruierte theoretische Polfigur des vielkristallinen
Kupfers mit der experimentell bestimmten verglichen. Die
Übereinstimmung ist gut.

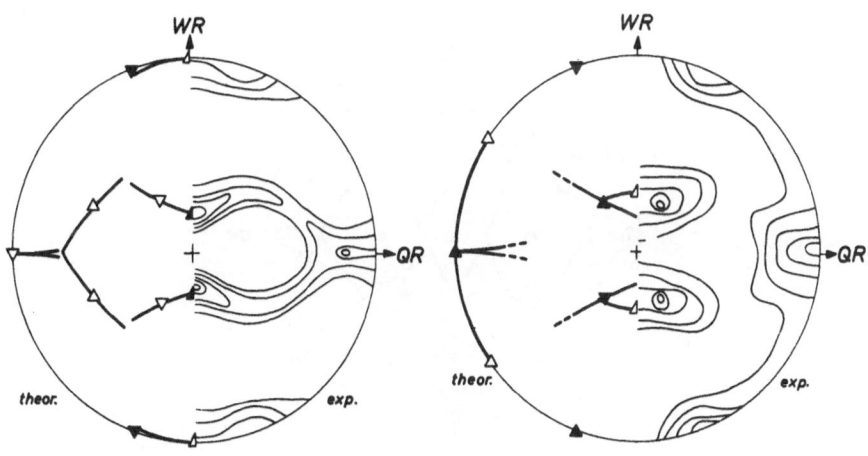

Abb.6. Experimentelle und theoretische {111}-Polfiguren des
     a) Kupfers   b) Silbers

**2.2.2.2 Kubisch flächenzentrierte Metalle mit niedriger Sta-
pelfehlerenergie**

Das Verhalten kubisch flächenzentrierter Metalle mit nie-

driger Stapelfehlerenergie soll am Beispiel des Silbers kurz
diskutiert werden. Silber ist hierfür besonders geeignet,
weil es das einzige reine Metall mit einer so niedrigen
Stapelfehlerenergie ist, daß mechanische Zwillingsbildung
beim Walzen in genügendem Ausmaß auftreten kann. Einflüsse
von Ordnungszuständen oder Ausscheidungen, wie sie bei Legierungen zusätzlich auftreten können, sind damit ausge-
schlossen. Als Beispiel für die untersuchten Silberein-
kristalle mit unterschiedlichen Ausgangsorientierungen soll
nur eine besonders charakteristische Orientierungsänderung,
nämlich die eines Kristalls mit (117)[77$\overline{2}$]-Ausgangsorien-
tierung diskutiert werden, um daraus Rückschlüsse auf die
Texturausbildung des Silbers zu ziehen.

2.2.2.2.1 (117)[77$\overline{2}$]-Ausgangsorientierung. Abb. 7 zeigt die
{111}⟨110⟩-Gleitsysteme dieses Kristalls.

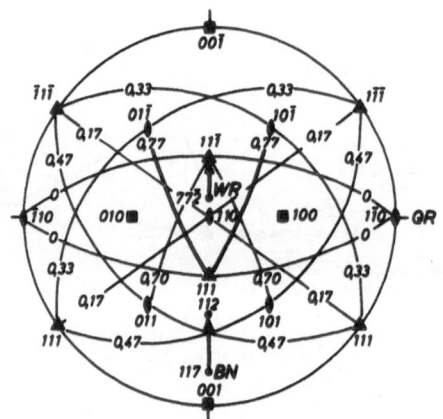

Abb.7. Gleitsysteme in einem (117)[77$\overline{2}$]-Silbereinkristall

Die Gleitsysteme mit der größten Schubspannung sind mit
einem Orientierungsfaktor von μ = 0,77 (111)[01$\overline{1}$] und (111)
[10$\overline{1}$] mit einer resultierenden [11$\overline{2}$]-Gleitrichtung. Mit
$\epsilon_{zz} = \epsilon_{xx}$ = a · 0,71 und ϵ_{yy} = 0 ist das für inkrementale
Formänderungen gültige Kriterium (9) erfüllt. Die gestri-
chelt eingezeichnete Orientierungsänderung von Walzrichtung und Blechnormale in Richtung auf eine hinsichtlich

Gleiten stabile (112)[11$\bar{1}$]-Orientierung erfolgt ohne Breitung. Die resultierende Gleitrichtung bleibt unverändert
in der Ebene Walzrichtung-Blechnormale. ϵ_z wird dabei grös-
ser als ϵ_x, damit ist dann auch $\langle$(10) erfüllt.

Außerdem ist mechanische Zwillingsbildung möglich. Einziges
nach (11) infrage kommendes Zwillingssystem ist (111)[11$\bar{2}$],
da nur seine Schiebungsebene (1$\bar{1}$0) in der Ebene Walzrichtung-Blechnormale liegt. Der Orientierungsfaktor für dieses Zwillingssystem ist mit μ_t = 0,81 größer als der der
primären Gleitsysteme mit μ = 0,77. Da jedoch die kritische
Zwillingsspannung τ_t größer als die kritische Gleitspan-
nung τ_s ist, wird die Matrixorientierung zunächst nur durch
Gleiten, und erst nach geringer Verfestigung auch durch
mechanische Zwillingsbildung geändert.

Nach einem Walzgrad von 50% (Abb. 8a) ist die Matrix in
Richtung auf die (112)[11$\bar{1}$]-Orientierung abgeglitten, sie
hat jetzt eine (114)[22$\bar{1}$]-Orientierung. Auch der schon
nach einer Verformung von etwa 30% primär entstandene

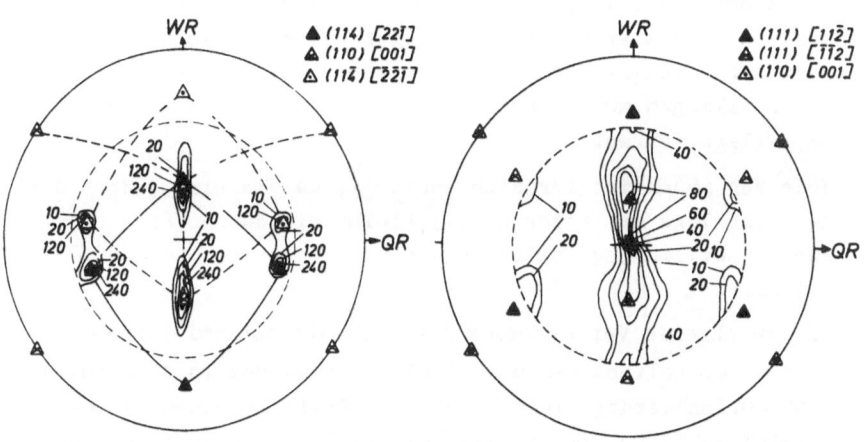

Abb.8. {111}-Polfiguren von Silbereinkristallen
a) Ausgangsorientierung (117)   b) Ausgangsorientierung (112)
   [77$\bar{2}$]; Walzgrad 50%   [11$\bar{1}$]; Walzgrad 80%

Zwilling hat seine Orientierung durch Gleiten auf seinen
Gleitsystemen mit den maximalen Schubspannungen und einer

in der Ebene Walzrichtung-Blechnormale liegenden resultierenden Gleitrichtung in eine (110)[001]-Lage geändert. Er
ist damit eine Zwillingslage der durch Gleiten und weitere
mechanische Zwillingsbildung abgebauten Matrixorientierung
geblieben. Es ist jedoch noch eine andere Belegung vorhanden, bei der es sich um Kristallbereiche in einer (11$\bar{4}$)
[$\bar{2}\bar{2}\bar{1}$]-Orientierung handelt. Sie ist eine Zwillingslage des
primären Zwillings, die durch sekundäre Zwillingsbildung
entstanden ist.

Durch Abgleitung und Verzwillingung aller bis jetzt entstandenen Lagen nähert sich die Orientierung des sekundären
Zwillings bei zunehmender Verformung immer mehr der Orientierung der (112)[11$\bar{1}$]-Matrix, die primären Zwillingslagen
einer (552)[$\bar{1}\bar{1}$5]-Orientierung. Wenn in diesem Fall Matrix-
und sekundäre Zwillingslagen die gleiche Orientierung haben, können letztere röntgenographisch nicht mehr nachgewiesen werden. Die sekundäre Verzwillingung ist dann also
eine "Rückverzwillingung". Sowohl bei dem bis jetzt diskutierten Silbereinkristall mit einer (117)[77$\bar{2}$]- wie auch
bei einem mit (112)[11$\bar{1}$]-Ausgangsorientierung, muß die
weitere Orientierungsänderung durch weitere primäre und
sekundäre mechanische Zwillingsbildung und Gleiten in die-
sen Zwillingen zwangsläufig in eine (111)[11$\bar{2}$] + (111)[$\bar{1}\bar{1}$2]-
Doppellage führen.

Jede von ihnen ist für sich instabil, da die eine jedoch die
Zwillingslage der anderen ist, bleibt sie durch "Hin- und
Rückverzwillingung" zwischen den Walzgraden von 60 - 80%
"stabil".

Die zunehmende Verfestigung der mechanischen Zwillingsbildung ermöglicht erneutes Gleiten, zunächst in eine (011)
[100]-Orientierung. In Abb. 8b, der Polfigur eines um 80%
gewalzten Silber-Einkristalls mit (112)[11$\bar{1}$]-Ausgangsorien-
tierung ist diese Orientierungsänderung deutlich erkennbar.
(011)[100] ist indessen nicht die stabile Endorientierung,
sie geht über in (011)[21$\bar{1}$] und deren andere Komponente
(011)[2$\bar{1}$1]. Unter den diskutierten Voraussetzungen ist diese
die einzige beim Silber stabile Orientierung.

Eine Gleitung von Halbversetzungen nach $\{111\}\langle211\rangle$, auf die hier nicht näher eingegangen werden kann, führt lediglich zu einer vorübergehenden Stabilität der $(011)[100]$-Orientierung.

2.2.2.2.2 Übertragung der Ergebnisse auf vielkristallines Material. Da beim Silber nur die $(011)[21\overline{1}]$-Orientierung stabil ist, wobei nach genügend hohen Verformungsgraden nur Reste von nach $(011)[100]$ orientierte Lagen vorhanden sind, setzt sich die Walztextur des vielkristallinen Silbers aus der ersteren und ihren Streulagen und der letzteren nur als Streulage zusammen. Abb. 6b zeigt eine Gegenüberstellung der theoretischen $\{111\}$-Polfigur mit der gemessenen. Die Übereinstimmung ist auch beim Silber gut.

Hier beim Silber, aber auch beim Kupfer, treten als Drehachsen zur Beschreibung der Streulagen genau die $\langle111\rangle$-Achsen auf, die bereits WASSERMANN [5] als Faserachsen zur Beschreibung der Walztexturen kubisch flächenzentrierter Metalle mittels beschränkter Fasertexturen verwendete. Eine weitere Bestätigung bringt eine Untersuchung von BUNGE [23], der auch in dreidimensionalen Orientierungsraum Drehachsen nahe $\langle111\rangle$ findet. Abweichende Resultate von LÜCKE [24] können evtl. auf einen zu geringen Walzgrad zurückgeführt werden. Die eigenen Untersuchungen zeigen nämlich, daß bei geringen und mittleren Walzgraden ebenfalls Drehungen um $\langle110\rangle$ erfolgen, die bei höheren Walzgraden jedoch in Drehungen um $\langle111\rangle$ übergehen.

2.2.3 Zusammenfassung

Zur Diskussion der Texturausbildung beim Walzen muß der Walzvorgang entweder als ein ebener Formänderungszustand oder als ein ebener Spannungszustand aufgefaßt werden.

Die Annahme einer ebenen Formänderung ist eine unzureichende Näherung, da die Breitung nicht vernachlässigt werden darf.

Die Annahme eines ebenen Spannungszustandes ist ebenfalls nur eine Annäherung, sie ist jedoch zur Berechnung der in den Verformungssystemen wirkenden Spannungen ausreichend. Das wird durch die Orientierungsänderung von Kupfer- und

Silber-Einkristallen mit unterschiedlichen Ausgangsorientierungen gezeigt. Es werden z.B. immer die Gleitsysteme betätigt, deren unter Annahme eines ebenen Spannungszustandes errechneten Orientierungsfaktoren einen Maximalwert haben.

Die von außen aufgebrachten Spannungen induzieren indessen nicht nur eine Schubspannung in den Verformungssystemen, sie bewirken auch eine Gitterdrehung und eine damit verbundene Formänderung. Deshalb genügt zur Voraussage, ob ein Verformungssystem betätigt werden kann, nur die Kenntnis der in den Verformungssystemen wirkenden Spannungen nicht.

Die Berechnung eines Orientierungsfaktors, der die Formänderung berücksichtigt, erlaubt die Taylor-Analyse. Dafür muß jedoch die Abhängigkeit der Formänderungen in Walz- und Querrichtung von der in Dickenrichtung bekannt und eindeutig sein. Es wurden deshalb die mathematischen Beziehungen zwischen den Formänderungen in den drei Hauptrichtungen des Bleches in Abhängigkeit von der Kristallorientierung abgeleitet.

Beim Gleiten ist die Beziehung nicht eindeutig, da beim Walzen beim Abgleiten auf einunddemselben Gleitsystem zwei unterschiedliche Gitterdrehungen und damit Formänderungen möglich sind.

Deswegen wurde ein mathematisch einfacherer Weg gewählt. Hierfür müssen zuerst die in den Verformungssystemen wirkenden Spannungen errechnet und danach die Formänderungskriterien berücksichtigt werden.

Bei kub. flz. Metallen führt dieser Weg zu einem befriedigenden Ergebnis. Bei Metallen dieser Struktur sind die allein texturbestimmenden Verformungsmechanismen Gleiten nach $\{111\}\langle 110\rangle$ und bei genügend niedriger Stapelfehlerenergie primäre und sekundäre mechanische Zwillingsbildung nach $\{111\}\langle 211\rangle$. Durch ihre Betätigung erfolgt eine Aufspaltung des Kristalls in Deformationsbänder, bzw. Matrix- und Zwillingslamellen; diese hat ebenfalls Einfluß auf die Texturausbildung, erlaubt andererseits aber auch die direkte Übertragung der an Einkristallen gewonnenen Ergebnisse auf die Texturausbildung polykristalliner Bleche.

Literatur

1. R.v. Mises: ZAMM 8(1928)161.

2. G.I. Taylor: J.Inst. Met. 62(1938)307.

3. G.Y. Chin, W.F. Hosford und D.R. Mendorf: Eingereicht bei Proc.Roy.Soc. (1968)-J. Inst.Met. 82(1953/54)655.

4. E.Schmid und G. Wassermann: Metallwirtsch. 9(1930)698.

5. G. Wassermann: Z. Metallkde. 54(1963)61.

6. H. Ahlborn, W. Heye u. G. Wassermann: Metall 7(1966)696.

7. W. Heye u. G. Wassermann: phys.stat.solid.18(1966)K107.

8. W. Heye u. G. Wassermann: Forschungsberichte Inst. f. Metallkde. u. Metallphys. Clausthal Nr. 1 - Scripta Met 2(1968)205.

9. W. Heye u. G. Wassermann: Z. Metallkde. 59(1968)617, 693.

10. G.E.G. Tucker: Acta Met. 12(1964)1093.

11. A. Johnson: Jb. Radioakt. Elektr. 11(1914)226.

12. C.S. Barrett: Structure of Metals, Mc Graw—Hill, New York 1952.

13. C.S. Barrett u. L.H. Levenson: Trans. AIME 135(1939) 327, 137(1940) 112, 145(1941)28.

14. C.S. Barrett u. F.W. Steadman: Trans. AIME 147(1942)57.

15. J.L. Walter u. E.F. Koch: Acta Met. 10(1962)1059, 11 (1963)923.

16. H. Hueck u. G. Wassermann: Z. Metallkde. 54(1963)32.

17. R.E. Reed u. C.J. Mc Hargue: Trans. AIME 239(1967)1604.

18. H. Ahlborn: Z. Metallkde. 56(1965)205, 411, 57(1966) 877—in: Recrystallization, Grain Growth and Textures, H. Margolin, Hrsg., Am. Soc. Metals, Metals Park, Ohio 1966, S. 374.

19. H. Hu in: Recovery and Recrystallization of Metals, L. Himmel, Hrsg., Interscience, New York 1964, S.311.

20. H. Ahlborn u. D. Sauer: Z. Metallkde. 59(1968)658.

21. M.R. Pickus u. C.H. Mathewson: J.Inst.Met. 64(1939)237.

22. W.R. Hibbard u. M.K. Yen: Trans.AIME 175(1948)126.

23. H.J. Bunge: Dieses Symposium, Kap. 1.2..

24. K. Lücke: Dieses Symposium, Kap. 1.3..

2.3 Diskussion der Verformungstexturen anhand der Kompatibilitätsbedingung im Vielkristall
Compatibility Condition in Polycrystals and its Importance for the Deformation Textures

von H.-P. Stüwe[+] und E. Aernoudt[++]

Abstract

It is assumed that ideal orientations of deformation textures should fulfill two conditions:
1. Each grain should experience the macroscopically correct strain under the macroscopic stress.
2. The orientation should automatically correct small fluctuations.

A comparison with the known textures of f.c.c. metals shows that orientations fulfilling both conditions usually do appear. Deformation by flat rolling is complicated by additional mechanisms of deformation.

Bei der plastischen Verformung eines vielkristallinen Werkstoffes unter technischen Bedingungen wird die Formänderung von außen vorgegeben. Die dazu erforderlichen Spannungen stellen sich von selbst ein. Sie lassen sich durch die üblichen Ansätze der Plastizitätsmechanik mit guter Genauigkeit berechnen. Gewöhnlich wird Isotropie des Werkstoffes vorausgesetzt; die Hauptachsen der Spannungen und der Formänderung sind dann einander parallel.

Die Erfahrung lehrt, daß bei einer solchen Verformung in der Regel jeder Kristallit die gleiche Formänderung erfährt wie der Gesamtkörper, die Verformung ist also homogen. Das bedeutet, daß auch jedem einzelnen Kristall seine Formänderung von außen aufgeprägt wird. Kristalle verformen sich vorwiegend durch kristallographische Gleitung. Die Gleitung findet statt, wenn im Gleitsystem eine bestimmte "kritische" Schubspannung herrscht. Im allgemeinen kann eine bestimmte

[+] Institut für Werkstoffkunde und Herstellungsverfahren, Technische Universität Braunschweig, Braunschweig, Bundesrepublik Deutschland

[++] Tréfileries Leon Bekaert, Zwevegem, Belgien

Formänderung nur erreicht werden, wenn sich mehrere Gleitsysteme gleichzeitig betätigen. Dann muß auch in jedem dieser Gleitsysteme die kritische Schubspannung herrschen.
Meist wird man davon ausgehen dürfen, daß die kritische
Schubspannung für alle Gleitsysteme den gleichen Wert hat.
Das Spannungsfeld, das einem Kristall eine bestimmte Formänderung aufprägt, muß also zwei Bedingungen erfüllen:

1. In mehreren Gleitsystemen muß die kritische Schubspannung herrschen (in den übrigen Gleitsystemen herrschen
   niedrigere Spannungen).
2. Die aktivierten Gleitsysteme müssen zusammen die geforderte Formänderung bewirken.

Bei vorgegebener Verformung hängt also das Spannungsfeld
von der Orientierung des Kristalles ab.

Bei der Verformung eines Vielkristalles mit regelloser
Orientierungsverteilung sollte danach in jedem Korn ein
anderes Spannungsfeld herrschen, das sich an den Korngrenzen unstetig ändern müßte. Weil das unmöglich ist, wird
sich in Wirklichkeit ein sehr kompliziertes inhomogenes
Spannungsfeld aufbauen. Die Spannungsfelder der in der Werk-
stoffoberfläche liegenden Körner liefern zusammen die Kräfte,
die makroskopisch gemessen werden. Über das Spannungsfeld
im Inneren des Werkstoffes ist nichts bekannt. Vielfach [1]
wird angenommen, daß sich der größte Volumenanteil in jedem
Korn unter einem anderen Spannungsfeld verformt, und daß der
Übergang zwischen den Spannungen benachbarter Körner in einer
relativ schmalen Schicht zu beiden Seiten der Korngrenze lokalisiert ist. Diese Annahme trifft sicherlich nicht immer
zu. Häufig wird vielmehr beobachtet, daß in verschiedenen
Teilen des gleichen Kornes unterschiedliche Gleitsysteme
betätigt werden. In solchen Fällen ist es sicher, daß auch
im Korninneren von Ort zu Ort verschiedene Spannungen ge-
herrscht haben.

Solange man nicht weiß, welche Spannungen im Korninneren
tatsächlich herrschen, weiß man auch nicht, welche Gleitsysteme die geforderte Verformung bewirken. Diese Unsicherheit belastet alle bisher vorgeschlagenen Theorien über die
Entstehung von Verformungstexturen.

Es gibt jedoch einen Spezialfall, in dem man die Spannungen

am Einzelkorn in guter Näherung kennt, nämlich dann, wenn alle Körner annähernd die gleiche Orientierung haben oder wenn sie mehrere Orientierungen einnehmen, die alle zu dem makroskopischen Spannungen symmetrisch liegen. Dann herrschen in jedem Korn die gleichen Spannungen, die auch makroskopisch am Werkstoff gemessen werden.

So ist es in einem Metall mit ausgeprägter Textur. Nur hier kann man unmittelbar sagen, welche Formänderung ein makroskopisch an den Werkstoff angelegtes Spannungsfeld bewirkt. Im allgemeinen werden die Hauptachsen dieser Formänderungen nicht den Hauptachsen der angelegten Spannung parallel sein. Dies zeigt sich sehr anschaulich beim plastischen Verdrehen von Rohren, die aus Kupferblechen mit einer scharfen Textur zusammengeschweißt wurden. Solche Rohre ändern beim Tordieren ihre Länge, obgleich keine Längsspannungen aufgebracht werden. Vorzeichen und Größe der Längenänderung lassen sich aus der Textur vorhersagen [2].

Es liegt nun die Vermutung nahe, daß als Endlagen der Verformungstextur nur solche Lagen in Frage kommen, bei denen – wie in der Plastomechanik angenommen – makroskopisches Spannungsfeld und makroskopische Formänderung gerade zueinander parallel laufen, d.h., die unter dem äußeren Spannungsfeld auch die "richtige" Formänderung bewirken.

Diese Vermutung wurde für fünf wichtige Verformungsarten an den kubisch flächenzentrierten Metallen geprüft, die sich durch Abgleitung auf {111}-Ebenen in ⟨110⟩-Richtungen verformen. Das Ergebnis zeigt die Tab.1. Man sieht zunächst, daß das hier beschriebene Formänderungskriterium nur von sehr wenigen Lagen erfüllt wird. Man sieht ferner, daß es im Scherversuch und im Zugversuch gerade die Lagen liefert, die auch tatsächlich beobachtet werden.

Bei den anderen Verformungsarten herrscht diese Übereinstimmung nicht. Hierfür scheint es drei verschiedene Gründe zu geben, die im folgenden angedeutet werden sollen (eine ausführliche Diskussion wird in [3] gegeben).

1. Die Würfellage (100)[001] erfährt beim Drahtziehen und beim Walzen die richtige Formänderung. Das gleiche gilt

Tabelle 1.

Verformungsart und Indizierung	untersuchte Lagen		Formänderungskriterium erfüllt	stabil	beobachtet
Scherversuch {Scherebene} (Scherrichtung)	{111} {100} {111} {110} {100} {110} {112}	⟨110⟩ ⟨011⟩ ⟨112⟩ ⟨110⟩ ⟨100⟩ ⟨001⟩ ⟨110⟩	ja ja ja nein nein nein nein		ja ja ja+) nein nein nein nein
Zugversuch (Zugrichtung)	⟨100⟩ ⟨111⟩ ⟨112⟩ ⟨110⟩		ja ja nein nein	ja ja ja nein	ja ja nein nein
Drahtziehen außerhalb des Drahtkernes (Drahtoberfläche) [Drahtachse]	(112) (112) (110) (100)	[$\bar{1}\bar{1}1$] [$11\bar{1}$] [001] [001]	ja ja ja ja	ja nein ja ja	ja nein ja nein
Stauchversuch (Stauchachse)	⟨100⟩ ⟨111⟩ ⟨112⟩ ⟨110⟩		ja ja nein nein	nein nein nein ja	nein nein nein ja
Walzen {Walzebene} ⟨Walzrichtung⟩	{110} {100} {110} {100} {112} {111} {110} {112}	⟨001⟩ ⟨001⟩ ⟨110⟩ ⟨011⟩ ⟨111⟩ ⟨112⟩ ⟨112⟩ ⟨110⟩	ja ja ja ja nein nein nein nein	ja (?) ja (?) nein nein ja nein ja nein	ja? (Ms) nein nein nein ja (Cu) nein ja (Ms) nein

+) schwach

für die Walzlage {110}⟨001⟩. Dies wären also geeignete Verformungstexturen. Wenn sie trotzdem nicht beobachtet werden, so liegt das an der Ausgangstextur der technisch üblichen Werkstoffe. Sorgt man durch experimentelle Kunstgriffe dafür, daß ein Draht oder ein Blech diese Textur bereits vor der Verformung zeigen, so bleibt sie bis zu hohen Verformungsgraden erhalten.

2. Einige andere Lagen erleiden zwar die richtige Formänderung, sind jedoch gegen Schwankungen nicht stabil. Das
bedeutet, daß ein Kristall, der die ideale Lage nicht
genau inne hat, bei der Verformung von dieser Ideallage
weiter weggedreht wird. Aus diesem Grund können im Stauchversuch die Lagen ⟨100⟩ und ⟨111⟩, beim Walzen die Lagen
{110}⟨110⟩ und {100}⟨011⟩ nicht als Endlagen auftreten.
Erzeugt man sie künstlich, so bleiben sie nicht erhalten.
Besonders erfolgreich ist dieser Gesichtspunkt bei der
Betrachtung der Lagen (112)[$\overline{11}$1] und (112)[11$\overline{1}$] in den
Außenschichten eines gezogenen Drahtes. Beide bewirken
die richtige Formänderung, aber nur eine ist stabil.
Entsprechend wird auch nur eine Lage beobachtet, was zu
einer sonst unverständlichen Asymmetrie der Verformungs-
textur führt [4].

3. Im Stauchversuch gibt es keine Lage, die sowohl die richtige Formänderung bewirkt, als auch stabil wäre. Die Natur hilft sich hier, indem die ⟨110⟩ Lage entsteht. Diese
Lage ist stabil. Ein Kreiszylinder aus einem Einkristall
dieser Orientierung nimmt beim Stauchen jedoch einen el-
liptischen Querschnitt an. Wir müssen also annehmen, daß
sich beim Stauchen kubisch flächenzentrierter Vielkri-
stalle in jedem Korn Querspannungen aufbauen, die eine
allseitige gleichmäßige Querschnittszunahme erzwingen.
Im Stauchversuch an kubisch flächenzentrierten Metallen
herrscht mithin – anders als beim Zugversuch – kein ho-
mogenes Spannungsfeld.

Auch beim Walzen von Blechen mit der Messingtextur {110}
⟨112⟩ bauen sich zwischen benachbarten Körnern Spannungen
auf, deren Orientierung sich in diesem Fall leicht einsehen
läßt. Freie Körner in dieser Lage würden nämlich außer der
richtigen Dickenabnahme und Längenzunahme eine Scherung erleiden, die in der Abb. 1 angedeutet ist. Zwischen benachbarten Körnern symmetrischer Orientierungen bauen sich also
Schubspannungen auf, die eine Abscherung in Walzrichtung
bewirken möchten. Die Scherebene steht senkrecht auf der
Blechebene.

Eine solche Scherung kann durch mechanische Zwillingsbildung

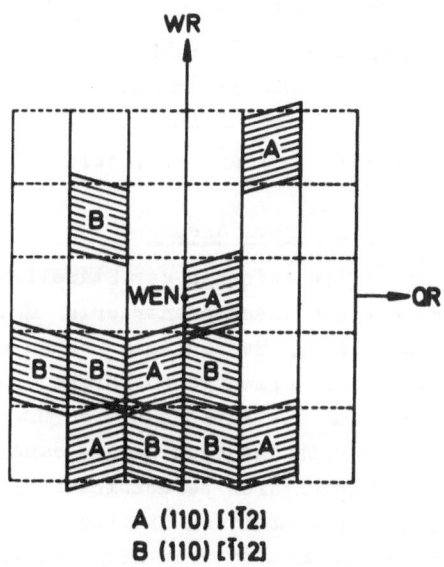

A (110) [1Ī2]
B (110) [Ī12]

Abb.1. Scherung beim Walzen. WEN = Walzebenen-Normale

bewirkt werden. So läßt sich verstehen, daß Metalle, die zur
Zwillingsbildung neigen (wie Silber und α-Messing) die {110}
⟨112⟩ Lage als Walztextur zeigen. Die Verformungszwillinge
sollten als Zwillingsebene die beschriebene Scherebene zeigen, was sich experimentell nachprüfen läßt.

Metalle, die keine Möglichkeit zur mechanischen Zwillingsbildung haben (wie Cu und Al) müssen entweder zwei zusätzliche Gleitsysteme betätigen, auf die makroskopisch jedoch keine Schubspannung wirkt, oder eine andere Walztextur
annehmen. Hierzu ist im Rahmen dieser Arbeit keine Voraussage möglich. Das Experiment zeigt, daß solche Metalle eine
komplizierte Textur mit weitem Streubereich annehmen.

Literatur

1. I.L. Dillamore und W.T.Roberts: Acta Met. 12(1964)281.
2. W. Rose und H.-P. Stüwe: Z. Metallkde. 59(1968)396.
3. G. Linßen, H.D. Mengelberg und H.-P. Stüwe: Z. Metallkde. 55(1964)600.
4. E. Aernoudt und H.-P. Stüwe: Veröffentlichung demnächst.

110

2.4 Texture Development under Conditions of Imposed Strain; The Influence of Stacking Fault Energy and Degree of Order
Entstehung von Walztexturen unter den Bedingungen aufgezwungener Formänderung; Einfluß von Stapelfehlerenergie und Ordnungsgrad

Zusammenfassung

Die für eine Korrelation zwischen der Stapelfehlerenergie
und den Texturen kubisch flächenzentrierter Metalle möglicherweise verantwortlichen Mechanismen werden untersucht.
Es wird eine Analyse der erzwungenen Formänderung benutzt,
die in erster Linie auf die Entstehung von Drahttexturen
angewendet wird. Die Betrachtungen werden dann verallgemeinert, um auch die Walztexturen berücksichtigen zu können.
Ferner werden Ergebnisse über den Einfluß einer Fernordnung
auf die Entstehung von Walztexturen mitgeteilt.

2.4.1 Introduction

The relationship between stacking fault energy and the
texture developed in f.c.c. metals and alloys has been
accounted for in two ways [1-4] which may be regarded as
being complementary rather than conflicting [5]. It is
suggested that for low stacking fault energy materials
crystal rotations result from deformation by a $\{111\}\langle112\rangle$
slip or twinning mode and, for high stacking fault energies
cross-slip predominates and crystal rotations resulting
from slip on $\{hhl\}\langle110\rangle$ systems control the development of
texture.

These viewpoints may be considered as extremes of the
possible range and as such they can only represent part of
the truth. The cross-slip theory almost certainly
approximates very closely to the situation prevailing in
b.c.c. metals [3] deforming by pencil glide on effectively
$\{hkl\}\langle111\rangle$ systems and the twinning theory accounts for some
of the crystal re-orientation observed in low stacking fault

[+] Birmingham University, Birmingham, England
[++] Rensselaer Polytechnic Institute, Troy, New York, U.S.A.

energy f.c.c. metals [1] but it is to be doubted whether either theory can account for the full range of crystals rotations contributing to texture development in f.c.c. metals of intermediate stacking fault energy.

In the present work the idea advanced by BISHOP [6] to account for the texture transition, that material variables may influence the choice of {111}⟨110⟩ slip systems, is revived and examined in the light of the known correlation between texture and stacking fault energy. In this viewpoint the influence of either stacking faults or of cross–slip would be to favour one or more of the combinations of five slip systems chosen from the available six or eight equally stressed slip systems. The effects which a preference for stacking fault formation or for cross–slip may have are examined in the first instance in relation to wire–textures. Later the ideas are generalised to consider rolling texture development and some results are presented on the effect of a further parameter influencing the choice of slip modes, the presence of long range order.

2.4.2 Experimental Background

The present analysis is suggested by two separate observations. ROBERTS, WILSON and GOODCHILD [7] and LAGNEBORG [8] have shown that in 18/8 stainless steel a tensile stress applied parallel to ⟨111⟩ promotes the formation of ε martensite, while tensile stressing parallel to ⟨100⟩ inhibits ε formation and yields a tangled dislocation structure more typical of a metal of high stacking fault energy, suggesting extensive cross–slip. ε martensite is thought to form from stacking faults, and the simple explanation of the experimental observations is that a tensile stress parallel to ⟨100⟩ constricts the slip dislocations and parallel to ⟨111⟩ extends the dislocations.

The second piece of experimental evidence is due to KOCKS[9], he performed work–hardening experiments on aluminium crystals subjected to different stress states. The states

of stress were chosen to activate either six or eight slip
systems (i.e. they were chosen from among the five
crystallographic distinct states of stress described by
BISHOP and HILL [10]) and it was found that the work –
hardening behaviour differed for the different states of
stress. It may be shown that the extent of sustained
work–hardening for the various stress states is
approximately in the order of increasing proportion of slip
systems whose dislocations are extended by the applied
stress. A possible interpretation of Kocks results is that
even in aluminium (γ/Gb$\simeq$20) the state of stress can, by
tending to extend the slip dislocations, inhibit cross-
slip.

These two observations together suggest the importance of
the imposed state of stress, and that the interaction
between the state of stress and a preference for either
cross–slip or stacking fault formation can influence the
choice of operative slip systems and hence of slip rotations.

2.4.3 Analysis of Slip–Rotations

BISHOP and HILL [10] showed that there are five
crystallographically distinct states of stress which are
capable of operating the five slip systems required to
impose an arbitrary shape change.

Under the five stress states either six or eight slip
systems sustain equal shear stresses and the choice of five
systems is to be made from the sets of either six or eight.

For the case of axisymmetric deformation as in wire drawing
the five stress states operating for fibre axes in different
regions of the unit triangle are shown in Fig. 1.
(Throughout this paper the notation of Bishop and Hill is
used). The equally stressed slip systems for the various
stress states are shown in Table 1 from which it may be
observed that there are two equally stressed slip
directions on either three planes (28,21) or four planes
(4,9,-1).

In order to have equal shear stresses in two $\langle 110 \rangle$ slip

113

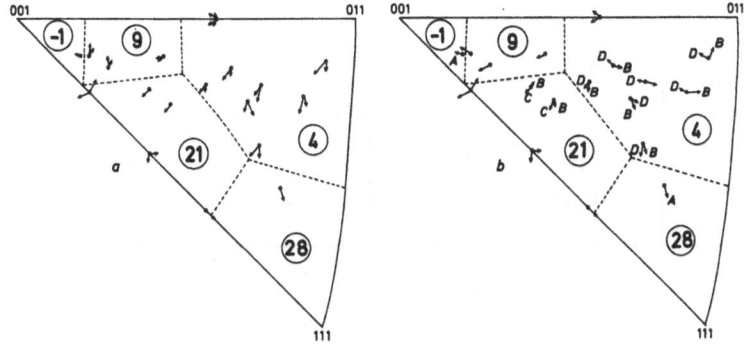

Fig.1a. Slip rotations favoured by stacking fault formation
obtained by selecting the maximum shear parallel to a plane
on which the dislocations are extended

Fig.1b. Slip rotations favoured by cross-slip. The notation
is described in the text

directions the maximum shear stress in a {111} plane must
be along a ⟨211⟩ vector, so that one of the Shockley
partials of each of the two slip dislocations is more
highly stressed than the other. The dislocations are thus
either extended or constricted by the applied stress. Both
slip systems on one plane are extended or constricted
together and a simple rule, consistent with Thompson's
rule is indicated in Table 1.

What principles, based on the information contained in
Table 1 and on a knowledge of work hardening behaviour, can
now be advanced to provide a basis for the choice of
operating slip systems assuming either that stacking fault
formation or cross-slip is favoured? A tendency to stacking
fault formation probably favours the maximisation of shears
parallel to a plane on which the dislocations are extended,
on the grounds that stacking faults form an effective forest
inhibiting intersecting slip. This principle is the only
obvious one, more contrived principles involving Lomer-
Cottrell locks now seem unlikely. The sense of slip
rotations using the principle of maximum shear parallel to a
plane of extension are shown in Fig. 1a.

Table 1.

Stress State	Slip Systems	Dislocations Extended or Constricted	Cross Related Pairs
4	a_2-a_3 $\quad$ c_2-c_3 $-b_2+b_3$ $\quad$ $-d_2+d_3$	extended constricted	a_2d_2, b_2c_2 a_3b_3, c_3d_3
28	b_1-b_2 $\quad$ c_2-c_3 $-d_1+d_3$	all extended	b_1d_1, b_2c_2 c_3d_3
21	$-a_1+a_2$ c_2-c_3 $\quad$ $-d_1+d_3$	constricted extended	 c_3d_3
9	$-a_1+a_2$ $\quad$ $-c_1+c_2$ $-b_1+b_3$ $\quad$ $-d_1+d_3$	constricted extended	a_1c_1 b_1d_1
-1	$-a_1+a_2$ $\quad$ $-b_1+b_2$ $-c_1+c_2$ $\quad$ $-d_1+d_2$	all constricted	a_1c_1, b_1d_1 a_2d_2, b_2c_2

A tendency to cross-slip readily could influence the choice of slip systems in a number of ways. For stress state (9) where there are two constricted systems in cross relation- ship it is probable that cross-slip favours the maximisation of shears with a common slip vector. For stress state (4) the cross-related systems are not both constricted; cross slip would favour either the maximisation of shear on one constricted system, using a theory of source hardening, or the minimisation of shears on systems not in cross- relationship with an active constricted system. The former principle (also considered in relation to stress state 21) gives the rotations indicated by arrows labelled B, and rotations predicted by the latter principle gives the arrows

labelled D. The rotation indicated by D is preferred to
that indicated by B since it can be justified mechanisti-
cally in terms of dislocation multiplication by the cross-
slip mechanism. A cross-slip source generating dislocations
on four, out of a required five, slip systems seems likely
to favour shear on those systems.

Under stress state 21 the total shear on constricted
system is not a variable and there is little to choose
between the two possible rotations shown as B and C in
Fig. 1b. The rotation indicated by the longer arrow (the
arrow length being proportional to the rate of rotation)
is likely to predominate.

Comparing the two cases in Figs. 1a and 1b we see that
material initially subjected to stress state 4 will in
both cases finish with the wire axis parallel to ⟨111⟩.
Stacking fault formation favours the rapid development
of this orientation while cross-slip causes a slower
development via a circuitous route. Similarly stress state
21 leads mainly to the development of the ⟨111⟩ orientation.

The main difference is seen in the region governed by
stress state 9, here cross-slip favours rotations in a sense
which reduces the probability of the orientation reaching
⟨100⟩. Under stress states (28) all rotations lead to ⟨111⟩
and under (-1) rotations lead to ⟨001⟩. Thus the main
difference predicted is that cross-slip will slightly
reduce the amount of ⟨100⟩ orientations in the final texture.
The composition of the final texture is predicted to be
75% ⟨111⟩ 25% ⟨100⟩ plus or minus a few percent depending
on whether cross-slip or stacking fault formation is
favoured. A rather larger proportion of material would
reach ⟨100⟩ if neither cross-slip nor stacking fault
formation were favoured.

With the additional variable of twin formation, favoured
for orientations under stress state 28 and causing reorientation to orientations near ⟨100⟩, the variation of
wire textures with stacking fault energy observed by
ENGLISH and CHIN [11] is fairly well accounted for on the

present analysis. This is not, however, the main objective
of the present contribution; here it is the aim to indicate
that slip rotations may be affected by stacking fault
energy in more subtle ways than proposed by the main
theories. The limited possibilities in the way of end
orientations for fibre textures should not obscure the
fact that the rotations shown in Figs. 1a and 1b differ
quite markedly and in a lower symmetry deformation process
such as rolling the differences may significantly affect
the end texture even if the more extreme effects of
twinning and gross cross-slip make no contribution.

2.4.4 Additional Experimental Work

In order to study the consequences of preventing twinning
and extensive cross-slip some preliminary experiments
have been carried out on a copper 25% gold alloy. Three
distinct procedures were adopted: (1) the alloy was
disordered and rolled 95%; (2) the alloy was initially
ordered and rolled 95%; (3) the alloy was initially ordered
and rolled in stages of 0.2 true strain, with re-ordering
anneals of 84 hours at 350°C after each reduction, to
a total of 95% deformation. In the latter material the
degree of order S as determined by X-Ray measurements
using 100, 200, 110 and 220 reflections was at no time less
than ~ 0.7. The variation of S with rolling reduction is
shown in the Fig. 2 for the initially ordered material;
it will be noticed that the degree of order falls more
rapidly for the 100 reflection than the 110. The
orientations contributing principally to these reflections
after some small strain will be (100)[001] and {110}⟨112⟩
which are the more stable of the orientations having (100)
and (110) sheet planes respectively. The difference in the
rate of decay of order does not therefore depend on the
accumulated shear strain since this is larger at a given
rolling strain for {110}⟨112⟩. This result may then be
taken as indicating the significance of the stress state
and the resulting combination of slip systems.

The textures, as judged from {111} pole figures, of the

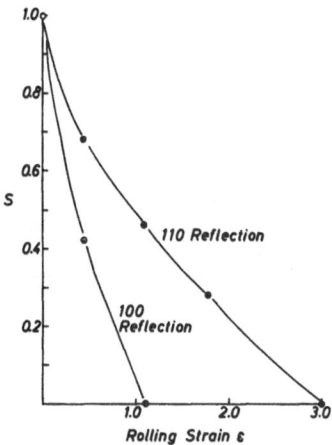

Fig.2. Degree of order S as a function of rolling
strain ε for Cu₃Au ordered prior to rolling. Note
that the variation of S is different for {100} and
{110} texture components

initially disordered and the initially ordered materials
are similar and the latter is shown in Fig. 3a. The
repeatedly reordered material shows only slight differences
as evidenced by Fig. 3b. Both pole figures are, however,
significantly different from those shown by copper and
70/30 brass after 95% deformation at room temperature.
It may be seen by comparison with any true random intensity
pole figures obtained from these materials that the texture
is less sharply developed in the Cu₃Au than in copper but
is about as strong in intensity as the texture of 70/30
brass. In fact the texture shows greater similarities with
that shown by brass than with that shown by copper. The
differences between Fig. 3 and a {111} pole figure of 70/30
brass arises from the lack of twinning as an available mode
in the former, as may be seen from the work of HEYE and
WASSERMANN [12].

The differences between Figs. 3a and b are too slight to
encourage speculation about possible differences in slip
modes but it may be remarked that in the initially ordered
material the presence of short range order will, after

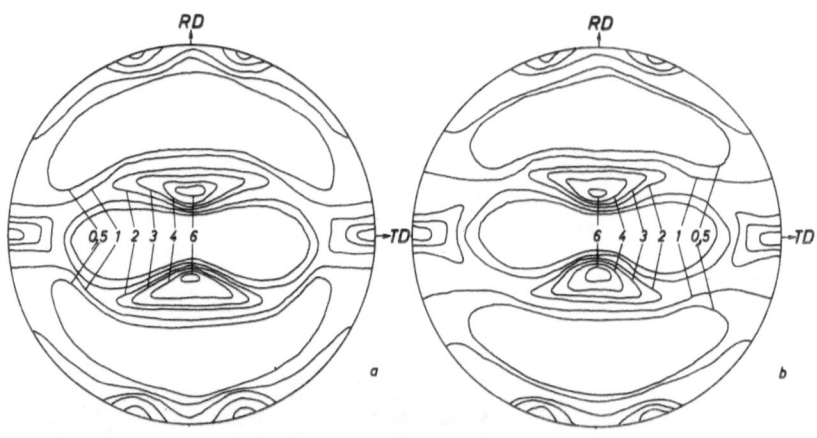

Fig.3a. {111} pole figure of Cu$_3$Au ordered prior to rolling 95%.

Fig.3b. {111} pole figure of Cu$_3$Au ordered repeatedly at intervals of 0,2 true strain. Rolled to a total of 95% reduction.

The intensity units are corrected true random multiples

initial disordering, play an important part. Short range order by confining dislocations to their primary slip system will affect the choice of slip systems in much the same way as stacking fault formation for half of the possible orientations (the similarities of slip line observations on disordered Cu$_3$Au and low stacking fault energy f.c.c. metals encourages this viewpoint [13], leading to textures closer in form to those of metals which fault readily than of those which do not.

The present results and analysis thus suggest that the rolling textures developed in aluminium and 70/30 brass are extremes of a range. Twinning and extensive cross-slip are probably not the only processes of importance, connecting slip and stacking fault formation may have influence on the choice of slip modes as indicated by the analogous effects of ordering.

Acknowledgement

We wish to thank Dr. W. T. Roberts for stimulating discussions.

References

1. G. Wassermann: Z. Metallkde. 54(1963)61.

2. F. Haeßner: Z. Metallkde. 54(1963)98.

3. I.L. Dillamore and W.T. Roberts: Acta Met. 12(1964)281.

4. H. Hu, R.S. Cline and S.R. Goodman: Recrystallization, Grain Growth and Textures, H. Margolin, Ed., Amer.Soc. Metals, Metals Park, Ohio 1966, p. 295.

5. I.L. Dillamore, E. Butler and D. Green: Met.Sc.J. 1968) 161.

6. J.F.W. Bishop: J. Mech. Phys. Solids 3(1954/55)130.

7. W.T. Roberts, D.V. Wilson and D. Goodchild, reported by I.L. Dillamore, W.T. Roberts and D.V. Wilson "Stainless Steel for the Fabricator and User" I.S.I. Spec.Rept., to be published.

8. R. Lagneborg: Acta Met. 12(1964)823.

9. U.F. Kocks: Acta Met. 8(1960)345.

10. J.F.W. Bishop and R.Hill: Phil.Mag. 42(1951)1298.

11. A.T. English and G.Y. Chin: Acta Met.13(1965)1013.

12. W. Heye and G. Wassermann: phys. stat. sol. 18(1966) K107.

13. N.S. Stoloff and R.G. Davies: Chalmers' Prog. Mat. Science 13(1966)3.

2.5 A Modified Cross-Slip Theory for the Rolling Texture of Face-Centered Cubic Metals and Alloys
Eine modifizierte Quergleitungstheorie der Walztextur kubisch flächenzentrierter Metalle und Legierungen

by T. Leffers[+]

Zusammenfassung

Die Entwicklung der Walztextur wird mit Hilfe eines Programms für einen Rechenautomaten simuliert. Dabei wird nur ein Gleitsystem, nämlich {111}⟨110⟩, berücksichtigt, doch können verschiedene Gleitsysteme dieses Typs auf unterschiedliche Weise kombiniert werden. Unter der Annahme, daß sich die Kristalle wie isolierte Einkristalle verformen, führt die Berechnung zum Messingtyp der Walztextur. Eine Textur vom Kupfertyp entsteht dagegen bei Verwendung eines Modells, das die TAYLOR'schen Bedingungen teilweise erfüllt. Um diese Ergebnisse zu deuten, wird ein Versetzungsmodell für die plastische Verformung von kubisch flächenzentriertem vielkristallinen Material vorgeschlagen. Der Übergang vom Messing- zum Kupfertyp wird auf zunehmende Quergleitung zurückgeführt, die als ein Katalysator für Mehrfachgleitung angesehen wird.

2.5.1 Introduction

Two different types of rolling texture, the copper type and the brass type, are found in face-centred cubic metals and alloys. The development of the copper-type texture is favoured by high stacking fault energy and high rolling temperature [1-3] and, as shown recently [4], by low deformation rate. The two different types of texture in f.c.c. materials must correspond to two different deformation mechanisms. The possibility of revealing the nature of this difference in mechanism is one of the most interesting aspects of deformation textures.

[+] Metallurgy Department, Danish Atomic Energy Commission, Research Establishment Risö, Roskilde, Denmark.

Numerous theories attempting to explain the two different
rolling textures have been advanced. Some of them introduce
mechanisms other than {111}⟨110⟩ slip to account for one
of the two types of texture (twinning [5], slip on planes
other than {111} [6], slip by partial dislocations [7]),
whereas other theories imply that the two rolling textures
are produced by different combinations of the {111}⟨110⟩
slip systems (the cross-slip theory [8,9], the theory of
dislocation interaction [10], the overshooting theory [11]).

All these theories suffer from the lack of reliable
calculated textures to support them. It has been attempted
to calculate the texture developed in different models [6,
9-11], but the scope of these calculations was limited to
finding the "stable end orientations". Also, only very
simple stress systems have been used for simulating the
rolling stresses. By using a computer, these limitations
have been overcome by the present author. Only {111}⟨110⟩
slip is considered in the calculations. Of the other
mechanisms proposed, slip on planes other than {111} and
slip by partials have never been shown to have any
statistical importance. Twinning does occur when the brass-
type texture is developed, but the twins cannot account for
the transition from copper-type to brass-type texture [12].

It turns out that both types of texture may be developed
by {111}⟨110⟩ slip. Initially it was found [13] that a
brass-type calculated texture was developed when the
individual grains were assumed to deform like isolated
single crystals, whereas the calculated texture was of the
copper type when the TAYLOR condition of continuity [14] was
fulfilled. A more detailed analysis [15] showed that the
agreement with the experimental copper texture was better
when the continuity condition as formulated by Taylor was
only partly fulfilled.

2.5.2 The Computer Programme

The "specimens" deformed in the computer consist of 100
grains of random orientation. The grains are taken one by
one and deformed stepwise. In each step, the {111}⟨110⟩ slip

system with the highest resolved shear stress is determined, and a certain amount of shear deformation (0.05) is performed on this system. The lattice rotation accompanying the deformation takes place in such a way as to fulfil two requirements: a) a string of material lying in the rolling direction before the slip event must retain this orientation when slip has taken place, and b) a plate of material lying in the rolling plane must also retain its orientation. Condition a) is the one normally used when dealing with the rotation of the slip direction in tensile experiments. It is not sufficient to define the lattice rotation unambiguously; therefore b) has been added. When all the crystals have been deformed to a predetermined reduction in thickness, their {111} and {100} poles are plotted in stereographic projection. All the poles in these pole figures are plotted in the first quadrant in order that the maximum density of poles may be obtained.

The different combinations of the {111}⟨110⟩ slip systems are obtained by using different stress systems to simulate the rolling stresses. The "basic" stress system consists of a tensile stress in the rolling direction and a compressive stress in the normal direction, the two stress components being of equal size numerically. When used without additional stresses this stress system is designated (I).

Fig. 1 illustrates the deformation of a cube-shaped crystal (Fig. 1a) when applying the different stress systems. Fig. 1b corresponds to system (I). The deformed crystal is a parallelepiped with four edges pointing in the rolling direction and two faces lying in the rolling plane in accordance with conditions a) und b). (I) leaves the problem of the continuity of the polycrystalline material unsolved. This problem is solved if the individual crystals follow the macroscopic deformation as proposed by TAYLOR [14]. The programm is written to permit the application of additional stresses together with the basic stress system. Each of the additional stresses represents a partial fulfilment of Taylor's condition; when they are used together, the condition is completely fulfilled. (I) leads

123

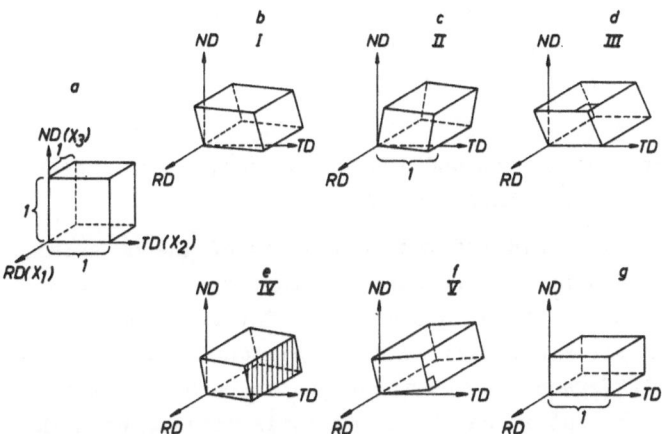

Fig.1. The deformation of a cube-shaped crystal using the
different stress systems. a) the undeformed crystal; b)
stress system (I); c) stress system (II); d) stress system
(III); e) stress system (IV); f) stress system (V); g) the
combined stress systems (II), (III), (IV), and (V). RD, TD
and ND stand for rolling direction, transversal direction
and normal direction respectively

to initial single glide like the deformation of single
crystals, whereas the additional stresses will provoke
multiple glide from the early stage of deformation (in this
connection multiple glide means change in slip system
between the individual slip events). The stress systems with
stresses added to the basic stresses are as follows:

(II) Stresses are added to impede changes in the
transverse dimension of the crystals. If, for instance,
a crystal shrinks in the transverse direction, a
transversal tensile stress proportional to the shrinkage
is added. The deformation caused by (II) is illustrated
in Fig. 1c.

(III) Shear stresses σ_{12} and σ_{21} are added in order to
preserve the right angles between the edges lying in
the rolling plane. The effect of (III) is shown in
Fig. 1d.

(IV) Shear stresses σ_{23} and σ_{32} are added to keep the

side faces (hatched in Fig. 1e) perpendicular to the
transverse direction.

(V) Shear stresses σ_{31} and σ_{13} are added to keep the
edges that do not lie in the rolling plane perpendicular
to the rolling direction (Fig. 1f).

In Fig. 1g (II), (III), (IV), and (V) are combined: The
transverse dimension and the orthogonality are preserved,
i.e. the Taylor condition is fulfilled.

Taylor only considered the continuity of the material.
BISHOP and HILL [16] pointed out that the stress continuity
should also be taken into account. Taylor's condition
cannot apply rigorously to the individual crystals because
of the difference in lattice orientation across the grain
boundaries, but it may be used statistically. When the
grains are considered one by one, as they are in this
programme, the effect of the orientation of the surrounding
grains upon the stresses seems to be random. It may therefore be simulated by random stresses. The computer programme
permits the addition of such random stresses to the above
stress systems. When used together with stress systems that
already include additional stresses, the random stresses
are given a mean value equal to the mean value of these
additional stresses. As demonstrated in [15] the random
stresses do not cause any drastic changes in the calculated
pole figures; they result in a decreased sharpness of the
calculated texture, thus increasing the similarity with
the experimental texture.

2.5.3 Results and Discussion

Many combinations of stress systems and random stresses and
different degrees of reduction have been tried. For obvious
reasons, only the pole figures for the crucial combinations
are given in this paper. A more comprehensive selection of
calculated pole figures is presented in [15].

As already mentioned, the model in which the grains are
considered to deform like isolated single crystals leads
to the development of a brass-type texture. This is

illustrated in Figs. 2 and 3, showing the calculated pole
figures for 50 and 80%[+]reduction for stress system (I)

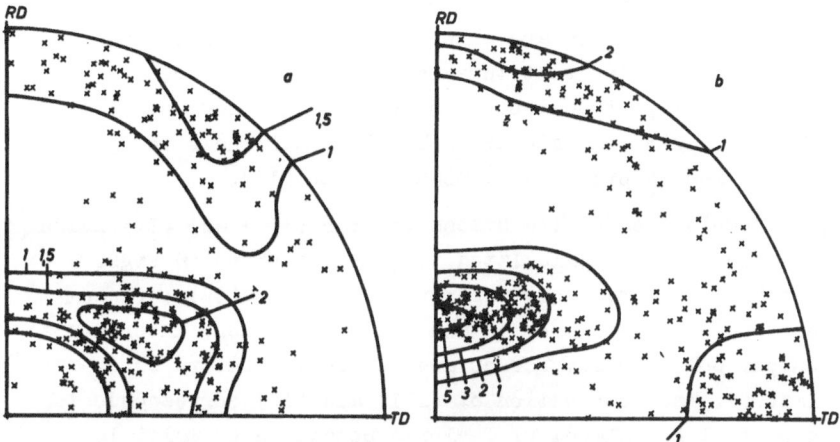

Fig. 2a and b. Calculated pole figures for (I) + random
stresses and experimental brass pole figures. 50% reduction.
a) {200}, b) {111}

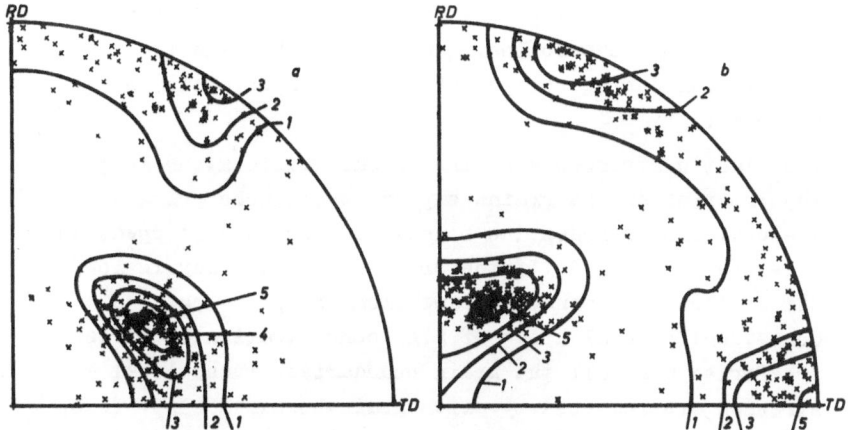

Fig. 3a and b. Calculated pole figures for (I) + random
stresses, 80% reduction, and experimental brass pole figures,
94% reduction. a) {200}, b) {111}

[+]An increase in reduction above 80% does not cause much
change in the calculated texture; it only increases the
running time of the programme.

with random stresses, i.e. the stress system with no formal restrictions on the change in shape that the crystals can undergo. In the figures, experimental pole figures for brass deformed 50 and 94% are superimposed on the calculated pole figures. In order that this model may apply to real polycrystals, it must be combined with the assumption that the continuity is maintained by alien slip near the grain boundaries as proposed by KOCHENDÖRFER [17].

The combination of the stress systems (III) and (IV) leads to a copper-type calculated texture, as shown in Figs. 4 and 5. Calculated pole figures for (III) + (IV) + random stresses are shown for 50 and 80%[+] reduction. Experimental copper pole figures for 50 and 94% reduction are super-imposed. The combination of (III) and (IV) corresponds to a partial fulfilment of Taylor's condition by multiple slip throughout the grains. The deformation caused by (III) + (IV) is illustrated in Fig. 6; it is seen that alien grain boundary-slip must be added.

The above models are purely crystallographic. On the dis-location scale the following model for the plastic deformation during rolling of f.c.c. polycrystals is proposed:

The elastic stresses make the primary dislocation loops in the interior of the grains expand. When these primary dislocations approach the boundaries they are stopped, and pile-ups are formed. If the cross-slip frequency is low, the dislocations cannot escape from the pile-ups. Consequently local areas of high concentration of stress are formed near all the grain boundaries. These local stresses are released by alien grain-boundary slip. Thus, for conditions unfavourable for cross-slip, i.e. the conditions under which the brass-type texture is developed, the continuity across all the boundaries is maintained by alien slip in the boundary regions. Therefore, the interior

---

[+] An increase in reduction above 80% does not cause much change in the calculated texture; it only increase the running time of the programme.

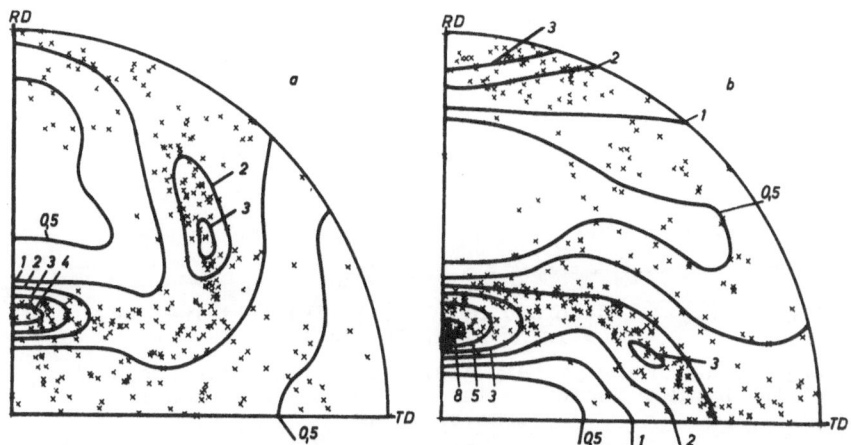

Fig. 4a and b. Calculated pole figures for (III) + (IV) + random stresses and experimental copper pole figures. 50% reduction. a) {200}, b) {111}

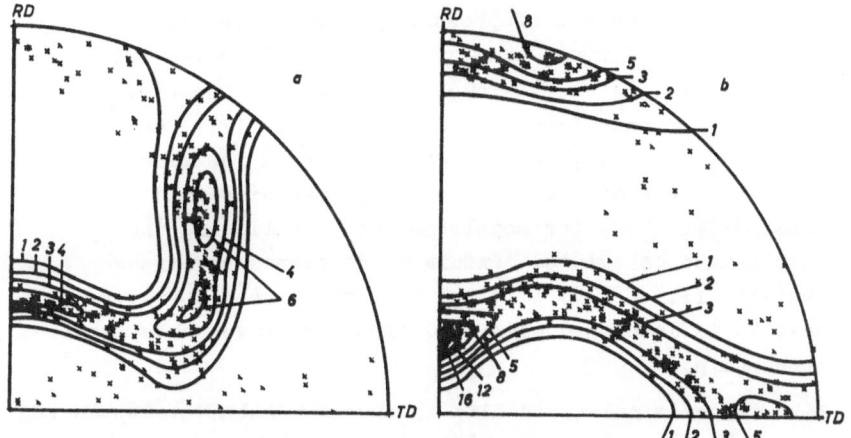

Fig. 5a and b. Calculated pole figures for (III) + (IV) + random stresses, 80% reduction, and experimental copper pole figures, 94% reduction. a) {200}, b) {111}

of the grains may deform in the single-crystal manner in accordance with the present results.

When the cross-slip frequency is high, the screw dislocations may escape from the pile-ups. Fig. 7 shows an expanding primary loop; if the rolling stresses consist of a tensile stress along RD and a compressive stress along TD,

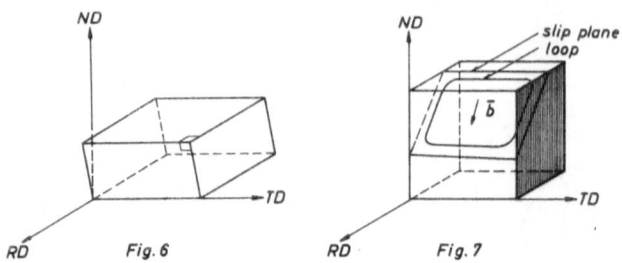

Fig. 6. Crystal deformed by (III) + (IV) (obtained by combining Figs. 1d and 1e)

Fig. 7. Expanding primary dislocation loop. The projection of the Burgers vector on the rolling plane is nearly parallel to RD

the primary dislocations approaching the side face (hatched in Fig. 7) are nearly screws. The pile-ups at this face are therefore dispersed by cross-slip, and the stresses from the misfit across the boundary are more evenly distributed. Instead of provoking local alien slip near the grain boundaries, they are transferred to the interior, causing multiple slip throughout the grain. One of the stress components from each of the systems (III) and (IV), namely σ_{21} and σ_{23}, is transferred through the side faces, i.e. the faces at which the pile-ups are dispersed by cross-slip. Thus, the model combines the finding that a copper-type calculated texture is developed by stress systems (III) + (IV) with the fact that the copper-type texture is developed when conditions are favourable for cross-slip.

Like the original cross-slip theory of SMALLMAN and GREEN [8] and DILLAMORE and ROBERTS [9,18], the present theory ascribes the transition from brass-type to copper-type texture to an increase in cross-slip frequency. The above authors considered slip on the cross-slip systems to be directly responsible for the copper-type texture, whereas, in the present model, cross-slip is assumed to cause multiple slip by preventing the formation of long pile-ups of screw dislocations. A recent estimate of the activation energy

for the texture transition in brass [4] supports the
assumption that the transition is caused by cross-slip.

References

1. F. Haeßner: Z. Metallkde. 53(1962)403.
2. H. Hu and S.R. Goodmann: Trans.Met.Soc.AIME 227(1963)627.
3. H. Hu and R.S. Cline: J. Appl. Phys. 32(1961)760.
4. T. Leffers: Scripta Met. 2(1968)447.
5. G. Wassermann: Z. Metallde. 54(1963)61.
6. F. Haeßner: Z. Metallkde. 54(1963)98.
7. H. Hu, R.S. Cline and S.R. Goodmann in: Recrystallization, Grain Growth and Textures, H. Margolin,Ed., Amer.Soc. Metals, Metals Park Ohio, 1966, p. 295.
8. R.E. Smallman and D. Green: Acta Met. 12(1964)145.
9. I.L. Dillamore and W.T.Roberts: Acta Met. 12(1964)281.
10. Y.C. Liu: Trans.Met. Soc. AIME 230(1964)656.
11. E.A. Calnan: Acta Met. 2(1954)865.
12. T. Leffers and A. Grum-Jensen: Trans. Met. Soc. AIME 242(1968)314.
13. T. Leffers: phys.stat.sol. 25(1968)337.
14. G.I. Taylor: J. Inst.Met. 62(1938)307.
15. T. Leffers: Risö Report No. 184 (1968).
16. J.F.W. Bishop and R. Hill: Phil. Mag. 42(1951)414 and 1298.
17. A. Kochendörfer: Plastische Eigenschaften von Kristallen und metallischen Werkstoffen, Springer, Berlin 1941.
18. See also I.L. Dillamore and N.S. Stoloff: This volume, chapter 2.4.

2.6 The Occurence of a "Brass-Type"-Texture in Aluminium
Das Auftreten einer Textur vom Messingtyp in Aluminium

by R. Sundberg[+)]

Zusammenfassung

Gegossene Aluminiumbänder können ein ausgeprägtes Gußgefüge
aus Fiederkristallen haben, deren Zwillingsgrenze parallel
zur Gußrichtung und senkrecht zur Bandoberfläche liegt. Wenn
diese Bänder unter verschiedenen Winkeln zur Zwillingsgrenze kaltgewalzt werden, entstehen unterschiedliche Texturen. Diese werden beschrieben und ihre Deutung wird diskutiert.

2.6.1 Introduction

This paper will report the results from cold rolling of
cast strips with a distinct texture. Some of the results
have been presented earlier [1].

It has been shown that the rolling texture of aluminium is
gradually changed from an orientation $\sim$(123)[41$\bar{2}$] to (112)
[11$\bar{1}$], as the reduction is increased. The "brass-type"-
texture (110)[$\bar{1}$12] can according to DILLAMORE and ROBERTS [2]
only occur at small reductions in aluminium. Some authors[3]
have found a "brass-type"-texture in pure aluminium with
small additions of iron. It is claimed that the effect
depends on high concentration of iron in solid solution [4].

Normally the recrystallization texture of aluminium is a
mixture of cube texture, (100)[001], and $\sim$ (123)[41$\bar{2}$].
According to [3], the recrystallization texture is $\sim$(123)
[41$\bar{2}$], when the rolling texture is of the "brass-type".
Some authors have tried to find a general connection
between the rolling and the recrystallization textures. A
mathematical method is given in [5]. A rolling texture
near $\sim$(123)[41$\bar{2}$] is transformed to the cube texture after
recrystallization and a "brass-type"-texture to $\sim$(113)[21$\bar{1}$].

[+)] Svenska Metallverken, Finspång, Sweden

2.6.2 Experimental Procedure

Pieces of a 6,5 mm thick cast aluminium strip with the
composition 0,08 % Si, 0,38 % Fe, 0,02 % Ti were rolled
in one direction in 20 passes to 0,5 mm. Samples were
rolled with an angle of 0, 10, 20, 30, 45, 60 and 90°
between the casting and the rolling direction. The rolled
strips were annealed at 380° C during 1 h.

The texture determinations were done with a Philips texture
goniometer. The samples were etched in NaOH to 1/3 of the
original thickness before testing, in order to remove the
surface texture, (100)[011]. The cast strip was tested
1 mm under the original surface. The results are presented
in (111)-pole figures. The intensities are multiples of
the intensity of a sintered sample.

2.6.3 Experimental Results

The cast strip has a sharp (112)[$\bar{1}$10] texture rotated 5°
around the transverse direction (TD), as shown in Fig.1.
It means that a (111)-plane is parallel to the casting
direction and perpendicular to the surface. Metallographic
studies have shown twin boundaries oriented in the same
way. The distances between the boundaries are normally
5-10 μm. Accordingly the strip has a feather crystal
structure of very high uniformity.

The strip rolled parallel to the twin boundaries has a
"brass-type"-texture, Fig. 2, which during annealing is
transformed to a (110)[001] texture. The rolling texture
is successively changed, as the angle between the twin
boundaries and the rolling direction is increased. One of
the two symmetric orientations of the "brass-type"-texture
gradually disappears. Fig. 3 shows the pole figure for the
sample rolled with an angle of 20°.

At 30° the disappearance is nearly complete, Fig. 4. The
annealing texture has in all cases as main component
(110)[001]. Higher angles between the twin boundaries and
the rolling direction result in textures more and more
like the "normal" rolling texture. At 45°, Fig. 5, the

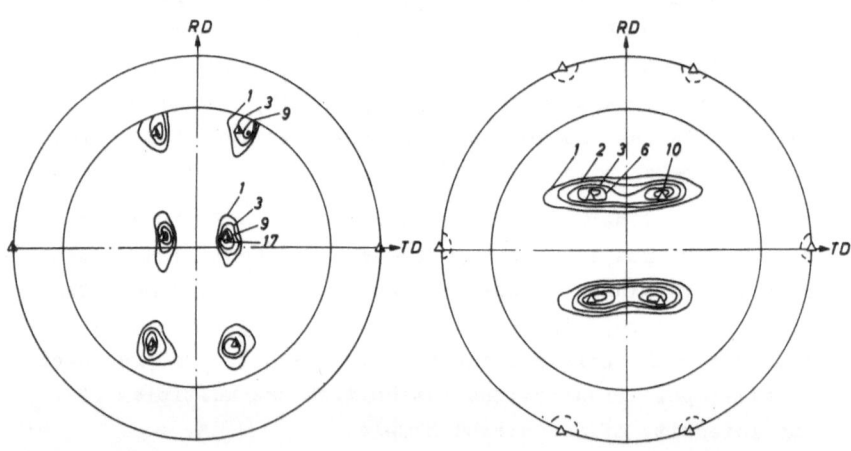

▲ *(112) [ī10] rotated 5° TD*   △ *(110) [ī12]*

Fig.1 (left). (111)-pole figure, as cast 6,5 mm strip

Fig.2 (right). (111)-pole figure, 0,5mm strip rolled with 92% reduction, rolling direction parallel to casting direction.
"Brass texture"

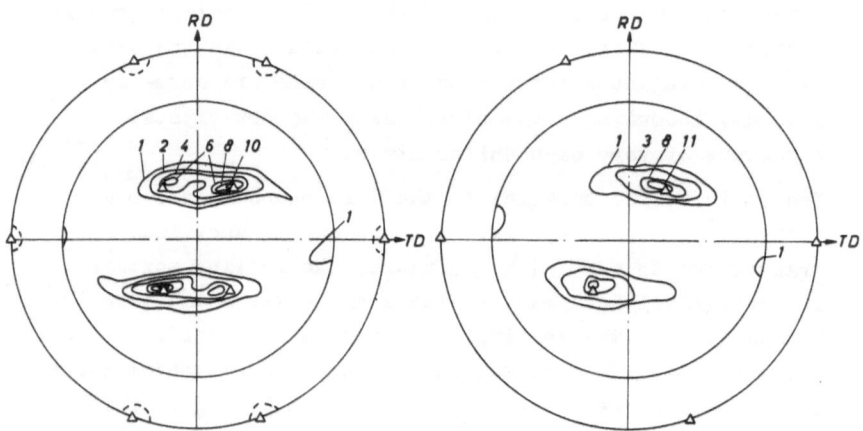

Fig.3 (left). (111)-pole figure, 0,5 mm strip rolled with 92% reduction, rolling direction 20° to casting direction

Fig.4 (right). (111)-pole figure, 0,5 mm strip rolled with 92% reduction, rolling direction 30° to casting direction

texture can be described as (011)[21$\bar{1}$] partially rotated
around the [11$\bar{1}$]-direction near the rolling direction. The
same type of texture was found in the 60°-samples. The
annealing texture is a mixture of cube texture and
complicated other orientations. In the sample, rolled
perpendicular to the twin boundaries, the texture is of
(112)[11$\bar{1}$]-type in the rolled condition, Fig.6, and a
mixture of cube texture and complicated other orientations
of rolling texture type after annealing. This means that the
texture of the cast strip is preserved after rolling.

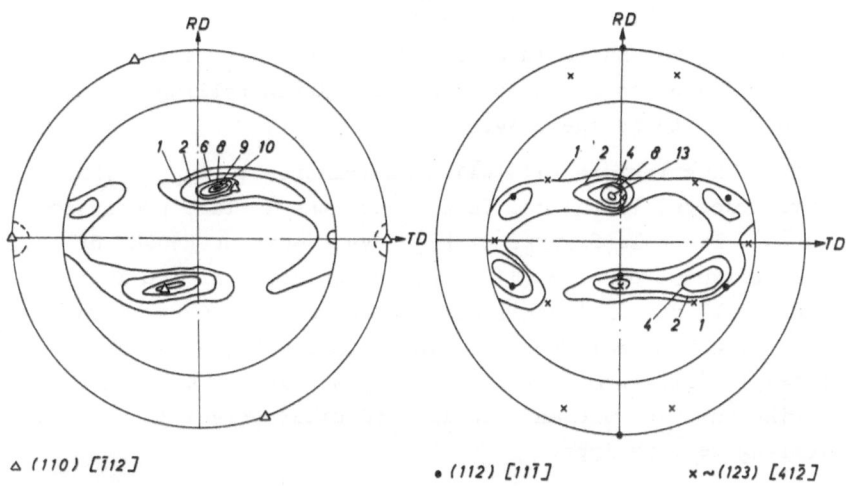

$\triangle$ (110) [$\bar{1}$12]

$\bullet$ (112) [11$\bar{1}$]   x ~(123) [41$\bar{3}$]

Fig.5. (111)-pole figure,
0,5 mm strip rolled with
92% reduction, rolling
direction 45° to casting
direction

Fig.6. (111)-pole figure,
0,5 mm strip rolled with
92% reduction, rolling
direction 90° to casting
direction

2.6.4 Discussion

The results presented in this investigation mean, if they
are interpreted according to DILLAMORE's and ROBERTS' [2]
theory for the formation of rolling texture in aluminium,
that cross slip is impossible when rolling parallel to the
twin boundaries. With increasing angle between casting
and rolling direction, cross slip becomes gradually more

and more probable.

Another way of interpretation is also possible. Lately
HEYE and WASSERMANN [6] have shown that the formation of
rolling textures can be described by slip on the $\{111\}\langle110\rangle$
glide systems. Single crystals of copper with the "stable
orientations" $(011)[21\bar{1}]$ and $(112)[11\bar{1}]$ are splitted by
rolling due to the formation of deformation bands. The
spread of orientation can be described by limited $\langle111\rangle$-
fibre textures, the fibre axes being those $\langle111\rangle$-poles
with high intensities. HEYE and WASSERMANN [7] pointed out
in a remark to [1] that deformation by rolling of a crystal
with the orientation $(112)[\bar{1}10]$ will give the orientation
$(011)[21\bar{1}]$ as end texture. Their theory predicts that the
$(112)[11\bar{1}]$ orientation is preserved during rolling, which
also seems to be the case.

The rolling texture $(110)[\bar{1}12]$ is transformed to $(110)[001]$
after recrystallization. This is not the texture predicted
in [5]. The pole figures of [3] have a weak component of the
same orientation. The $(110)[001]$ orientation is the main
recrystallization texture as long as the rolling texture
is of the "brass-type". When the normal rolling texture
$(112)[11\bar{1}]$ is formed, the annealing texture is a mixture
of the cube texture and complicated other orientations of the
rolling texture type.

From a practical point of view the results mean that the
feather crystal structure of ingots can influence the
rolling and annealing texture of aluminium very much. This
is also reported in many cases. The scatter in earing is
much worse when an ingot with feather crystals is used
according to [8]. If the feather crystals are oriented with
the twin boundaries approximately parallel to the rolling
direction and perpendicular to the rolling plane, the
earing after annealing will be of 45° type. The amount of
cube texture will be low. The result means that annealed
sheet rolled from ingots with a feather grain structure
normally will have increased 45° earing. The results
reported in [3] about the influence of iron may also be an
effect of feather crystals.

References

1. R. Sundberg: Z. Metallkde. 59(1968)202.
2. I.L. Dillamore and W.T. Roberts: Acta Met. 12(1964)281.
3. W. Bunk: Z. Metallkde. 56(1965)645.
4. W. Bunk: Z. Metallkde. 57(1966)34.
5. F. Lihl and W. Pexa: Z. Metallkde. 58(1967)465.
6. W. Heye and G. Wassermann: Z. Metallkde. 59(1968)617.
7. W. Heye and G. Wassermann: Z. Metallkde. 59(1968)667.
8. M.F. Jordan and J.C. Blade: Metall 13(1959)193.

2.7 Texturen von Kupfereinkristallen nach Walzen bei Raumtemperatur und 77°K
Textures of Copper Single Crystals rolled at RT and 77°K

von R. Bauer, H. Mecking und K. Lücke[+)]

Abstract

Textures of copper single crystals with various initial
orientations were investigated after rolling up to very high
degrees of deformation at RT and 77°K. In particular, the
texture development at various depths from the surface of
the crystals was examined. Results were explained by the
stress condition in the roll gap. Furthermore, measurements
indicated that the "end-orientations" developed at the
various layers across the thickness of the strip depend
on initial orientation and rolling temperature.

Bei der theoretischen Deutung der Entstehung und Stabilität
der Walztexturen wird häufig der von TUCKER [1] angenommene
effektive Spannungszustand zugrundegelegt, nach dem die Verformung unter reinem Druck senkrecht zur Walzebene und
reinem Zug in Walzrichtung stattfindet. Wie unter [2] beschrieben, bedeutet diese Annahme jedoch nur eine grobe
Näherung. Die tatsächlichen Verhältnisse sind in Abb. 1
erläutert. Dort sind schematisch die Richtungen der maximalen Schubspannung in den verschiedenen Bereichen des
Walzspaltes bei breitungslosem Walzen dargestellt. (Die
gestrichelten Linien kennzeichnen von rechts nach links
Eintrittsseite, Lage der Fließscheide und Austrittsseite).
Die Walzballen übertragen auf das Walzgut Druckkräfte und
–infolge der Reibung – auch Tangentialkräfte. Dabei sind
die aus den Tangentialkräften resultierenden, in der Walzebene wirkenden Schubspannungen an der Ober- und Unterseite
des Walzgutes gegenläufig. In der Probenmitte müssen demnach die parallel zur Walzebene wirkenden Schubspannungen
aus Symmetriegründen verschwinden. Dort, aber auch nur dort,

[+)] Institut für Allgemeine Metallkunde und Metallphysik,
Technische Hochschule Aachen, Aachen, Bundesrepublik
Deutschland

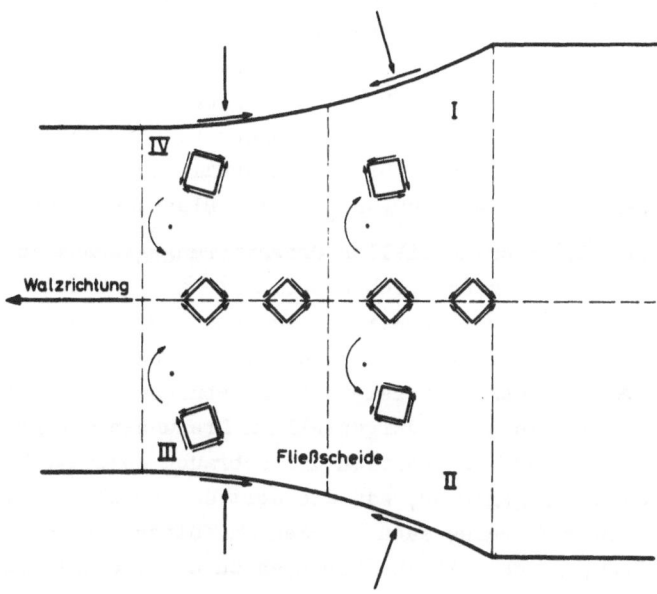

Abb.1. Auf das Walzgut wirkende Kräfte und die daraus resultierenden Richtungen maximaler Schubspannungen in den verschiedenen Bereichen eines Walzspaltes (schematisch).

Der Drehsinn des Spannungstensors gegenüber dem in der Probbenmitte vorliegenden Tucker'schen Spannungszustand ist in den einzelnen Quadranten (I-IV) durch Pfeile angegeben

liegt der spezielle Tucker'sche Spannungszustand mit reinem Druck in Normalrichtung und reinem Zug in Walzrichtung vor; nur hier liegen demnach die Richtungen maximaler Schubspannung um 45° gegen diese Hauptspannungen geneigt. Unter dem Einfluß der an der Probenoberfläche wirkenden Tangentialspannungen dreht sich der Spannungstensor vom Probenkern zur Oberfläche in dem durch Pfeile angezeigten Sinne (Abb.1), wobei sich der Drehsinn in der Probenmitte umkehrt. In der Fließscheide wechseln die Tangentialkräfte und damit auch die Drehrichtung des Tensors ihre Vorzeichen, so daß sich das gesamte Spannungsfeld im Walzspalt hinsichtlich des Drehsinns in vier Quadranten aufteilen läßt. Dabei sind die Quadranten I und III bzw. II und IV jeweils gleichberechtigt.

Eine etwas außerhalb der Probenmitte liegende Schicht durchläuft während der Umformung einen Spannungszustand, in dem
der Spannungstensor zunächst im einen, dann im anderen Sinne
um die Querrichtung gekippt ist. Dabei wird der Kippwinkel
mit dem Abstand von der Blechmitte größer. Diese Inhomo-
genitäten des Spannungstensors wurden bereits an anderer
Stelle [2] als Ursache für die an gewalzten Einkristallen
gemessenen Texturänderungen über die Bleckdicke angesehen.

Die dort an einer speziellen Orientierung gewonnenen Ergebnisse sollen hier durch Experimente an Einkristallen anderer Orientierungen erweitert werden, wobei auch der Bereich extrem hoher Walzgrade in die Untersuchung einbezogen wird. Bei der Diskussion der Ergebnisse wird ausführ-
lich von den in Abb.1 dargestellten Drehungen des Spannungstensors um die Querrichtung Gebrauch gemacht. Dabei
wird davon ausgegangen, daß die dort dargestellten Verhältnisse näherungsweise auch für den anisotropen Einkristall
zutreffen, obwohl die Ausführungen zu Abb.1 exakt nur für
isotropes Material gültig sind.

Es wurden Einkristalle benutzt, deren Ausgangsorientierungen
Hauptkomponenten der Textur des polykristallinen Kupfers
entsprachen, für die also nach den Texturtheorien eine Stabilität vorausgesagt wird. Diese wurden bis zu sehr hohen
Verformungsgraden streng reversierend auf trockenen Walzen (Ø 15 mm) gewalzt. Die Probenausgangsdicke von 20 mm
erlaubte die Texturaufnahme in einzelnen Schichten auch bei
hohen Walzgraden. Die Abtragung der Schichten erfolgte durch
einseitiges Abätzen der Proben. Die Texturen wurdem mit
Hilfe eines automatisch korrigierenden und schreibenden
Texturgoniometers [3] als (111)-Polfiguren gemessen.

In Abb. 2 ist die Texturentwicklung eines Kupfereinkristalles mit einer (112)[11$\overline{1}$]-Ausgangsorientierung nach Walzen
beim Raumtemperatur dargestellt. Über dem steigenden Verformungsgrad ist der relative Abstand s=2h/d (h = Abstand
der Probenschicht von der Blechmitte, d = Blechdicke) der
untersuchten Schicht von der Blechmitte aufgetragen. Die
Polfiguren der oberen Reihe entsprechen der Oberfläche
(s = 1), die mittleren der Zwischenschicht (s = o,5) und

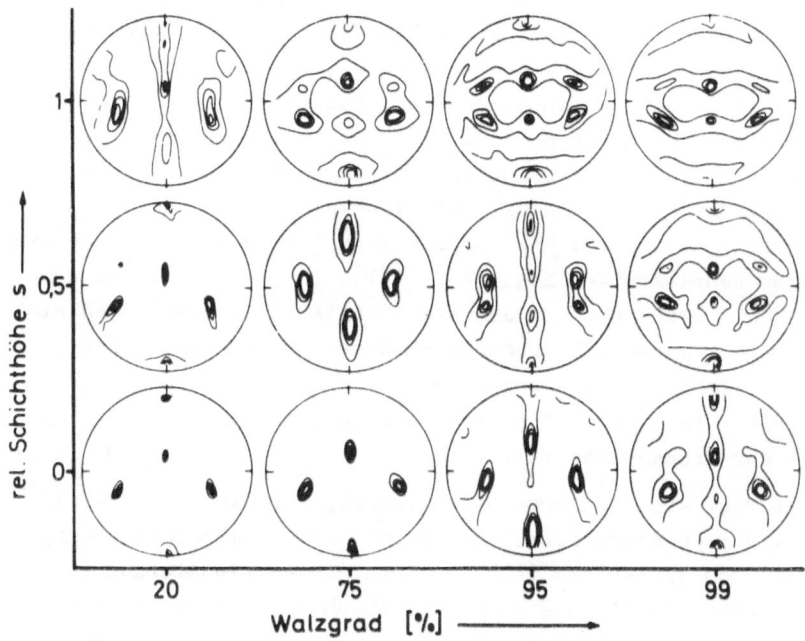

Abb.2[+]). Texturentwicklung in verschiedenen Abständen s von der Probenmitte eines bei 295°K gewalzten Einkristalls mit einer (112)[11$\bar{1}$]-Ausgangsorientierung.

Die Höhenlinien bedeuten (ebenso in Abb.3-5) die 1-, 3-, 6- bzw. 10-fache Intensität einer texturlosen Pulverprobe

die unteren der Mitte der Probe (s = 0). Abb. 2 zeigt deutlich, daß sich die Textur über die Blechdicke ganz erheblich ändert.

In der Probenmitte bleibt die ursprüngliche Lage, abgesehen von einer Polverschmierung, erhalten. Zu den Streuungen bei Walzgraden oberhalb 85% sei bemerkt, daß hier Proben und Schichten sehr dünn werden. Infolge der endlichen Eindringtiefe des Röntgenstrahles ist eine saubere Trennung der Schichten nicht mehr möglich, so daß die in der Kernschicht gemessenen Polfiguren durch die Polverteilungen der Zwischenschicht verfälscht werden. Während sich also in der Probenmitte die Ausgangsorientierung nicht wesentlich ändert,

---

[+]) Bei allen Polfiguren ist die Walzrichtung vertikal und die Querrichtung horizontal.

erkennt man in der Zwischenschicht deutlich Drehungen um die
Querrichtung, und zwar schmieren die Polverteilungen im
Sinne einer Hin- und Rückdrehung um die Querrichtung aus,
also im gleichen Sinne, wie der Spannungstensor in der
Zwischenschicht (s $\neq$ 0) um die Querrichtung pendelt. Außerdem dreht sich die Orientierung insgesamt einsinnig um die
Querrichtung. In dieser Schicht entstehen über die "schräge
Würfellage" sowohl die Ausgangslage (112)[11$\overline{1}$] als auch die
dazu komplementäre Lage (112)[$\overline{11}$1]. In der Oberfläche werden die Drehungen und die Komplementärlagen schon nach etwas geringeren Walzgraden beobachtet. Danach ist offensichtlich die Drehung der Orientierung umso größer, je
größer die Drehung des Spannungstensors gegenüber der Lage
in der Probenmitte ist.

Die Wahl einer anderen Orientierung, für die eine Stabilität vorausgesagt wird, nämlich die (110)[$\overline{1}$12]-Lage, bringt
ähnliche Ergebnisse (Abb.3). Auch diese Orientierung ist
gegenüber Verformung unter dem Spannungszustand der Kern-

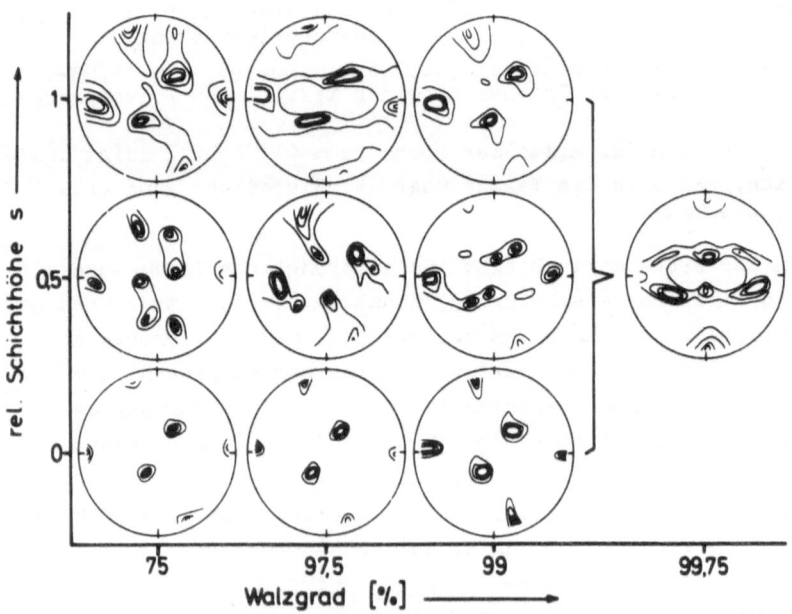

Abb.3. Texturentwicklung in den verschiedenen Schichten
eines (110)[$\overline{1}$12]-Einkristalles, Walztemperatur = 295° K

schicht stabil. Die Polfiguren der Zwischenschicht zeigen
wieder, zumindest bis 97,5% Walzgrad, Streuungen und Orientierungsänderungen im Sinne einer Drehung um die Querrichtung. Die Instabilität dieser Orientierung kann also ebenfalls auf die in der Zwischenschicht auftretende Rotation
um die Querrichtung zurückgeführt werden.

Diese unterschiedliche Texturentwicklung über die Blechdicke spielt auch bei Walzen unter tiefer Temperatur (77°K)
eine wesentliche Rolle, wie aus Abb.4 zu entnehmen ist. Die
Ausgangsorientierung des untersuchten Einkristalls ist wieder (112)[11$\bar{1}$] wie in Abb.2. Hier tritt die Zwillingsbildung
als zusätzlicher Verformungsmechanismus [4] auf. Aus diesem
Grund ist hier die Orientierung im Kern nicht stabil (im Gegensatz zur Raumtemperatur-Verformung). In den Zwischenschichten beobachtet man wieder Rotationen um die Querrichtung.
Schliesslich stellt sich hier bei hohen Verformungsgraden die
Messingtextur ein, im Gegensatz zur Kupfertextur bei Raumtem-
peraturwalzen.

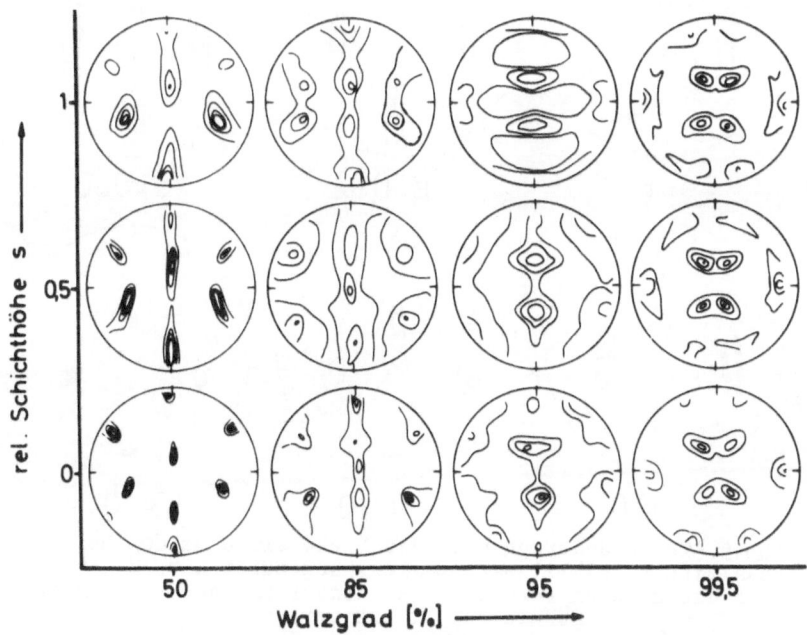

Abb.4. Texturentwicklung in den verschiedenen Schichten
eines bei 77°K gewalzten Kristalles der Ausgangsorientierung (112)[11$\bar{1}$]

Natürlich können in der Oberfläche auch qualitative Abweichungen von dem hier angenommenen Spannungszustand auftreten,
da dort infolge der Wechselwirkung Walze–Walzgut komplizierte Verformungsvorgänge ablaufen können. Die bei polykristallinen Materialien in dünnen Oberflächenschichten gefundenen
spezifischen Oberflächentexturen [5–9] werden meistens mit
diesen Wechselwirkungen erklärt. Solche Effekte mögen auch
hier in der Oberfläche eine Rolle spielen, sind aber (wie
auch bereits unter [2] gezeigt wurde) nur von untergeordnetem Einfluß auf das Verhalten der einzelnen Schichten
in der Tiefe des Bleches. Das folgt aus einigen Stich-
versuchen, in denen die Reibungsverhältnisse, die Walzendurchmesser sowie die Ausgangsdicke der Probe gegenüber den
im Vorhergehenden beschriebenen Versuchen geändert wurden.

Abb. 5 zeigt als Beispiel die Texturänderung über die Pro-

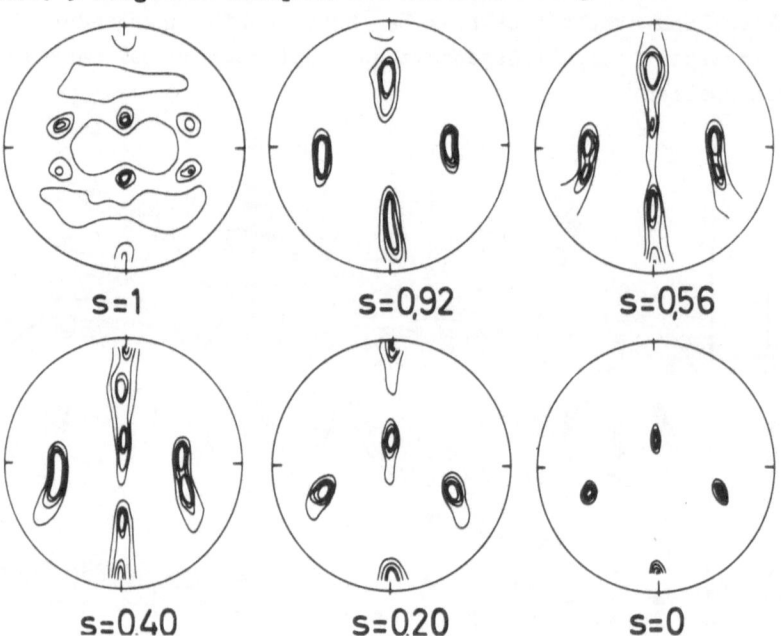

Abb.5. Texturänderung über die Dicke eines um 60% bei 295°K
gewalzten (112)[11$\bar{1}$]-Kristalles. Gegenüber Abb. 2–4 wurden
hier die äußeren Parameter (Walzendurchmesser, Probendicke,
Schmierung) stark verändert

bendicke bei einem Walzgrad von 60% für eine Probenausgangsdicke von 3 mm (statt 20 mm), einen Walzendurchmesser von

75 mm (statt 15 mm) und Ölschmierung (statt trockener Walzen). Die Ergebnisse unterscheiden sich außer in der Oberflächenschicht qualitativ und auch quantitativ nur wenig von den bereits besprochenen. In der sehr dünnen Oberflächenschicht hat sich die Komplementäranlage bereits gebildet, während die anderen Schichten trotz der extrem geänderten Walzbedingungen das für die vorhergehende Abbildung bei vergleichbaren Verformungsgraden typische Texturbild zeigen.

Zusammenfassend kann aus den vorliegenden Untersuchungen geschlossen werden:

1.) Bei der Frage nach der Stabilität einer Orientierung dürfen nicht nur die Formänderungskriterien herangezogen werden, sondern es müssen auch die Spannungszustände betrachtet werden, unter denen diese Formänderungen stattfinden. Beim Walzen erfahren nämlich alle Probenschichten, abgesehen von kleinen Randzonen, die gleiche Formänderung. Diese wird aber offenbar in den verschiedenen Probenschichten durch Betätigen unterschiedlicher Gleitsysteme erreicht, wie die Texturunterschiede zeigen.

2.) Einkristalle der Orientierung (112) $[11\overline{1}]$ und $(110)[\overline{1}12]$ (Kupferendlagen) besitzen eine stabile Orientierung gegenüber Gleiten nur, wenn die Verformung unter dem speziellen biaxialen Spannungszustand, wie er in der Blechmitte vorliegt, stattfindet+). Gegenüber Verformung unter dem Spannungszustand der Zwischenschichten $(s \neq 0)$ sind diese Lagen nicht stabil.

3.) Die Endlagen sind in der gesamten Probe erst dann stabil, wenn sich die jeweils komplementären Lagen gebildet haben. Diese Tatsache legt den Schluß nahe, daß eine endgültige Stabilität gegenüber der Walzverformung erst dann erreicht ist, wenn sich eine zur Querrichtung symmetrische Textur eingestellt hat.

+) HEYE und WASSERMANN [11] folgern aus ihren Texturmessungen, die offenbar in der Kernschicht vorgenommen wurden, daß auch dort die obigen Orientierungen bei hohen Verformungsgraden instabil werden. Nach unseren Messungen handelt es sich dabei jedoch um einen Effekt der instabilen Zwischenschicht. Diese nimmt nämlich mit wachsender Verformung immer größere Teile des Probenvolumens ein, so daß die Kernschicht für eine getrennte Beobachtung zu dünn wird.

Literatur

1. G.E.G. Tucker: J.Inst.Metals 82(1953)655 - Acta Met. 12(1964)1093.

2. H. Mecking, H.D. Mengelberg und K. Lücke: Veröffentlichung demnächst.

3. K. Lücke, G. Burmeister, R. Alam und H.D. Mengelberg: Veröffentlichung demnächst.

4. Vgl. hierzu auch W. Heye: Dieses Symposium, Kap. 2.2.

5. G.v.Vargha u. G. Wassermann: Metallwirtsch.12(1933)511.

6. K. Lücke: Z. Metallkde 45(1954)584.

7. I. L. Dillamore und W.T. Roberts: J.Inst.Metals 92 (1963/64)193.

8. P.J. Regenet und H.P. Stüwe: Z. Metallkde 54(1963)273.

9. F. Haeßner, G. Masing und H.P. Stüwe: Z. Metallkde. 47(1956)743.

10. W. Bunk, K. Lücke und G. Masing: Z. Metallkde. 45(1954) 584.

11. W. Heye und G. Wassermann: Z. Metallkde 59(1968)617.

2.8 Röntgenographisches Studium von Texturen bei gesinterten Magnesiumröhren
X-Ray Study of Sintered Magnesium – Tubes

von J. Šedivý[+)] and V. Šatánek[++)]

Abstract

The tubes were fabricated by sintering and extruding
magnesium powder at ~ 500°C. The samples were cut from
different parts of the tubes and examined by the conventional
X-ray-transmission method and the back-reflection technique.
Different fibre textures have been found; some of them are
of a new type.

2.8.1 Herstellung der Proben

Dünnwandige Magnesium-Hüllrohre für Uranstäbe im Reaktor
wurden röntgenographisch auf ihre Textur untersucht. Die
Rohre waren aus Magnesiumpulver mit einer mittleren Teilchengröße von 1 bis 2 μm durch Strangpressen bei etwa
500°C hergestellt worden (Verformungsgrad 95 bis 99%). Sie
hatten einen Durchmesser von 30 mm bei 1.0 bis 1,5 mm Wand-
dicke.

Die Wandung der Rohre war für eine unmittelbare Durchstrahlung ungeeignet, sie wurde daher auf etwa 0,3 mm Dicke ab-
geätzt und zwar einseitig oder doppelseitig, so daß Proben
von den beiden Oberflächen wie aus der Mitte der Wandung
erhalten wurden.

2.8.2 Röntgenographische Methoden

Von den vorhandenen Proben wurden eine Reihe von Aufnahmen
auf ebenem Film in Durch- sowie Rückstrahlanordnung mit
Cu K_α-Strahlung hergestellt (Ni-Filter, Abstand Probe-
Film 3,5 oder 1,0 cm), außerdem auch normale DS-Aufnahmen
auf zylindrischem Film.

[+)] Lehrstuhl für Festkörperphysik der Mathematisch-Physi-
kalischen Fakultät, Karlsuniversität, Prag, ČSSR
[++)]Forschungsinstitut für Pulvermetallurgie, Šumperk, ČSSR

Tab.1. Orientierungsbestimmung in der Mittelzone der Rohre mit dem höchsten Verformungsgrad. (hkl) = Beugungslinie, [uvw] = Faserachse

(hkl)	θ	δ[++)	[uvw]	(hkl)	θ	δ	[uvw]
100	16° 5'	0°/34°30'/59°20'	410/6$\bar{1}$0/120 o	210	45°32'	75°	$\bar{1}$60
002	17°10'	90°	⊥[001]	211	48°27'	73°	$\bar{1}$60
101	18°20'	23°/62°30'	210/120 o	114	49°44'	33°/90°	120 x /1$\bar{1}$0
102	24°10'	45°/68°	610/120 o	105	52°26'	73°	120 o
110	28°48'	0°/90°	120 x /1$\bar{1}$0	204	54°30'	53°	120 o
103	31°40'	54°/72°	610/120 o	300	56°24'	27°	120 o
200	33°40'	54°	120 o	213	59°18'	72°	1$\bar{5}$0
112	34°24'	27°/90°	120 x /1$\bar{1}$0	302 / 115	62°9'	0°/90°	120 o /1$\bar{1}$0
201	35°12'	55°	120 o	006	62°22'	90°	⊥[001]
004	36°18'	90°	⊥[001]	205	67°41'	45°	120 o
202	38°37'	55°	120 o	106	68°35'	65°	120 o
104 / 113	40°41'	73°/90°	120 o /1$\bar{1}$0	214	71° 3'	62°	$\bar{1}$60
203	45°19'	56°	120 o	220	73°45°	90°	1$\bar{1}$0

+) Die mit o bezeichnete Richtung [120] ist gleichwertig mit [1$\bar{1}$0]. Die mit x gekennzeichnete Richtung [120] entspricht [210]. [$\bar{1}$60] ist [6$\bar{1}$0] gleichwertig; man kann nämlich z.B. die Ebene (214) auch mit (124) angeben.

++) Zur Definition von δ s. unter [1].

2.8.3 Auswertung der Aufnahmen

Die Ergebnisse sind in Tab. 1 zusammengestellt. Abb.1 zeigt
ein typisches Röntgenbild aus der Mitte eines Rohres mit
99 % Verformungsgrad. Im einzelnen geht aus der Tabelle
Folgendes hervor:

1.) Die Lage der Schwärzungen auf dem (002) Durchstrahl-
Kreis zeigt, daß [001] senkrecht zur Rohrachse liegt.

2.) Die Lage der Schwärzungen auf den übrigen Durchstrahl-
Ringen ergeben, daß auch bestimmte, in der Basisebene liegende Richtungen [uvw] parallel zur Rohrachse ausgerichtet
sein können.

Abb.1. Durchstrahlaufnahme mit den Beugungsringen
(100), (002), (101) und (102)

Wenn man die Textur der Rohre so betrachtet und beschreibt,
als ob es sich um Stangen handelte (deren Radius also gleich
der Rohrwanddicke wäre), so ergibt sich eine Ringfasertextur mit der hexagonalen Achse als Faserachse in Radialrichtung. Daneben treten jedoch auch mehr oder weniger ausgeprägte Abweichungen hiervon auf. Es sind nämlich bestimmte, in der Basisebene liegende Richtungen parallel zur Rohr-
Längsrichtung ausgerichtet, die als Faserachsen beschrieben
werden können.

Die längs als Faserachsen vorkommenden Richtungen [120],
[1$\bar{1}$0] und [210] entsprechen den üblichen Gleitrichtungen der
partiellen Versetzungen, die bei der plastischen Verformung

von Einkristallen aus Magnesium zu den Hauptgleitrichtungen [110], [100] und [010] führen.

Die weiteren als Faserachsen vorhandenen drei Richtungen [410], [610], [6$\overline{1}$0], die sich der Gleitrichtung [100] nähern, können - ähnlich wie die Hauptgleitrichtung - auch aus den Gleitrichtungen der partiellen Versetzungen [120], [210], [1$\overline{1}$0] zusammengesetzt werden. Diese drei Orientierungen traten erst auf den Aufnahmen der gut ausgeprägten Texturen bei den höchsten Verformungsgraden in der angegebenen Reihenfolge auf.

2.8.4 Vergleich mit der Literatur

In der Literatur liegen nur Ergebnisse über gezogene Drähte und stranggepreßte Stangen aus Magnesium und einigen Legierungen, nicht aber über Rohre vor.

GLOCKER [1] gibt als Deformationstextur für Magnesiumdraht eine Ringfasertextur [001] an, ferner als Zugtextur [210] und bei Temperaturen oberhalb 450°C [320].

Bei TAYLOR [2] sind neben einer Ringfasertextur [001] die Faserachsen [210] und [110] angegeben. Bei höheren Temperaturen oberhalb 225°C erfolgt die Gleitung in der Richtung [110], aber in der Ebene (101) oder (102).

WASSERMANN und GREWEN [3] erwähnen für Magnesium neben einer Ringfasertextur [001] auch die Faserachse [210] für 300°C. Bei Magnesiumlegierungen tritt neben der Faserachse [110] oberhalb 450°C auch die Richtung [210] als Faserachse auf.

2.8.5 Beispiele praktischer Anwendung

Im allgemeinen dienten diese Texturstudien zur Kontrolle des Herstellungsvorganges im Zusammenhang mit den unterschiedlichen Formen der Preßwerkzeuge, der Verformungsgeschwindigkeit und ähnlichem.

So wurde erstens die Änderung der Textur entlang des Rohres untersucht. Dabei ergab sich, daß die Textur am Anfang des Rohres einem niedrigeren Verformungsgrad entspricht als die in der Mitte und am Ende des Rohres.

Zweitens wurde die Schärfe und Stärke der Texturen in Abhängigkeit von der Probenlage im Rohrquerschnitt studiert. Hier bewährte sich die Rückstrahlmethode, während die Durchstrahlmethode nur kleine Schwärzungsunterschiede zwischen den einzelnen Schichten zeigte.

Die inneren Schichten des Werkstoffes wiesen die in der Tab.1 ausgewertete Textur auf.

Die an der Oberfläche des Rohres (bis zu 0,15 mm Tiefe) vorliegende Orientierung entsprach einem niedrigerem Orientierungsgrad, wobei die Faserachse mehrere Richtungen einnehmen kann. Diese Tatsache hängt mit den möglichen Richtungen der einwirkenden Kräfte in den äußeren Schichten des Rohres während des Pressens zusammen.

Die Aufnahmen der Schicht an der inneren Wand des Rohres (Schichtdicke etwa 0,15 mm) deuten auf keine oder eventuell nur auf eine schwache Vorzugsorientierung hin.

2.8.6 Zusammenfassung

Die Textur von gesinterten Magnesiumrohren mit hohem Verformungsgrad, die durch Strangpressen von Pulver bei 500°C hergestellt worden waren, ist eine Ringfasertextur, bei der die hexagonale Achse [001] senkrecht zur Rohrachse steht. Sie entspricht somit der Textur, die auch bei Drähten und Stangen vorhanden ist. Daneben treten mehrere Fasertexturen mit Faserachsen parallel der Rohrlängsrichtung auf. Zu den bisher beobachteten Faserachsen [210] bzw. [120] kommen weitere vier Richtungen [1$\bar{1}$0], [410], [610] und [6$\bar{1}$0], die alle der Basisebene angehören. Die vorliegende Textur kann also als mehrfache Fasertextur bezeichnet werden. Die von manchen Autoren oft noch angegebene Faserachse [110] kam hier nur bei den Proben mit dem niedrigsten Verformungsgrad vor.

Literatur

1. R. Glocker: Materialprüfung mit Röntgenstrahlen, Springer, Berlin 1958.
2. A. Taylor: X-Ray Metallography, J. Wiley, New York 1961.
3. G. Wassermann und J. Grewen: Texturen metallischer Werkstoffe, Springer, Berlin 1962.

3 GLÜHTEXTUREN
ANNEALING TEXTURES

3.1 Nucleation in Recrystallisation
Keimbildung bei der Rekristallisation

by R.W.Cahn[+]

Zusammenfassung

Der Bericht beginnt mit einem kurzen, historischen Überblick über das Polygonisationsmodell der Keimbildung und
dem Beweis, auf dem dieses Modell beruht. Hierauf folgt eine
Diskussion der Orientierungen der Keime in Bezug auf die
der Matrix; besondere Berücksichtigung erfahren dabei noch
offene Fragen. Sodann wird die kritische Verfestigung, die
für die Keimbildung erforderlich ist, diskutiert. Ab-
schließend wird ein Überblick über die mit der Keimbildung in zweiphasigen Legierungen in Zusammenhang stehenden
Tatsachen gegeben.

3.1.1 Introduction

This short discussion of a long-standing problem should
be read in conjunction with my survey of Recrystallisation
Mechanisms [1]. I will take the following generalisations,
concerning nucleation of strain-free grains on annealing
deformed metals, as being soundly established and accepted:

a) Uniform "laminar" glide, even to very large extensions,
leaves a metal crystal immune to nucleation. The presence
of localised misorientations within the deformed crystal
lattice is an essential prerequisite for recrystallisation
generally and nucleation specifically.

b) Nuclei form in regions of maximum local strain gradient,
which are also regions of maximum misorientation. The two
necessarily go together, because high strain gradient

[+] School of Applied Sciences, University of Sussex,
Brighton, England

implies a high concentration of dislocations of one sign
(whether random or in sub-boundaries), which implies a large
misorientation.

c) In polycrystals, o n e "nucleation" mechanism is strain-
induced boundary migration, which may be on a coarse scale
[2] or on a fine scale [3]. I like to call this the "bulge
mechanism". The driving force for this stems from an imbalance
in dislocation density (hence of misorientations) on the two
sides of the boundary. Plainly this mechanism is inapplicable
to deformed single crystals.

d) Another mechanism is the growth of a small, highly
misorientated subgrain. It becomes more perfect as it grows
and the orientation difference across its boundary increases
until it becomes a "viable" nucleus; there is no satisfactory criterion for identifying when this stage has been
reached. The early stages of growth may happen by thermally
activated rearrangement of dislocations (polygonisation) or
by subgrain rotation [4]; I do not think that on the available
evidence one can decide generally between these alternatives,
and I propose to leave this question open.

3.1.2 Historical Note

The polygonisation model of nucleation (of which HU's [4]
subgrain rotation model is a variant) was independently
conceived by BECK [5] and CAHN [6], subsequently somewhat
amended by COTTRELL. At that time it seemed the only feasible
way of interpreting the non-random orientations of the new
grains produced by recrystallisation of deformed aluminium
crystals, as first established many years ago by BURGERS and
LOUWERSE [7]; the polygonisation model in its original form
implied a specific kind of local rotation of the lattice,
consistent with deductions from asterisms in Laue photographs.
Soon afterwards, it was found [8] that these asterisms were
caused by deformation bands[+) and that indeed there was some
preferential nucleation in such bands; later work on other
metals has confirmed this systematically, and nucleation is
widespread where deformation bands intersect [9], that is, at
sites of high deformation gradient.

+)Fußnote s.S. 152.

The original view that annealing textures are attributable to oriented nucleation became untenable some years ago, and with it vanished the original justification for the polygonisation model of nucleation. This does not, however, invalidate the model! The model, as used now, no longer postulates, as once it did, a particular type of misorientation, i.e. a rotation about an axis in the slip plane, normal to the slip direction; the misorientation can now be arbitrary, and so is the geometrical form of the nucleus.

3.1.3 Nucleus Orientations

While the existence of annealing textures can certainly not be attributed to orientated nucleation, this is not to say that nucleation is random. Whether, in any particular instance, nuclei form by the polygonisation process or by the bulge mechanism, plainly there must be some restriction to the available orientations. This consideration raises the following questions:

1.) What factors favour the bulge mechanism over the polygonisation mechanism, or vice versa?

2.) Where the bulge mechanism is preferred, when does one find coarse-scale bulge formation as found by BECK and SPERRY [2], and when is the process fine (grain-boundary microbulges no larger than individual subgrains)? In the first case, the annealing texture m u s t arise from major components of the deformation texture. In the second case, this is not necessarily so, because individual microbulges can easily be grossly misorientated relative to the gross orientation of the parent grain.

Fußnote von S. 151.

+) The deformation bands being referred to here are regions in the form of thin lamellae in which the lattice is continously rotated from the mean orientation of the rest of a deformed single crystal. This kind of band is described, for instance, in my paper in J. Inst.Metals 79(1951)129. The term "deformation bands" has also been applied to describe larger regions of deviating orientation, for example in rolled or wire-drawn crystals.

3.) If either the polygonisation process or the microbulge process operates, then the new grains will be quite severely misorientated with respect to the deformed grains (since high misorientation e n c o u r a g e s nucleation). Even allowing for effective growth selection, still this selection is limited to the available nucleus orientations, which cannot (at least in the presence of a really sharp deformation texture) be even approximately random. Just how far does the nucleus distribution depart from randomness, and what effect, if any, does this have on the annealing textures?

Nobody has yet attempted to establish the texture of a population of nuclei when these are still small: this would be very hard to do, since nuclei do not all form simultaneously. However, HAESSNER and his coworkers [11-13] have developed a technique of determining a deformation texture by multiple selected-area diffraction in a thin foil. The same technique could be applied to measuring the orientation of nuclei or very small grains, where recrystallisation has only begun, for comparison with the final annealing texture of the completely recrystallised material. Experiments on zinc monocrystals recrystallising from small twin lamellae, described by CAHN [15] represent an early attempt to obtain information of this type, although by means of a cruder technique. In this case the orientation distribution of the nuclei resembled that of the mature grains.

4.) A special variant of the preceding question is: Where nucleation takes place exclusively a t grain boundaries not by the bulge mechanism [10], but simply by a special form of nucleation by polygonisation, is the orientation of the resulting grains random or related in some statistical way to either or both of the bounding parent grains? The answer to this question is currently under investigation at Sussex University.

5.) The special question remains: when a moderately deformed single crystal is recrystallised, a wide range of new orientations is found, even though the crystal remains

perfect enough to give quite good Laue patterns. Is this
finding consistent with the polygonisation model? The
answer here seems to be that it is, because it is precisely
the s m a l l e s t subgrains that turn into nuclei. This
emerges from the excellent study of nucleation from
subgrains in aluminium and nickel recently published by
MICHELS and RICKETTS [14] – a study which incidentally
constitutes the most critical confirmation to date of the
polygonisation model. To quote from this paper: "The
observation that recrystallised grains can grow out of
very small regions of the cold-worked structure by the
rapid migration of high-angle boundaries (after zero or
very short incubation times) indicates the possibility that
major components of the annealing texture can originate
in regions which contribute little or nothing to the
deformation texture". What we do not know yet is whether
these very small subgrains, preferred as nuclei, are also
always the most severely misoriented ones; but this is very
probably so.

One special form of nucleation is not discussed here,
because I have recently fully treated this elsewhere [1,15].
Suffice it to say here that grains nucleated at t w i n s
either all have the orientation of the twins, in which case
there can of course be no growth selection, or they have a
range of orientations and are then probably formed by the
polygonisation mechanism; in this case there i s growth
selection and a texture is formed.

3.1.4 Critical Strain for Nucleation and Growth

It has long been understood that for any particular metal
sample, depending on composition, grain size and deformation
mode, there is a critical minimum strain below which it
cannot be induced to recrystallise at any temperature. It
has normally been asssumed tacitly that the limiting
condition is nucleation, that is, if nuclei can form then
they can also grow large. The truth or otherwise of this
supposition is interesting because, on both nucleation
models, there is no sharp dividing line between nucleation

and growth, but it is normally assumed that the early
stages of sub-boundary migration (or bulge migration) are
difficult because the misorientation is in the difficult
intermediate range. Very small-angle sub-boundaries move
very easily, and large-angle grain boundaries move easily;
when a subgrain or microbulge is preparing to become a
fully viable nucleus, the sub-boundary is at its most
immobile. On this basis, one would expect the critical
strain for nucleation to be higher than that for growth.
This point has been examined by HAMELIN and GOUX [16] who
found, for low-purity aluminium single crystals of various
controlled orientations, that the critical strain for
nucleation was more than twice as great as for growth of
artificially nucleated (scratched) crystals. MONTUELLE,
in a discussion of the same paper, confirmed the conclusion
by means of an elegant experiment. One would like to see
this work repeated with purer aluminium and careful
statistical study (in view of the small number of nuclei
one can hope to find in weakly deformed monocrystals).

I have carried out (unpublished) experiments many years ago
in which I sought to establish different critical strains˙
for nucleation and growth in strained p o l y crystalline
99.99% aluminium, but was not able to find any difference:
any strain large enough to permit an artificial nucleus to
grow also permitted spontaneous nucleation. A similar
conclusion was reached by DESALVO and NOBILI [17], who
examined the recrystallisation of S.A.P.; again, and
irrespective of oxygen content, if a sample would not
recrystallise spontaneously it would not support the growth
of an artificial nucleus either.

The foregoing suggests that polycrystals and monocrystals
behave differently, and indirectly confirms again, what
has long been known, that nucleation at grain boundaries
(by whatever mechanism) is easier than in grain interiors.

3.1.5 Nucleation in Two-Phase Alloys

A good deal has been published in the last five years on
the recrystallisation of metals containing dispersed

phases of various fineness, from very fine S.A.P. [17,18] (ref. [1], fig. 9) and thoria-disperse nickel [19] at one extreme to coarse Al/Al$_3$Fe alloys at the other extreme [20]. Other studies include one by HAESSNER et al [21] on copper containing dispersed boron carbide, the study by GAWNE and HIGGINS [22] on mild steel containing spheroidised carbides, a report by RYAN [23] on chromium containing a titanium carbide dispersion, a paper by DOHERTY and MARTIN [24] on Al/CuAl$_2$ alloys, and a series of studies of alloys in the Cu/Co and Ni/γ' systems which contain coherent precipitates [25-28].

The whole subject of the role of second-phase dispersion on recrystallisation is being treated elsewhere in this volume by HORNBOGEN and KREYE [29]. My only concern here is with the influence of particles on the process of nucleation itself. The most detailed analysis of the effect of the principal geometrical variables was made in the papers on Al/CuAl$_2$ [24] and Al/Al$_3$Fe [20], and from these one important conclusion emerges: A sufficiently fine dispersion, with a mean separation of less than 3 - 4 microns, hinders both nucleation and migration of boundaries, while a coarse dispersion aids nucleation and does not significantly hinder migration. HAESSNER et al [25] state that a separation of 20 microns in Cu/boron carbide is large enough to meet this latter condition, and carbide spheres about 0,5 micron in diameter (and presumably separated by several times that distance) accelerated nucleation in iron [22]. Two other recent studies also show acceleration for coarse dispersion [30,31]. A critical separation of about 4 microns appears to be consistent with all the available information. C o h e r e n t precipitates, however, which dissolve as a boundary passes and then reprecipitate, appear to exert a large drag on the boundaries and always seriously retard recrystallisation.

MOULD and COTTERILL [20] and DOHERTY and MARTIN [24] interpret their findings in terms of the size of a viable nucleus (i.e. the critical size of a subgrain when it first becomes capable of rapid growth). If each subgrain encounters

a particle before or when it reaches critical size, its
growth will be strongly affected. This happens when the
mean particle separation becomes about twice the critical
subgrain size, which critical size both sets of authors
take to be 1 - 2 microns. A particularly interesting
observation was made by DOHERTY and MARTIN [24]: For a
mean particle spacing of 1.2 microns (well below the
critical value), nucleation could only take place by the
microbulge mechanism of strain-induced boundary migration
(see. Fig. 7. of reference [1]). A systematic study of
strain-induced boundary migration as a function of
dispersion morphology would now be of great interest, to
complement the equally urgent study of the range of
conditions, in single-phase materials, under which the
microbulge mechanism is favoured over the polygonisation
mechanism.

The acceleration of nucleation in two-phase alloys with
coarse dispersions is presumably due to the large localised
strain gradients at dispersions, and consequential large
local dislocation concentrations [32], provided that the
subgrain boundaries are not arrested before they become
large enough. - One novel observation has been reported by
WEBSTER [19]. When thoria-disperse nickel is heavily cold-
rolled, the thoria particles break away from the matrix
and the metal thus becomes full of voids. These voids
inhibit subsequent grain growth more than the original
unbroken thoria dispersion did. There is no evidence that
fine pores can aid recrystallisation by promoting nucleation,
and there is indeed some evidence that under suitable
circumstances pores can considerably inhibit recrystallisation [33,34] (ref. [1], Fig. 13).

Acknowledgements

I am grateful to Dr. F. Haeßner, Dr. G. T. Higgins, and Dr.
N.E. Ryan for helpful advice and information in advance of
publication.

References

1. R.W. Cahn in: Recrystallisation, Grain Growth and Textures, H. Margolin, Ed., Amer.Soc.Metals, Metals Park, Ohio 1966, p. 99.

2. P.A. Beck and P.R. Sperry: J. Appl.Phys. 21(1950)150.

3. J.E. Bailey and P.B. Hirsch: Proc.Roy.Soc. 267A(1962)11.

4. H. Hu in: Recovery and Recrystallisation of Metals, L. Himmel, Ed., Interscience, New York 1963, p. 311.

5. P.A. Beck: J. Appl.Phys. 20(1949)633.

6. R.W. Cahn: Proc.Phys.Soc. (London) 63A(1950)323.

7. W.G. Burgers and P.C. Louwerse: Z. Phys. 61(1931)605.

8. R. W. Cahn: J. Inst. Metals 79(1951)129.

9. T. H. Schofield and A.E. Bacon: Acta Met. 9(1961)653.

10. R.A. Vandermeer and P. Gordon: Trans.Met.Soc. AIME 215(1959)577.

11. F. Haeßner, U. Jakubowski and M. Wilkens: Acta Met. 14(1966)454.

12. F. Haeßner and W. Hemminger: Z. Metallkde. 58(1967)104.

13. F. Haeßner and D. Keil: Z. Metallkde. 58(1967)220.

14. L.C. Michels and B. G. Ricketts: Trans.Met.Soc. AIME 239(1967)1841.

15. R.W.Cahn in:Deformation Twinning, AIME Metallurgical Society Conferences 25, Gordon and Breach, New York 1964, p. 20.

16. A. Hamelin and C. Goux in: Ecrouissage, Restauration, Recristallisation, 7e Colloque de Saclay, Presses Universitaires de France, 1963, p.167 – Mém.Sci.Rev.Mét. 60(1963)45.

17. A. Desalvo and D. Nobili: J. Materials Sci. 3(1968)1.

18. D. Nobili and R. de Maria: J. Nucl.Mat. 17(1965)5.

19. D. Webster: Trans.Met.Soc.AIME, 242(1968)640.

20. P.R. Mould and P. Cotterill: J. Materials Sci. 2(1967) 241.

21. F. Haeßner, E. Hornbogen and M. Mukherjee: Z. Metallkde. 57(1966)171.

22. D.T. Gawne and G.T. Higgins: This volume, chapter 4.3.

23. N.E. Ryan: (Australian) Aeronautical Research Laboratories, Melbourne, Report ARL/Mat 64, (1967).

24. R.D. Doherty and J.W. Martin: J. Inst.Metals 91 (1962/63)332.

25. F. Haeßner, E. Hornbogen and M. Mukherjee: Z. Metallkde. 57(1966)270.

26. L.E. Tanner and I.S. Servi: Mat.Sci. and Eng. 1(1966) 153.

27. V.A. Phillips: Trans.Met.Soc.AIME 236(1966)1302.

28. V.A. Phillips: Trans.Met.Soc.AIME 239(1967)1955.

29. E. Hornbogen and H. Kreye: This volume, chapter 4.1.

30. C. Antonione, G. della Gatta and G. Venturello: Trans. Met.Soc.AIME 230(1964)700.

31. A.T. English and W.A. Backofen: Trans.Met.Soc.AIME 230(1964)396.

32. P. Guyot and E. Ruedl: J. Materials Sci. 2(1967)221.

33. A.B. Middleton, L.B. Pfeil and E.C. Rhodes: J.Inst. Metals 75(1949/50)595.

34. M.L. Bhatia and R.W. Cahn: To be published.

3.2 Growth Selectivity in Single Crystals
Wachstumsauslese bei Einkristallen

by K. T. Aust[+]

Zusammenfassung

Untersuchungen der durch Rekristallisation und Kornwachstum gebildeten Texturen haben gezeigt, daß die Orientierungen der neuen Kristalle häufig eine den Koinzidenz-Korn-
grenzen ähnliche Beziehung zur Matrix aufweisen, aus der sie
entstanden sind. Eines der wichtigsten Probleme für das
Verständnis der Glühtexturen ist deshalb mit der Frage
verbunden, warum diese speziellen oder Koinzidenz-Groß-
winkelkorngrenzen in einem rekristallisierten Material auftreten.

In Einkristallen aus zonengereinigten Metallen mit einer
Zellstruktur (striation-structure) wurden künstliche Keime
erzeugt und deren Wachstum verfolgt. Dabei zeigte sich, daß
Vorzugsorientierungen vom Koinzidenztyp entstehen. Dies ist
auf die speziellen Eigenschaften derartiger Grenzflächen,
nämlich ihre niedrigere Energie und höhere Beweglichkeit
unter bestimmten Bedingungen, zurückzuführen. Es wird auch
der Schluß gezogen, daß eher das Kriterium der Korngrenzen-
als das der Gitter-Koinzidenz zur Auswahl der speziellen
Korngrenzen führt, die in Gegenwart von Verunreinigungen
hohe Beweglichkeit besitzen.

An important application of research on interfacial
properties is the problem of preferred orientation in
annealed materials. By suitable control of crystal orientations in a polycrystalline material it is possible to
control the directionality of mechanical or magnetic
properties. Studies of preferred orientations obtained by
recrystallization and grain growth have indicated that the
orientations of the new crystals are often related to the
matrix material in which they formed by orientation

[+] Department of Metallurgy and Materials Science,
University of Toronto, Toronto, Canada

relationships that are similar to those of the coincidence site type.

One of the important problems, therefore, is the understanding of preferred orientations in annealed metals, is why these special or coincidence types of large angle grain boundaries are obtained in a recrystallized material. This same problem was stated more explicitly by BROOKS [1] over 15 years ago: "An important problem which has received little attention is the detailed elucidation in atomic terms of the rather remarkable orientation relationships which appear to exist in the migration of boundaries, especially during recrystallization and grain growth".

The purpose of the present paper is to provide some understanding of this important problem in the field of materials science and technology. It will be shown that preferred orientations of the coincidence type are developed in annealed metals as the result of the special properties of such interfaces, namely their lower energy and higher mobility under certain conditions.

3.2.1 Experimental Procedure

The experimental techniques employed in these studies are similar to those used by RUTTER and AUST [2,3] in the previous work on the migration of individual grain boundaries in bicrystals of zone refined lead. The materials used were (1) zone refined lead with and without various small concentrations of added solute, (2) zone refined aluminium and (3) zone refined copper and normal-purity copper.

The starting specimens were single crystals, about 0,5 to 1 cm. square in cross-section and by 5 to 10 cm. in length, grown from the melt in a horizontal graphite container. A crystal grown in this manner usually contains lineage or striation substructure, consisting of an array of low angle boundaries which partition the crystal into regions misoriented from each other by a few degrees. This substructure constitutes a source of driving energy, estimated to be between 10^3 to 10^4 ergs per cu. cm., for the migration of a grain boundary into the striated crystal.

New grains were introduced into each crystal by plastic deformation in compression of a localized region at one end of the specimen. Recrystallization occurs in the deformed region, introducing a large number of new, striation-free grains. These grains are able to grow into the melt-grown crystal on annealing at sufficiently high temperatures. The striations are removed by the passage of the grain boundary, as the substructure energy is utilized to produce the grain boundary motion.

Generally, only one or two of the recrystallized grains persist after growth has occurred for about 1 cm. out of the deformed region. The orientation relationships between the striated crystal and the recrystallized grain or grains which grew most successfully in this manner in each specimen were determined from Laue back-reflection x-ray photographs.

3.2.2 Experimental Results

3.2.2.1 Effects of Grain Boundary Mobility in High Purity Lead

Fig.1, taken from the earlier work of AUST and RUTTER [3], shows the measurements of the velocity of grain boundary migration at $300^{\circ}C$ as a function of solute (tin) concentration in zone refined lead. The upper curve pertains to the "special" boundaries and the lower curve to the random boundaries. The addition of tin to the zone refined lead produced a rapid decrease in the speed of migration of the random boundaries, but had a much smaller effect on the velocity of the special boundaries. If this difference in migration rate between special and random grain boundaries, when tin is present, is related to the development of preferred orientations observed in annealed metals, then the following may be expected to occur. Random orientation relationships should be obtained in the pure lead when annealed at $300^{\circ}C$, and special orientation relationships should be observed at the higher tin concentrations where

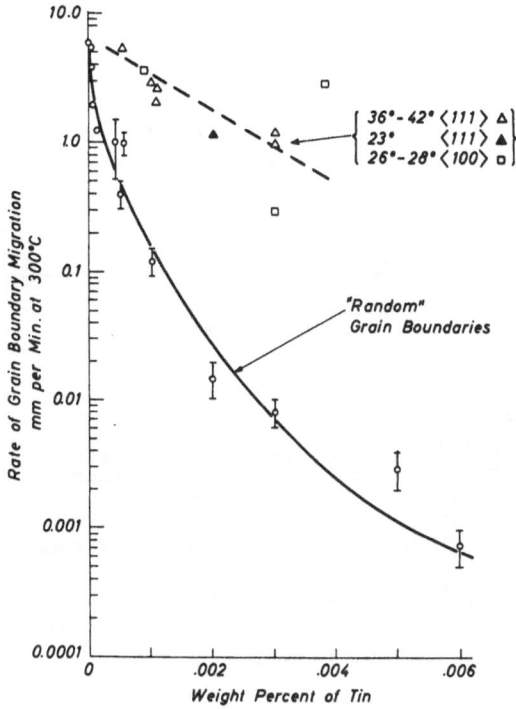

Fig.1. Rate of grain boundary migration at 300°C vs.
tin concentration in zone refined lead, for "random"
and "special" grain boundaries [3]

the difference in rates of boundary migration is as large as
a factor of 100.

The experiment then simply consisted of introducing many
recrystallized grains into one end of a striated crystal,
by means of artificial nucleation and growth, and
determining which grains are most successful in growing into
specimens of zone refined lead with various tin additions
after annealing at 300°C. Fig. 2 shows the stereographic
plots of the axes of rotation which relate each dominant
recrystallized grain to the striated crystal into which
it grew by the smallest amount of rotation θ [3]. The range
of angles or orientation differences is indicated below Fig.2
with the stereographic triangles.

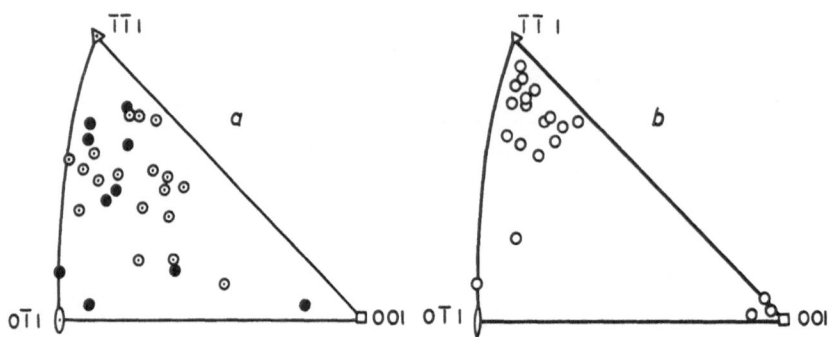

Fig.2a and b. Rotation axes of dominant artificially
nucleated grains in striated crystals of zone refined lead:
a) with tin additions less than 2 atom p.p.m.; angles of
        rotation 15 to 60° about axes shown,
b) with tin additions from 10 to 80 atom p.p.m.; angles of
rotation primarily 30 to 50° about axes within 12° of ⟨111⟩
        and 26 to 28° about axes within 5° of ⟨100⟩ [3]

Data points in the triangle of Fig.2a are for zone refined
lead specimens containing less than about 2 atom p.p.m. of
tin and annealed at 300°C. It appears that essentially a
random distribution of orientation relationships is
obtained; the distinctive feature of the data in Fig.2a is
that the orientation relationships correspond to those of
large angle grain boundaries. The stereographic triangle of
Fig.2b shows the data obtained for zone refined lead
containing tin solute in the concentration range of 10 to
80 atom p.p.m., also after annealing at 300°C. It is
evident that a preferred orientation is observed (Fig.2b),
with the axes of rotation clustering near ⟨111⟩ and the
angles of rotation between 30° and 50°. There are also a
significant number of cases in which the new, recrystallized
grains and the starting, striated crystals are related by
rotations of 26 to 28° about a common ⟨100⟩ axis.

These observations indicate that the difference in boundary
migration rate between special and random boundaries, which
results with tin additions to the high purity lead (Fig.1),
is a significant factor in determining whether preferred
orientations will appear or not.

Preferred orientation relationships are no longer observed
in the striated crystals of zone refined lead when the tin
concentration is increased to about 100 to 2000 atom p.p.m.,
or when either silver or gold is added in the concentration
range from 0,1 to 1 atom p.p.m.[4]. Although the solute
concentrations in lead are quite different for Ag or Au as
compared to Sn, the experiments covered approximately the
same range of boundary migration rates for these solutes.
It was observed that the addition of Ag or Au to the lead
produces a much more drastic reduction in the rate of
boundary migration than a similar tin addition [5]. This
observation indicates that the interaction between Ag or
Au solute atoms and a grain boundary is much larger than that
between Sn atoms and the boundary.

The results shown in Fig. 1 and 2 suggest that the
interaction between tin atoms and the boundary can result
in a selective segregation, depending upon the type of grain
boundary and the concentraion of tin present. However, the
interaction between Ag or Au atoms and the boundary is large
enough that sufficient solute segregation occurs to all the
grain boundaries, so that very little s e l e c t i v e
segregation occurs; consequently, no special boundaries are
observed after artificial nucleation and growth in these
specimens.

During a study of the mobility of grain boundaries in zone
refined lead, using bicrystals having controlled orientations,
it was observed that the temperature-dependence of boundary
motion is smaller for large-angle coincidence (or special)
grain boundaries than for large-angle non-coincidence (or
random) boundaries [6]. This is illustrated in Fig.3, which
shows a plot of log rate of boundary migration against the
reciprocal of the absolute annealing temperature for
coincidence (38° about $\langle 100 \rangle$) and non-coincidence (41° about
$\langle 100 \rangle$) boundaries. The mobilities are similar at temperatures
near the melting point of the lead, but the mobility of a
coincidence boundary is greater than that of a non-
coincidence boundary at lower temperatures. This observation,
shown in Fig.3, suggested an experiment for testing the

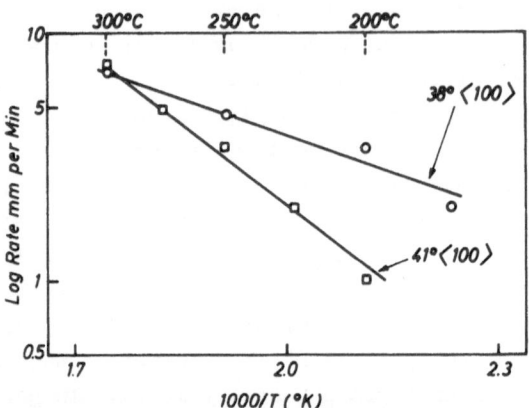

Fig.3. Log rate of boundary migration vs. 1/T for 38°
and 41° ⟨100⟩ boundaries in zone refined lead [6]

importance of the temperature-dependence of boundary
mobility in the development of preferred orientation in an
annealed high-purity metal. If the temperature-dependence
of boundary mobility is an important factor, then randomly
oriented grains would be expected after annealing at the
higher temperatures (e.g. at 300°C), and preferred orientations of the coincidence type should be observed after
annealing at lower temperatures (e.g. at 175°C.).

The results of the experiments on artificial nucleation and
growth into striated crystals of zone refined lead, after
annealing at 300°C and 175°C are given in Fig. 4 [7]. The
300°C anneal resulted in random orientation relationships
(Fig. 4a) while the 175°C anneal gave preferred orientation
relationships of the coincidence site type (Fig. 4b) as
proposed by KRONBERG and WILSON [8], FRANK [9] and DUNN [10].
The coincidence orientations shown in Fig. 4b are 32, 38
and 47° about ⟨111⟩, 26 1/2, 39 and 50 1/2° about ⟨110⟩ and
37° about ⟨100⟩.

The fact that preferred orientations are observed in high-
purity lead after annealing at 175°C, but not after
annealing at 300°C, may be rationalized in terms of the
greater mobility of the coincidence boundaries over the
non-coincidence boundaries at the lower temperature and the

167

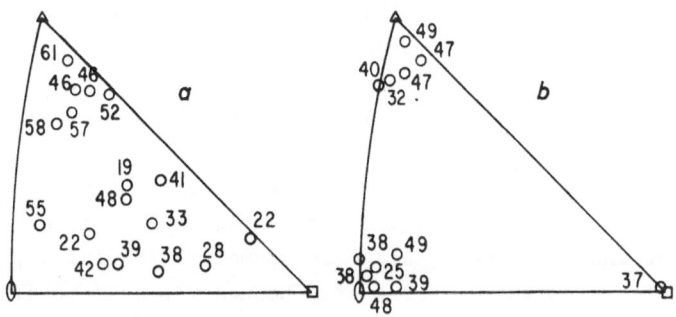

Fig.4a and b. Stereographic plots of the axes of
rotation which relate, by the smallest amount of
rotation, the most successful growing grain to the
striated crystal in specimens of zone refined lead.
The amount of rotation is indicated beside each axis

a) Specimens annealed at 300°C
b) Specimens annealed at 175°C[7]

similar mobilities at the higher temperature. These results
clearly demonstrate the importance of the temperature
dependence of grain boundary mobility in the development of
preferred orientation in an annealed high-purity metal.

3.2.2.2 Effects of Grain Boundary Energy in High Purity Lead

Additional studies on striated lead crystals indicated that
the boundary mobility is not the only important factor
leading to preferred orientations in an annealed metal [7].
It was found that many of the coincidence relationships
obtained by annealing the zone refined lead at lower
temperatures (175°C) were introduced when twinning occurred
during grain boundary migration. As a result of twinning,
the coincidence boundaries replaced either non-coincidence
boundaries or boundaries of lower density of coincidence
sites at the growth front. For example, eight of the twelve
coincidence relationships observed after annealing at 175°C
(Fig.4b) were identified as originating by twinning during
grain growth in the recrystallized end of the specimen [7].

The driving energy for this twinning process is believed
to be due to the lower interfacial energy of "ideal"
coincidence boundaries as compared to other large angle

boundaries, in agreement with the FULLMAN-FISHER [11]
theory of annealing twins. This theory states that a twinned
grain will persist in a specimen when the energy of the
boundary between the twin and the matrix grain is less than
that of the boundary between the original growing grain and
the matrix grain by an amount which exceeds the energy of
the coherent twin boundary. In the experiments on zone
refined lead annealed at 175°C after artificial nucleation,
the lower-energy coincidence grain boundaries replaced at
the growth front the higher-energy boundaries as a result
of twinning. These observations indicate the importance of
relative grain boundary energies in the introduction of
suitably oriented grains into a metal specimen.

It was pointed out previously that random orientation
relationships are obtained for zone refined lead containing
0.1 to 1 atom p.p.m. of Ag or Au. However, twinning during
subsequent grain boundary migration can produce coincidence
relationships in Pb-Ag or Pb-Au specimens. For example,
Fig.5 shows the orientations observed before and after

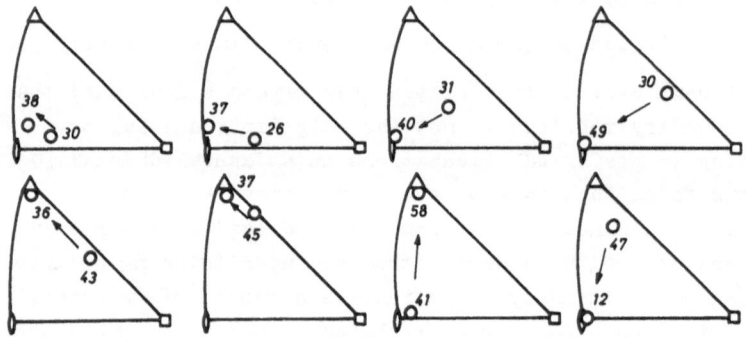

Fig.5. Rotation axes and angles before and after
twinning in specimens of zone refined lead containing
0.1 to 1.3 atom p.p.m.Au[12]

twinning in lead containing 0.1 to 1.3 atom p.p.m.Au [12].
The axis and angle of rotation relating the first growing
grain to the striated crystal is shown at the first position
in each stereographic triangle. The axis of rotation relating
the twin of the first growing grain to the striated crystal

is given at the second position after the arrow. In three
of six specimens considered here (Fig.5), a new twin
of the original twin formed; and the orientation
relationship between this second twin and the striated
crystal is depicted after the second arrow. The new grains
produced by twinning are related to the striated crystals
by rotations about ⟨111⟩, ⟨110⟩ and ⟨100⟩ axes. In every
case where an annealing twin formed during boundary
migration, a large angle random boundary is replaced by a
large angle coincidence boundary. When twinning occurred
a second time, the coincidence boundary was replaced at the
growth front by another coincidence boundary having a higher
density of coincidence sites.

It has been suggested [13] that, in addition to the
decrease in grain boundary free energy which results from
twinning, the new boundary formed may also have a greater
mobility than the initial grain boundary. Although no
difference, within a factor of two, was found among the
mobilities at $300^{o}C$ of the grain boundaries observed before
and after twinning in the Pb-Ag or Pb-Au specimens [12],
slight mobility differences may still exist. However, the
free energy decrease appears to be the major difference
among the grain boundaries in the zone refined lead
specimens containing very small Ag or Au additions.

In the experiments on the migration of grain boundaries in
zone refined lead containing tin as solute, boundaries of
the coincidence or near-coincidence type were frequently
observed migrating out of the deformed and recrystallized
region of the specimen (Fig.2b). Whenever they appeared,
these boundaries exhibited a migration rate up to 100 times
faster than other boundaries (Fig.1). It appears, therefore,
that the appearance of preferred orientations in the Pb-Sn
specimens is largely the result of the mobility differences.
The great advantage in mobility possessed by a coincidence
boundary in lead containing a suitable tin addition would
virtually assure its continued existence in competition with
other grains of less favourable orientation.

3.2.2.3 Growth Selectivity in High Purity Aluminium and Copper

Additional experiments involving artificial nucleation and growth into striated single crystals were conducted using zone refined aluminium. This material was characterized by a ratio of electrical resistivity at room temperature to that at $4.2^{\circ}K$ of 9000, indicating a purity of the order of 99.9999 per cent. The activation energy for the migration of large angle random boundaries in the aluminium was only 13 K.cal per gm. atom in the temperature range from 560 to $640^{\circ}C$ [14].

The orientation relationships between the striated aluminium crystals and the predominant recrystallized grains obtained after annealing ($560-640^{\circ}C$) were found to be random. However, during subsequent migration of these large angle random boundaries it was sometimes observed that a twin of a recrystallized grain would form and grow into the striated crystal [15]. An example of such a case is depicted in Fig.6; a recrystallized grain A which was introduced by artificial nucleation and growth, and its twin A^1, which formed during the growth of grain A, have advanced into the striated crystal S.

Fig.6. Macrophotograph showing growth of grain A and its twin A^1 into an aluminium crystal S, X 7 [15]

It was noted that where an annealing twin was formed, a large angle random or non-coincidence boundary was generally

replaced by a large angle coincidence boundary [15]. The
results of the orientation determinations are given in the
separate stereographic triangles of Fig.7 for each example
of twinning. The single axis of rotation relating the first
growing grain A to the striated crystal S by the smallest
amount of rotation is shown at the position before the
arrow; the axis of rotation relating the twin of grain A,

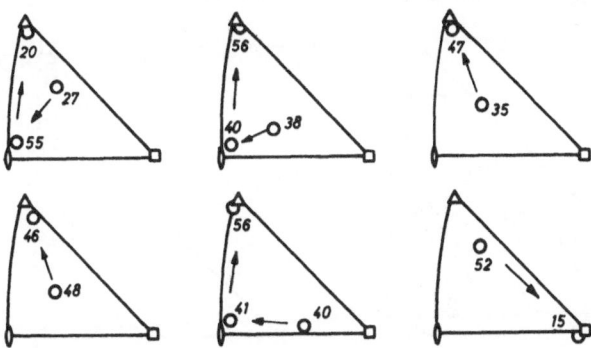

Fig.7. Rotation axis and angles before and after twinning
       in specimens of zone refined aluminium [15]

namely grain A^1, to the striated crystal is given at the
position after the arrow. The amount of rotation is
indicated at each axis position in Fig.7. The coincidence
boundaries introduced by twinning correspond to orientation
relationships having a h i g h density of coincidence
sites, e.g. 39° and 50° about $\langle 110 \rangle$, 38° and near 60° about
$\langle 111 \rangle$. In the previous work discussed using zone refined lead
(section 3.2.2.2), the formation of annealing twins resulted
in the appearance of coincidence grain boundaries having low
as well as high densities of coincidence sites. These
observations are consistent with the experimentally measured
larger ratio of twin boundary energy to grain boundary
energy in aluminium [13] as compared to lead [16]. In one
case shown in Fig.7, twinning produced a medium-angle grain
boundary and, therefore, a lower energy interface separating
grains A^1 and S.

Similar studies on striated single crystals of copper also
indicated the importance of impurities and twinning in the

selective growth of coincidence grain boundaries. For
example, Fig.8 gives results obtained by FERRAN, CIZERON
and AUST [17] for copper of two different purities, after
local deformation at room temperature followed by annealing
between 1000 and 1070°C. For zone refined copper (resistivity
ratio of 2000 to 2500), the orientation relationships
between the successful recrystallized grains and the striated
crystals appear random (Fig.8a). However, in the less-pure

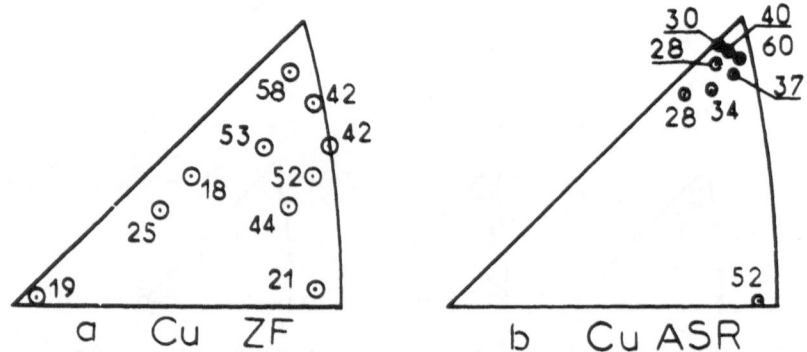

Fig.8a and b. Rotation axes and angles for a) zone refined
copper, b) ASR copper [17]

ASR copper (resistivity ratio of 800 to 1000), the orientations are grouped preferentially near ⟨111⟩ (Fig.8b). The
orientations observed in the ASR copper are similar to the
predicted high density coincidence relations of 28, 32, 38
and 60° about ⟨111⟩, and 50 1/2° about ⟨110⟩. In addition,
it was established that several of these relationships resulted from twinning during grain boundary migration [17].
These observations indicate that selective growth in copper
is influenced by the purity of the material and by twinning,
in agreement with the results obtained for lead and aluminium.

3.2.3 Discussion

3.2.3.1 Lattice and Boundary Coincidence

The preferred orientation relationships observed in the
studies of selective growth into striated crystals of
Pb, Al and Cu appear to be similar to those of the
l a t t i c e coincidence site model proposed by KRONBERG
and WILSON [8], and later extended by FRANK [9] and DUNN [10].

At certain angular relationships between the crystals, sites
of the two adjacent crystals coincide at the boundary. An
example is shown in Fig. 9, which represents a 38° (or 22°)
boundary about a ⟨111⟩ axis. It is a property of
relationships such as that, if one of the two lattices is
extended to overlap the other, there is a three dimensional
array (or lattice) of coincidence. The only a t o m s that
are common to the two lattices are those at the boundary
itself. A series of "ideal coincidence" relationships of
this kind can be developed for rotations about ⟨111⟩, ⟨100⟩
and ⟨110⟩ axes.

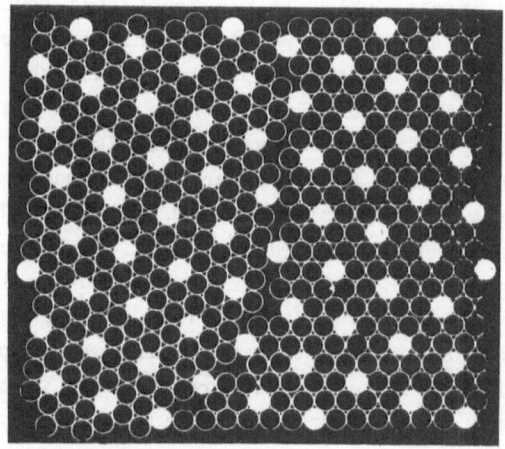

Fig.9. Model of a 38° (or 22°) ⟨111⟩ "ideal" coinci-
     dence grain boundary (after FRANK [9])

A departure from this geometrically exact relationship
should destroy the coincidence lattices. However, as shown
by BISHOP and CHALMERS [18], it is possible to retain
boundary coincidence at the expense of the precise
periodicity of coincidence sites that characterizes the
"ideal coincidence" relationship defined above. The
boundary coincidence, but not lattice coincidence, persists
and gradually changes in a predictable way as the
transition is made from one ideal coincidence orientation to
another. For example, any angle between the ideal

coincidence relations of 32° and 38° about $\langle 111 \rangle$ can be
represented by the appropriate ratio of dislocations and
boundary coincidence atoms [18]. The maintainence of
coincidence in the plane of the boundary requires some
distortion of the crystals; this distortion increases with
increasing departure from an ideal coincidence relationship.
It can be shown that a n y boundary can be represented as
a special case of the transition between ideal coincidence
angles [18].

Direct confirmation of the existence of coincidence grain
boundaries was obtained by BRANDON, RALPH and co-workers
by means of field ion emission studies in tungsten [19]
iron [20] and iridium [21]. It was found that boundaries
lying along planes of high density in the coincidence
lattice appear to be quite narrow, i.e., less than 2 atom
diameters in width. In boundaries whose grains are rotated
several degrees off a high density coincidence rotation,
the fit is observed to be poorer and the boundary wider [19].
The field ion microscope studies also suggest that deviations
from the exact lattice coincidence relation are accommodated
by dislocation networks in a manner analogous to the
structure of low angle boundaries [19,20]. This is an
alternative way of stating that the coincidence spacing is
no longer uniform; the distortion required is identical with
that introduced by the appropriate dislocations.

3.2.3.2 Role of Solutes and Boundary Mobility in Preferred
Orientations

The difference in migration rate between "special" and
"random" boundaries when solute is present (Fig.1), and
the resulting development of preferred orientation
relationships (Fig.2) can be accounted for in terms of the
boundary coincidence model as follows. When the two crystals
are exactly at the ideal coincidence relationship, there is
little if any elastic strain outside the "core" region of
the boundary. Solute atoms would, therefore, be adsorbed
only in the core region. Because of the rather "loose"
structure of this region, diffusion of solute atoms is
relatively easy, and they are, therefore, able to migrate

with the boundary without retarding it seriously. On the
other hand, departure from the ideal orientation
relationship is equivalent to the addition of an extra, more
widely spaced, dislocation array, which gives rise to a
strain field extending an equivalent distance into the
crystals. This strain field would interact with solutes in
a region where the lattice is only slightly distorted.
The much slower diffusion of solute atoms (compared with
self diffusion) would lead to substantial drag on the
boundary, especially under the conditions of low driving
force that are under discussion. This effect would arise
gradually with progressive departure from ideal coincidence.
The observation that the special properties, such as high
mobility, of "ideal coincidence" boundaries are still
observable some degrees away from the ideal relationship
lends strong support to the concept that boundary
coincidence, rather than lattice coincidence, is important.

It has been suggested [22,23] that a grain boundary
approximating to, but deviating somewhat from the ideal
coincidence relationship, will have the highest mobility.
At present, there is no conclusive evidence to support
this suggestion, although the results shown previously in
Fig.3 appear to argue against it. In addition, a summary
of the temperature-dependence data for the migration of
⟨100⟩ tilt grain boundaries in zone refined lead is given
in Fig.10 [6]. A lower apparent activation energy for

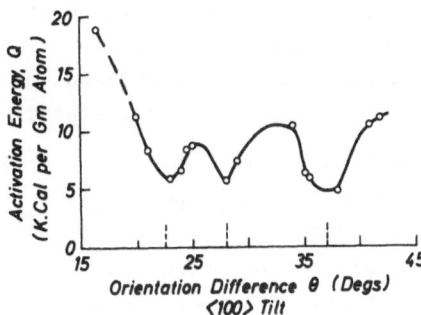

Fig.10. Measured activation energies vs. orientation
difference, θ, for ⟨100⟩ tilt boundaries in zone
refined lead [6]

boundary motion is found for grain boundaries with the ideal coincidence relations, namely 23, 28, and 37° about ⟨100⟩. These results provide strong support for the interpretation of large angle boundaries in terms of the coincidence model. However, it is not definitely known whether the differences in migration behaviour, such as depicted in Fig.3 and 10, are due solely to differences in boundary structure, or due to the possible interaction of residual impurity in the zone refined lead with the different boundary structures. It is known that the addition of a suitable solute to the lead will accentuate the differences in migration behaviour among large angle boundaries (e.g. Fig.1 and references [3] and [24].

FORTES and RALPH [21] employed the field ion microscope to study segregation of oxygen to grain boundaries in iridium. They found that for a volume concentration of about 500 p.p.m. of oxygen, a level of six times this amount at a grain boundary that deviates from a high density or ideal coincidence relation. However, no oxygen segregation was found at a coincidence boundary having a high density of coincidence sites. For an ideal coincidence boundary, appreciable impurity segregation is not expected since the number of distorted sites should be very low, whereas for a non-ideal coincidence boundary a large number of distorted sites exists which provide the driving force for impurity segregation. These field-ion microscope observations provide direct experimental confirmation of the interpretation given bei AUST and RUTTER [3,24] for the observed difference in boundary migration between special and random boundaries in the presence of solute atoms.

It should be noted that these studies of growth selectivity in striated crystals of Pb, Al und Cu were conducted with a relatively small driving energy for boundary migration. The question arises whether these results can apply to preferred orientations developed during the recrystallization of heavily deformed materials. There is considerable evidence that these findings are also applicable to driving energies at least 1000 times greater than those used in the striated crystal experiments. For example, RATH and GORDON [25] in a study of single boundary migration in deformed crystals of aluminium employed a driving force estimated to be 10^3 times that present in the striated single crystals. Rath and Gordon found lower apparent activation energies for the migration of coincidence boundaries (near 37° about $\langle 100 \rangle$ and 38° about $\langle 111 \rangle$) than for random boundaries in zone refined aluminium containing 0.008 and 0.016 wt. pct. of copper.

In addition, FROIS and DIMITROV [26] observed the preferred orientation relationship defined by a rotation of 40° about $\langle 111 \rangle$ between primary recrystallized grains and a highly deformed matrix when the copper content of zone refined aluminium is between 2 and 50 atom p.p.m. They also found that the coincidence boundaries near 40° about $\langle 111 \rangle$ are apparently characterized by a high speed of migration and a low activation energy, in agreement with the results of AUST and RUTTER [24] for dilute alloys of tin in lead. Finally, RATH and HU [27] showed that most anisotropically-growing grains are related to the matrix crystal in zone refined aluminium by a 40° or 20° rotation about a common $\langle 111 \rangle$ axis, and 35° or 25° about $\langle 100 \rangle$. All of these orientation relationships are within 2° of predicted high density coincidence values. Also, in most cases the pure tilt boundary segments have a higher migration rate than do either the mixed or the pure twist types [27]. A possible explanation of this latter observation is that the "core" region of the twist boundary has no free volume, while the tilt ones do. Hence, the easy diffusive motion of solutes (as discussed earlier) should apply only to the tilt case.

3.2.3.3 Role of Twinning and Interfacial Energy in Preferred Orientations

The experiments on selctive growth in Pb, Al and Cu have shown that ideal coincidence grain boundaries may be introduced into a metal as a result of twinning during grain boundary migration. Also, in some cases for lead [12] and copper [17], twinning occurred a second time during boundary motion, thereby replacing a coincidence boundary

having a lower density of coincidence sites with a higher
density coincidence boundary. The twinning observations
indicate that interfacial energy differences among large
angle grain boundaries may be as high as about 10 pct.
depending upon the density of coincidence site atoms.

Direct evidence that coincidence boundaries are lower energy
interfaces was obtained by RUTTER and AUST [28] using
tricrystal specimens of zone refined lead. The results
obtained are given in Fig.11, where the relative energies
of ⟨111⟩ symmetrical tilt boundaries are expressed in terms
of the energy of a large angle non-coincidence boundary. The
orientation relationships of the ⟨111⟩ boundaries were
maintained to within $0,5^{\circ}$ of the boundary angles shown;

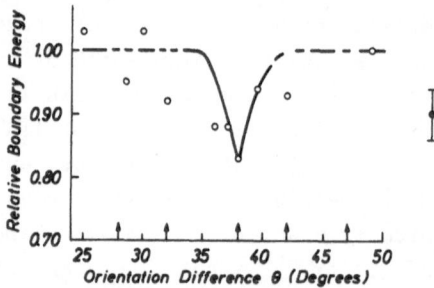

Fig.11. Relative boundary energy vs. θ for ⟨111⟩ tilt
boundaries in zone refined lead. The predicted
coincidence boundary angles are indicated by the
arrows [28]

the possible error in the relative energy measurement is
given in Fig.11. The data here suggest that the 28° and
32° ⟨111⟩ boundaries have lower energies, although the
error in the relative energy measurement obviates any
definite conclusion. However, it is evident in Fig.11 that
the interfacial energy of a 38° ⟨111⟩ coincidence boundary
is about 15 pct. lower than that of a high angle non-
coincidence boundary in zone refined lead. This result
provides additional support for the coincidence description
of grain boundaries and also of certain preferred
orientation relations.

It is known that the frequency of annealing twins increases

with decreasing ratio of twin boundary energy to grain
boundary free energy [11,16]. Consequently, the introduction
of i d e a l coincidence boundaries by twinning may also
increase when this ratio decreases. For example, a greater
frequency of twinning and i d e a l coincidence grain
boundaries might be expected in a high purity metal as
compared to an impure metal, and in metals such as Cu or
Pb where the interfacial energy ratio is lower than in a
metal such as aluminium. In addition, coincidence boundaries
very close to the ideal type are observed when twinning
occurs (e.g. Fig. 5 and 7). This result is in agreement
with the boundary coincidence description given by BISHOP
and CHALMERS [18], which predicts that the grain boundary
energy should increase rapidly with departure from ideal
coincidence.

It seems likely, however, that whenever a strong mobility
dependence exists, it will be the dominating influence
in growth selectivity. In this case, a r a n g e of
orientations around the ideal coincidence is obtained since
the effects of solute interaction, leading to mobility
differences, should develop gradually with increasing
deviation from the ideal coincidence orientations [18].

Acknowledgement

The author would like to thank the National Research
Council of Canada for a grant (NRC A 4611) during the
preparation of this paper and for a travel grant to attend
the meeting in Clausthal, Germany.

References

1. H. Brooks in: Metal Interfaces, Amer.Soc.Metals,
   Cleveland 1952, p. 20.
2. J.W. Rutter and K.T. Aust: Acta Met. 6(1958)375.
3. K.T. Aust and J.W. Rutter: Trans.Met.Soc.AIME 215(1959)119.
4. K.T. Aust and J.W. Rutter: Trans.Met.Soc.AIME 218(1960)50.
5. J.W. Rutter and K.T. Aust: Trans.Met.Soc.AIME 218(1960)682.
6. J.W. Rutter and K.T. Aust: Acta Met. 13(1965)181.
7. K.T. Aust and J.W. Rutter: Trans.Met.Soc.AIME 224(1962)111.

8. M.L. Kronberg and F.H. Wilson: Trans. AIME 185(1949)501.

9. F.C. Frank: Seminar at General Electric Research Laboratory, September 1958.

10. C. G. Dunn: Annual AIME meeting, San Francisco, February 1959.

11. R.L. Fullman and J.C. Fisher: J. Appl.Phys.22(1951)1350.

12. K.T. Aust and J.W. Rutter: Trans.Met.Soc.AIME 218(1960)1023.

13. J.E. Burke and D. Turnbull in: Chalmers' Progr.Metal Physics 3(1952)220.

14. K.T. Aust and J.W. Rutter in: Ultra-High-Purity Metals, Amer.Soc. Metals 1962, p. 115.

15. K.T. Aust: Trans.Met.Soc.AIME 221(1961)758.

16. G.F. Bolling and W.C. Winegard: J. Inst. Metals 86(1958)492.

17. G. Ferran, G. Cizeron and K.T. Aust: Acad.Sci., Paris 257(1963)3595 – Mém.Sci. Rev.Mét. 54(1967)1067.

18. G.H. Bishop and B. Chalmers: Scripta Met. 2(1968)133 – also annual AIME meeting, Los Angeles, February 1967.

19. D.G. Brandon, B. Ralph, S. Ranganathan and M.S. Wald: Acta Met. 12(1964)813.

20. R. Morgan and B. Ralph: Acta Met. 15(1967)341.

21. M.A. Fortes and B. Ralph; Acta Met. 15(1967)707.

22. J.C.M. Li in: Recovery and Recrystallization of Metals, L. Himmel, Ed., Interscience, New York 1963, p. 160.

23. P. Gordon and R.A. Vandermeer in: Recrystallization, Grain Growth and Textures, H. Margolin, Ed., Amer.Soc. Metals, Metals Park, Ohio 1966, p. 205.

24. K.T. Aust and J.W. Rutter: Trans.Met.Soc.AIME 215(1959)820.

25. B.B. Rath and P. Gordon: Tech.Rept. No. 5, U.S. Army Research Office, October 1962.

26. C. Frois and O. Dimitrov: C.R.Acad.Sci., Paris 252(1961) 1465.

27. B.B. Rath and H. Hu: Trans.Met.Soc. AIME 236(1966)1193.

28. J.W. Rutter and K.T. Aust: Referred to by K.T. Aust in: Surfaces and Interfaces, I – Chemical und Physical Characteristics, Syracuse University Press, Syracuse, N.Y. 1967, p. 435.

3.3 The Mechanism of Grain Boundary Migration
Der Mechanismus der Korngrenzenbewegung

by H. Gleiter[+]

Zusammenfassung

Elektronenmikroskopische Untersuchungen (Al-Cu-Mischkristal-
le) zeigen, daß die Oberflächen der Körner, die eine Korngrenze bilden, stufenförmige Atomanordnungen haben. In Kristallen mit kubisch flächenzentrierter Struktur werden diese
Stufen durch die {111}-Ebenen beider Kristalle gebildet. Die
Korngrenzenbewegung wird durch die Emission von Atomen der
Stufen des kleiner werdenden Korns in die Korngrenze hinein
und durch die Absorption der gleichen Anzahl von Atomen an
den Stufen des wachsenden Korns vollzogen.

Dieser Vorgang bewirkt, daß die Stufen sich auf den Oberflächen beider Körner bewegen und damit wandert die Korngrenze. Ausgehend von diesem Modell wird die Beweglichkeit
einer Korngrenze bei reinen Kristallen, festen Lösungen und
Kristallen mit Einlagerungen berechnet. Die Ergebnisse der
Berechnung sind in Übereinstimmung mit den experimentellen
Beobachtungen.

Grain boundary migration is defined as the movement of a
grain boundary in the direction perpendicular to its tangent
plane. Although in many papers observations on grain
boundary migration are described very little has been
published on the atomistic mechanism. Its the aim of this
paper to give a summary of recent experimental and
theoretical results on the atomistic mechanism of grain
boundary migration.

Transmission electron microscopy was used to investigate
the process of grain boundary migration. The alloy used
for this purpose was a solid solution of Al-Cu (0.39 wt% Cu).
In order to obtain grain boundary migration the alloy was
heavily deformed, recrystallized and subsequently annealed
at 270°C. At this temperature grain boundary migration

[+] Harvard University, Gordon McKay Laboratory, Cambridge,
Mass. U.S. A.

occurs between the nearly dislocation free grains. During
the migration process the alloy was quenched to room
temperature. The structure of the boundaries was investigated
by transmission electron microscopy. The results of the
electron microscopic observations are summarized in the
following paragraphs:

1.) In a migrating grain boundary there exists a pattern of
    lines. These lines will be called grain boundary lines
    (g.b.l.). The pattern is composed of two or more line
    systems. Up to 5 line systems can be observed at one
    point of a boundary. (Fig. 1) The lines intersect each
    other.

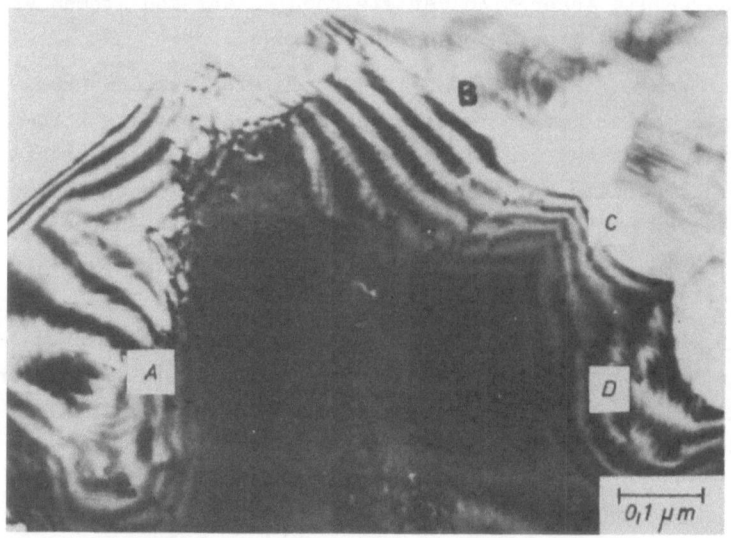

Fig.1. Electron micrograph showing different sets of grain
boundary lines. Near B, C and D the intersection of the
two sets of lines is visible. The displacement in a
thickness fringe caused by these lines can be seen near A
(Al – Cu, 0.39 wt%, 99.9% deformed, recrystallized 5 min
270°C/20°C)

2.) The g.b.l. form a pattern of parallel lines or have the
    form of a spiral (Fig.2). In the center of the spiral
    always a dislocation with a screw component was
    observed.

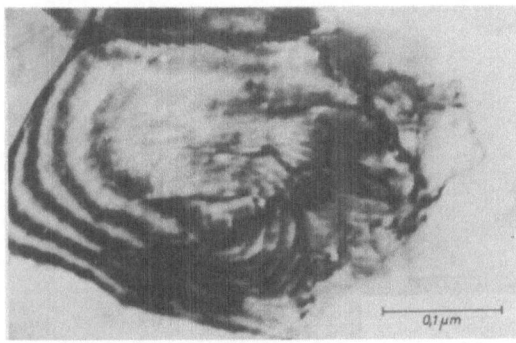

Fig.2. Grain boundary line having the form of a spiral

3.) At the points, where a g.b.l. crosses a thickness fringe,
a displacement in the fringe or a jump in the thickness
fringe occurs (Fig.1 and 3).

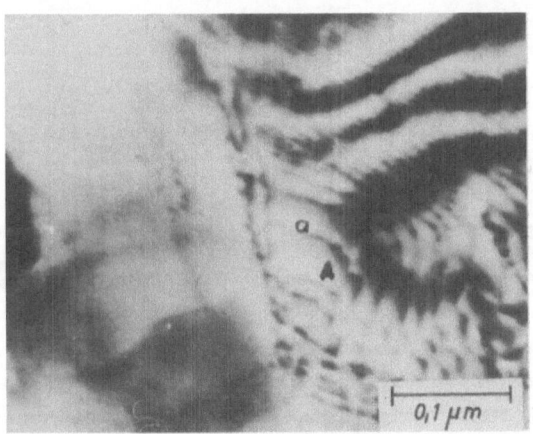

Fig.3. Jump in the thickness fringe of a boundary at
a grain boundary line

4.) Often the g.b.l. contains steps. These steps are aligned
in the direction of another line system in the same
boundary (Fig. 4).

5.) During grain boundary migration the g.b.l. move very
rapidly along the boundary. The different line systems
move in different directions. The spirals in the

Fig. 4. Steps in grain boundary lines. The steps are aligned
in the direction of a weak second line system

boundary rotate. Grain boundary migration without the
movement of the g.b.l. was never observed.

From the observation reported in paragraph 5 it follows that
there is a relationship between these lines and grain
boundary migration. In order to understand this relationship
the atomistic structure of the lines must be known. The only
model consistent with all the above summarized observations,
is a step-model for the g.b.l.. (A detailed discussion of the
explanation of the observations is given in the original
publication [1].) From the jump in the thickness fringes,
the displacement of the fringes and the spirals, the height
of the steps can be determined. In many cases a step height
of about 3 Å was measured. From field ion microscopy it
follows that the thickness of a boundary is about 8...15 Å.
Therefore we know that the step height is smaller than the
boundary thickness. From the form of the g.b.l., the

relationship between the crystal lattice of the adjacent
grains and the steps was obtained in the following way:
By electron diffraction the orientation of the two grains
and the spacial form of the g.b.l. was determined. If the
direction of the two tangents on this spacial line is
measured then the lattice plane forming this g.b.l. can be
calculated. By this method it was found that the steps are
formed by the {111} - planes of the grains. Therefore we
obtain the following grain boundary model (Fig.5): The
surfaces of both grains forming a grain boundary consists
of atomic steps formed by the {111} planes of the crystals.
Because each grain has 4 different {111}-planes, we obtain
4 different sets of steps on each grain surface (Fig.6).

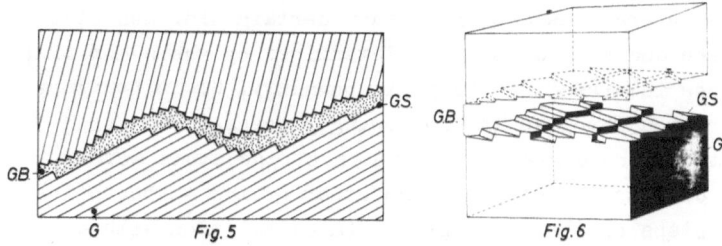

Fig.5. The formation of steps in the surface of a grain by
ending {111} planes. (G = grain, G.S. = grain surface, G.B.
= grain boundary) Each {111} plane is symbolized by one
plate

Fig.6. Grain boundary model: On the surface of both grains
there is a pattern of intersecting steps due to the
different {111} planes of the grains

These steps intersect each other and form steps at the
intersection points. (Comp. paragraph 4).

From the observation in the electron microscope it follows
that the steps move along the grain boundary during grain
boundary migration. The movement of the steps will only
occur, if atoms are added at the steps or are emitted from
the steps (Fig. 5). If we add atoms at the steps of the
lower grain then these steps move from the left to the
right side and the whole grain surface moves towards the
other grain. The only sources for these atoms are the steps

on the upper grain. These steps move by the emission of
atoms from left to the right and therefore the whole surface
of this grain migrates in the upper direction. These
simultaneous movement of the steps in different directions
was observed in the microscope during grain boundary
migration. Therefore this observation tells us that grain
boundary migration occurs in the following way: The steps
of the shrinking grain emit atoms into the boundary while
the steps of the growing grain absorb these atoms. By this
process the surfaces of the two grains move perpendicular
to the tangent plane and therefore the whole boundary
migrates. Based on this model, the mobility of a boundary as
a function of the grain orientation can be calculated. In
order to do this calculation, we consider the shrinking
grain as a plane with a certain source density and the
growing grain as a plane with a certain sink density. The
source and sink density is determined by the numer of $\{111\}$-
planes intersecting the grain boundary. Grain boundary
migration involves the following processes: Emission of the
atoms from one grain – diffusion of the atoms across the
boundary to the steps of the other grains – absorbtion at
the steps of the other grain. (A calculation with a more
detailed model will be published). From this simplified
model we expect that the mobility of the boundary is a
maximum when the sink and source density is a maximum and it
decreases as the source and sink density decreases. Because
the source and sink density is directly related to the
number of $\{111\}$ planes intersecting the boundary, we get a
relationship between the grain orientation and the mobility.
The results of a more sophisticated calculation are shown
in Fig. 7 and Fig. 8. The different mobility of a boundary
at different grain orientation can therefore be described
in terms of the step structure of a boundary proposed by
this grain boundary model. The mobility of a grain
boundary in solid solutions and in alloys containing
particles can be calculated by considering interaction
between the steps and the solute atoms or the particles [4].

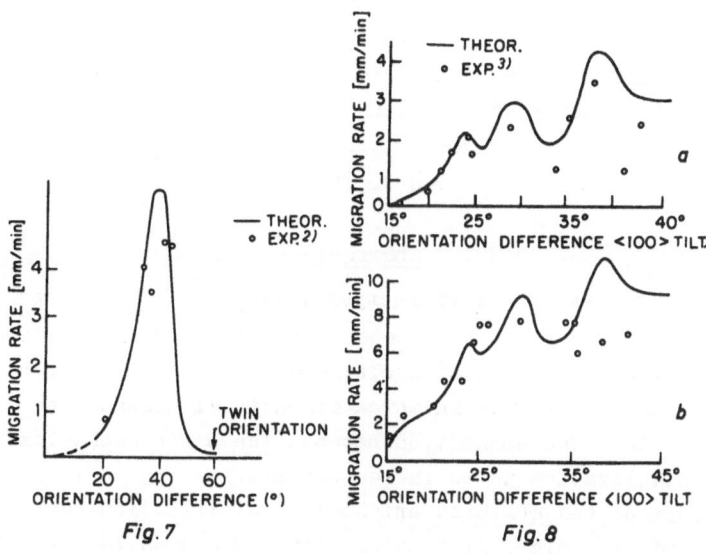

Fig.7. Calculated and observed mobility of a grain boundary in aluminum; (experimental data after LIEBMANN et. al [2])

Fig.8. Calculated and observed mobility of a grain boundary in high purity lead (experimental data after RUTTER and AUST [3]

      a) 200°C   b) 300°C

References

1. H. Gleiter: Acta Met., in press.

2. B. Liebmann, K. Lücke and G. Masing: Z. Metallkde. 47(1956)57.

3. F. Rutter and K. Aust: Acta Met. 13(1965)181.

4. Papers about these results will be published in Acta Met. and Trans. Met. Soc. AIME.

3.4 Ansätze zu einer statistischen Theorie der Rekristalli-
sationstexturen und ihre Anwendung auf die Wachstumsauslese
in Einkristallen
Attempts for a Quantitative Theory of Recrystallization
Textures and their Application to the Growth Selectivity of
Single Crystals

von G. Ibe[+]

Abstract

A quantitative theory of recrystallization textures requires
the knowledge of the basic mechanism of recrystallization –
nucleation and growth of nuclei – and the technique for
measuring the texture in three-demensional orientation
coordinates. The connection between the basic mechanism of
recrystallization which is responsible for the texture
development and measured orientation distribution can be
obtained with simplified assumptions in a quantiative
manner by a statistical theory. This theory permits the
calculation of simple textures from the nucleation and
growth parameters or vice versa the calculation of the most
important parameters of nucleation and growth from the
measured orientation distribution, but there are presently
only a few direct measurements of nucleation and growth of
nuclei. As an example, results from a study on the growth
selectivity in single crystals will be discussed.

3.4.1 Einleitung

Um zu einer quantitativen Beschreibung der Texturentstehung
bei der primären Rekristallisation zu gelangen, ist es
zweckmäßig, den Zusammenhang der Erscheinungen in drei
Gruppen zu unterteilen.

Die erste Gruppe umfaßt die Grundvorgänge der Rekristalli-
sation: Keimbildung $N(t,r,\rho,\epsilon)$ und Keimwachstum $v(t,\vec{r},\rho,\epsilon)$
mit ihrer Abhängigkeit von Zeit (t), Ort (r), Richtung $(\vec{r})$,

---

[+] Leichtmetall-Forschungsinstitut der Vereinigten Alumi-
niumwerke AG, Bonn, Bundesrepublik Deutschland

Orientierung (ρ) und Verformung (ε). Diese Grundgrößen und
ihre Abhängigkeiten werden, soweit erforderlich, im folgenden als gegeben vorausgesetzt. Die Physik der Grundvorgänge
soll hier nicht betrachtet werden.

Die dritte Gruppe betrifft die nach beendeter Rekristallisation vorliegenden Texturen. Um zu eindeutigen Zusammenhängen zu gelangen, müssen wir ihre Messung in dreidimensionalen Orientierungskoordinaten voraussetzen, etwa in den
drei Koordinaten der Drehung φ um eine räumliche Drehachse
mit den Polarkoordinaten ω, θ. Die Gesamtheit dieser Koordinaten erfüllt einen Orientierungsraum, Vorzugsorientierungen sind als mehr oder weniger kugelsymmetrische Häufungen
von Orientierungspunkten (Rotationen) in diesem Raum erkennbar. Da die Vorzugsorientierungen (Ideallagen) hier
durch die als bekannt vorausgesetzten Grundvorgänge gegeben
sind, kann man die Beschreibung der Orientierungsverteilung
auf die eindimensionale radiale Häufigkeitsverteilung H(ρ)
als Funktion eines Orientierungsabstandes ρ von der Ideallage beschränken. Dieser Abstand ρ läßt sich leicht als
größter Abstand einander entsprechender Pole von Keim und
Ideallage aus der Polfigur entnehmen [1,2], wie Abb. 1 für
{100}-Pole zeigt.

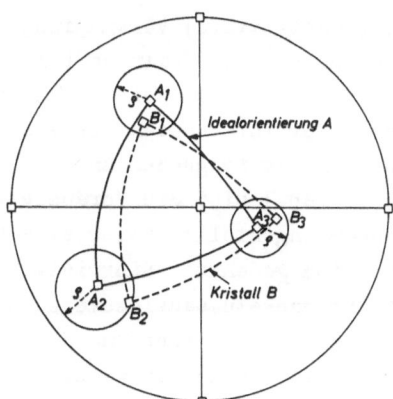

Abb.1. Zur Definition des Orientierungsabstandes ρ

Die zweite Gruppe schließlich stellt die bisher noch weit-

Tab.1. Zwei Beispiele zu einer statistischen Theorie der
Rekristallisationstexturen unter vereinfachenden Annahmen:
Keimbildung nur zur Zeit $t = 0$

Grundvorgänge			Texturbildung durch	
			Wachstums-Auslese	homogene Rekristall.
Keim-bil-dung	Dichte	Anzahl	N	n
		Geschwind.	$\delta(t = 0)$	$\delta(t = 0)$
	Ortsverteilung		punktförm.	statistisch
	Orientierungsver-teil.		regellos (bevorzugt)	regellos (bevorzugt)
Keim-wachs-tum	Dimension		linear	3-dim.
	Orient.-Abhängig-keit		$v(\rho)$	$v(\rho)$
	Anisotropie		experim.	-

gehend fehlende (quantitative) Verknüpfung der Grundvorgänge (I) mit den gemessenen Texturen (III) dar. Diese Verknüpfung ist mit Hilfe geometrisch-statistischer Überlegungen unter gewissen vereinfachenden Voraussetzungen bereits
durchführbar und soll im folgenden an zwei Beispielen erläutert werden. In Tab.1 sind die Voraussetzungen der beiden Beispiele zusammengestellt. Beispiel A bezieht sich auf
die Rekristallisation gedehnter Einkristalle nach der bekannten Technik der Wachstumsausleseversuche [3,4]. Je Probe
nehmen N durch zusätzliche Verformung an der Probenspitze
erzwungene Keime von $t = 0$ an der Wachstumskonkurrenz in
Längsrichtung der stabförmigen Einkristalle teil. Die jeweils schnellstwachsenden Keime vieler Proben ergeben eine
Orientierungsverteilung, die im Prinzip die vereinfachte
Textur einer linearen Rekristallisation darstellt.

Beispiel B stellt die homogene Rekristallisation einer dreidimensionalen Probe mit statistischer Verteilung der Keime
im Orts- und im Orientierungsraum dar, bildet also gegenüber
A einen anderen Grenzfall möglicher Verhältnisse.

3.4.2 Wachstumsauslese in Einkristallen, Statistik der größten Geschwindigkeiten

Die N Keime in jeder Probenspitze seien regellos orientiert
mit einer Abstandsverteilung gegen jede beliebige Ideallage A [1,2]

$$W(\rho) = k \cdot \rho^3; \quad k = 1,42 \text{ für } \{100\}; \quad k = 1,38 \text{ für } \{111\} \qquad (1)$$

Die Wachstumsgeschwindigkeiten haben um die Ideallagen A_1
den Verlauf

$$v_i = v_{io} \cdot \exp\left(-\frac{\rho_i^3}{s_i^3}\right) \qquad (2)$$

Für eine einzige Ideallage A überlebt unabhängig vom genauen
Verlauf der Geschwindigkeit stets der Keim mit dem kleinsten
Abstand ρ von A die Auslese. Die sich ergebende differentielle Orientierungsverteilung setzt sich aus zwei Faktoren
zusammen:

$$dH(\rho) = g_o(\rho) \cdot f_1(\rho) \, d\rho \qquad (3)$$

Dabei bedeutet $g_o(\rho)$ die Wahrscheinlichkeit, k e i n e n
Keim im Bereich 0 bis ρ anzutreffen, und $f_1(\rho)d\rho$ die Wahr-
scheinlichkeit, w e n i g s t e n s   e i n e n Keim im Be-
reich ρ bis $\rho+d\rho$ zu finden. Mit Hilfe der in guter Näherung gül-
tigen Poisson-Verteilung $\varphi(z)$ für $z = 0$ bzw. $z = 1$ um die
Erwartungswerte $N \cdot W(\rho)$ bzw. $N \cdot dW(\rho)$ für jeweils einen
Keim kann man $g_o(o)$ und $f_1(\rho)d\rho$ bestimmen und erhält durch
Integration von Gl. (3) die integrale Orientierungshäufigkeit

$$H(\rho) = 1 - \exp\left(-N \cdot k \cdot \rho^3\right) \qquad (4)$$

Wichtig ist der Vergleich mit der regellosen Verteilung durch
den Häufungsfaktor

$$Q(\rho) = H(\rho)/W(\rho) \qquad (5)$$

Nach Gl.(4) folgt aus $\rho \to 0$

$$Q(\rho \to 0) = N \ , \tag{6}$$

die Anzahl der an der Konkurrenz beteiligten Keime. Der Verlauf von $H(\rho)$ und $Q(\rho)$ ist für verschiedene N aus Abb.2 entnehmbar.

Bei zwei Ideallagen A_1 und A_2 ist die gegenseitige Konkurrenz von Keimen verschiedener Ideallagen zusätzlich zu berücksichtigen. Man erhält daher für die differentielle Orientierungshäufigkeit

$$dH(\rho_1) = g_o(\rho_1) \cdot g_o(\rho_2) \cdot f_1(\rho_1)d\rho$$
$$dH(\rho_2) = g_o(\rho_1) \cdot g_o(\rho_2) \cdot f_1(\rho_2)d\rho \tag{7}$$

stets bezogen auf gleiche Geschwindigkeit nach Gl. (2). Solange mit wachsendem ρ noch $v_1 > v_{20}$ ist, nehmen nur Keime nahe A_1 an der Konkurrenz teil und man erhält wieder Gl.(4). Von dem Abstand $\rho_{1krit.}$ für $v_1 = v_{20}$ ab konkurrieren beide Ideallagen und man erhält durch Einsetzen der entsprechenden Wahrscheinlichkeiten in die Gl.(7) die integralen Orientierungshäufigkeiten:

$$H(\rho_1) = 1 - \frac{1}{2} \left[\left(\frac{v_{20}}{v_{10}}\right)^{Nks^3} + \left(\frac{v_{10}}{v_{20}}\right)^{Nks^3} \cdot \exp(-2Nk\rho_1{}^3) \right]$$
$$H(\rho_2) = \frac{1}{2} \left(\frac{v_{20}}{v_{10}}\right)^{Nks^3} \cdot \left[1 - \exp(-2Nk\rho_2{}^3) \right] \tag{8}$$

Dieser Verlauf ist für zwei Beispiele in Abb.3 dargestellt. Mit größerem N erfolgt wieder wie vorher ein steilerer Anstieg der Kurven, aber auch eine stärkere Differenzierung der relativen Anteile H_1 und H_2 für $\rho \to \infty$.

Diese Überlegungen lassen sich nun auf beliebig viele hinsichtlich A_i, v_{io} und s_i unterschiedliche Ideallagen ausdehnen.

Eine erste Anwendung erfolgte auf die Ergebnisse von Wachstumsausleseversuchen an gedehnten Einkristallen aus Reinst-

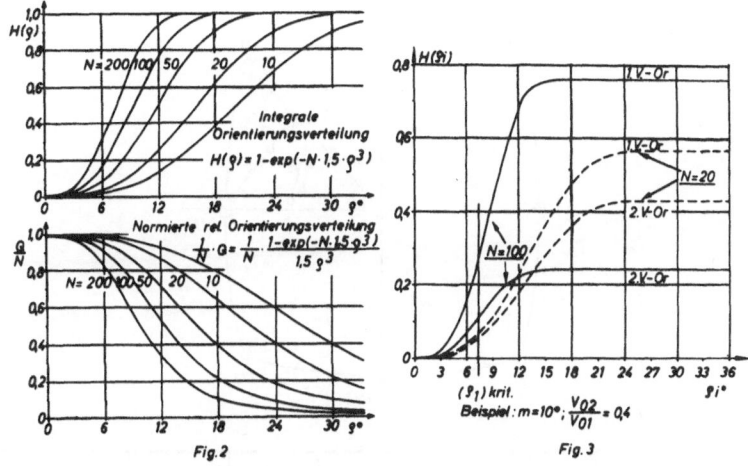

Fig.2   Fig.3

Abb.2. Orientierungshäufigkeit H(ρ)und normierter
Häufungsfaktor Q(ρ)/N nach Wachstumsauslese für
eine Ideallage schnellsten Wachstums

Abb.3. Orientierungshäufigkeit H(ρ_i)nach Wachstumsauslese für zwei unterschiedlich bevorzugte Ideallagen und zwei verschiedene Keimzahlen N

aluminium [5], wobei unter Umkehrung der abgeleiteten Formeln aus den gemessenen Orientierungsverteilungen H(ρ_i)
die Keimzahl N und die maximalen Geschwindigkeiten v_{io} der
verschiedenen Komponenten der Ideallage $40^\circ \langle 111 \rangle$ in Abhängigkeit von den Versuchsbedingungen bestimmt wurden (Abb.4).
Man erkennt deutlich den sehr plausiblen Gang von N mit dem
Durchmesser d und der Dehnung ε der Proben sowie die Abhängigkeit der ungleichen Maximalgeschwindigkeiten v_{io} der verschiedenen $\langle 111 \rangle$-Drehachsen von der mehr oder weniger sym-
metrischen Lage der Probenachse.

Daraus geht offensichtlich hervor, daß die skizzierten Zusammenhänge die Texturentstehung bei der Wachstumsauslese
in guter Näherung beschreiben können.

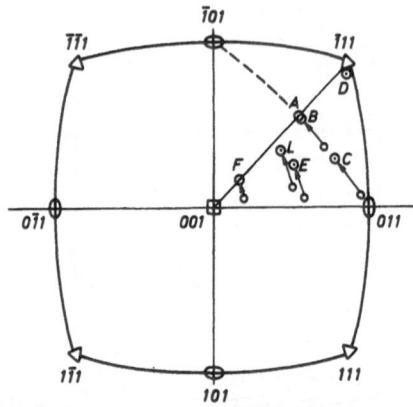

Serie	n	d (mm)	ϵ(%)	N	v_{io}/v_{10}			
					[$1\bar{1}1$]	[$\bar{1}\bar{1}1$]	[111]	[$\bar{1}11$]
A–01	99	1,4	20	31	1	0,42	0,45	0,18
A–02	17	1,4	40./.50	126	1	0,55	0,48	0
A–04	24	1,9	20	53	1	0,56	0,52	0
B–14	66	1,4	20	37,5	1	0,33	0,43	0,21
C–08	82	1,4	20	72	1	0,51	0,19	0,19
D–19	12	1,4	20	25	1	0,79	0,62	0
Ergebnisse von Direktmessungen 3 :								
L	$s\approx10^{\circ}$	1,5	20	–	1	0,15	0,03	0,01

Abb.4. Ermittlung der Keimzahlen N und der Wachstumsgeschwindigkeit v_{io} der verschiedenen $40^{\circ}\langle111\rangle$-Ideallagen aus den Orientierungshäufigkeiten $H(\rho_i)$ nach Wachstumsauslese in Al-Einkristallen (nach [5]) unter verschiedenen Versuchsbedingungen

3.4.3 Homogene Rekristallisation in Einkristallen, Statistik der kleinsten Ankunftszeiten

In Beispiel B der homogenen Rekristallisation müssen wir die Zusammenhänge gleichzeitig im Orientierungsraum und im Ortsraum betrachten, beide vereinfacht dargestellt durch den Orientierungsabstand ρ und den Ortsabstand r. Der Abstand r

bezieht sich dabei auf einen beliebigen Aufpunkt in der Probe, der bei der Rekristallisation die Orientierung des zuerst dort ankommenden Keimes annimmt. Die Orientierungsverteilung in der Probe ergibt sich also aus der Häufigkeitsverteilung, mit der Keime bestimmten Orientierungsabstandes ρ mit der kürzesten Ankunftszeit τ in einem beliebigen
Aufpunkt eintreffen.

Im Volumen V seien N Keime mit der mittleren Dichte n regellos verteilt, die Orientierungen seien ebenfalls regellos
nach Gl.(1).

Für eine Vorzugsorientierung maximaler Wachstumsgeschwindigkeit gelte für v wieder die Gl.(2) um die Ideallage A.
Die differentielle Häufigkeitsverteilung ist wieder durch
2 Faktoren gegeben:

$$dH(\rho,r) = g_0(\tau) \cdot f_1(\rho,r)d\rho \; dr \qquad (9)$$

Dabei bedeutet $g_0(\tau)$ die Wahrscheinlichkeit, daß k e i n
Keim mit einer kleineren Ankunftszeit als τ auftritt, die
durch die Koordinaten ρ und r gegeben ist. ρ und r sind
jetzt durch τ miteinander gekoppelt:

$$r = \tau \cdot v = \tau \cdot v_0 \cdot \exp\left(-\frac{\rho^3}{s^3}\right) \qquad (10)$$

$f_1(\rho,r)d\rho \; dr$ stellt jetzt die Wahrscheinlichkeit dar, daß we-
nigstens ein Keim g l e i c h z e i t i g in den Bereichen
ρ bis $\rho+d\rho$ und r bis r+dr auftritt und ergibt sich als Fall
z = 1 aus der Poisson-Verteilung $\varphi(z) = e^{-\lambda} \cdot \lambda^z \cdot 1/z!$ um
den Erwartungswert λ, einen beliebigen von N Keimen gleichzeitig in diesen Bereichen zu finden

$$\lambda = N \cdot dP(\rho) \cdot dP(r) = N \cdot (3k\rho^2 d\rho) \cdot \left(\frac{1}{V} \cdot 4\pi r^2 dr\right) \quad (11)$$

wobei $dP(\rho)$ und $dP(r)$ die Erwartungswahrscheinlichkeiten
dieser Bereiche sind.
Für z = 1 folgt dann gerade

$$\lambda = f_1(\rho,r)d\rho \; dr \qquad (12)$$

$g_0(\tau)$ bestimmt sich etwas umständlicher als Fall z = 0 der

Verteilung $\wp(z)$ um den Erwartungswert λ, eine beliebigen
Keim gleichzeitig im Bereich 0 bis ρ und 0 bis r zu finden.
Wegen der Gl.(10) ist jetzt aber r nicht mehr von ρ unab-
hängig, und es gilt für den Erwartungswert

$$\lambda = N \cdot P(\rho,r) = N \cdot \int\limits_{\rho=0}^{\infty} \int\limits_{r=0}^{r_m} dP(\rho) \cdot dP(r) \qquad (13)$$

wobei $dP(\rho)$ und $dP(r)$ wieder die differentiellen Erwartungswahrscheinlichkeiten darstellen. Durch r_m nach Gl.(10) ist
die Integration hierbei auf das Gebiet unterhalb der Kurven
τ = const. in der ρ-r-Ebene der Abb.5 beschränkt. Mit dem

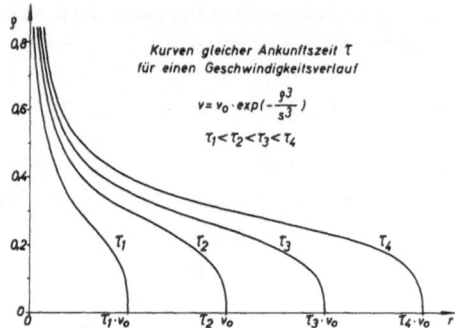

Abb.5. Abhängigkeit des Ortsabstandes r vom Orientierungsabstand ρ für gleiche Ankunftszeiten τ_i bei
homogener Rekristallisation mit einer Ideallage

Erwartungswert nach Gl. (13) ist dann

$$g_0(\tau) = \wp(z = 0) = e^{-\lambda} \qquad (14)$$

Einsetzen von Gl. (12) und (14) gestattet die Berechnung
der integralen Orientierungshäufigkeit

$$H(\rho) = \int\limits_{\rho=0}^{\rho} \int\limits_{r=0}^{\infty} dH(\rho,r) = 1 - \exp\left(-3\,\frac{\rho^3}{s^3}\right) \qquad (15)$$

für eine Ideallage. Die Verteilung gibt somit den durch die
dritte Potenz verschärften Verlauf der Geschwindigkeitsver-
teilung wieder. Hier hat die Keimdichte n keinen Einfluß auf
die Texturschärfe. Abb.6 zeigt den Verlauf $H(\rho)$ für ver-

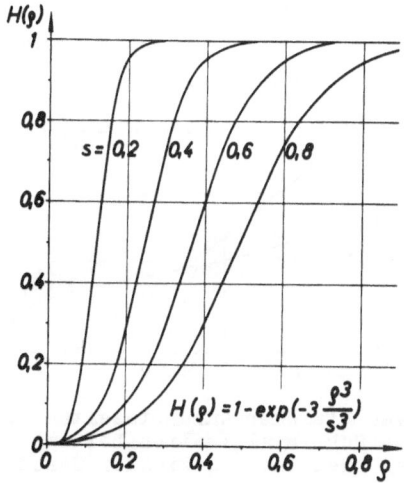

Abb.6. Orientierungshäufigkeit H(ρ) nach homogener Rekristallisation mit einer Ideallage für verschieden breite Geschwindigkeitsverteilungen

schiedene Geschwindigkeitsparameter s. Wie man sieht, geht die Maximalgeschwindigkeit v_o ebenfalls nicht in die Verteilung ein.

Für zwei Idealorientierungen unterschiedlichen Geschwindigkeitsverlaufs nach Gl. (2) gilt jetzt für die differentiellen Häufigkeiten

$$dH(\rho_1, r_1) = g_o(\tau_1) \cdot g_o(\tau_2) \cdot f_1(\rho_1, r_1) \, d\rho \, dr$$
$$dH(\rho_2, r_2) = g_o(\tau_1) \cdot g_o(\tau_2) \cdot f_1(\rho_2, r_2) \, d\rho \, dr \tag{16}$$

Die Faktoren $g_o(\tau_i)$ und $f_1(\rho_i, r_i) d\rho \, dr$ bestimmen sich genau wie für eine Ideallage, wobei jedoch trotz ungleichen Geschwindigkeitsverlaufes ($v_{10} \neq v_{20}$, $s_1 \neq s_2$) stets $\tau_1 = \tau_2 = \tau_d$ einzuhalten ist. Das gilt besonders für die Ermittlung der Erwartungswerte $N \cdot P(\rho_i, r_i)$ analog zu Gl.(13), wie Abb.7 verdeutlicht. Nach Einsetzen in Gl. (16) und Rückführung auf jeweils ein Variablenpaar (ρ_i, r_i) mit Hilfe der Gl.(10) läßt sich Gl. (16) integrieren und ergibt schließlich für die integralen Orientierungshäufigkeiten den Verlauf

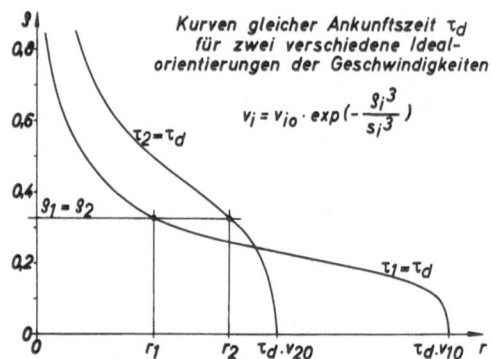

Abb.7. Kurven gleicher Ankunftszeiten $\tau_1 = \tau_2 = \tau_d$
für homogene Rekristallisation bei zwei unterschiedlichen Ideallagen maximaler Geschwindigkeit

$$H(\rho_1) = \frac{v_{10}^3 \cdot s_1^3}{v_{10}^3 \cdot s_1^3 + v_{20}^3 \cdot s_2^3} \cdot (1 - \exp[- 3\, \frac{\rho_1^3}{s_1^3}])$$

$$H(\rho_2) = \frac{v_{20}^3 \cdot s_2^3}{v_{10}^3 \cdot s_1^3 + v_{20}^3 \cdot s_2^3} \cdot (1 - \exp[- 3\, \frac{\rho_2^3}{s_2^3}]) \tag{17}$$

woraus die relativen Anteile für $\rho_i \to \infty$ folgen. Wie Gl.(17)
zeigt, haben die beiden Komponenten den gleichen relativen
Verlauf wie für einzelne Ideallagen nach Gl. (15). Die Maximalgeschwindigkeiten v_{10} und v_{20} bestimmen nur die Anteile,
die für große ρ erreicht werden.

Diese Rechnung kann im Prinzip auf den allgemeinen Fall beliebig vieler unterschiedlicher Ideallagen ausgedehnt werden, doch läßt sich das Prinzip der gegenseitigen Konkurrenz viel klarer für 2 Ideallagen zeigen, wie aus den beiden Beispielen der Abb.8 für unterschiedliche v_{io} und s_i
hervorgeht.

Die hiermit gegebene, noch stark vereinfachte statistische
Theorie der Rekristallisationstexturen läßt sich im Prinzip
durch Berücksichtigung der Zeitabhängigkeit von N und v

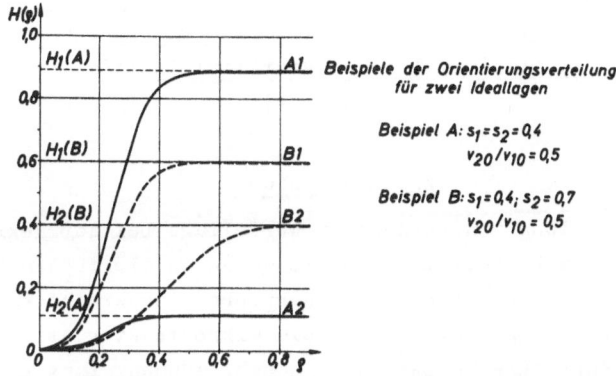

Abb.8. Orientierungshäufigkeiten H(ρ_i) nach homogener Rekristallisation für zwei Beispiele mit zwei unterschiedlichen Ideallagen

sowie etwaiger nichtregelloser Orientierungs- und Ortsverteilungen der Keime und insbesondere der polykristallinen Struktur der Matrix verallgemeinern, doch werden die Rechnungen dazu sehr schnell unübersichtlich, so daß für diese Abhängigkeiten nach brauchbaren Näherungen gesucht werden muß.

Literatur

1. C.G. Dunn: Phys. Rev. 66(1944)215.
2. G. Ibe und K. Lücke: Acta Met., im Druck.
3. B. Liebmann, K. Lücke und G. Masing: Z.Metallkde. 47(1956)57.
4. H. Yoshida, B. Liebmann und K. Lücke: Acta Met. 7(1959)51.
5. W. Dietz, A.C. Fraker, G. Ibe und K. Lücke: Z.Metallkde. im Druck.

3.5 Reorientation in Recrystallization: Origin of Cube Grains
in Copper - A Study of Nucleation in Recrystallization
Orientierungsänderung durch Rekristallisation: Ursprung von
Körnern in Würfellage in Kupfer - Eine Untersuchung der
Keimbildung bei der Rekristallisation

by H. Hu[+)]

Zusammenfassung

Die Keimbildung von Körnern in Würfellage bei der Rekristallisation stark gewalzten Kupfers wurde in elektronenmikroskopisch durchstrahlbaren Querproben (senkrecht Walz-
und Querrichtung) untersucht. Die Mikrostruktur stark ge-
walzter Kupferbleche besteht aus sehr dünnen, fast flachen
und stark gelängten strukturellen Elementen, ähnlich Strei-
fen, die im allgemeinen in der Walzebene liegen und parallel
zur Walzrichtung ihre größte Ausdehnung haben. Eine Vergrößerung der polygonisierten Elemente oder Subkörner durch
Koaleszenz, die eine Orientierungsbeziehung zur benachbarten Matrix durch eine gemeinsame Drehachse haben, führt zur
Bildung eines rekristallisierten Kornes. Die Orientierungsbeziehung steht eher in Zusammenhang mit der Verformung als
mit der Korngrenzenbeweglichkeit. Im anfänglichen Stadium
der Rekristallisation entstehen sowohl Keime der Würfellage
als auch solche anderer Orientierungen. Ein selektives
Wachstum dieser Körner auf Kosten der verformten Matrix bestimmt die Rekristallisationstextur.

3.5.1 Introduction

Recrystallization of a deformed metal involves the
nucleation and growth of new grains widely different in
orientation from the surrounding matrix. These new grains
grow at the expense of the matrix by high-angle boundary
migration, resulting in the formation of recrystallization
texture. During the past one-quarter of a century, many
authors have discussed the relative validity of the two

[+)] Edgar C. Bain Laboratory for Fundamental Research,
United States Steel Corporation, Monroeville, Pa., U.S.A.

mechanisms, <u>oriented nucleation</u> and <u>oriented growth</u>, for the formation of annealing textures. A detailed and critical review of the experimental data in connection with this topic was presented by BECK and HU [1] only three years ago. Further elaboration of the same subject in the present paper is unnecessary. It is more appropriate to provide new information on the general problem of texture formation in recrystallization.

Since recrystallization is a result of nucleation and growth, an ultimate understanding of recrystallization textures depends on an understanding of the nature and crystallographic aspects of both these processes. In connection with the growth of recrystallized grains, extensive studies have been made of the effects of orientation and solute elements on boundary mobility and growth selection. References to these works can be found in recent review articles and contributions [2-4]. Most of these studies were conducted under either variable or low driving forces. Growth anisotropy and the effect of driving force on boundary mobility and orientation selection have also been studied recently [5,6].

In contrast to these wide studies of the growth process, very little is known about the mechanism or the orientation aspects of nucleation. Understandably, the study of nucleation in recrystallization is extremely difficult, for nuclei are by nature small. However, since the advent of transmission electron microscopy the structure and orientation of very small domains can be examined in more detail. For example, it is now well known that cells form during deformation, and that the growth or coarsening of polygonized cells or subgrains leads to the formation of a recrystallized grain. Thus, not only is the classical process of nucleation by random thermal fluctuation ruled out [7-9], but also the concept of "invisible growth" of established nuclei is unnecessary. The main questions remaining to be resolved about nucleation are:

1. Do polygonized subgrains coarsen by simple <u>boundary migration</u> [10,11] or by a cooperative process such as <u>coalescense</u> [12,13] and

2. What are the crystallographic aspects of nucleation?

This paper presents results of a recent study of the origin of cube grains in heavily rolled copper during the early stages of recrystallization.

For a study of nucleation in a polycrystalline metal, the formation of cube-oriented recrystallized grains in copper is attractive, because the recrystallization texture of copper is remarkably sharp, and the cube orientation is one of the simplest to identify by diffraction. Various workers [14-18] have studied the recrystallization of heavily-rolled copper by electron transmission, using thin foils prepared parallel to the plane of the sheet. A serious flaw in these studies is that the structure of the as-rolled or partially annealed specimen is ill-defined. Frequently, a nearly dislocation-free recrystallized grain is observed amid a matrix poorly defined in structure, as shown in Fig. 1. The lack of a cell structure in highly

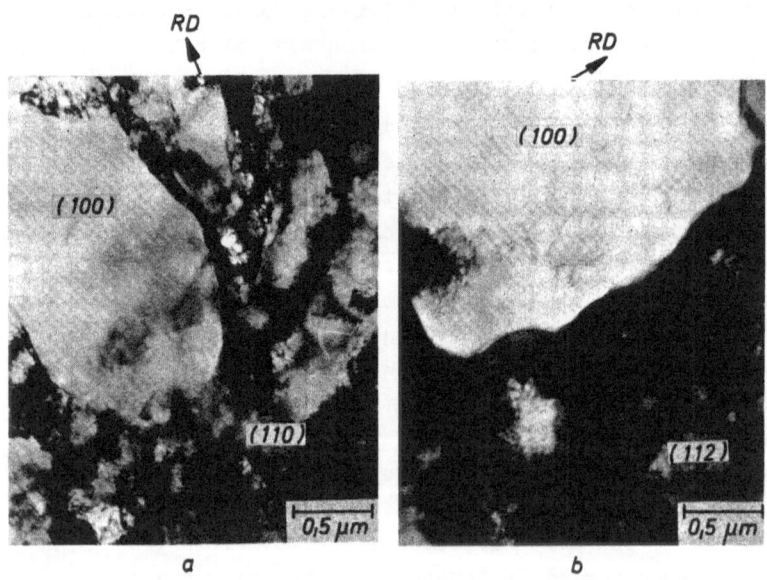

Fig.1a and b. Structure of partially recrystallized area. Electrolytic copper rolled 99.5%, annealed at 100°C for a) 25 min., and b) 625 min. Rolling plane section. Note cube grains formed near (110)[1$\bar{1}$2] and (112)[11$\bar{1}$] matrix with ill-defined structure

deformed copper was suspect; it could be due to the fact
that the average thickness of these cells was considerably
smaller than the thickness of the foil [19,20]. It has long
been known that cube texture fails to develop, in Ni-Fe for
example, if the rolled strip is etched to a very small
thickness (1 μm or so) prior to annealing [21]. The necessity
for examining the strip in cross-section was obvious. This
should not be a problem, as thin foils have been prepared
successfully from fine wires [22,23], and from the cross-
sections of heavily rolled sheet of low-carbon martensite[24]
by nickel plating the sample prior to sectioning and
thinning. In the more recent work [24] it was pointed out
that examinations on two or three principal cross-sections
of structurally anisotropic materials are necessary before
an interpretation of the structure is attempted.

3.5.2 Material and Technique

Part of the copper strips prepared for an earlier
investigation was used in the present study. The copper was
an electrolytic grade (99,92% pure), containing 0.025%
oxygen as the major impurity. It was processed to a uniform
penultimate grain size of 0.06 mm with nearly random
orientation before final cold rolling. The processing
details were described previously [25]. Two strips were
selected, one rolled 80% and the other 99%, both being
0,50 mm thick. Part of each was then rolled to 0.25mm thick,
thus adding two more levels of deformation, 90 and 99,5%
reduction. The purpose of the late additional rolling was
two-fold:

1. To provide both thin and thick specimens in very high
   and relatively low deformation ranges, and
2. To compensate for any recovery that might have occurred
   in the previously rolled strips.

However, it was found that both of these considerations
were insignificant.

Examinations of longitudinal and transverse cross-sections
of the strips were made possible by plating copper on both
faces of the strip to a total thickness of about 5mm, then

thinning slices cut from the plated sample in the usual manner. Information on the technique was given in an earlier note [26].

3.5.3 Results and Discussion

3.5.3.1 Structure as Revealed in the Cross-sections of the Strip

Fig.2 shows the structure of an "as-rolled" specimen of the strip rolled 99%. In contrast to the messy appearance of the matrix as seen in the rolling plane section of the

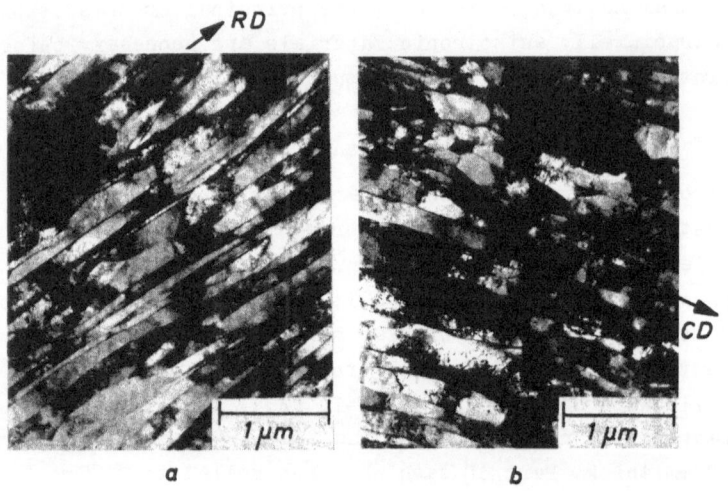

Fig.2 a and b. Structure of "as-rolled" specimen. Electro-
    lytic copper rolled 99%
        a) Longitudinal section
        b) Transverse section

strip (Fig.1), the structure of the longitudinal and transverse cross-sections consists of very narrow and elongated cells throughout the entire thickness of the strip. Some polygonization is evident, as electrolytic plating was conducted at 40 – 45°C for many hours. The heavily rolled copper, therefore, consists of very thin, nearly flat, elongated structural elements, like ribbons, mostly parallel

to the rolling plane of the strip and elongated in the
direction of rolling. However, large curvatures are very
common in local regions. These thin structural elements do
not individually represent the initial grains, as the
average thickness of the initial grains after rolling is
several times greater than the average thickness of the
elements. With such a cell structure, it is not surprising
that thin foils prepared parallel to the rolling plane of
the strip cannot reveal the details.

These thin, flat structural elements are developed by high
degrees of deformation. At smaller reductions, the cells
are coarser and more irregular, as shown in Fig.3, which
represents the structure of an "as-rolled" specimen of the
strip rolled 80%. There is apparently a reduction of cell
thickness with increasing deformation.

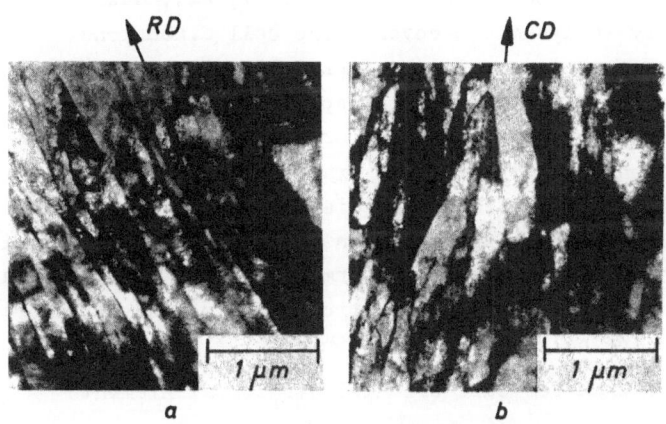

Fig.3 a and b. Structure of "as-rolled" specimen.
Electrolytic copper rolled 80%.
   a) Longitudinal section
   b) Transverse section

3.5.3.2 Cell Thickness and Frequency Distribution vs
Deformation

For the four strips used in the present investigation,
measurements of cell thickness were made over 200 structural
elements on several specimens of each strip. Fig.4 shows

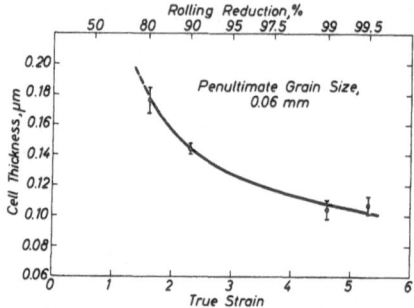

Fig.4. Cell thickness in rolled copper as a function
of deformation

the average cell thickness as a function of deformation.
The structural elements decrease gradually in thickness
with increasing deformation. From 99 to 99,5% reduction,
the decrease in cell size is practically nil, due
apparently to dynamic recovery. The cell dimensions
observed in the present investigation are somewhat finer at
comparable strains than those reported by EMBURY, KEH and
FISHER [23] for cold-drawn copper wires. This may be due
to differences in the initial grain size or purity of the
materials used. The frequency distribution of cell thickness
at the various rolling reductions is shown in Fig.5. It is
clear that with increasing deformation, more fine cells are
produced. At 99 or 99.5% reduction, most of the structural
elements are only 500 – 1000 Å thick.

3.5.3.3 Search for Cube-oriented Structural Elements

An extensive search was made for cube-oriented structural
elements in each thin foil. Using a small aperture, which
covered an area of approximately 1 μm in diameter, SAD
(selected-area-diffraction) patterns were taken at many
locations over the specimen. Diffraction patterns were also
examined continuously along and across elongated structural
elements by scanning the electron beam across the entire
thickness of the strip. Cube orientation was occasionally
observed in the less heavily deformed strips (rolled 80
and 90%). In the very heavily rolled strips (99 or 99.5%

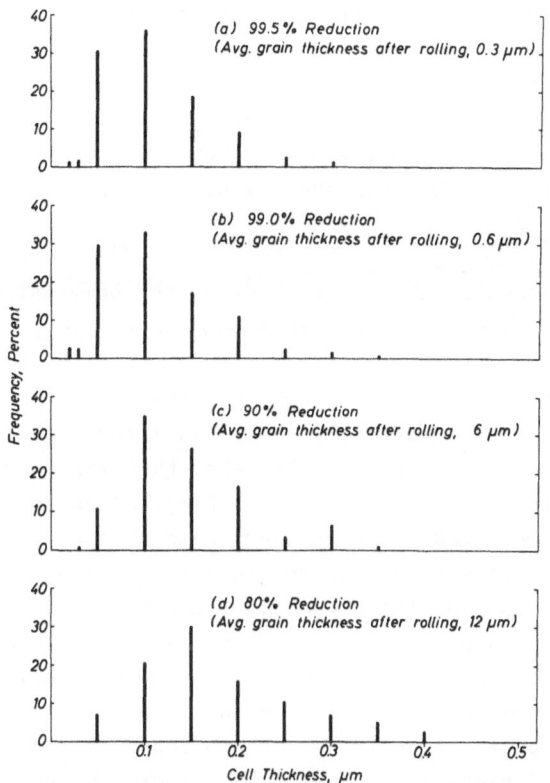

Fig.5. Frequency distribution of cell thickness in
rolled copper at various reductions

reduction), cube orientation was not detected[+]. This
observation is in agreement with the results reported by
two independent groups [27,28], who determined the deformation texture of copper rolled 95% by a large number of SAD
patterns taken from thin foils parallel to the rolling
plane of the strip. It was pointed out [1] that these
results did not necessarily prove the absence of cube-
oriented crystallites in heavily rolled copper. X-ray

[+] During continuous examination by scanning the electron
beam over an annealed specimen, cube pattern appears only
from local areas where subgrain coarsening has occurred,
or a cube grain has formed.

measurements [25] of the (200) intensity in the rolling
plane and in the plane perpendicular to the rolling
direction indicated detectable amount of cube orientation
in heavily rolled copper strips. In light of the present
observation, cube and nearly-cube crystallites would have
to be represented by cells or subgrains too small to be
detected by the SAD technique employed.

3.5.3.4 Structure Evolution During Nucleation[+)]

Evidence indicates that coarsening of the polygonized cells
or subgrains leads to the formation of a recrystallized
grain. The growth process is largely a sequence of
coalescence of subgrains rather than the migration of a   .
particular subgrain boundary. These features are quite
similar to those observed in the microband region of a
rolled silicon-iron single crystal [12]. As shown in Fig.
6a, the polygonized cells in the middle of the
photomicrograph have joined together through the
"disintegration" of the low-angle boundaries. At a somewhat
later stage, when the remaining dislocations of the
disintegrated cell boundary disappeared completely, the
boundary curvature would mislead one to believe that
coarsening of these cells was accomplished by boundary
migration. Structural features indicating coalescence
between coarsened subgrains are shown in Fig. 6b. A
recrystallization nucleus is thus being developed. BOURELIER
and MONTUELLE [18] showed similar features suggesting
coalescence among subgrains during recrystallization of
high-purity copper. The recrystallization nucleus shown
in Fig.6b, and many other similar nuclei observed, had a
non-cube orientation.

[+)] In the present investigation, the early stages of
recrystallization in strips rolled 99.5 and 99% were
studied more in detail than in those rolled 90 and 80%, as
the main interest was in the nucleation of cube grains.
Although several series of isothermally annealed specimens
were prepared, e.g., at 100°C for up to 625 min., at 150°C
up to 15 min. and at 175°C up to 5 min., one could always
find local regions where structural changes were more or
less advanced irrespective of the annealing treatment,
including the "as-rolled" specimen.

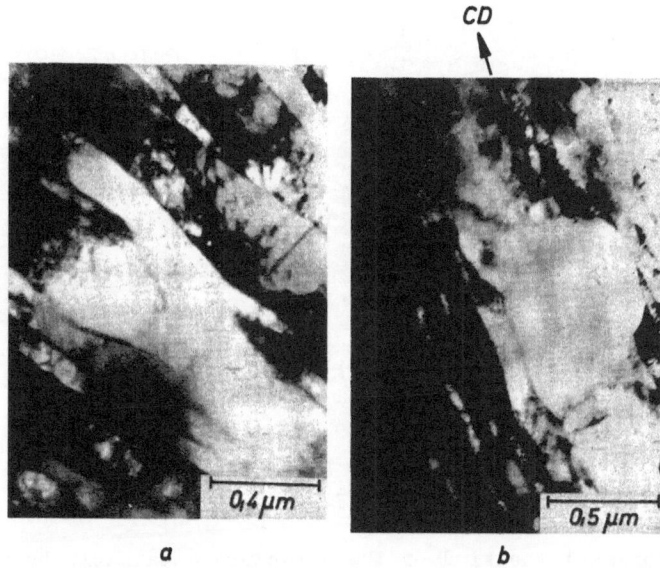

Fig.6 a and b. Structure evolution during nucleation.
Electrolytic copper rolled 99.5%, annealed 25 min. at 100°C
a) Polygonized cells or subgrains coarsen by coalescence.
Longitudinal section
b) Coarsened subgrains join by coalescence. Transverse
section

Fig.7 shows the microstructure and the SAD pattern of a
nearly cube-oriented recrystallized grain in a very early
stage of formation. A twin boundary is clearly visible.
The structure of the twin portion of the grain is worthy
of particular attention. As indicated by the structure of
the matrix immediately adjacent to the grain, the
recrystallization nucleus was developed in a region of very
fine and elongated structural elements. While the cube grain
proper has already developed to an advanced stage and the
mechanism of its formation cannot be deduced from its
internal structure[+], the twin portion of the grain does

[+] See further discussion about its formation in connection
with the twin in the next section.

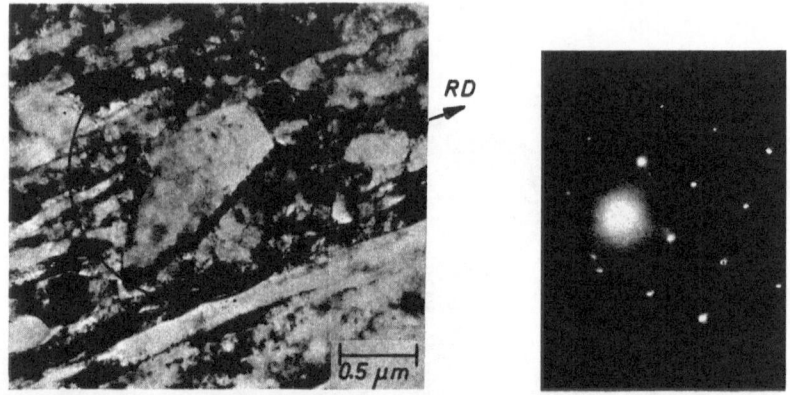

RD

Fig.7. A nearly cube-oriented grain and its twin during
recrystallization. Electrolytic copper rolled 99%.
Longitudinal section.

not represent the kind of the structure attainable by simple
migration of one cell boundary. The coalescence nature of
its formation by the merging of a number of cells and of
coarsened subgrains is clearly indicated.

3.5.3.5 Orientation Aspects of Nucleation near $(110)[1\bar{1}2]$
Matrix

Referring back to the microstructure shown in Fig. 7,
diffraction patterns were taken with various apertures. With
the smallest aperture on the recrystallized grain alone, the
pattern indicated a $\{100\}$ plane in the plane of the foil.
With an intermediate aperture to include the grain, its
twin and a portion of the matrix, as indicated by the circle,
extra spots appeared as shown. With a still larger aperture,
the pattern was essentially the same. The extra spots were
due to the twin and the matrix. These orientations are
plotted in a stereographic projection shown in Fig.8. If
the direction of cell elongation in this microscopic area
is considered to be exactly parallel to the rolling
direction (RD), the matrix orientation would be $(110)[1\bar{1}2]^{+)}$,

_____

+) All such expressions, (hkl) [uvw], used in the present
   paper, refer to rolling plane and rolling direction of
   the strip.

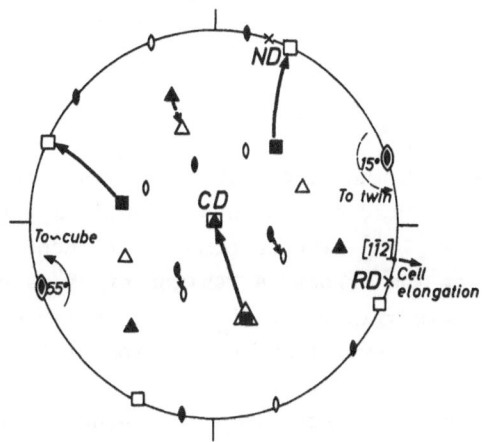

Fig.8. Stereographic projection showing orientation
relationship during nucleation near (110)[1$\bar{1}$2] matrix.
Filled symbols: ~ (110)[1$\bar{1}$2]. Open symbols: ~ (100)
[001]

and the grain orientation would be approximately (140) [4$\bar{1}$0],
which is about 14° away from (100) [001]. However,
considering the likely inaccuracy in assigning the rolling
direction according to local microstructure, this grain
could very well be only a few degrees away from the exact
cube orientation. It certainly should be within the spread
of the cube texture, particularly at an early stage of
recrystallization. On the other hand, if the grain is
assumed to be exactly in the (100)[001] orientation, the
adjacent matrix orientation would be approximately (134)
[4$\bar{3}$1], which may be considered as a spread from the (110)
[1$\bar{1}$2] main orientation. For nucleation studies, it is
necessary to realize the fact that local orientations can
rarely be expressed by simple "ideal orientations". This is
much more so in the study of nucleation than in the study
of textures, because nucleation always occurs in regions of
high curvature. The orientations of these regions are likely
to be irrational or high-indexed. For convenience in
representation, the symbols ~(110) [1$\bar{1}$2], ~(100) [001], etc.,
are used to indicate their approximate nature. The formation

of nearly cube-oriented grains adjacent to ~(110) [1T2] or
~(112) [11T] matrix (the latter will be discussed in a
following section) was most frequently observed, in
agreement with results reported recently by other
investigators [15,16].

In connection with the orientation relationship between the
grain and the matrix, one can see immediately from Fig. 8
that there is a [110] axis in common for these orientations.
As indicated by the arrows, a roation of 55° around this
common [110] axis will bring the matrix orientation to the
orientation of the grain; and a rotation of 15° in the
opposite direction will result in a twin orientation to the
grain. It can also be noted that the twinning plane agrees
with the twin trace in the microstructure[+]. This range of
orientation (70°) is spread out across 15 or 20 elongated
structural elements. The average disorientation between
adjacent elements is, therefore, about 4°. Since these
rotational relationships are in opposite directions, one
concludes that the matrix orientation above and below the
grain with its twin, as shown by the photomicrograph in
Fig.7, must be the same. This agrees with the observation
that the diffraction patterns taken with an intermediate
and a large apertures were the same. Among the fine
structural elements between the two matrix regions, there
must be a twin boundary. This implies that this annealing
twin was not produced by the so-called "growth accidents[29]"
during the migration of a grain boundary, as this twin
boundary already existed in the deformed metal. Owing to the
disorientation between the structural elements, and to the
general direction of the twin trace being at an angle to
the direction of cell elongation, this twin boundary must
be a distorted one in the deformed metal. To arrive at a
microstructure with features shown in Fig.7, some
cooperative orientational adjustments of the structural
elements would appear necessary.

[+] To bring the orientation plot, Fig.8, in exact correspondence with the microstructure, rotate the projection
28.5° counter-clockwise with respect to the micrograph
shown in Fig.7.

Difficulties in association with the formation of this cube
grain by cell-boundary migration is illustrated
schematically in Fig.9. Small arrows indicate the relative
orientations of the structural elements, or cells, before

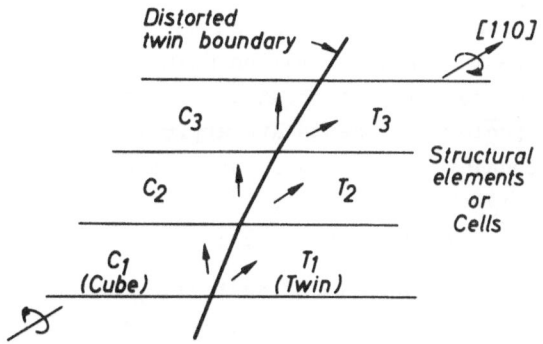

Fig.9. Schematic representation showing orientations
of structural elements and distorted twin boundary
before nucleation

nucleation. As a consequence of the disorientation between
the cells, the twin boundary will not be a smooth boundary
but distorted. Assuming C_1 is the cube-oriented cell, and
that it grows by boundary migration into cell C_2, the twin
boundary between C_2 and T_2 will be altered unfavorably
energetically because C_1 is not twin related to T_2. The
only way to avoid this difficulty will be for C_1 and T_1 to
grow simultaneously by migrating their boundaries into C_2
and T_2. The microstructure shown in Fig.7, however, does
not suggest that this was the case, because the internal
structures of the cube grain and its twin are distinctly
different. Accordingly, one must conclude that, although the
cube grain has already assumed a smooth appearance within
itself, its formation was, nonetheless, by a similar process
occurring at the stage in the twin portion of the crystal,
namely, coarsening of the cells by coalescence.

The most interesting finding in the orientation analysis
shown in Fig.8 is the significance of the [110] rotation
axis. This [110] axis corresponds to the slip direction of
the most favored slip systems for the orientations

concerned[+]). This would imply that screw dislocations were
most responsible for the disorientations produced between
the elongated structural elements – a condition most
favorable for coalescence to occur, as a twist boundary
can be untwisted without a transfer of matter. It is only
necessary to rearrange the atoms in the boundary [13].
However, since this [110] rotation axis is not in the normal
direction (ND) to which the boundaries between the
elongated structural elements are approximately perpendicular, these boundaries are, therefore, low-angle boundaries
of the mixed type. According to dislocation theory, such
boundaries are highly immobile [31].

3.5.3.6 Orientation Aspects of Nucleation near $(112)[11\bar{1}]$
Matrix

As mentioned earlier, a nearly cube-oriented grain adjacent
to an $\sim(112)$ $[11\bar{1}]$ matrix was frequently observed (e.g.
Fig. 1b). An analysis of the orientation relationship in
this case is shown in Fig. 10. To facilitate illustration,
the approximately common [110] axes were plotted to coincide;
the deviation is thus shown by the poles of the rolling
planes, or the normal directions (ND), and the rolling
directions (RD). If a mean position is assumed for these
directions, the deviation is only $\pm$ 5°. The characteristic
features are again similar to those in Fig.8. The near cube
and its twin orientations are related to the matrix
orientation by rotations around a common [110] axis, which
is the slip direction of the most favored slip systems.
VERBRAAK [32] also used this slip direction as a rotation
axis to describe the orientation spreads associated with
the $(112)[11\bar{1}]$ textures in rolled copper single crystals
and polycrystals. However, in the present case, the twin
orientations can be obtained by rotations of 20° from

[+]) For the $\sim(110)$ $[11\bar{2}]$ matrix it is the slip direction of
one of the duplex slip systems [30]. The Schmid factor
for these slip systems is 0.82, based on a biaxial stress
system. For the $\sim(100)$ [001] crystallites, it is the
slip direction of one of the four symmetrically oriented
and most stressed slip systems, having a Schmid factor
of 0.81.

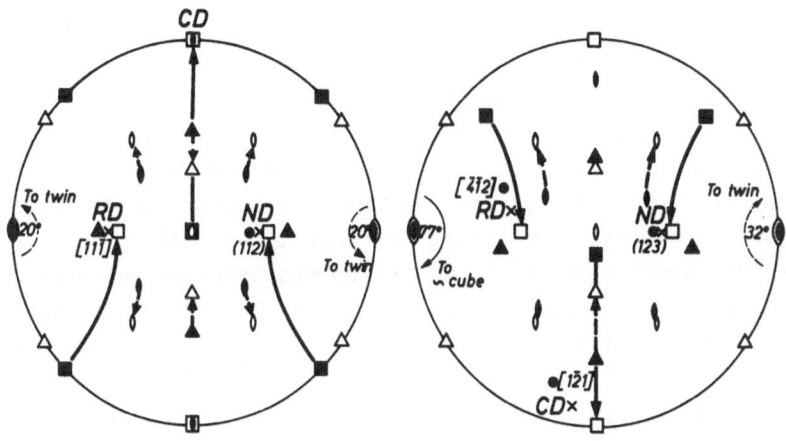

Fig.10. Stereographic
projection showing
orientation relationship
during nucleation near
(112) [11$\bar{1}$] matrix.
Filled symbols: ~ (112)
[11$\bar{1}$]. Open symbols:
~ (100) [001]

Fig.11. Stereographic
projection showing
orientation relationship
during nucleation near
(123) [$\bar{4}$$\bar{1}$2] matrix.
Filled symbols: ~ (123)
[$\bar{4}$$\bar{1}$2]. Open symbols:
~ (100)[001]

~(112) [11$\bar{1}$] in either direction. Beyond these, one can
reach the near cube orientations by further rotation.

It should be emphasized that these orientation analyses,
Fig.8 and 10, do not imply that nearly cube-oriented grains
can be nucleated w i t h i n a matrix of (110) [11$\bar{2}$] or
(112) [11$\bar{1}$] by such large-angle rotations. On the contrary,
they are probably nucleated in a region adjacent to these
matrices where the orientation spread is related to the
matrix orientation in the manner just described. The
orientation of recrystallized grains so nucleated would
not be limited to the cube orientation only; the whole range
of orientations should be possible. From this range of
orientations, the oriented-growth mechanism governs the
selection, hence the recrystallization texture. This conclusion was reached in an earlier study of reorientation
during recrystallization of titanium [33], based on X-ray
measurements.

3.5.3.7 Orientation Aspects of Nucleation near (123)[4̄1̄2] Matrix

Cube grains nucleated adjacent to a (123) [4̄1̄2] matrix were not observed during this investigation. Neither were such cases reported in the work of HINKEL, et al. [16]. An orientation analysis of this case is shown in Fig.11. It is seen that by plotting the [110] slip directions of the most favored slip systems in both crystals to exactly coincide, the deviation between the poles representing the rolling and the transverse, or cross directions (CD) is very large ($\pm$ 10° or 11° from the mean position), although that in the ND is only $\pm$ 5°. For the range of orientations described by such [110] rotations, even the one closest to the cube orientation will not be recognized as such in the SAD pattern. This may serve as an interpretation for the off-cube orientations observed by LIU [34] and by VERBRAAK [32] in the recrystallization texture of copper single crystals rolled in the (358) [5̄2̄3] orientation, which is very near to the (123) [4̄1̄2] orientation. Among the range of orientations provided, those closer to the off-cube orientations would be more favorably oriented for growth, as they have a [111] axis approximately in common with the matrix [1]. From Fig.11 one sees also that with a [110] rotational relationship, the ~(123) [4̄1̄2] and ~(100) [001] orientations are nearly twin-related, off by only 6°.

3.5.3.8 Other Possible Nucleation Sites for Non-cube Grains

Since the primary interest of the present work was limited to the origin of cube grains, very little information was collected in connection with the formation of n o n -cube grains. The possibility for nucleation of non-cube grains is evidently present even in regions where a [110] rotational relationship exists with an adjacent matrix. Fig.12 is one of the examples observed during the examination of orientation variations in high-curvature regions. As shown in the micrograph, from location 1 to location 2 there is a large change in structure. The orientation changes accordingly, as shown by the SAD patterns. The assignment of the true rolling direction

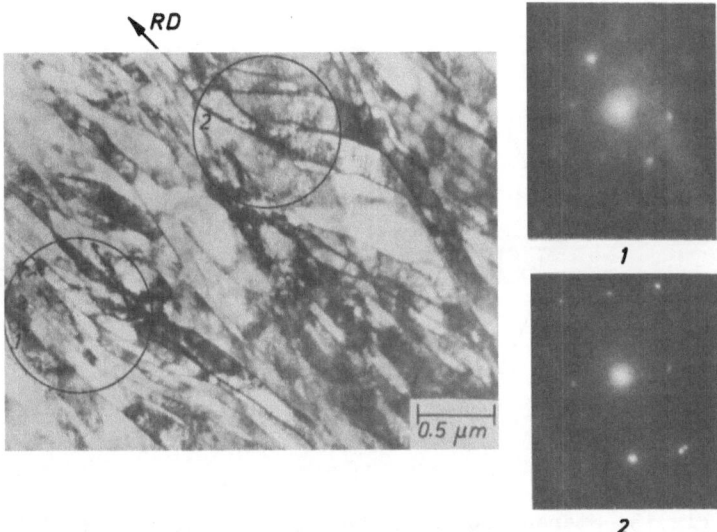

Fig. 12. Structure and orientation variation. Electrolytic
copper rolled 90%. Longitudinal section

from this micrograph is rather difficult. However, if a
compromise RD is chosen as indicated, the orientation of
location 1 is ~(110) [001] and that of location 2 is ~(123)
[4̄1̄2]. These orientations can be related by rotations
around an approximately common [110] axis, as shown in
Fig. 13. This common [110] axis, again, corresponds to the
slip direction of the most favored slip systems in both
crystals. With this orientation relationship, one cannot
bring the poles to even an approximate cube position by any
amount of rotation. It is interesting, but puzzling, to note
that cube grains are favorably oriented for growth in all
four crystallographically equivalent (123) [4̄1̄2] components
of the deformation texture, but the nucleation of cube grains
appears to be associated with this deformation texture only
very remotely. This, if true, seems to suggest that
nucleation and growth are basically different processes.
While growth is controlled by high-angle boundary mobility,
nucleation is closely related to the mechanism of
deformation.

218

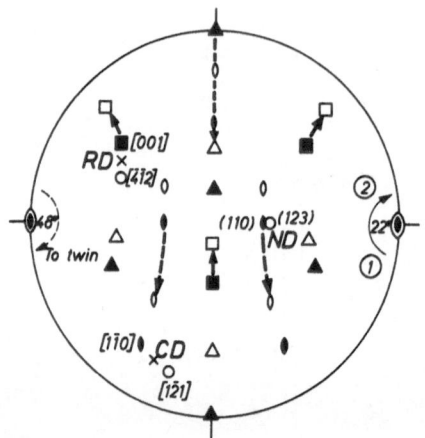

Fig.13. Stereographic projection showing orientation relationship between locations 1 and 2 in Fig.12. Filled symbols: 1 ~ (110) [001]. Open symbols: 2 ~ (123) [4̄1̄2]

During the present investigation, many non-cube grains[+] were observed. In fact, at very early stages of recrystallization, there were more non-cube grains than cube grains. This is in contrast to the observations of HINKEL et al. [16], who reported that 95% of the grains formed during recrystallization in high purity (99.999% pure) polycrystalline copper rolled 99% had the cube orientation. These authors examined thin foils of the rolling plane section only. Consequently, the early stages of nucleation could not be observed, because of the highly ill-defined structure. With progressive recrystallization, cube grains would gain the advantage for growth over non-cube grains as they come in contact with the structural elements corresponding to other deformation texture components of the matrix [1,35,36]. The frequency of cube grains would thus become higher in regions where recrystallization is more

---

[+] Understandably, these may include grains having off-cube orientations, when the deviation is sufficient to yield a non-cube SAD pattern.

advanced, as can be seen in Fig.14. In connection with this
micrograph, the orientation of one of the large grains was
not positively identified. Its SAD pattern showed one row
of spots identical with the corresponding row of spots for
the other two cube grains. It is, therefore, very likely
that this grain had a nearly cube orientation.

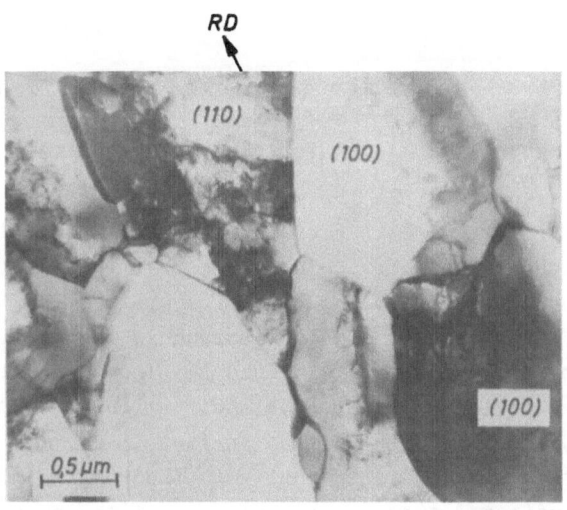

Fig.14. Structure of partially recrystallized area
showing several cube grains formed near (110) [1Ī2]
matrix. Electrolytic copper rolled 99.5%, annealed
625 min. at 100°C. Rolling plane section

3.5.3.9 A Schematic Model for Cube Grain Nucleation

Fig.15 is a schematic representation of the orientation
spread across a series of structural elements. The individual
structural elements are represented by the short rods; and
the rod axis represents the [001] direction. Each successive
element is thus disoriented a few degrees by rotation around
the common [110] axis in the same direction. The vertical
position represents the exact cube orientation. In the less
heavily rolled strips, the cube-oriented material has not
been fragmented by deformation, hence the probability for
its detection is high. With increasing deformation, it
breaks up into finer structural elements as a result of

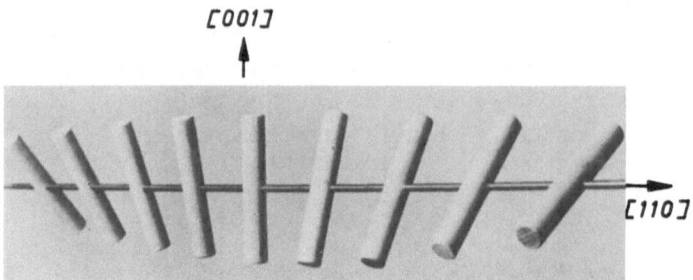

Fig.15. A schematic model for cube-grain nucleation

change in orientation. As the disorientation between adjacent elements may be as high as 4 or 5°, the cube-oriented, together with its neighboring nearly cube-oriented elements, could escape detection by selected-area diffraction. On the other hand, since the orientation spread is represented by the discrete orientations of individual structural elements, it will be a matter of chance for one particular element to be in the exact cube orientation. In fact, X-ray measurements of the (200) intensity in the rolling plane of the sheet indicated a continuous decrease with increasing deformation [25]. According to the subgrain coalescence mechanism for nucleation, cube grains can be formed even without the presence of a cube-oriented structural element.

Based on the critical-size requirement for growth by boundary migration, it is difficult to rationalize that very small subgrains are energetically more favored for growth than large ones. Furthermore, according to the theory of dislocations, low-angle boundaries of the general type should be highly immobile [31]. Experimental observations indicated that even for simple tilt boundaries the mobility decreases with increasing disorientation [37]. There have been numerous examples showing that boundaries of very low mobilities are difficult to move even in the presence of adequate driving forces [2,38-40]. For cube-texture formation in copper, it is also difficult to explain on the basis of low-angle boundary migration why cube texture is

less pronounced in strips rolled less heavily, where cube-
oriented structural elements are coarser and more abundant
than in strips rolled very heavily (see next section).

The orientation relationship during nucleation as observed
in the present investigation is, in essence, a confirmation
of the nucleation model proposed by BURGERS [41,42] for
deformed aluminum single crystals. According to his model,
nuclei were developed from the most heavily deformed
fragments in the vicinity of glide planes. These fragments
were rotated with respect to the main body of the crystal
around a [112] axis lying in the active slip plane and
perpendicular to the active slip direction (disorientations
due to edge dislocations), or around a [111] axis oblique
to the slip plane (disorientations due to edge and screw
dislocations). The present observation on heavily rolled
copper indicates a rotational relationship around a [110]
axis, corresponding to the most favored slip direction.
Hence, screw dislocations would be responsible for the
disorientations produced. However, as pointed out earlier,
both edge and screw dislocations would be required to
account for the disorientations subtended by the elongated
structural elements, as the rotation axis is not normal to
the plane of the boundary between these structural elements.
Consequently, these boundaries are of the general type, i.e.,
they contain both edge and screw dislocations.

Based on the findings of the present investigation a main
deviation from Burgers' nucleation theory is that a full
range of orientations is represented by the structural
elements where nucleation by subgrain coarsening through
coalescence may occur. From these nuclei having a range of
orientations, selective growth determines the
recrystallization texture. The orientation relationship in
nucleation is intimately connected with deformation, but is
not necessarily the same for favorable growth by boundary
migration.

In connection with the "martensitic shear" mechanism for
cube grain nucleation in copper as proposed by BURGERS and
VERBRAAK [43], a full discussion has been given in an

earlier publication [1]. The observation that cube or nearly
cube grains are frequently nucleated adjacent to an ~(110)
[1̄12] matrix, or in a region between two ~(110) [1̄12]
matrices as seen in Fig. 7, is inconsistent with the specific
model (martensitic shear between two twin-related (112)[111]
orientations) proposed by these authors.

3.5.3.10 Recrystallization Texture of the Strip

Fig. 16 shows transmission Laue patterns of the
recrystallized strips after annealing at 300°C for 15 min.

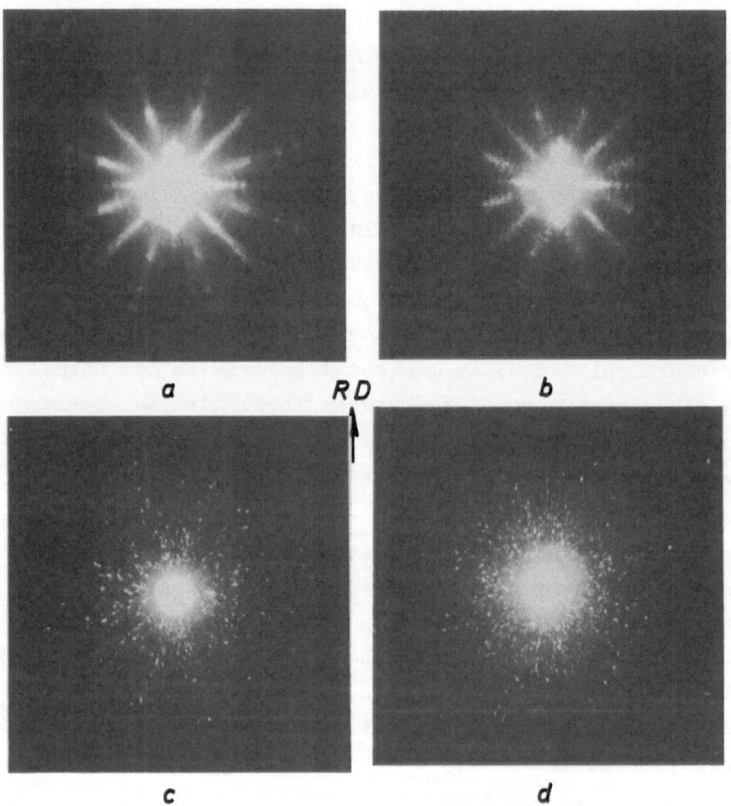

Fig.16 a – d. Transmission Laue patterns showing recrystallization texture of electrolytic copper rolled a) 99.5%,
b) 99%, c) 90%, and d) 80%, after annealing at 300°C for
15 min

The amount and sharpness of the cube texture increase with prior rolling reduction as expected. There is very little difference between the cube textures of the strips rolled 99.5 and 99%. This is consistent with the similarity of their cell structure (Figs. 4 and 5). The recrystallized grains, however, appears to be finer in the more heavily rolled strip, suggesting a higher rate of nucleation with increasing deformation. It is clear that the amount and the size of cube-oriented structural elements in the deformed metal are irrelevant to cube texture formation. This well-known fact cannot be explained on the basis of oriented-nucleation for annealing texture formation, and of subgrain growth by boundary migration for nucleation of recrystallized grains.

3.5.4 Summary and Conclusions

The nucleation of cube-oriented grains during recrystallization of heavily rolled copper has been studied in detail by application of transmission electron microscopy to cross-sections of the strip. The structure of heavily-rolled strips of copper consists of very thin, nearly flat, and elongated elements, like ribbons, mostly parallel to the rolling plane and elongated in the direction of rolling. The thickness of these structural elements decreases with increasing deformation. At 99% or higher reductions, most of these structural elements are only 500 to 1000 Å thick. In the "as-rolled" or partially annealed specimens, cube-oriented elements were occasionally found in strips rolled 80 and 90%. In strips rolled 99 and 99,5%, cube orientation was not detected in extensive surveys by selected area diffraction along and across the elongated structural elements over the entire thickness of the strip. Although this does not necessarily indicate the absence of cube orientation, its volume fraction would certainly n o t increase, but decrease with increasing deformation; and its presence would have to be represented by very small structural elements. It is difficult to rationalize that these very small polygonized structural elements or subgrains are energetically more favored for growth by boundary migration over the large ones. Evidence indicates that coarsening of polygonized subgrains leads to the formation of a recrystallized grain, and such coarsening occurs largely by coalescence of subgrains. The nucleation of a cube grain, or a grain of any other orientation within the spread subtended by these subgrains, can be achieved by coalescence without the actual presence of that particular orientation. During early stages of recrystallization, both cube and non-cube grains are nucleated. The nucleation of a recrystallized grain appears to be related in orientation to the adjacent matrix. This orientation relationship has a close connection with the deformation process rather than with boundary mobility. Consequently, the orientation of recrystallization nuclei would have to be within the range of orientations produced by deformation. Once a boundary of sufficient angle of misfit, hence mobility, is attained, growth competition will come into play. The formation of recrystallization texture will thus be controlled continuously by the mechanism of oriented growth.

While the present results indicate the existence of an orientation relationship for nucleation which appears to prevail in the examples encountered, the actual situation in the recrystallization of a highly deformed metal is undoubtedly more complicated. Much extensive investigations, such as a statistical study of the correlation between microstructure and orientation, the orientation distribution of recrystallization nuclei, and the orientation relationship between nuclei and their adjacent matrix are needed before a complete understanding of the nucleation process in recrystallization and its role in texture formation can be obtained.

Acknowledgment

This work was initiated and completed while the author was visiting at the Institut für Metallkunde und Metallphysik, Technische Universität Clausthal during 1967–1968. He wants to thank Professor Dr. G. Wassermann for his interest and encouragement, and many members of the Institut, in particular, Dr. H. Ahlborn and Dipl.-Ing. D. Sauer for their assistance during the course of this work. The skillful preparation of the thin foils for electron microscopic examinations by Miss C. Pommerehne is greatly appreciated.

References

1. P.A. Beck and H.Hu in: Recrystallization, Grain Growth and Textures, H. Margolin, Ed., Amer. Soc. Metals, Metals Park, Ohio 1966, p. 393.

2. K.T. Aust and J.W. Rutter in: Recovery and Recrystallization of Metals, L. Himmel, Ed.,Interscience, New York 1963, p. 131.

3. P. Gordon and R.A. Vandermeer in: Recrystallization, Grain Growth and Textures, H. Margolin, Ed., Amer.Soc. Metals, Metals Park,Ohio 1966, p. 205.

4. G. Ibe and K. Lücke in: Recrystallization, Grain Growth and Textures, H. Margolin, Ed., Amer.Soc. Metals, Metals Park, Ohio 1966, p. 434.

5. B.B. Rath and H. Hu: Trans.Met.Soc.AIME 236(1966)1193.

6. B.B. Rath and H. Hu: To be published.

7. E. Orowan: Dislocations in Metals, AIME, New York 1953, p. 183.

8. R.A. Oriani: Acta Met. 8(1960)134.

9. J.E. Bailey: Phil.Mag.,5(1960)833.

10. R.W. Cahn: Proc.Phys.Soc. (London) A 63(1950)323.

11. J.L. Walter and G.F.Koch: Acta Met. 11(1963)923.

12. H.Hu in: Recovery and Recrystallization of Metals, L. Himmel, Ed., Interscience, New York 1963, p. 311.

13. J.C.M. Li: J. Appl. Phys. 33(1962)2958.

14. E. Votava: Acta Met. 9(1961)870.

15. Sh. Horiuchi, T. Okubo and J. Gokyu: Trans. Japan Inst. Met. 7(1966)257.

16. H. Hinkel, G. Hasse and F. Granzer: Acta Met. 15(1967) 1875.

17. M. Hatherly and K. Brown: J. Austr.Inst.Met. 11(1966) 264.

18. F. Bourelier and J. Montuelle: Mém. Scient. Rev. Mét. 65(1968)65.

19. H.Hu: Trans.Met.Soc. AIME 230(1964)572.

20. H. Hu and R.S. Cline: Trans. Met. Soc. AIME 242(1968) 1013.

21. J.F.H. Custers and W.G. Rathenau: Physica, 8(1941)759.

22. R.C. Glenn and W.R. Duff: Trans. Amer. Soc. Metals 58(1965)428.

23. J.D. Embury, A.S. Keh and R.M. Fisher: Trans.Met.Soc. AIME 236(1966)1252.

24. R.C. Glenn, G. Langford and A.S. Keh: Submitted to Amer. Soc. Metals.

25. S.R. Goodman and H.Hu: Trans.Met. Soc. AIME 242(1968)88.

26. H. Hu: Submitted to Z. Metallkde.

27. K. Lücke, H. Perlwitz and W. Pitsch: phys.stat.sol. 7(1964)733.

28. F. Haeßner, U. Jakubowski and M. Wilkens: phys.stat.sol. 7(1964)701 - Mat.Sci. Eng. 1(1966)30.

29. **J.E. Burke and D. Turnbull** in: **Progr.Chalmers' Met. Phys.** 3(1952)220.

30. H. Hu, R. S. Cline and S.R. Goodman in: Recrystallization Grain Growth and Textures, H. Margolin, Ed., Amer.Soc. Metals, Metals Park, Ohio 1966, p. 295.

31. A.H. Cottrell, Dislocation and Plastic Flow in Crystals, Clarendon Press, Oxford 1953, p. 188.

32. C.A. Verbraak: Acta Met. 6(1958)580 - Thesis Technische Hogeschool, Delft, Holland 1959.

33. H. Hu and R.S. Cline: Trans Met. Soc. AIME 242(1968)1013.

34. Y.C. Liu: Trans. AIME 9(1957)836.

35. P.A. Beck: Phil.Mag.Suppl. 3(1954)245.

36. I.L. Dillamore: Acta Met. 12(1964)1005.

37. D.W. Bainbridge, C.H. Li and E.H. Edwards: Acta Met. 2(1954)322.

38. T.J. Tiedema, W. May and G. Burgers: Acta Cryst. 2 (1949)151.

39. P. Lacombe and A. Berghezan: Métaux & Corrosion 24(1949)1.

40. C.D. Graham, Jr., and R.W. Cahn: Trans.AIME 206(1956) 517.

41. W.G. Burgers and P.C. Louwerse: Z. Phys. 61(1931)605.

42. W.G. Burgers and T.J. Tiedema: Proc.Kon.Ned-Akad.v. Wet. 53(1950)1525.

43. W.G. Burgers and C.A. Verbraak: Acta Met. 5(1957)765.

3.6 Secondary Recrystallization
Sekundäre Rekristallisation

by J. L. Walter[+)]

Zusammenfassung

Unter sekundärer Rekristallisation versteht man das Wachstum einiger weniger Körner auf Kosten einer stabilen Matrix
aus feinen Körnern. Die Steuerung von Korngröße und Orientierung (Textur) durch sekundäre Rekristallisation hat große
praktische Bedeutung.

Der Beitrag behandelt den heutigen Stand des Wissens über
die sekundäre Rekristallisation und berücksichtigt besonders
die dabei entstehenden Texturen (kubisch flächenzentrierter
und kubisch raumzentrierter Werkstoffe) und ihre möglichen
Beziehungen zur primären Rekristallisationstextur.

Einige Betrachtungen werden auch über die primäre Rekristallisation selbst angestellt, weil Korngröße, Textur und
Dispersion zweiter Phasen von ihr abhängen. Außerdem spielt
die Stabilität der primär rekristallisierten Matrix eine
Rolle. Besondere Bedeutung hat bei der sekundären Rekristallisation die treibende Kraft. Es wird hier vor allem
die durch die Unterschiede in den Grenzflächenenergien
Gas-Metall entstehende besprochen, weil sie bei hochreinen
Materialien für die Texturbildung wesentlich ist.

3.6.1 Introduction

Secondary recrystallization is the discontinuous growth of
certain grains at the expense of the smaller matrix grains
surrounding them. Such growth often results in formation
of a preferred orientation that differs from any preferred
orientation in the matrix. Secondary recrystallization is
interesting for that which is still unknown about it and
for its industrial applications, with particular reference

[+)] General Electric Company, Metallurgy and Ceramics
Laboratory, Schenectady, N.Y., USA

to iron-silicon alloys. In this case, the preferred
orientation brought forth by secondary recrystallization
imparts desirable magnetic properties. In other cases, the
growth of large grains by secondary recrystallization imparts
poor mechanical properties and is to be avoided.

3.6.2 Grain Boundary Migration

The conditions for secondary recrystallization are attained
by normal grain growth. The rate of grain boundary
migration during grain growth (normal as well as secondary)
may be expressed by an equation relating velocity of
migration, V, to grain boundary mobility, M, and the
driving force, P. The driving force is expressed in terms
of the two principal radii of curvature, ρ_1 and ρ_2, and the
grain boundary energy γ_b. In sheet material, differences in
the gas-metal interfacial energies between two grains at the
sheet surface ($\Delta\gamma_s = \gamma_{s_1} - \gamma_{s_2}$) may add to the driving force
for migration. In addition, the driving force may be
reduced by the drag exerted by the grove formed where the
boundary intersects the surface [1]. The drag is expressed
as θ_c, the angle between the grain boundary and the normal
to the surface. θ_c is related to the ratio γ_b/γ_s. So, for
sheet specimens of thickness a, the velocity for boundary
migration is

$$V = M \left(\gamma_b/\rho_1 + 2\Delta\gamma_s/a - 2\theta_c\gamma_b/a \right). \tag{1}$$

θ_c replaces ρ_2, the curvature in the radial cross section.

3.6.3 Normal Grain Growth

HILLERT [2] substituted grain radii for radii of curvature
and obtained the relation

$$V = \alpha M\gamma_b / 4R_A \tag{2}$$

where R_A is the average grain radius. The term α is a
geometrical factor approximately equal to 1 for a three-
dimensional grain structure and 1/2 for a two-dimensional
grain structure. In steady state normal growth, the
diameter of the largest grains would be about 1.5 times the

average grain diameter.

Fig.1 relates velocity of grain growth (velocity parameter) to the grain size parameter S = 2R/a [3]. In a three-dimensional system, one where the grain diameters are less than the sample thickness, the velocity decreases as the

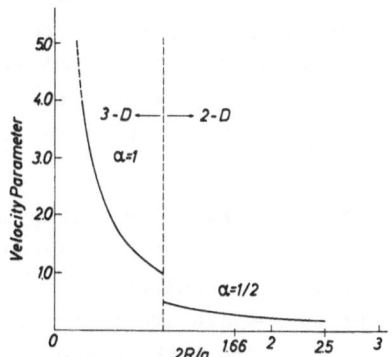

Fig.1. Rate of normal grain growth (velocity parameter) versus grain size parameter (after DUNN and WALTER [3].

grain size increases. When the grain diameter equals the sample thickness, there is a substantial reduction in growth velocity for two reasons. First, the geometrical shape factor α in eq. 2 becomes 1/2 which decreases the velocity by a factor of 2. Secondly, the boundary grooves formed at the surface further reduce the velocity by decreasing the driving force (the θ_c term in eq. 1). The sudden decrease in rate of grain growth on reaching the two-dimensional configuration describes a "thickness effect" [4]. In a system free of second phases, the limiting grain diameter will be not greater than 2.5 times the sheet thickness, as indicated in Fig.1.

3.6.4 Normal Grain Growth in Presence of Second Phase

When second phase particles are present, the driving force is reduced by the grain boundary-particle interaction. The particle drag term is [5]

$$3f\gamma_b / 4r_o \ \text{ or } \ Z\gamma_b$$

giving
$$V = M[\alpha\gamma_b / 4 \ R_A - Z\gamma_b / 2], \tag{3}$$

The inclusion of the $Z\gamma_b$ term in Fig.1 also causes a reduction in the velocity of grain growth depending on the volume fraction of second phase present. In the case of two-dimensional system, addition of a particle drag term of $Za = 0.1$ reduces the velocity of grain growth to zero at a limiting grain diameter of about 1.67 times the sheet thickness.

Thus, normal grain growth does not continue indefinitely but ceases when the grains are slightly larger than the sheet thickness. Also, the effect of the second phase particles is greater as the average grain sizes increases. Continuing growth requires that the effectiveness of the second phase particles be diminished, thus increasing the limiting size.

3.6.5 Secondary Recrystallization

Secondary recrystallization occurs only when the matrix grains remain relatively constant in size during the period when the secondary grains are growing. If the matrix grains grow during this period the driving force for growth of the secondaries may be decreased depending on the rate of growth of the primary grains.

The necessary stable matrix structure for secondary recrystallization is obtained either by normal grain growth to a limiting grain size or by primary recrystallization. In the latter case, the grain size may equal or exceed the limit for normal grain growth depending on the frequency of nucleation of primary grains. Or, if the primary recrystallization texture is a strong single-orientation texture, there may be little growth of the matrix grains during normal growth or secondary recrystallization.

Table 1 gives some calculated values of driving force for

Table 1. Driving Force for Growth of a Large Secondary Due Only to Grain Boundary Energy and a Zener Drag (a)

Material	Thickness, cm	Matrix type	R_M, cm	Z, cm^{-1}	P(b)	P	P, erg-cm^{-3}
Pt	0.0057	2-D	0.0091	20	$(110 - Z)\gamma_b$	$90\,\gamma_b$	43,000
Pt	0.010	2-D	0.016	20	$(62 - Z)\gamma_b$	$42\,\gamma_b$	20,000
Pt	0.020	2-D	0.032	20	$(31 - Z)\gamma_b$	$11\,\gamma_b$	5,300
3 % Si-Fe	0.0025	2-D	0.0088	-	-	$114\,\gamma_b$	68,000
3 % Si-Fe	0.03	2-D	0.034	14	$(29,5 - Z)\gamma_b$	$15.5\,\gamma_b$	9,300
Ag	0.05	3-D	0.001	$1/R_M$	$(2000 - 1000)\gamma_b$	$1000\,\gamma_b$	300,000
3 1/4 % Si-Fe	0.036	3-D	0.002	500	$(1000 - 500)\gamma_b$	$500\,\gamma_b$	300,000
3 % Si-Fe	0.036	3-D	0.0025	400	$(800 - 400)\gamma_b$	$400\,\gamma_b$	240,000

(a) $\Delta\gamma_s$ and θ_c terms are omitted.

(b) Values for γ_b are taken as approximately equal to 480 erg-cm^{-2} for Pt (17), 600 erg-cm^{-2} for 3 % Si-iron (17) and 300 erg-cm^{-2} for Ag

growth of secondary grains due only to grain boundary
energy and the drag caused by second phase particles [3].

In the two dimensional case the drag force, $Z\gamma_b$, is
relatively small, and because the matrix grains have
diameters greater than the sheet thickness, the driving
forces from grain boundary energy are also small compared
with the driving forces for growth of a secondary into a
three-dimensional matrix.

The rate of growth of a secondary depends on the driving
force and mobility of the boundary. The driving force
depends on the energy of the boundaries between the matrix
grains and the secondary (γ_{s-m}) and the boundary curvature.
The curvature depends on a) the relative size of the
secondary grain with respect to the matrix grains and b)
the ratio $\gamma_{m-m} / \gamma_{s-m}$ where γ_{m-m} is the average energy of the
matrix grain boundaries. The mobility of the boundary
depends on its angle of disorientation and on the possible
concentration of solute atoms in the boundary.

Thus, in a three-dimensional system, with a weak matrix
texture, there will be a high average angle of disorientation, high average mobility and high average driving force
for the secondary. Under these conditions, a potential
secondary could grow if it were at least twice the size
of the matrix grains. If the matrix texture is a strong
single-orientation texture with a low average grain
boundary energy, then, for growth of the secondary to occur,
$R_{sec} > \dfrac{\gamma'_b}{\overline{\gamma}_b}$ R_m where γ'_b is the energy of the boundary

between the matrix grains and the secondary grain and $\overline{\gamma}_b$
is the average matrix grain boundary energy. The sharper
the texture (the lower $\overline{\gamma}_b$) the larger must be the minimum
size of the potential secondary for growth to occur. The
orientation for highest driving force is the average
orientation of the matrix but since the mobility is probably
quite low for such an orientation, the rate of growth of
any potential secondary will be small. Thus, although the
driving force may be lower, a potential secondary must be
in deviating orientation for growth to occur.

In the case of the two-dimensional grain structure, differences in surface energy become important. An additional term, $\Delta\gamma_s = \bar{\gamma}_s - \gamma_s'$, must be added to the driving force. $\bar{\gamma}_s$ is the average surface energy of the matrix grains and γ_s' is the surface energy of a potential or actual secondary and varies according to the crystallographic orientation of the surface plane. The relative contributions of surface and grain boundary energies to the driving force of a secondary grain in a weak-texture matrix is

$$\frac{\text{G.b. energy}}{\text{Surface energy}} = \frac{\dfrac{\gamma_b}{R_m} - \dfrac{\gamma_b}{R}}{2\,\Delta\gamma_s\,/\,a} \tag{4}$$

where a is the thickness of the material [3].

The ratio increases as the secondary grain size increases; that is, the driving force from the surface energy difference is most important when the secondary grain is just slightly larger than the matrix grains. Also, if the ratio of matrix grain size to thickness remains constant, the ratio of energies is independent of thickness.

3.6.6 Illustrations of Secondary Recrystallization

3.6.6.1 Secondary Recrystallization in a Stable Two-Dimensional Matrix

We shall first examine some illustrations of secondary recrystallization and the textures obtained in a stable two-dimensional matrix. In this case, it must be remembered, the differences in surface energy between specifically oriented grains and the matrix grains may act as a driving force for boundary migration.

McLEAN and MYKURA [6] observed secondary recrystallization in thin sheets of high purity platinum and measured the macroscopic grain boundary velocity. The average radius of the matrix grains was 1.6 times the sheet thickness prior to secondary recrystallization. Values of $\gamma_b / \gamma_s = 0.24$ were

234

obtained from measurements of the equilibrium groove angles
of grain and twin boundaries at the surface. Effective
driving forces of 48,000 ergs/cm^3 were calculated with
impurity drag values of 8,000 to 10,000 ergs/cm^3.

The orientations of the secondary grains are shown in Fig.2
plotted as normals to the rolling plane [6]. The texture

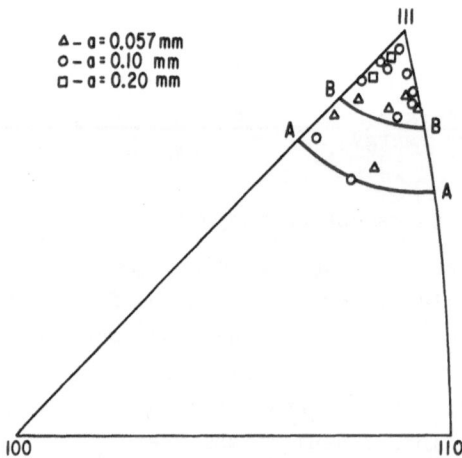

Fig.2. Orientation of secondaries in platinum sheet;
a = sheet thickness (after McLEAN and MYKURA [6])

is near (111) with considerable spread. The secondaries
grew into a matrix with only slight preferred orientation.
The arc AA indicates the limiting deviation from exactly
(111) to be expected if, in addition to the surface
energy driving force, a θ_c term is included. If the particle
drag value is added in, further reducing the effective
driving force, the limiting deviation is as indicated by the
arc BB. The greatest difference in surface energy and, hence,
the largest driving force; is obtained for those grains with
(111) planes precisely parallel to the plane of the sample.
If there are no retarding forces, then grains farther from
the ideal (111), those with smaller surface energy
differences, will also grow. From this description of the

relationship between driving force and crystallographic orientation, it is possible to deduce some information concerning the primary matrix. For instance, examination of the orientations of the secondaries in the thinnest sheet shows no secondaries near (111). But, these would be the grains with the highest driving force. Therefore, it is concluded that there were no primary grains with orientation close to (111) in the 0.057 mm thick sheet.

3.6.6.2 Secondary Recrystallization in High-Purity bcc **Metals**

In all of the cases of secondary recrystallization in high-purity bcc materials, the matrix grain structure is two-dimensional with negligible effect of impurities on rates of growth of secondaries [7 - 10]. The two-dimensional grain structure makes it possible to obtain information concerning the orientations of the matrix grains, particularly in the thicker sheets. Thus, the precise correlation between the orientations of the secondaries and the matrix texture may easily be seen.

A given matrix may have more than one set of potential secondary grains depending upon specimen composition, annealing temperature, and annealing atmosphere. In 3% silicon iron there are three potential sets of secondaries in the primary matrix; one set has the (100) plane parallel to the rolling plane [7, 10-12]. The second set has (110) parallel to the rolling plane [8,9,12,13] and, recently reported, a set having (111) planes parallel to the rolling plane [14]. What happens depends on which crystallographic plane has the lowest surface energy under the conditions of material purity and annealing atmosphere.

The relationship between secondaries and potential secondaries may be illustrated by the results of DETERT [11]. Specimens of 3 % Si-Fe were cold rolled to reductions of 95, 50, and 13% of thickness. When annealed at $1100^{\circ}C$ in hydrogen, (100) secondaries were obtained. The distributions of the cube edge directions for both matrix and secondary grains are shown in Fig.3. The (100) pole densities of

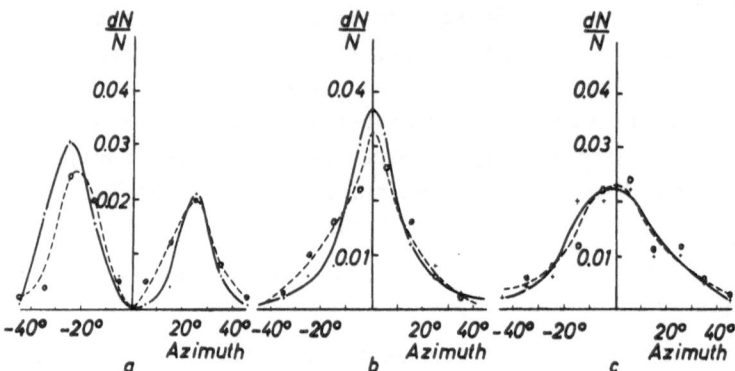

Fig.3. Distribution of cube edge directions about the
rolling direction .... (100) matrix grains ————(100)
secondaries. a) 95 %, b) 50 % and c) 13 % reduction of
thickness (after DETERT [11])

secondaries were 250 times random for the three reductions.
The fraction of matrix grains with (100) planes parallel
to the rolling plane was 1-2 % and their size was no
different than the size of the other, differently-oriented,
matrix grains. It is clear that the orientations of the
secondaries are directly related to the orientations of the
matrix grains having (100) planes nearly parallel to the
plane of the sheet.

Secondary grains having (110) planes parallel to the sheet
surface may also be obtained in 3 % silicon iron if the
oxygen content of the material is low and is kept low by
annealing in vacuum at high temperatures [8]. Fig.4 shows
the course of growth of (110) secondaries in 0.025 mm thick
foil. After 6 minutes, growth of a number of (110) potential secondaries could be recognized from thermal groove
markings. Instance of growth of 6- and 5-sided (110)-
oriented grains with migration away from centers of
curvature were observed. All secondaries had (110) planes
within 5° of the surface. Growth of a (110) grain with 1°
tilt into a (110) grain with 4° tilt was observed with the
direction of migration away from the center of curvature [8].
The small difference in tilt provides a small driving force;
the impurity drag force must be very small.

10Sec. 1Min. 4Min. 6Min. 8Min.12Min.

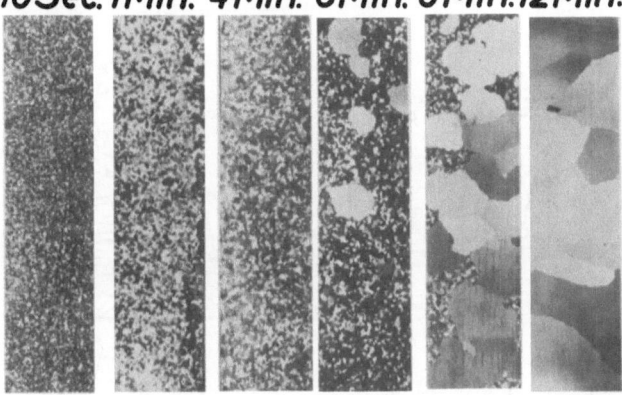

Fig.4. Course of secondary recrystallization in high-purity
3 % silicon-iron foil annealed in vacuo at 1200°C (after
DUNN and WALTER [3])

Data on both the matrix texture and the secondary
recrystallization texture were obtained to show the
correlation between the orientation of the secondaries and
the matrix grains. Only primary grains with (110) planes
within 5° were included. The orientation distribution of
the (110) primaries and secondaries as a function of the
angle between the [001] direction and the rolling direction
is shown in Fig.5. The distribution is seen to be the same.

A matrix texture which produces either the (110) [001] or
(100) [001] secondary recrystallization texture in high-
purity 3 % silicon iron is shown in Fig.6. The matrix has a
(110) pole concentration parallel to the rolling plane and a
fairly strong concentration of [001] directions parallel to
the rolling direction. It might be expected that the
mobility of the boundary between a (110) [001]-oriented
secondary and the matrix grains might be lower than the
mobility of the boundary of a (110) [113] grain, for
instance. There is, however, no evidence for a mobility
effect in this case, for there is a 1 to 1 correspondence of
[001] directions of the secondary grains to the [001]
directions of the (110) oriented matrix grains.

The selection of (110) grains within 5° of the rolling plane

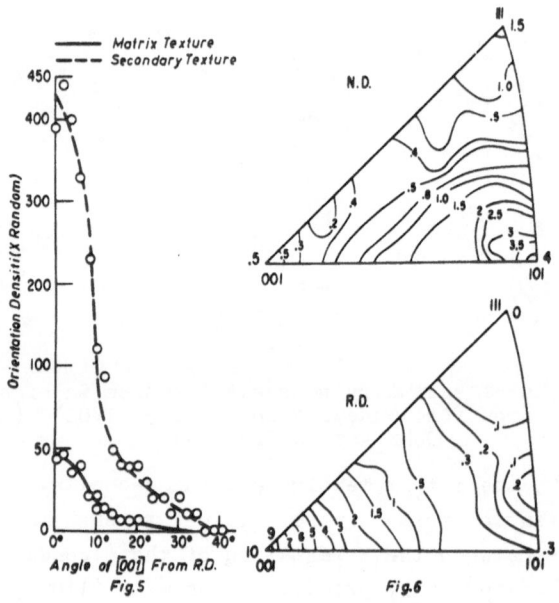

Fig.5. Orientation densities for {110} secondaries and {110}
matrix grains versus angular position of [001] directions for
high-purity 3 % Si-Fe annealed in vacuo at 1200°C (after
WALTER and DUNN [8])

Fig.6. Matrix texture produced in high-purity 3 % Si-Fe
during 5 minute anneal at 1200°C in vacuo (after WALTER and
DUNN [8])

(though the spread of the matrix texture is greater than 5°)
is the result of the higher driving force resulting from
greater differences in surface energy for these grains. If
there were no primaries within 5° of the rolling plane, the
spread of the secondaries might be greater but the surface
energy difference would be less, the driving force would be
diminished and so would the growth rate.

The spread of (100) planes as a function of sheet thickness
was studied by KRAMER and FOSTER [15] under conditions for
growth of (100)-oriented grains by surface energy. There was
no significant effect of thickness on the spread.

There might even be occasions where the surface energy differences would be so high as to give too great a spread of secondary grain orientations. In this case, the addition of impurities, as second phases, to cause a drag on the boundaries might reduce the tilt of the desired low index planes. An example of a possible effect of large differences in surface energy on spread of secondary grains is given by the work of MILLER and WILLIAMS [16]. They found that rolled zinc, when annealed in air at 410°C grew very large secondary grains with basal planes (0001) within 22° of the plane of the surface. This is a much larger spread of surface energy driven secondary recrystallization textures than has been reported previously. In the absence of a strong primary matrix texture, a large difference in energy between near (0001) surfaces and all other orientations could result in such a large spread. It appeared that the growth rate was greater for those grains with basal planes nearest the plane of the sheet.

Another satisfactory matrix for both (100) [001] and (110) [001] secondaries in iron and 0.6 % silicon-iron is shown in Fig.7. In this case there is a near (111) pole concentration parallel to the rolling plane and a rolling direction concentration near [011].

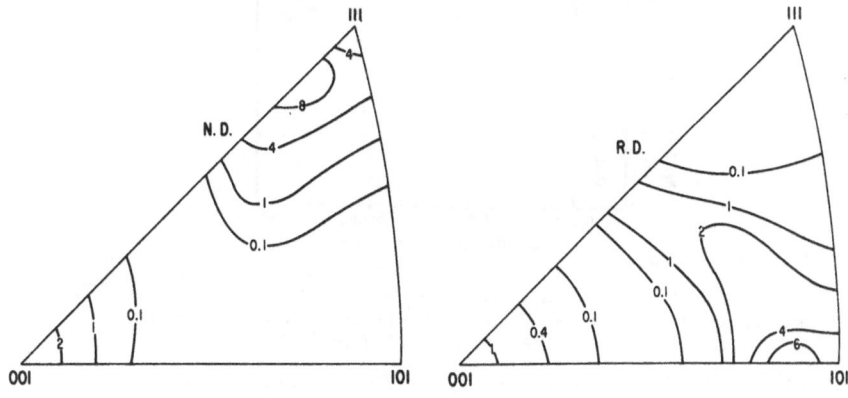

Fig.7. Matrix texture produced in high-purity 0.6 % Si-Fe during 2 hour anneal at 875°C in dry hydrogen (after DUNN and WALTER [12])

This might be an ideal starting texture for growth of
secondaries with (111) planes parallel to the plane of the
surface. Such a texture has been recently described by
MEE [14]. Mee was able to effect growth of secondaries with
(111) planes within 5° of the plane of the sheet by
annealing 0.03 cm thick samples of commercial purity 3 %
silicon iron in vacuum or hydrogen. Mee observed a distinct
impurity effect on the relative values of the surface
energies. The (111) oriented secondaries grew most readily
in the commercially pure silicon iron which contained
appreciable amounts of sulfur and phosphorus compared with
vacuum melted high purity silicon iron. Further, the addition of H_2S to the hydrogen atmosphere favored growth of (100)
oriented secondaries instead of (111) oriented secondaries.
The effect of H_2S additions on reducing the (100) surface
energy relative to either (110) or (111) has been studied
by KOHLER [17]. Fig. 8 shows the effect of amount of H_2S
in the hydrogen atmosphere on growth of (100) oriented
secondaries at 1200°C. The critical concentration of sulfur
in the silicon iron is between 0.5 ppm and 6 ppm.

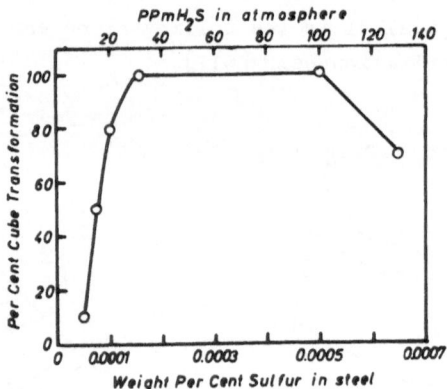

Fig.8. Effect of amount of H_2S gas in hydrogen
atmosphere at 1200°C on formation of cube texture
in 3 % Si-Fe (after KOHLER [17])

Growth of (110) secondaries by surface energy occurs only
when the surface is nearly free of foreign atoms, such as
sulfur or oxygen. WALTER and DUNN [8] determined that
oxygen in excess of about 6 ppm in the silicon iron would

Table 2. Driving Force, Boundary Velocity, and Boundary Mobility from Some Examples of Growth of Secondaries in a 2-D Matrix

Material	Thickness, cm	Temp, °C	Homologous Temp, T/T_{MP}	Atmos	P, erg-cm^{-3}	V, cm-sec^{-1}	M, cm^{4}erg^{-1}sec^{-1}
Pt	0.010	1500	0.86	Vacuum	20,000	4.2×10^{-6}	2.1×10^{-10}
3 % Si-Fe	0.0025	1200	0.81	Vacuum	72,000	1.8×10^{-3}	2.5×10^{-8}
3 % Si-Fe	0.03	1200	0.81	Hydrogen	10,400	1.0×10^{-3}	9.6×10^{-8}

result in growth of (100) secondaries in vacuum whereas, with smaller oxygen contents, only (110) secondaries grew indicating that (110) surface had the lowest energy. In the presence of oxygen, chemisorbed at the surface, the (100) surface had the lowest energy [18].

Finally, some calculated values for driving force and boundary velocity for growth of secondaries into a two-dimensional matrix are given in Table 2 [4]. These driving forces are small compared with driving forces for growth of secondaries in three-dimensional samples (which will be discussed later). This shows the necessity for having material free of impurities to take advantage of the differences in surface energies for grain growth.

3.6.6.3 Secondary Recrystallization in an Impurity Stabilized **Matrix**

3.6.6.3.1 Secondary recrystallization to the (110) [001] texture in silicon iron. The matrix stability of commercial 3% silicon iron is provided by impurities. MAY and TURN-BULL [19] studied the effect of MnS additions to high-purity 3% SiFe. Fig.9 shows that the grain size of undoped material increases in a normal way with increasing temperatures of anneal. In the doped material the matrix grain growth was slow between 700 and 925°C at which temperature secondary recrystallization occurred. Above 925°C, the matrix grains grew more rapidly; that is, the matrix lacked stability with the result that the final secondary grain size was smaller and the texture less sharp.

FIEDLER [20] studied the effects of various annealing treatments designed to change the amount and distribution of MnS particles and concluded that particles are required for matrix stabilization. A variety of other second phases such as VN [21] SiN [22] and Al_2O_3 [23] has also been studied.

There is no single matrix texture from which one obtains the (110) [001] secondary recrystallization texture. Mostly, the primary recrystallization textures are distinguished by the fact that the (110) [001] component is a very small fraction of the total texture. For good mobility and high

rate of growth of the secondaries, the matrix should have a
markedly different orientation than the secondaries.
According to FIEDLER the (110) [001] component is the off-
spring of a (111) ⟨112⟩ type component obtained by primary
recrystallization after the penultimate reduction [24]. The
secondary grains in the commercial silicon iron are not as
sharply oriented with respect to the sheet surface as are
the secondaries in the high purity silicon iron. The degree
of spread is indicated by Fig. 10 taken from the work of

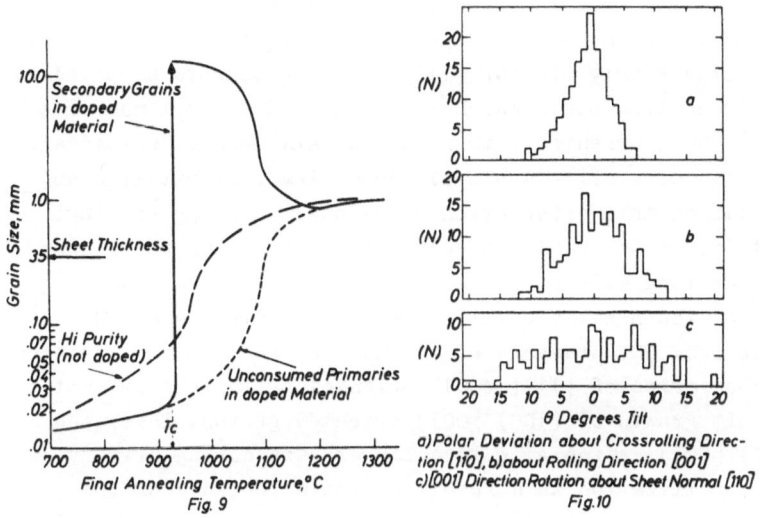

Fig.9. Grain size as a function of temperature (1-hr.
anneals) in 3 % Si-Fe with a final 50 % cold reduction to
0.36 mm thickness (after MAY and TURNBULL [19])

Fig.10. Histograms of distribution of crystal {110} poles
about the [110] and [001] axes of the (110) projection and
of crystal [001] directions rotated about the sheet normal
(after JAMES, JONES and LEAK [25])

JAMES, JONES and LEAK [25]. The planar deviations about the
cross rolling direction and the rolling direction are at
least twice the deviation of the (110) planes in the high-
purity silicon iron. This would seem to rule out any gas-
metal interfacial energy assist in the early growth of these

grains to become secondaries. In view of the unprecise
relationship between the matrix texture and the secondary
texture, this leaves the important question of the
nucleation of the secondaries unanswered.

A possible explanation for the selection of the (110) [001]
secondaries was suggested by BROWN [26] and may be found in
the later work of TAGUCHI and SAKAKURA [27]. They observed
secondary recrystallization in (100) [001]-oriented single
crystals of Si-Fe rolled to 70% reduction only when needle-
like AlN particles were present in correct quantity and
orientation. On rolling, the crystals reoriented to form
deformation bands and transition bands as described by
WALTER and KOCH [28]. The primary recrystallization texture
consisted mainly of four (113) ⟨301⟩ components supplied by
the transition bands which also supplied a small number of
(100) [001] oriented primaries. The AlN needles formed along
specific crystallographic planes during precipitation and
maintained this orientation relationship during rolling;
that is, as the [001] direction rotated during rolling so
did the long axis of the needles. After rolling, the long
axis of the needles still corresponded with the [001]
direction. The presence of the needles restricted migration
of boundaries of {113} ⟨301⟩-oriented grains but did not
inhibit growth of (100) [001]-oriented grains. Thus, the
(100) [001] grains grew to a larger size during primary
recrystallization as a result of the difference of the
coincidence of orientation relationships between AlN and
the parent crystal. These then grew to become secondaries.

3.6.6.3.2 Secondary recrystallization in silver. ROSI et al
[29] prepared sheet specimens of 99.999 % silver. The primary
recrystallization texture of the rolled sheet was comprised
of four components of the {113} ⟨112⟩ type. The matrix grain
diameter averaged 2×10^{-3} cm, small enough to provide a
relatively high driving force of 250 γ_b for normal grain
growth. The lack of normal grain growth and the relatively
low values of migration rate (0.05 mm/min.) and mobility of
the secondaries suggest an impurity effect [3]. The
retardation of grain growth near exposed surfaces led Rosi

et al to assume that an impurity (oxygen) was inhibiting migration.

The secondary grains had the (110) ⟨112⟩ type orientation which was the same as the rolling texture. The large angle of disorientation between the secondary grains and the matrix grains insures a high mobility and high growth rate for these grains unless this boundary is a coincidence boundary or other special type of boundary.

The sharpness of the matrix texture also insures that the potential secondaries must have been larger than the average matrix grains as a result of primary recrystall- ization and normal grain growth. Otherwise, the potential secondaries, in deviating orientation, would have been eliminated by the other matrix grains. It is not unreason- able to expect that the potential secondaries arose as a result of recrystallization in situ of the deformation component and by strain-induced grain growth. Some calcu- lated values of driving force, boundary velocity and mobil- ity are shown in the Table 3.

3.6.6.4 Secondary Recrystallization with Stabilization Supplied by Impurities Segregated at Grain Boundaries

New work provides evidence for secondary recrystallization in systems where matrix stability is provided not by a second phase but by impurity atoms in solution. By segre- gating at the boundary, slowly diffusing impurity atoms may create a drag on the boundary. The impurity drag (P_i) can be approximated by CAHN's [30] expression

$$P_i = \alpha V \, C_o / (1 + \beta^2 v^2) \tag{4}$$

where C_o is the impurity concentration and α and β are para- meters that depend on the interaction energy of the impurity and the grain boundary and upon impurity diffusion coeffi- cients.

DE SALVO, NOBILI and ZIGONI [31] studied the effect of thermal cycling on rate of growth and incubation time for secondary recrystallization in zone refined zinc. The material was of two purities: Ingot (1) contained ~1 ppm

Table 3. Driving Force, Boundary Velocity, and Boundary Mobility from Some Examples of Growth of Secondaries in a 3-D Matrix

Material	Thickness, cm	Temp, °C	T/T_{MP}	Atmos	P, erg-cm^{-3}	V, cm-sec^{-1}	M, cm^4 erg^{-1} sec^{-1}
Ag	0.05	806	0.87	Argon	3×10^5	8.8×10^{-5}	2.9×10^{-10}
3 1/4 % Si-Fe	0.036	900	0.65	Hydrogen	3×10^5	6.0×10^{-5}	2×10^{-10}
3 1/4 % Si-Fe	0.036	980	0.69	Hydrogen	3×10^5	5.5×10^{-4}	1.8×10^{-9}

of impurities while the ingot (2) contained 20-30 ppm of
impurities. The zinc was rolled to 90% reduction and recrystallized at room temperature. The primary recrystallization
texture of zinc is generally characterized by a uniform
scatter of the hexagonal axis around the sheet normal and is,
therefore, not a strong single orientation texture.

Samples of the two different purities were then repeatedly
cycled between room temperature and an upper temperature,
holding four times shorter than the incubation time for
secondary recrystallization. Fig.11 shows the effect of
number of cycles on a) growth rate of the secondaries and
b) incubation time for secondary recrystallization.
Cycling has a greater effect on the induction period for
secondary recrystallization in the less pure zinc. The effect
of cycling on growth rate of the secondaries does not differ
greatly as a function of purity. In the absence of second
phases in the zinc, and without the strong matrix texture,
it is supposed that the matrix inhibition derives from
impurities segregated at the grain boundaries. Once past the
incubation period, however, the driving force is
sufficiently high to cause the boundary to move so rapidly
as to prevent segregation of impurities at the moving
interface.

A more definitive example of segregated solute inhibition
of normal grain growth is given by the recent work of
GRENOBLE and FIEDLER [32]. They investigated the effect of
small additions of sulfur and nitrogen on matrix stability
and secondary recrystallization in vacuum melted 3 % silicon
iron. The manganese content was less than 10 ppm and it is
expected that, at the temperature of secondary recrystallization, $1000^{o}C$, the sulfur will be in solution. The material
was rolled to 0.03 cm thick and decarburized at $800^{o}C$ in
wet hydrogen.

Fig.12 shows the effect of additions of nitrogen alone and
sulfur alone and the effect of simultaneous additions of the
two elements on normal growth and on secondary recrystallization. When either sulfur or nitrogen are present in small
amounts, only normal grain growth occurs. When both are
present, there is sufficient inhibition to normal grain

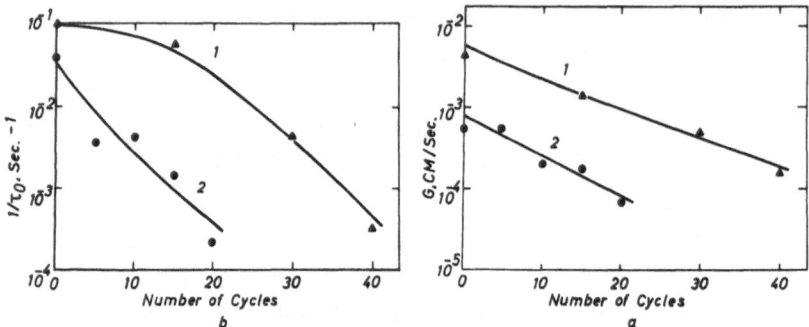

Fig.11a and b. Effect of number of thermal cycles on a) rate of growth of secondaries and b) incubation time for secondary recrystallization of zinc at 200°C (after DE SALVO et al [31])

Sulfur Only

Nitrogen Only

Nitrogen+Sulfur (1)

Nitrogen+Sulfur (2)

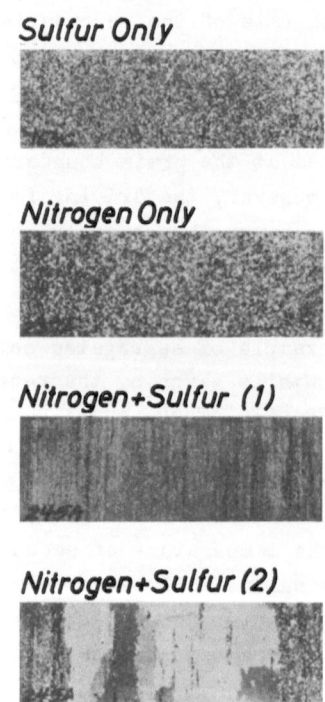

Fig.12. Effect of additions of nitrogen and sulfur singly and together on normal growth and secondary recrystallization in 3% Si-Fe. Sulfur only: 0,008% S, < 3 PPM N; nitrogen only: 0,002% S, 96 PPM N; nitrogen + sulfur (1): 0,008% S, 40 PPM N; nitrogen + sulfur (2): 0,008% S, 40 PPM N

growth to allow secondary grains to grow readily.

There is no ready explanation for the need for both elements to be present to inhibit normal grain growth and allow secondary recrystallization to occur. Presumably, one or the other, or both, segregate to the grain boundaries. There is a characteristic grain boundary etching effect associated with the presence of the two elements. Many of the grain boundaries are more heavily delineated compared with samples from which the nitrogen had been removed.

3.6.6.5 Secondary Recrystallization in a Strong Single-Orientation Matrix

The best known example of a single orientation matrix texture is the cube texture that occurs in heavily rolled f.c.c. metals. Secondary recrystallization from the cube texture has been studied for copper, [33,34] nickel–iron [35,36] and stainless steel [37]. The cube texture and the texture at the very early stage of secondary recrystalliza-tion in stainless steel are shown in Fig.13. The material had been annealed for 16 hours at $1000^{o}C$, long enough for secondary recrystallization to begin. In this case, the orientation described as {520} [001] by GOODMAN and HU [37] begins to appear as indicated by the triangles. The orien-tation may be related to the cube texture by rotations of about 22° around the ⟨001⟩ axis in the rolling direction. A number of other orientations are obtained in the f.c.c. metals and these may be related to the cube texture by similar rotations about the rolling plane normal and the cross direction. There is also a component related by approximately 40° rotations around a ⟨111⟩ axis of the cube texture. This latter component becomes predominant at higher temperatures. Goodman and Hu suggested that the grains related by rotations about ⟨001⟩ axis had the higher growth rate at low temperatures but lower growth rates at high temperatures than did the grains related by rotations about the ⟨111⟩ axis. This growth rate relationship is shown schematically in Fig. 14.

It was further suggested that the growth rate kinetics are related to effects of impuritites on the boundary mobility,

the mobility being affected differently at different
temperatures depending on the boundary structure and kinds of
impurities segregated at the boundaries.

It seems that means now exist to begin to sort out some of
the factors involved in secondary recrystallization. Such
information as rate of growth of matrix grains, the matrix
grain size at incubation of the secondaries, and rates of
growth as well as orientations of secondaries is needed.

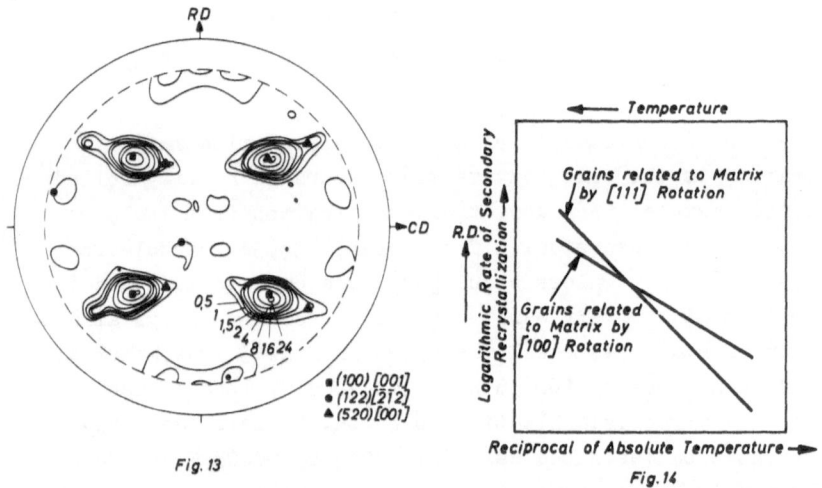

Fig. 13

Fig. 14

Fig.13. (111) pole figure showing texture changes at an early
stage of secondary recrystallization in 18-8 stainless steel
(type 304). Specimen annealed at 1000°C for 16 hr (after
GOODMAN and HU [37])

Fig.14. Schematic illustration of temperature dependence of
rate of growth for grains related to matrix by ⟨111⟩ and
⟨100⟩ rotations (after GOODMAN and HU [37])

Also needed is information concerning the presence of grains
in the matrix of the proper orientation to become
secondaries.

References

1. W.W. Mullins: Acta Met. 6(1958)419.

2. M. Hillert: Acta Met. 13(1965)227.

3. C.G. Dunn and J.L. Walter, Recrystallization, Grain
   Growth and Textures, ASM, 1966, p.461.

4. P.R. Beck, J.C. Kramer, J.L. Demer, and M.L. Holzworth:
   Trans. AIME 175(1948)372.

5. C.S. Smith: Trans. AIME 175(1949)15.

6. M. McLean and H. Mykura: Acta Met. 13(1965)1291.

7. J.L. Walter and C.G. Dunn: Trans.Met.Soc. AIME 218 (1960)914.

8. J.L. Walter and C.G. Dunn: Trans.Met.Soc. AIME 218 (1960)1033.

9. J.L. Walter: J.Appl.Phys. 36, part 2(1965)1213.

10. K. Foster, J.J. Kramer and G.W. Weiner: Trans.Met.Soc. AIME 227(1963)185.

11. K. Detert: Acta Met. 7(1959)589.

12. C.G. Dunn and J.L. Walter: Trans.Met.Soc. AIME 224 (1962)518.

13. C.G. Dunn and J.L. Walter: Trans.Met.Soc.AIME 221(1961) 413.

14. P. B. Mee: Trans.Met.Soc.AIME 242(1968)2155.

15. J.J. Kramer and K.Foster: Trans.Met.Soc.AIME 233(1965) 1244.

16. W.A. Miller and W.M. Williams: J.Inst.Metals 93(1964/65) 125.

17. D. Kohler: J.Appl.Phys. 31(1960)4085.

18. J. L. Walter and C.G. Dunn: Acta Met. 8(1960)497.

19. J.E. May and D. Turnbull: Trans AIME 212(1958)764.

20. H.C. Fiedler: Trans.Met.Soc.AIME 230(1964)95.

21. H.C. Fiedler: Trans.Met.Soc.AIME 221(1961)1201.

22. D. Fast: Philips Res.Reports 11(1956)490.

23. A. Markuzewicz, J. Groyecki, J. Lassota, and A. Zawada: Trans.Met.Soc.AIME 236(1966)196.

24. H.C. Fiedler: J.Appl.Phys. 29(1958)361.

25. D. W. James, H. Jones, and G.M. Leak: Trans.Met.Soc. AIME 236(1966)36.

26. J. R. Brown: J.Appl. Phys. 29(1958)359.

27. S. Taguchi and A. Sakakura: Acta Met. 14(1966)405.

28. J.L. Walter and E.F. Koch: Acta Met. 10(1962)1059.

29. F.D. Rosi, B.H. Alexander, and C.A. Dube: Trans.AIME 194(1952)189.

30. J.W. Cahn: Acta Met. 10(1962)789.

31. A. DeSalvo, D. Nobili, and F. Zigrani: J.Inst.Metals 94(1966)175.

32. H.E. Grenoble and H.C. Fielder: Private communication.

33. M.L. Kronberg and F.H. Wilson: Trans.AIME 185(1949)501.

34. M. Sharp and C. G. Dunn, Trans.AIME 194(1952)42.

35. G. Wassermann: Z.Metallkde. 28(1936)262.

36. G.W. Rathenau and J.F.H. Custers: Philips Res.Reports 4 (1949)241.

37. S.R. Goodman and H.Hu: Trans.Met.Soc.AIME 236(1966)710.

3.7 The Recrystallization of Heavily Rolled Copper Bi-Crystals
Rekristallisation von Kupfer-Bikristallen

by J. Eady and M. Hatherly[+]

Zusammenfassung

Eine kürzlich durchgeführte Arbeit hat gezeigt, daß das Rekristallisationsverhalten von stark gewalzten Einkristallen
von der Ausgangsorientierung abhängt. Es wird hier über eine
Erweiterung der Untersuchungen auf Bikristalle aus Kupfer
berichtet, durch die auch der Einfluß einer Korngrenze untersucht werden konnte. Dabei wurde gefunden, daß sich die
einzelnen Kristalle zum Teil so verhielten, als seien sie
isolierte Einkristalle (z.B. Texturen), während sie in anderer Hinsicht (z.B. Kinetik) ein abweichendes Verhaltenb
zeigten. Polykristalline Proben mit 5 Körnern ergaben ähnliche Ergebnisse wie Bikristalle.

Recent work [1,2] on the recrystallization of heavily cold-
rolled single crystal specimens of copper has established
that, although the annealing kinetics vary considerably with
initial orientation, they are independent of the rolling
texture (Fig.1). The recrystallized grain size also varies
and is independent of the annealing kinetics. Because of
these observations it seems unlikely that the annealing
textures of polycrystalline specimens can be explained by
any theory based directly on single crystal results. This
paper describes an extension of the single crystal studies
to bi-crystals in order to examine the effect of a single
grain boundary on the recrystallization behaviour.

Seeded bi-crystal specimens, each 2" long by 3/4" wide by
1/2" thick, and having a plane boundary normal to the rolling
surface and parallel to the rolling direction, were prepared
by a modified Bridgman technique. These were cold-rolled to
99% reduction in thickness and (111) pole figures of each of

[+] School of Metallurgy, University of New South Wales,
Kensington, Australia

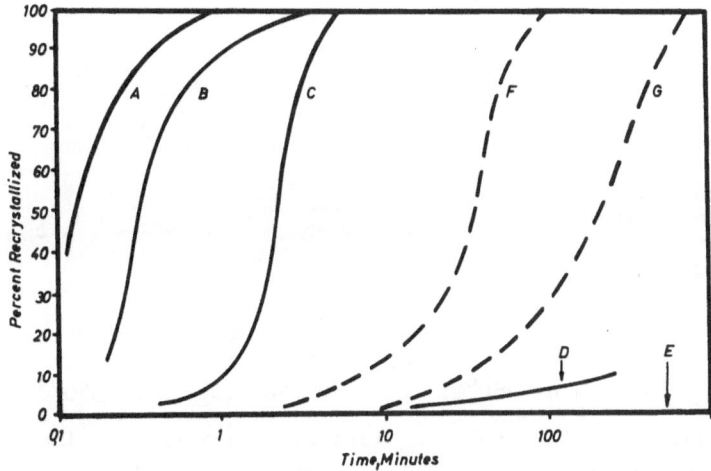

Fig.1. Isothermal recrystallization curves for cold-rolled
copper crystals: (after BROWN [1])

A:polycrystalline specimen annealed at 300°C; B:(100) [011]
crystal annealed at 300°C; C:(110)[001] crystal annealed at
300°C; D:(100) [001] crystal annealed at 300°C; E:(110) [1$\bar{1}$2]
crystal annealed at 300°C; F:(110) [1$\bar{1}$2] crystal annealed at
450°C; G:(100) [001] crystal annealed at 325°C

the individual crystals were obtained before and after
annealing. The kinetics of recrystallization were determined
for each crystal by a statistical grain counting method. The
techniques, specimen material and equipment were identical
to those used in the earlier single crystal work. A typical
analysis of a rolled specimen is shown in Table.1

Table 1. Typical Bi-crystal Composition

% P	0.007	% As	0.001	% Fe	0.001
% Pb	0.002	% Bi	0.001	% Sn	0.001
% Sb	0.001	% Fe	0.001	% Mn	0.001
		% Ni	0.001	% Zn	0.001

It has been found that in many respects the individual
crystals of the bi-crystal specimens behaved as if they were
isolated single crystals. In particular, it has been observed
that if one half of the specimen recrystallized rapidly there

was no migration of the recrystallized material into the other half (Fig.2).

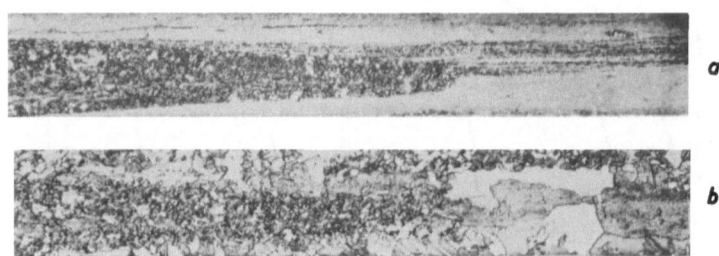

Fig.2 a and b. Microstructure of bi-crystal specimen after annealing at 350°C. X100

a) 3 minutes at 350°C;   b) 90 minutes at 350°C

The rolling and the annealing textures obtained from the component crystals were similar to those reported for equivalent single crystals. In other respects, however, this similarity was not observed.

1. The recrystallization kinetics of the component crystals differed from those obtained for the corresponding single crystals.

2. In nearly all cases the rate of recrystallization varied with the distance from the original grain boundary.

3. The microstructure was different, in several cases, from that of the previously studied single crystal.

These results, which have been observed in a considerable number of specimens, can be illustrated in detail by the example of a specimen made up of the two orientations, (112) [110] and (110)[001]. The rolling and annealing textures obtained from the respective parts of this bi-crystal were similar to those previously found in the relevant single crystal specimens (Fig.3). However, that side of the specimen having the orientation (1T2) [110] (Side A) was completely recrystallized after twenty minutes at 300°C. whilst the (110) [001] side (Side B) was fully recrystallized after 100 minutes at 300°C. (Fig.4). Single crystal studies have shown that a single crystal of copper with the former orientation required 1,000 minutes at 300°C. for complete recrystallization while a crystal having the latter orientation needed four minutes at 300°C.

255

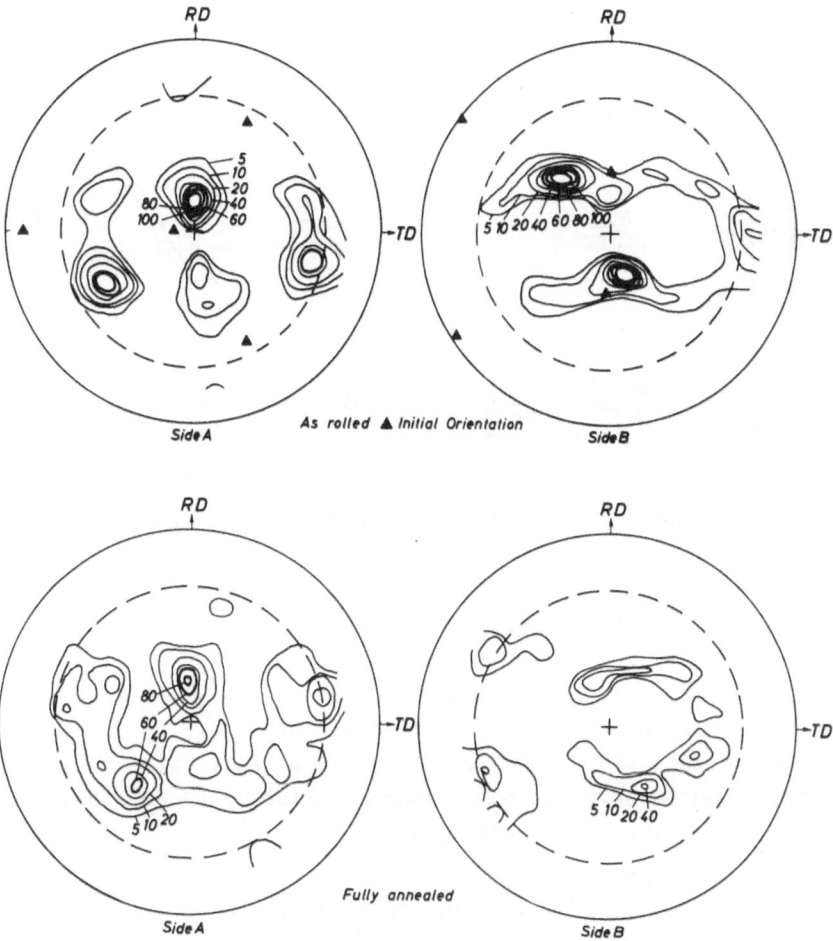

Fig.3. (111) pole figures of rolled and fully annealed bi-
crystal
Side A (1T2) [110]   Side B (110) [001]

Fig.4 also illustrates the variation in the rate of recrystallization usually found across the width of the
specimen; the three curves shown for each crystal refer to
approximately equal volumes in the regions indicated. It will
be seen that there is a marked effect arising from the
presence of the original boundary. The nature of this

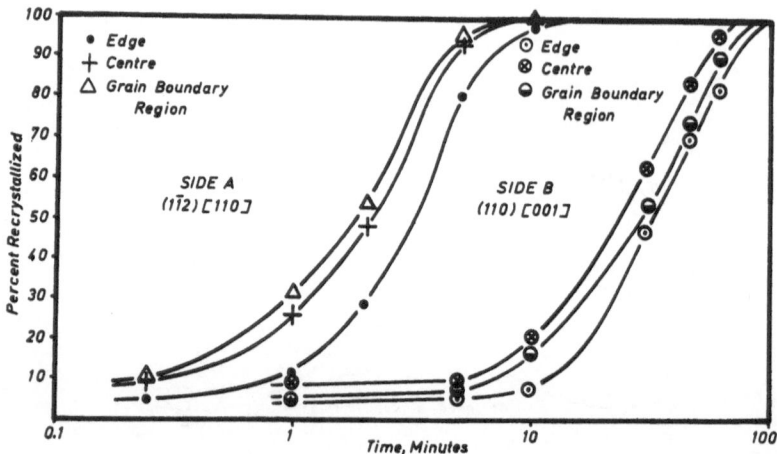

Fig.4. Isothermal recrystallization curves for the bi-crystal (1Ī2) [110] – (Ī10) [001]. Annealed at 300°C

variation was not constant; in some cases recrystallization commenced near the original grain boundary but in others, this was the last region to recrystallize. Although no pronounced differences in texture could be detected in the various regions of each crystal of the specimen, it is believed that the changed kinetics can be attributed to variations in orientation arising from the more complicated deformation process in bi-crystals.

The only exceptions to this behaviour have been a group of specimens prepared so that the individual crystals were slightly different in orientation; e.g. by 3° in both rolling plane and direction. The rolling and annealing textures were identical on the two sides of these specimens. The recrystallized grain size in the crystals were identical and the annealing kinetics so similar that it was extremely difficult to detect the position of the original boundary.

The observation that one half of a heavily rolled bi-crystal can recrystallize rapidly without initiating recrystall- ization in the second half is surprising for it would have been expected that in suchcases the recrystallized grains

would have grown across the boundary and replaced the
neighbouring strained material. It is possible that this
effect is due to the concentration of the small amounts of
impurities present at the single grain boundary or to a lack
of grains at this boundary that are oriented favourably for
rapid growth into the unrecrystallized half of the specimen.
Subsequent work has shown that these explanations are
inadequate and a more detailed study of the structural
changes in and near the deformed grain boundary is necessary.

A specimen has also been examined which consisted originally
of the five crystals shown in Fig.5. Although the rolling and

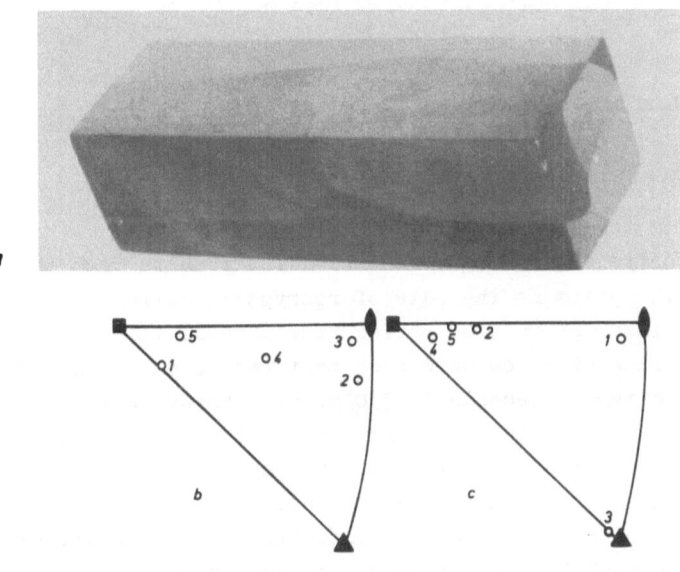

Fig.5 a-c. Orientation of individual crystals in polycrystalline specimen.
a) Polycrystalline specimen, b) orientation of rolling
plane, c) orientation of rolling direction

annealing textures were determined at several positions pole
figures characteristic of polycrystalline material were not
observed. Two pole figures from the annealed strip are
shown in Fig.6; neither of these corresponds to a strong
cube texture but in one there is clear evidence of a cube

258

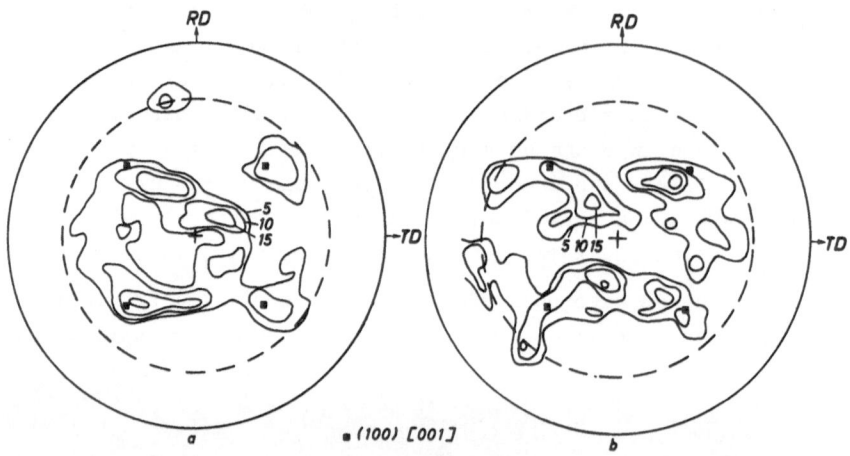

Fig.6 a and b. (111) pole figures of annealed polycrystalline specimen.
a) Top surface
b) Bottom surface

texture component. In agreement with the results from the
bi-crystal specimens the rate of recrystallization and the
recrystallized grain size varied from crystal to crystal but
the annealing kinetics were such that recrystallization was
complete after 15 seconds at 350°C. From these results and
the observations made on bi-crystals it seems that, although
the rate of recrystallization within the individual grains
of the specimens is changed by the presence of a grain
boundary, the migration of recrystallized grain boundaries
across original boundaries does not occur readily.

References

1. K. Brown: Ph.D thesis, University of N.S.W., 1967.
2. M. Hatherly and K. Brown: J. Aust.Inst.Met. 11(1966)264.

3.8 Relationships between the Cold Rolling Textures of Iron
Single Crystals and their Primary Recrystallization Textures
Beziehungen zwischen den Kaltwalztexturen von Eisen-Einkri-
stallen und ihren primären Rekristallisationstexturen

by R. Penelle and P. Lacombe[+]

Zusammenfassung

Zwischen der Walz- und der Rekristallisationstextur kalt-
gewalter Eisen-Einkristalle besteht eine Orientierungsbeziehung durch eine $30°$-Drehung um eine gemeinsame $\langle 110 \rangle$ oder
$\langle 112 \rangle$ Achse. Die primäre Rekristallisationstextur von vielkristallinen Blechen kann nicht – wie die Walztextur – durch
Überlagerung der bei Einkristallen beobachteten Orientierungen dargestellt werden.

Die Untersuchung zeigt, daß die Theorie des orientierten
Wachstums die Ergebnisse am besten zu deuten vermag, während
orientierte Keimbildung ausscheidet.

3.8.1 Introduction

The two main theories put forward in order to explain the
formation of recrystallization textures are, the oriented-
nucleation mechanism [1] and the oriented-growth mechanism
[2]. Supporting this last theory, IBE and LÜCKE [3] have
recently shown that in the case of an Fe-3% Si alloy the
migration rate of recrystallization grain boundaries is at
its maximum when there is a rotation orientation
relationship of about $30°$ around a $\langle 110 \rangle$ axis common to the
new crystal and to the deformed matrix.

TAOKA, TAKEUCHI and FURUBAYASHI [4] have shown that the
rotation axes, relating the recrystallization texture to
the cold rolling textures of Fe-Si single crystals are
situated between the $\langle 110 \rangle$ and $\langle 112 \rangle$ poles, moreover HU [5]
demonstrated a rotation relationship around a $\langle 100 \rangle$ pole.

The aim of this work was to determine which mechanism

[+] Centre de Recherches Métallurgiques de l Ecole des
Mines de Paris, Paris, France

explains best the annealing textures of cold rolled iron single crystals, and to attempt an explanation of the annealing textures in polycrystalline iron in terms of the various orientations found for single crystals.

3.8.2 Experimental Determination of the Recrystallization of Cold-rolled Iron Single Crystals

3.8.2.1 Experimental Procedure

The single crystals, prepared by the strain-anneal method [6], were rolled 80% at room temperature and annealed for 30 mn at $840^\circ C$ in an argon atmosphere.

The orientations of the recrystallized specimens were determined using the Schulz reflection method.

3.8.2.2 Influence of the orientation on the recrystallization characteristics of the rolled single crystals

During rolling at room temperature, iron single crystals behave very differently according to their orientation. It was found that crystals with orientations, after rolling, close to (001) [110] are more easily deformed, and should then store less energy, than those with the orientation ($\bar{1}11$) [211] [7].

Furthermore, it was found by electron microscopy that the dislocation density is higher in ($\bar{1}11$) [211] crystals than in (001) [110] crystals. HU [8] confirmed these results with silicon-iron single crystals.

The recrystallization rate of such crystals after identical reduction should then be different during the anneal which follows rolling. It was indeed found that ($\bar{1}11$) [211] crystals recrystallize more quickly than (001) [110] crystals; these observations are in agreement with the results of HIBBARD and TULLY [9].

3.8.2.3 Recrystallization Textures of Single Crystals with Orientations close to (001)[110] after Rolling

After the recrystallization anneal, the mean orientation and the corresponding rotation axis for each recrystallized specimen were determined and plotted on two standard stereographic (001) projections, Figs. 1 and 2.

The first figure shows that the poles of the planes parallel to the rolling plane are about 20° from the (001) pole, whereas the rolling directions are about 20° from [010], [Ī00] and [0Ī0].

The orientation of each recrystallized sample is related to that of the rolled crystal by rotation of about 30° around one of the rotation axes plotted in Fig.2.

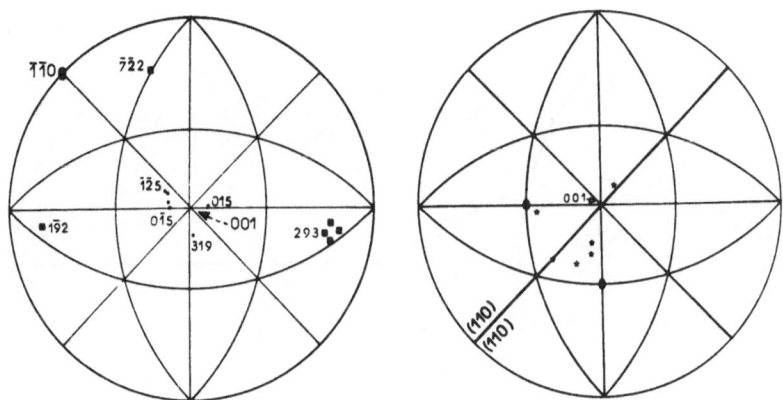

Fig.1. Recrystallization textures of crystals close to (001)[110] after rolling

Fig.2. Rotation axes corresponding to Fig.1

For all figures: ● or ○ = pole of rolling plane;   ☐ or □ = rolling direction; ✻ or ★ = rotation axis

Although the distribution of these axes is heterogenous, they are nevertheless close to the [Ī11] zone and fall between the [0Ī1] and [101] poles.

In conclusion, the primary recrystallization texture can be written:

$$\{114 \rightarrow 014\} \quad \langle 225 \rightarrow 115 \rangle$$

Finally the comparative study of the pole figures of the rolled and annealed samples shows that the components of the primary recrystallization texture are absent in the cold rolled single crystals; this result is evidently contrary to the hypothesis of oriented nucleation proposed by BURGERS.

3.8.2.4 Recrystallization Textures of Single Crystals with
Orientations Close to (T̄12)[110] after Rolling

In order to distinguish the different components of the recrystallization texture they are plotted on four separate
standard (001) projections. Each projection shows for a series
of results the poles of the surface planes of the rolled and
annealed crystals as well as the corresponding rolling
directions and the rotation axes relating the rolled
orientations to the recrystallized orientations, Figs. 3 - 6.

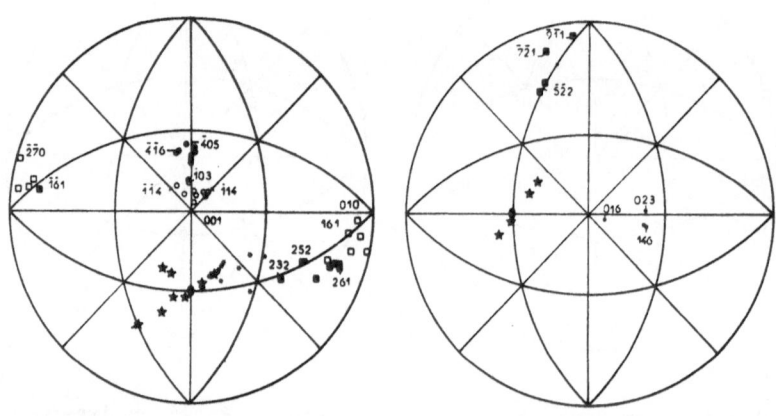

Fig.3. Recrystallization
textures of crystals close to
(T̄12) [110] after rolling.1st
and 2nd components

Fig.4. Recrystallization
textures of crystals close
to (T̄12) [110] after rolling.
3rd component

It was found that the orientation of each recrystallized
sample is related to that of the corresponding rolled crystal
by a mean rotation of $30°$ around one of the experimentally
determined axes; these axes are situated on the four $\langle 111 \rangle$
zones without noticeable grouping around the $\langle 110 \rangle$ poles.

The five main components found for the recrystallization
texture are the following:

1) (001)→T̄T̄4→T̄03→T̄14) [161→010→T̄6̄1→2̄70]
   The rotation axes fall between the [101] and [112] poles
   of the [T̄11] zone, Fig. 3.

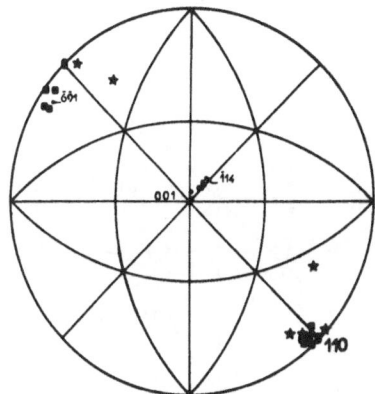

Fig.5. Recrystallization
textures of crystals close
to (T12)[110] after
rolling. 4th component

Fig.6. Recrystallization
textures of crystals close
to (112)[110] after
rolling. 5th component

2) (T02→405→4T6) [261→252→232]

   The rotation axes fall between the [1T2], [101] and [2T1]
   poles, Fig.3.

3) (023→146) [7Z1→522]

   These orientations are symmetrical about the (110) plane
   with those found in 2). The rotation axes fall between
   the [12̄1] [0T1] and [TT2] poles, Fig.4.

4) (001→T14) [110].

   The rotation axes fall between the [2T1] [110] and [121]
   poles, Fig.5.

5) (011) [Z3̄2→T2̄2→223]

   The rotation axes fall between the [T10] and [T21] poles,
   Fig.6.

It should be noted that the (001→T14) [110] orientation
is related to (T12) [110] by a 35° rotation around the
rolling direction and corresponds to one of the main
components of the recrystallization texture of a polycrystalline sheet.

These results are illustrated by the following example,Fig.7.
After a 30 mn anneal at 840°C, the rolled crystal with an
orientation close to (T12) [110] gives three components R_1,

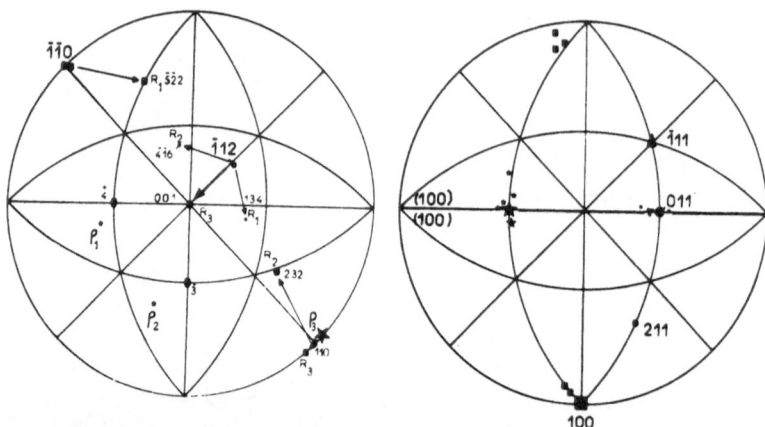

Fig.7. Recrystallization
textures of a crystal close
to (Ī12) [110] after rolling

Fig.8. Recrystallization
textures of crystals close
to (Ī11) [211] before or
after rolling

R_2 and R_3 simultaneously. The rotation axes o_1, o_2, o_3, relating, the three components to the "deformed parent crystal", correspond to the [Ī32], [3Ī2] and [110] poles.

The R_1 orientation, or (134) [Ī2̄2] is related to the rolled crystal by a $31°$ rotation around o_1.

The R_2 orientation, or (4̄Ī6) [232] is related to the rolled crystal by a $33°$ rotation around o_2, and finally R_3, or (001)[110] by a $35°$ rotation around the rolling direction.

The experimentally determined R_1, R_2 and R_3 orientations did not exist in the pole figure of the rolled single crystal. This was a general observation.

3.8.2.5 Recrystallization textures of single crystals with orientations close to (Ī11)[211] before or after rolling

After plotting in a stereographic projection the orientations of the recrystallized samples and their corresponding rotation axes, Fig. 8, it was found that all crystals have an approximate (011)[100] orientation. This orientation is related to (Ī11)[211] by a $35°$ rotation around the [0Ī1] axis. These last results are in agreement with those of several authors [10,11] concerning silicon-iron single crystals.

3.8.3 Conclusion

This study of the recrystallization textures of cold-rolled iron single crystals showed existence of an orientation relationship between the recrystallized crystals and the deformed matrix. The orientation of the new recrystallized crystals is related to the single crystal by a rotation about 30° around a common ⟨110⟩ or ⟨112⟩ axis.

Furthermore, the various components of the primary re-crystallization texture of a polycrystalline sheet do not in general result from the simple superposition of the various orientations obtained with single crystals, as was found for the rolling texture [12]. It is thus likely that some of the ⟨110⟩ and ⟨112⟩ axis relate the rolling and the recrystall-ization textures of a polycrystalline sheet and that the others relate the rolling textures of single crystals to those of the resulting polycrystalline sheet after anneal.

The comparative study of poles figures of rolled and annealed samples showed that none of the orientations found in the recrystallized crystals preexisted in the deformed metal as BURGERS and LOUWERSE [1] assumed in their oriented nucleation hypothesis.

This result is in agreement with those found by HAESSNER et al in copper [13] and MERLINI and GUINIER [14] in Aluminium.

BECK's oriented growth theory seems then best adapted to these results.

References

1. W.G. Burgers and P.C. Louwerse: Z. Phys. 61(1931)605.
2. P.A. Beck: Trans.AIME 191(1951)475.
3. G. Ibe and K. Lücke in: Recrystallization, Grain Growth and Textures, H. Margolin, Ed., Amer.Soc.Metals, Metals Park, Ohio 1966, p.434.
4. T. Taoka, S. Takeuchi and E. Furubayashi: Trans.Met. Soc. AIME, 239(1967)13.
5. H. Hu: Trans.Met.Soc.AIME 221(1961)130.
6. R. Penelle et D. Pollnow: Colloque International sur le fer de très haute pureté, propriétés physiques et chimiques, CNRS – Mém. Scient.Rev.Mét. LXV(1968)351.

7. R. Penelle, J. Mion et P. Lacombe: C.R. Acad.Sci., Paris 256(1963)1514.

8. H. Hu in: Recovery and Recrystallization of Metals, L. Himmel, Ed., Interscience, New York 1963, p. 311.

9. W.R. Hibbard, Jr. and W.R. Tully: Trans.AIME 221(1961) 336.

10. Yu. Avramov, B.V. Molotilov, G. Naumann and N.M. Samarina: Fiz.Metal.Metallov. 21(5)(1966)740.

11. C.G. Dunn and P.K. Koh: Trans.AIME 206(1956)1017.

12. R. Penelle et P. Lacombe: Acta Met. 16(1968)443.

13. F. Haeßner, U. Jakubowski and M. Wilkens: phys.stat.sol. 7(1964)701.

14. A. Merlini et A. Guinier: Symposium de Métallurgie Spéciale, Ed; C.E.N. Saclay et P.U.F. Paris 1957, p. 73.

3.9 Das Wachsen einer Textur bei der Sekundärrekristallisation von Si-Eisen
Texture Development During Secondary Recrystallization in Iron-Silicon

von E. H. Möbius[+)]

It is of greatest techniqual interest to study the development of a texture during annealing, but the present experimental techniques are inadequate. This is especially true, when the conditions of recrystallization in practice are simulated in the laboratory.

In connection with an investigation on the production of cube texture in iron-silicon by secondary recrystallization a method was developed, which enables the study of recrystallization step by step. The texture results will be presented and discussed.

Die Texturen von kaltgewalztem und geglühten Si-Eisen, die bei der Sekundärrekristallisation bei Temperaturen über 1000°C entstehen, sind darum von besonderem Interesse, weil sie magnetisch genutzt werden können. Es entstehen die sogenannte Goss- und Würfellage.

In letzter Zeit hat man sich zunehmend gefragt, welche Ursachen einmal zur Goss- und zum anderen zur Würfellage führen. WALTER und DUNN [1] konnten zeigen, daß - wie MULLINS[2] vermutete - die von Glühatmosphäre und Legierung bestimmte Oberflächenspannung wirksam ist. Dieser Faktor reicht jedoch nicht aus, um alle Ergebnisse zu erklären. Ein weiterer Einfluß ist zu erkennen, wenn man die Ergebnisse der folgenden drei Versuche, deren Aussage zum Schluß zusammengefaßt wird, betrachtet.

3.9.1 Versuch 1

Diesem lag der Gedanke zugrunde, das Wachsen der Textur in Schritten zu verfolgen, um Hinweise zu erhalten, welche Einflüsse die Sekundärrekristallisation zur Würfellage

---

[+)] Röchling'sche Eisen- und Stahlwerke, Völklingen/Saar, Bundesrepublik Deutschland

begünstigen und welche hemmend wirken. Um in rascher Folge
Glühung an Glühung reihen zu können, wurde ein Vertikalofen
mit Schleuse gemäß Abb.1 benutzt. In diesen Ofen wurde jeweils ein Stapel von Ronden, durch ein stabiles körniges
Oxyd voneinander isoliert und von einem Käfig getragen, über
die Schleuse eingeführt. Geglüht wurde bei 1250°C unter Vakuum über jeweils 3 h. Anschließend wurden die Ronden rasch

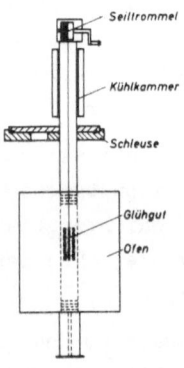

Abb.1 Schleusenofen

abgekühlt, über die Schleuse ausgeführt, ausgemessen, erneut eingesetzt und weitere 3 h geglüht. Bei dieser Art des
schrittweisen Glühens zeigten die Ronden nach eben erreich-
ter Endtemperatur von 1250°C und auch noch nach 3 h Glühzeit bei 1250°C ein gleichmäßiges, relativ feinkörniges Gefüge. Nach jeder weiteren dreistündigen Glühung wurde ein
strahlenförmiges Wachstum weniger Kristalle aus der Mitte
der Ronden zum Rand hin beobachtet. Abb.2 zeigt solche Ronden von 26 mm ⌀ und 0,3 mm Dicke verschiedener Schmelzen
und unterschiedlicher Vorbehandlung nach 21 h Glühzeit.

Natürlich ist nicht zu verkennen, daß die beschriebene Glüh-
weise mit großer Erhitzungs- und Abkühlungsgeschwindigkeit
auch nicht annähernd mit einer stationären Schlußglühung
verglichen werden kann.

Die harmonischen Koeffizienten der Drehmomentkurve – entsprechend dem Vielfachen des Drehmomentwinkels mit a_2 und
a_4 bezeichnet – wurden nach jedem Glühabschnitt bestimmt
(s.Abb.3). Der Gang von a_2 und a_4 ist stetig.

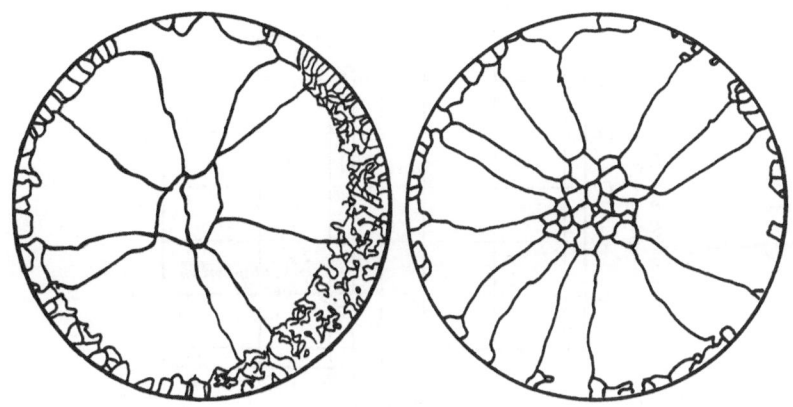

Abb.2. Gefüge geglühter Ronden in schematischer Darstellung,
V = 2x

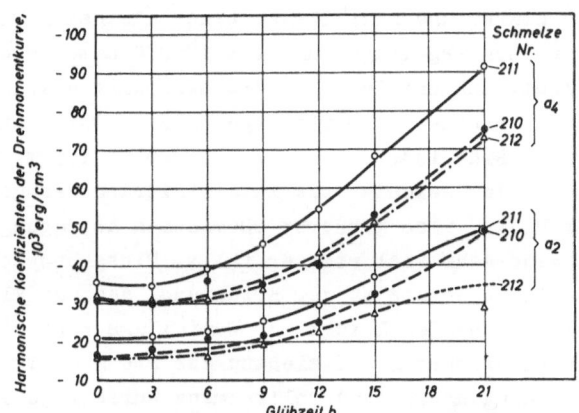

Abb.3. Drehmoment von in Schritten geglühten Ronden

Die kontinuierlich geglühten Ronden der gleichen Schmelzen
zeigten normale sekundäre Rekristallisation; sie verläuft
im Vergleich zur schrittweisen Glühung langsamer (s.Abb.4).

3.9.2 Versuch 2

Im Ausgangszustand lag ein kaltgewalztes Band mit einer Dicke

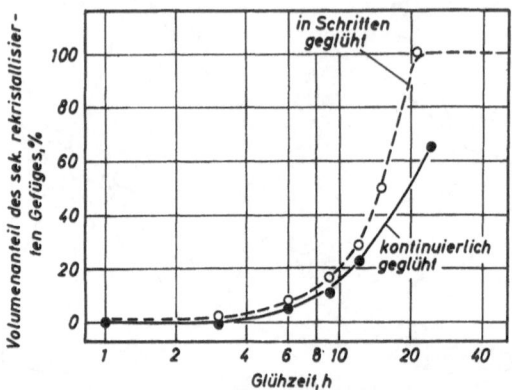

Abb.4. Volumenanteil des sekundärrekristallisierten
Gefüges von Ronden mit 26 mm ∅ und 0,3mm Dicke

von 0,3 mm vor. Das Band wurde vor der Schlußglühung unterschiedlich stark, und zwar bis zu einer kleinsten Dicke von
0,15 mm chemisch abgetragen und bei 1250°C unter Hochvakuum
schlußgeglüht. Bekannt ist, daß die sekundäre Rekristallisation des Si-Eisens zur Würfellage umso schwerer abläuft,
je dicker das Band ist. Das Versuchsergebnis zeigt, daß der
Volumenanteil des sekundär rekristallisierten Gefüges mit
geringerer Dicke, also größerer chemischen Abtragung, erst
mäßig und dann sehr viel stärker fällt. Diese Beobachtung
steht also im Gegensatz zu der Erfahrung, daß auf kleinere
Dicken gewalzte Bänder leicht zur Würfellage rekristallisie-
ren. Die harmonischen Koeffizienten der Drehmomentkurve lassen für den Ausgangszustand vollkommene Würfellage erkennen.
Mit zunehmend größerem chemischen Abtrag wird die Würfellage
gestört; a_4 fällt und a_2 steigt an. Dargestellt ist in Abb.5
das Meßergebnis einer Schmelze. Andere Schmelzen zeigten ein
von diesem nur sehr wenig abweichendes Ergebnis.

3.9.3 Versuch 3

Um beide Phänomene besser deuten zu können, muß an eine Beobachtung erinnert werden, die PHILIP und LENHART [3]
machten. Sie untersuchten die Kinetik der sekundären Rekristallisation von Si-Eisen zur Goss-Lage und fanden, daß

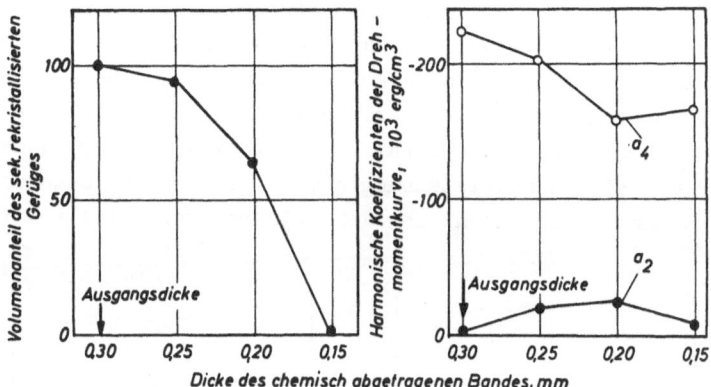

Abb.5. Einfluß der chemischen Abtragung auf die Sekundärrekristallisation

die Keime der sekundären Rekristallisation zumeist zwei oder
drei Körner unterhalb der Oberfläche des Bandes und nur ge-
legentlich auch in der Oberfläche lagen. Bei letzteren war
nicht sicher festzustellen, ob sie von der Oberfläche aus
ins Innere oder vom Inneren zur Oberfläche bereits gewach-
sen waren. Nie fanden sie Keime der sekundären Rekristallisation in der Mitte der Banddicke.

3.9.4 Deutung

Die drei Phänomene, strahlenförmiges Wachstum der Körner
der Ronde, Erschwerung der sekundären Rekristallisation beim
chemischen Abtragen und die Lage der Keime bei der Entstehung der Goss-Lage können gedeutet werden, wenn man die
Keime oder Körner, von denen die sekundäre Rekristallisation
ausgeht, hinsichtlich ihrer Lage innerhalb des verbleibenden
Spannungsfeldes nach der primären Rekristallisation untersucht.

Beim ersten Versuch liegen die Keime der sekundären Rekristallisation im Zentrum der Ronde. Beim 2. Versuch liegen
sie in der oberflächennahen Zone, die beim chemischen Abtrag verlorengeht. Die Ergebnisse der 3. Untersuchung be-
stätigen diesen 2. Befund. Es ist nun auffallend, daß die
Keime der sekundären Rekristallisation in den drei be-

schriebenen Fällen stets innerhalb des Gebietes zu beobachten sind, in dem eine, wenn auch nur geringe, Zugspannung
herrscht (siehe Abb.6). Das kann darauf zurückgeführt werden, daß eine Zugspannung die Diffusion stark erhöht, also
den Platzwechsel der Atome erleichtert. Eine Auswahl hinsichtlich der kristallographischen Orientierung der Keime
in Bezug auf die Hauptspannungen des verbliebenen Spannungsfeldes kann hinzutreten. Bei der primären Rekristallisation

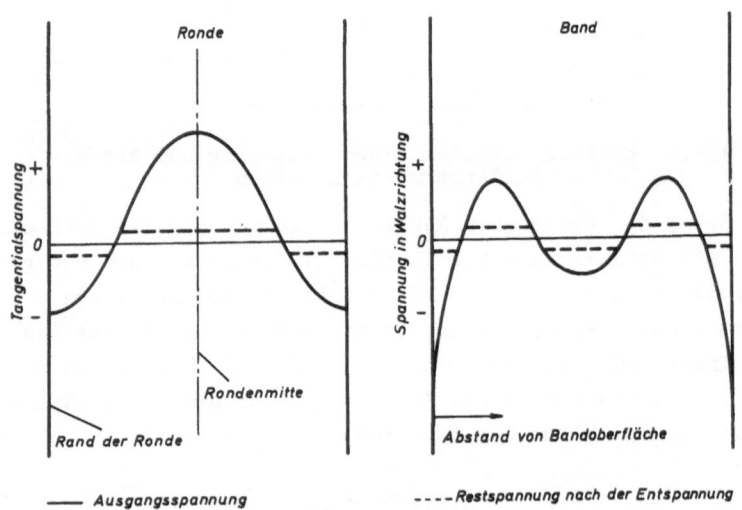

Abb.6. Spannungen in Ronde und Band (schematisch)

von Kupferdrähten wurde die dabei auftretende zyklische
Würfellage von STADELMAIER und BROWN [4] in diesem Sinne gedeutet. Auf Grund der Anisotropie des E-Moduls ist die Keimbildungsarbeit minimal, wenn die Hauptspannungsrichtung mit
der [100]-Richtung zusammenfällt. Läßt man diese Auswahl-
regel auch für die sekundäre Rekristallisation zu, so ist
z.B. wegen der unterschiedlichen Spannungsgradienten in den
drei Ordinaten des Bandes verständlich, das bei der Würfellage nicht eine Streuung um Walzrichtung und Querrichtung,
sondern bevorzugt um die Walzebenennormale beobachtet wird.
Bei Eisen beträgt das Verhältnis der Moduln $E_{[111]}$ zu $E_{[100]}$
2,15 und bei Kupfer 2,85. Die Höhe der Anisotropie des
E-Moduls scheint danach bei Eisen ausreichend, um die Keim-

bildung zu beeinflussen. Die Energie nimmt, wenn man eine
eindimensionale Spannung von einer [100]-Richtung in eine
[111]-Richtung legt, bei Eisen um 54 % und bei Kupfer um
63 % zu [4]. Die absolute Größe des möglichen Energiege-
winnes wird durch die Höhe der verbliebenen Restspannung
nach der primären Rekristallisation bestimmt. Nimmt man an,
daß sie in der [111]-Richtung 1 kg/mm^2 beträgt, so ist die
freie Energie bei gleicher Dehnung und 1250°C grob geschätzt

$$\text{in der } [111]\text{-Richtung } 6900 \text{ erg/cm}^3 \text{ und}$$
$$\text{für die } [100]\text{-Richtung } 3200 \text{ erg/cm}^3.$$

Diese freie Energie ist vom Quadrat der Spannungen, also
nicht von deren Vorzeichen abhängig. Die Beobachtung, daß
die Keime der sekundären Rekristallisation bevorzugt in dem
Gebiet mit Zugspannung zu beobachten sind, ist darum ein
davon unabhängiger zusätzlicher Einfluß.

Die besprochenen Beobachtungen können also gedeutet werden,
wenn man einerseits der verbleibenden geringen Restspannung
einen Einfluß auf die Diffusion und damit auf die Platzwechsel bei der Keimbildung einräumt und zum anderen die Vorstellung von STADELMAIER und BROWN [4] über die Abhängigkeit der Keimbildungsarbeit von der Anisotropie des E-Moduls
auf die sekundäre Rekristallisation überträgt.

Literatur

1. J.L. Walter und C.G. Dunn: Trans. AIME 215(1959)465.
2. W.W. Mullins: Acta Met. 6(1958)414.
3. T.V. Philip und R.E. Lenhart: Trans.AIME 221(1961)493.
4. H. Stadelmaier und B.F. Brown: Z. Metallkde. 47(1956)1.

4 TEXTUREN UND ZWEITE PHASEN

TEXTURES AND SECOND PHASES

4.1 Beeinflussung der primären Rekristallisation durch Teil-chen

Influence of Second Phase Particles on Recrystallization

von E. Hornbogen[+] und H. Kreye[++]

Second phase particles have an influence on recrystallization because of the following reasons:

a) an incoherent interface (of particles, which are present before deformation) can accelerate the formation of re-crystallization nuclei,

b) precipitation of particles before recrystallization hinders rearrangement of dislocations and retards the formation of recrystallization nuclei,

c) precipitation of particles at grain boundaries and at the fronts of recrystallization reduces their velocity,

d) in case of high particle drag, dislocation rearrangement and annihilation, formation of low and high angle boundaries and boundary velocity are determined by the kinetics of particle growth (recrystallization in situ)

e) discontinuous precipitation (i.e. formation of a new phase or transformation of a metastable into a more stable phase) increases the velocity of a recrystallization interface,

f) immediately below the equilibrium temperature, recrystall-ization precedes precipitation. At lower temperatures the influence of particles on the beginning of recrystallization and the precipitation process itself increases with falling temperature (b-e)

There are the following possibilities that primary re-crystallization textures can be influenced by particles:

1) Recrystallization in situ does not change the deformation texture.

2) Coarse dispersion of non-coherent particles in a matrix lattice with strong anisotropy of grain boundary velocity

---

[+] Institut für Metallphysik, Universität Göttingen, Göt-tingen, Bundesrepublik Deutschland

[++] Max-Planck-Institut für Metallforschung, Stuttgart, Bun-desrepublik Deutschland

leads to optimum conditions for growth selectivity and therefore to strong, well developed recrystallisation texture.

3) Segregation, precipitation at the recrystallization front or discontinuous precipitation lead to greater isotropy of boundary velocity and therefore to disappearance of texture during recrystallization.

These aspects will be explained by results from the literature and own investigations on alloys of Al, Cu, Ni and Fe, which form precipitates.

4.1.1 Einführung und Literaturübersicht

Es gibt folgende Fälle, in denen Teilchen einer zweiten Phase die Rekristallisation beeinflussen.

a) Die Teilchen sind vor der Erzeugung des defekten Zustandes (Verformung, Bestrahlung) vorhanden.

b) Der defekte Zustand wird in einer übersättigten Legierung erzeugt. Die Teilchen bilden sich erst beim Anlassen im gleichen Temperaturbereich in dem auch Rekristallisation erfolgen kann.

Entsprechendes wie für b) gilt auch dann, wenn Übersättigung und Defekte gleichzeitig erzeugt werden, wie beim mechanischen Auflösen von kohärenten Teilchen oder bei Erzeugung neuer Atomarten und Defekte bei Bestrahlung.

Wird eine Legierung, die schon Teilchen einer zweiten Phase enthält, plastisch verformt, so bestehen zwei grundsätzliche Möglichkeiten für das Verhalten der Teilchen.

a) Die Versetzungen des Grundgitters können durch die Teilchen laufen, diese abscheren und so bei Betätigung vieler Gleitsysteme und nach hohen Verformungsgraden praktisch zu Auflösung der Teilchen führen.

b) Die Teilchen besitzen mechanische Eigenschaften, die nicht erlauben, daß sie sich plastisch verformen. Die Kristalle des Grundgitters müssen um die Teilchen herumfließen, die ihre Form nur durch Bruchbildung ändern können [1].

Es soll im Folgenden immer vorausgesetzt werden, daß nur ein kleiner Volumenanteil (< 10%) der zweiten Phase auftritt. Gasblasen und Löcher verhalten sich in vieler Hinsicht wie nichtkohärente Teilchen und können deshalb in die Betrachtung einbezogen werden.

Es liegen schon eine große Zahl von Untersuchungen über die

Wirkung von vorhandenen oder sich ausscheidenden Teilchen
vor. Ihre Wirkung ist nicht ohne weiteres zu verstehen. Es
gibt sowohl Fälle in denen der Ablauf der Rekristallisation
durch Teilchen stark beschleunigt als auch stark verzögert
wird. Außerdem wird gefunden, daß Teilchen sowohl die Beibehaltung der Walztextur, als auch das Entstehen von neuen
Rekristallisationstexturen bewirken können. Die Umordnung
von Versetzungen zu einer wanderungsfähigen Reaktionsfront
wird beschleunigt durch Erhöhung der Versetzungsdichte, sowie durch das Vorhandensein von Senken wie ursprüngliche
Korngrenzen, nicht-kohärente Phasengrenzflächen und von
Kristallblöcken, die bei der plastischen Verformung um
große Winkelbeträge verkippt werden (Verformungsbänder). Segregation und Ausscheidung an den Versetzungen sowie Ausscheidung zwischen den Versetzungen verringern die Geschwindigkeiten von deren Umordnung. Die Kraft auf eine
Reaktionsfront kann von dem Energieunterschied zwischen de-
fektem und defektfreiem Mischkristall oder zwischen übersättigter und entmischter Legierung stammen. Der erste Vorgang ist als (diskontinuierliche) Rekristallisation der
zweite als diskontinuierliche Ausscheidung bekannt. Den Kräf-
ten, die die Bewegung der Rekristallisationsfront bewirken,
sind diejenigen entgegengesetzt, die durch Segregation von
Atomen in die Korngrenze oder durch dispergierte Teilchen
(der stabilen Phase) bewirkt werden, Abb. 1 und 2.

Diskontinuierliche Rekristallisation tritt nicht auf, wenn
die Bildung von Rekristallisationsfronten durch Segregation
oder Ausscheidung von Teilchen an Versetzungen verhindert
wird, oder wenn keine Möglichkeit zur Bewegung solcher Fron-
ten infolge der rücktreibenden Kräfte besteht. Das Ausheilen
der Defekte (Korngrenzen, Versetzungen) erfolgt dann gleichzeitig mit dem Auflösen kleiner Teilchen durch Umlösung.
Dieser Prozess führt ebenfalls, aber sehr langsam, zu einem
rekristallisierten Gefüge und wird als kontinuierliche oder
in situ-Rekristallisation bezeichnet.

4.1.1.1 Bisherige Arbeiten an speziellen Legierungen

In der Technik interessieren die oben erwähnten Probleme aus
sehr verschiedenen Gründen. Zwei sehr wichtige Legierungs-

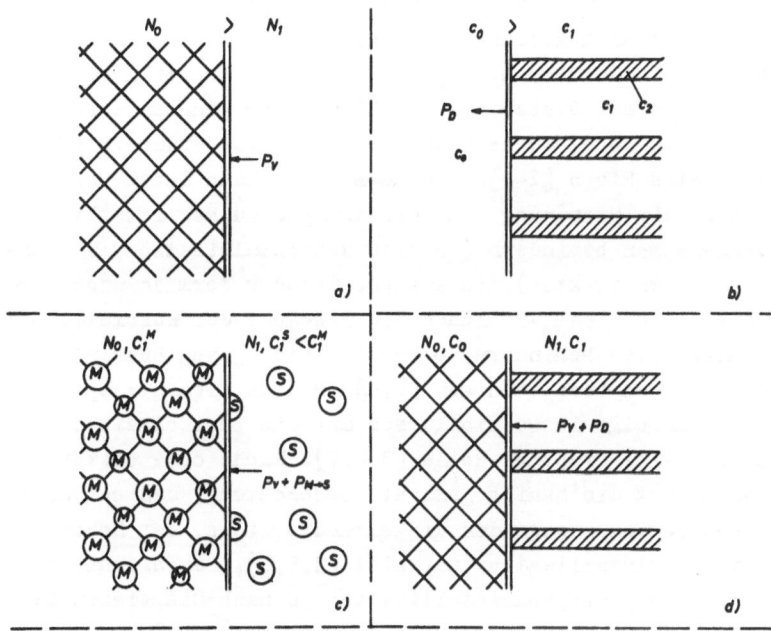

Abb.1a - d. Schematische Darstellung der Kräfte, die auf eine
Rekristallisationsfront wirken können:

a) Differenz der Versetzungsdichte p_V (disk.Rekristallisa-
tion)
b) Differenz der Konzentration p_D (disk. Ausscheidung)
c) Differenz der Versetzungsdichte und Umwandlung von metastabiler (M) in stabile Phase (S)
d) Differenz der Versetzungsdichte und Entmischung des homogenen Mischkristalls

gruppen, die Stähle und die Al-Legierungen sind fast immer
heterogen, so daß Teilcheneffekte bei ihrer Rekristallisation zu erwarten sind. Weiterhin ist die Kenntnis der Wirkung von Teilchen z.B. wichtig, wenn:

a) bestimmte Rekristallisationstexturen erzeugt werden sollen,

b) durch Kombination von Aushärtung und Verfestigung hohe
Festigkeit erzeugt werden oder umgekehrt

c) günstige Bedingungen zum Weichglühen heterogener Legierungen gefunden werden sollen.

In den letzten Jahren ist eine größere Zahl von Arbeiten auf
diesem Gebiet erschienen. Im folgenden sollen zunächst einige Ergebnisse dieser Arbeiten erwähnt werden, die an Al-,

Ni-, Cu- und Fe-Legierungen gewonnen worden sind.

4.1.1.1.1 Al-Legierungen. Es gibt eine große Zahl von Arbeiten die darüber berichten, daß die Rekristallisation in Al mit feiner Dispersion von Al_2O_3 sehr langsam erfolgt. Viel untersucht wurde auch die Rolle des Verunreinigungselementes Eisen [2-6]. Dieses Element kann durch Segregation die Umordnung von Versetzungen zu Rekristallisationskeimen behindern (in situ Rekristallisation, Beibehalten der Walztextur), in ausgeschiedener Form je nach Verteilung der $FeAl_3$-Teilchen die Bewegung der Rekristallisationsfronten behindern, oder die Bildung von Rekristallisationskeimen beschleunigen [2,6]. Dritte Elemente spielen dabei infolge ihres Einflusses auf die Löslichkeit des Eisens eine wichtige Rolle [3,4,7]. Nicht ohne weiteres zu deuten ist die häufig gemachte Beobachtung, daß bei mittleren Verformungsgraden diskontinuierliche, bei hohen in situ-Rekristallisation auftritt [3,5,6]. Beschleunigung oder Verzögerung der Rekristallisation je nach Dispersion der nicht kohärenten ϴ-Teilchen wurde auch in Al-Cu-Legierungen beobachtet [8]. Wegen des Auftretens mehrerer metastabiler Phasen sind die Vorgänge beim Anlassen verformter übersättigter Mischkristalle der Al-Cu-Legierungen kompliziert und nur elektronenmikroskopisch zu ermitteln [9].

4.1.1.1.2 α-Fe-Legierungen. Alle Arbeiten über α-Fe-Legierungen und Stähle deuten darauf hin, daß ihr Rekristallisationsverhalten nur erklärt werden kann, wenn bekannt ist, ob die Atome in Lösung oder als Teilchen bestimmter Dispersion vorhanden sind. Durch Ausscheidung von Cu an Versetzungszellwänden kann disk. Rekristallisation stark verzögert werden, oder in situ Rekristallisation unter Beibehaltung der Walztextur erreicht werden [10]. Zunehmende Konzentration an Sauerstoff, der infolge seiner geringen Löslichkeit in α-Fe in Form von großen FeO-Teilchen vorhanden ist, bewirkt bis zu 0,1 Gew % O eine Beschleunigung der Rekristallisation [11]. An Legierungen mit Mn und O konnte gezeigt werden, daß eine sehr starke Verzögerung der Rekristallisation bei Konzentrationen und Wärmebehandlungen auftrat, bei denen beide Elemente vor der Verformung in Lösung waren. Nach Bildung von MnO-Teilchen wurde eine Beschleunigung gefunden [12]. Entsprechendes tritt auch in Fe-C-Legierungen mit C-Gehalten innerhalb und außerhalb der Löslichkeit von Fe_3C auf [13]. Im ersten Fall ist die Zahl der Rekristallisationskeime mit zunehmendem C-Gehalt durch Segregationseffekte erniedrigt, sie erhöht sich aber, sobald Fe_3C-Teilchen vorhanden sind. Ein Vergleich einer Fe-N-Legierung, die im homogenen Zustand und nach vorheriger Ausscheidung von Fe_4N rekristallisiert wurde lieferte für den letzteren Fall unter Bedingungen, bei denen sich Teilchen auflösen, eine Veränderung der Rekristallisationstextur, während die Kinetik der Rekristallisation nicht sehr stark beeinflußt wurde [14].

4.1.1.1.3 Ni-Legierungen. Die γ'-Phase (kfz.-geordnet) scheidet sich in Ni-Si- und Ni-Al-Legierungen als stabile, in vielen Ni-Legierungen vom Nimonictyp und in aushärtbaren austenitischen rostfreien Stählen als metastabile Phase aus. Diese Teilchen können durch starke plastische Verformung mechanisch aufgelöst werden [15,16]. Stabile γ'-Teilchen bilden sich in verformten Legierungen in sehr feiner Verteilung und verhindern die Umordnung der Versetzungen zu Rekristallisationskeimen, so daß die Walztextur vollständig erhalten werden kann [16]. Es gibt in diesen Legierungen eine kritische Versetzungsdichte für den Beginn der diskontinuierlichen Rekristallisation, die mit der Übersättigung zunimmt [17]. Beim Laufen einer Rekristallisationsfront lösen sich kleine Teilchen in der Rekristallisationfront, während größere durchlaufen werden. Durch beide Vorgänge wird die Bewegung der Reaktionsfront verzögert [15].

4.1.1.1.4 Kupferlegierungen. Kupfer mit verschiedenen Dispersionen von SiO_2, die durch innere Oxidation erzeugt werden können, zeigt Verzögerung oder Beschleunigung je nach Abstand der Teilchen [18]. Der Vorgang der Beschleunigung der Rekristallisation und Verschärfung der Rekristallisationstextur wurde an groben Dispersionen von B_4C in Cu untersucht [19]. Wie in Fe-FeO-Gemischen [11] wirken die Grenzflächen der Teilchen als Keimstellen. Das Gefüge zeichnet sich durch eine sehr scharfe Textur aus. Cr scheidet sich aus Cu als krz.-Mischkristall aus. Unter den Bedingungen einer kürzlich durchgeführten Arbeit an einer Legierung mit 1 Gew. % Cr war es für den Ablauf der Rekristallisation unerheblich, ob die Ausscheidung vor oder nach der Verformung stattgefunden hatte. Nach allen Behandlungen blieb ein Anteil der Walztextur erhalten [20]. Cu-Be-Legierungen zeigten die Besonderheit, daß trotz etwas verschiedener Walztextur des homogenen und des ausgeschiedenen Zustandes die Rekristallisationstexturen fast gleich waren (α-Ms-Typ). Nur bei niedrigen Anlaßtemperaturen ($<350^{\circ}$C) wird die Walztextur beibehalten [21].

Es ist nicht leicht, trotz der großen Zahl von Daten allgemeingültige Zusammenhänge zu finden. Es sollen dazu zunächst einige dafür wichtige Grundphänomene behandelt werden.

4.1.2 Allgemeine Erscheinungen

4.1.2.1 Reihenfolge von Ausscheidung und Rekristallisation

In verformten übersättigten Mischkristallen verzögern die sich ausscheidenden Teilchen die Umordnung der Versetzungen, während die im Kristallgitter vorhandenen Defekte die Keimbildung bestimmter Phasen stark beschleunigen können [22]. Nur in dem Falle, daß Rekristallisation abgeschlossen ist bevor Keimbildung beginnt, beeinflussen sich beide Reaktionen nicht. Um einen Überblick zu gewinnen, unter welchen Bedingungen dieser Fall oder die gegenseitige Beeinflussung von Ausscheidung und Rekristallisation auftritt, ist es sinnvoll, die Temperaturabhängigkeit des Beginns beider Reaktionen zu betrachten (Abb.2). Für eine bestimmte Versetzungsdichte N, unter Annahme konstanter Größe der Rekristallisationskeime und unter Vernachlässigung der Temperaturabhängigkeit der Segregationseffekte in Mischkristallen, gilt für die Zeit bis zum Beginn der Rekristallisation t_R:

$$t_R(T) \sim \exp \frac{Q_R(N)}{RT} \tag{1}$$

Q_R, die Aktivierungsenergie für Bildung von wanderungsfähi-

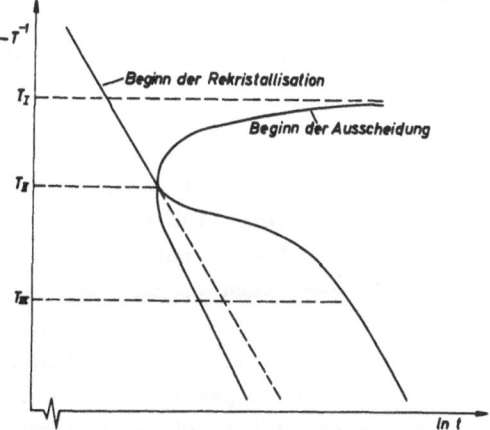

Abb.2. Beginn von Ausscheidung und Rekristallisation
einer Legierung, die unterhalb T_I heterogen wird,
$T < T_{II}$ Wechselwirkung von Ausscheidung und Rekristallisation, $T < T_{III}$ keine Bildung und Bewegung
von Rekristallisationsfronten

gen Rekristallisationsfronten, hat die Größenordnung von Q_{SD}
(SD Selbstdiffusion), nimmt mit zunehmender Versetzungsdichte ab und durch Segregation an Versetzungen zu. Demgegenüber ändert sich die Aktivierungsenergie für Keimbildung
ΔG_K von ∞ bei der Gleichgewichtstemperatur für den betreffenden Mischkristall bis zu sehr kleinen Werten bei großer
Unterkühlung [22]. Die Zeit bis zum Beginn der Ausscheidung
t_A ist erst dann allein von der Aktivierungsenergie für
Diffusion Q_D abhängig, wenn bei hoher Übersättigung
$\Delta G_K \rightarrow 0$:

$$t_A(T) \sim \exp \frac{\Delta G_K(T) + Q_D}{RT} \qquad (2)$$

Für eine Legierung mit bestimmter Versetzungsdichte und Konzentration ergibt sich eine Temperatur T_{II} bei der $t_R = t_A$
ist und damit eine Wechselwirkung im gesamten Volumen der
Legierung auftritt. Wir unterscheiden folglich drei Temperaturbereiche im Rekristallisationsverhalten (T_I Gleichgewichtstemperatur Abb.2).

1) $T > T_I$ Rekristallisation, die nur durch Segrega-
                     tion beeinflußt ist,

2) $T_I > T > T_{II}$: Ausscheidung erfolgt nach Beendigung der
                     Rekristallisation im rekristallisierten

Gefüge.

3) T < T$_{II}$: Ausscheidung erfolgt im verformten Gefüge, die
Rekristallisation wird durch Teilchen beein-
flußt.

Der Übergang bei T$_{II}$ ist natürlich nicht scharf, da zum Ablauf der Rekristallisation Zeit notwendig ist. In diesem
Übergangsbereich tritt das Verhalten von Bereich 2 und 3 an
verschiedenen Orten in der Probe gemischt auf. Unterhalb von
T$_{II}$ wird ΔG_K durch die dann noch vorhandenen Gitterbaufehler
beeinflußt, in den meisten Fällen stark herabgesetzt. Die
t_A = f(T) Kurve ändert sich dadurch verglichen zu der für
den defektfreien Kristall gültigen Kurve, die Ausscheidung
wird im allgemeinen beschleunigt, während Beginn und Ablauf
der Rekristallisation verzögert wird.

4.1.2.2 Bewegung der Rekristallisationsfront

Vorausgesetzt, daß eine als Rekristallisationsfront geeignete Großwinkelkorngrenze vorhanden ist, so wird auf diese
unterhalb von T$_{II}$ eine beschleunigende Kraft durch die erhöhte Energie des gestörten Gitters und eine rücktreibende
Kraft durch die ausgeschiedenen stabilen Teilchen ausgeübt.
Das gleiche gilt für den Fall, daß nicht verformbare Teilchen vor der plastischen Verformung vorhanden waren. Die
Wechselwirkung einer Korngrenze mit Teilchen ist zuerst von
ZENER [23], später von ASHBY [24] behandelt worden (Abb.3).

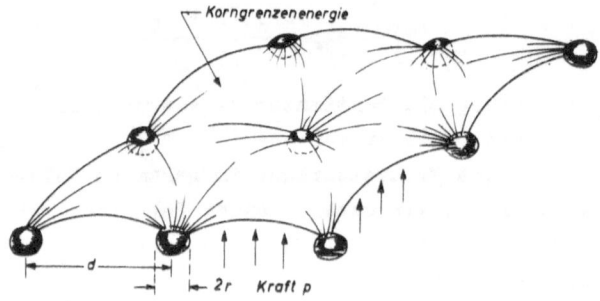

Abb.3. Größen bei der Wechselwirkung einer Korngrenze mit Teilchen in räumlicher Darstellung (nach
ASHBY und LEWIS [24])

Zener berechnete die Festhaltekraft eines kugelförmigen
Teilchens mit Radius r wobei er annahm, daß die Fläche der
Korngrenze um πr^2, die Energie um $E_{max} = \gamma_{KG} \; \pi r^2$ verringert
wird (γ_{KG} = Energie der Korngrenze). Diese Energie muß aufgebracht werden, um die Korngrenze loszureißen. Dazu ist
eine Vorwärtsbewegung um r notwendig. Die maximale Festhaltekraft pro Teilchen P_T ist:

$$\frac{E_{max}}{r} = \pi r \; \gamma_{KG} = P_T \qquad (3)$$

Die Zahl n der Teilchen mit Radius r, die bei einem Volumenanteil f von der (ebenen) Korngrenze geschnitten werden, ist
aus geometrischen Gründen:

$$n = \frac{3f}{2\pi r^2} \; , \; \text{die gesamte Festhaltekraft } P_T \cdot n = p_T \qquad (4)$$

Die treibende Kraft.p auf die Korngrenze ist umso größer je
kleiner ihr Krümmungsradius ρ werden kann: $p \simeq 2 \; \gamma_{KG}/\rho$.
Unter der Annahme, daß $\rho \simeq d$ ist (d = Abstand der Teilchen),
ergibt sich aus der Bedingung, daß die treibende gleich der
rücktreibenden Kraft ist, ein kritischer Teilchenabstand,
unterhalb dessen die Korngrenze unbeweglich wird:

$$d_{krit} = \frac{4}{3} \frac{r}{f} = \frac{2\gamma_{KG}}{n \cdot P_T} = \frac{2\gamma_{KG}}{p_T} \qquad (5)$$

Die Voraussetzung ist, daß die Wechselwirkung unabhängig
von der Natur des Teilchens ist. Aus dieser Beziehung folgt
dann im Einklang mit der Erfahrung, daß kleine Teilchen bei
gegebenen Volumenanteil die Korngrenzenbewegung stärker
hemmen als große. Diese Beziehung setzt aber eine gleichmäßige Verteilung der Teilchen voraus. Falls der Volumenanteil f b e v o r z u g t an der Korngrenze ausgeschieden ist, gilt die angenommene Beziehung zwischen d, r und f
nicht mehr. Es tritt eine größere Festhaltekraft auf.

ZENER's [23] Ansatz ist von ASHBY [24] verbessert worden.
Es muß nämlich berücksichtigt werden, ob die Korngrenzenenergie im Innern des Teilchens (γ'_{KG}) größer oder kleiner

als im Grundgitter (γ_{KG}) ist. Für $\dfrac{\gamma'_{KG}}{\gamma_{KG}} < 1$ wird die Korn-

grenze durch das Teilchen hindurch laufen, während sie es

bei $\dfrac{\gamma'_{KG}}{\gamma_{KG}} < 1$ umschlingt. Der erste Fall sollte z.B. auftre-

ten in Ni–Ni$_3$Al Legierungen ($\mu_{Ni} > \mu_{Ni_3Al}$, μ = Schubmodul),
der zweite Fall ist jedoch der häufigere (Abb.4b,c). Auch
wenn die verschiedenen Arten der Wechselwirkung berücksichtigt werden, ist die Festhaltekraft $P_T = \pi r\, \gamma_{KG}$ noch eine
gute Näherung [24].

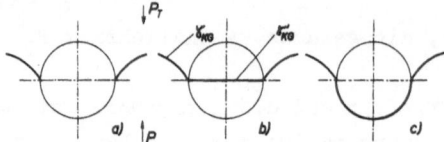

Abb.4a – c. Drei Möglichkeiten der Wechselwirkung von
        Korngrenzen mit Teilchen:
a) unabhängig von Eigenschaften des Teilchens (ZENER)
    b) Korngrenze dringt ein $\gamma'_{KG} < \gamma_{KG}$,
c) Korngrenze umschlingt Teilchen $\gamma'_{KG} > \gamma_{KG}$ (nach
       ASHBY und LEWIS [24])

Eine weitere Möglichkeit der Wechselwirkung ist das Mit-
schleppen von Teilchen durch wandernde Korngrenzen. Die
Atome müssen dabei quer durch das Teilchen diffundieren. Deshalb findet man diese Erscheinung entweder, wenn die treibende Kraft auf die Korngrenze und damit deren Geschwindigkeit sehr klein ist, oder wenn der Diffusionskoeffizient der
Atome im Innern der Teilchen sehr groß ist. Letzteres ist
der Fall, wenn es sich um flüssige Einschlüsse handelt. Mit-
schleppen führt durch Akkumulation zu sehr ungleichmäßiger
Verteilung von Teilchen in der Umgebung von Korngrenzen [25].

4.1.2.3 Rekristallisation in situ

Die Aktivierungsenergie zur Umordnung (Klettern) der bei
der Verformung entstandenen Versetzungen zu wanderungsfähigen Korngrenzen liegt für reine Metalle bei $Q_R \approx Q_{SD}$

(Aktivierungsenergie für Selbstdiffusion). Durch bevorzugte
Ausscheidung von Teilchen an Versetzungen kann dieser Wert
um mehr als eine Größenordnung erhöht werden, so daß praktisch keine Umordnung mehr stattfinden kann ($t_R \rightarrow \infty$). Wenn
weiterhin vorausgesetzt wird, daß vorhandene Korngrenzen
durch bevorzugte Ausscheidung dort festliegen, kann in der
Legierung keine Rekristallisation durch Bildung und Bewegung
einer Reaktionsfront auftreten. In einem solchen Fall wird
das Ausheilen der Defekte (Versetzungen, Korngrenzen) ge-
steuert durch die Umlösung der ausgeschiedenen Teilchen. Bei
diesem Vorgang bleibt der Volumenanteil f konstant, während
die größeren Teilchen auf Kosten der kleineren wachsen. Dies
führt zu einer Zunahme des mittleren Teilchenradius $\bar{r}$ und
damit des Teilchenabstandes $\bar{d}$:

$$\bar{r} = (\frac{3f}{4\pi})^{1/3} \bar{d} \tag{6}$$

Die Beziehung zwischen r und der isothermen Auslagerungszeit
t ist von WAGNER [26] für Teilchen in Flüssigkeiten abgeleitet worden. Diese Beziehung kann für das Wachstum kohärenter Teilchen in Kristallen in guter Näherung angewandt
werden [27]. Da aber Volumendiffusion (D_V) vorausgesetzt
wird, kann sie nicht für stark defekte Kristalle gelten. In
diesem Falle muß berücksichtigt werden, daß ein wesentlicher
Anteil des Stofftransportes über Versetzungen oder Korngrenzen mit Diffusionskoeffizienten $D_P \gg D_V$ erfolgt. Wagner's
Gleichung ist für den Fall erweitert worden, daß Diffusion
durch Versetzungskanäle zwischen den Teilchen eine Rolle
spielt, deren Zahl n (pro Teilchen) während der Umlösung
konstant bleiben soll [28]. Für kleine Teilchen ergibt sich
dann ein starker Einfluß der Versetzungen während bei großen
Teilchen der Anteil der Volumendiffusion überwiegt.

$$\bar{r} = [\ \frac{8}{9}\ \frac{D_V\ \sigma\ \bar{v}^2\ c\infty}{RT}\ t + (\ \frac{4}{5}\ )^3\ (\ \frac{25}{8}\ \frac{nb^2\ D_P\ \bar{v}^2\ \sigma\ c\infty}{\pi\ RT}\ )^{3/5}\ \cdot$$

$$t^{3/5}\ ]^{1/3} \tag{7}$$

$\bar{r}$ = mittlerer Teilchenradius
D_V Diffusionskoeffizient der gelösten Atome in der Matrix

σ = Grenzflächenenergie Teilchen – Matrix
$\bar{v}$ = Molvolumen der Ausscheidung
c = molare Gleichgewichtskonzentration der gelösten Atome
RT = Gaskonstante, absolute Temperatur
n = Zahl der Versetzungskanäle die von einem Teilchen weg-
    führen
b = Burgersvektor
D_p = Diffusionskoeffizient der gelösten Atome entlang der
    Versetzungskanäle
t = Auslagerungszeit

Elektronenmikroskopische Untersuchungen [3,9] deuten darauf
hin, daß die Umordnung der Versetzungen kontrolliert ist
durch das Auflösen kleiner Teilchen, die Versetzungen oder
Versetzungsgruppen festhalten, d.h., die Größe der ver-
setzungsfreien Bereiche entspricht dem Teilchenabstand. Der
mikroskopische Vorgang wird am besten durch Subkornkoaleszenz
beschrieben [29]. Sobald sich ein Teilchen, das sich an einem
Versetzungsknoten befindet, auflöst, erniedrigt sich dort
die Aktivierungsenergie auf Q_{SD}. Versetzungen können dann
sehr viel leichter wandern als diejenigen, an denen sich
noch Teilchen befinden. Falls sich zwischen zwei Subkörnern
eine Grenze befindet, die frei von Teilchen ist, so wird
diese durch Herauswandern der Versetzungen ausheilen, während das kleinere Subkorn zum Ausgleich des Winkelunter-
schiedes etwas rotiert. Dieser Prozess setzt sich fort, sobald neue Knoten durch Auflösen der Teilchen gelöst werden.
Es entstehen immer größere defektfreie Bereiche, die immer
größere Winkel zueinander bilden (Abb.5). Das Wachstum setzt
sich fort, wobei in die oben genannte Gleichung schließlich
der Diffusionskoeffizient und Querschnitt für Großwinkelkorngrenzen eingesetzt werden muß, wenn die Versetzungsnetze
den Charakter von Großwinkelkorngrenzen annehmen.

Dieser Prozeß führt schließlich zu einem rekristallisierten
Gefüge, das sich nur in der Verteilung der Korngrößen und
der Teilchen von dem durch diskontinuierliche Rekristallisation entstandenem Gefüge unterscheidet. Dieser Vorgang,
bei dem sich nie bewegliche Rekristallisationsfronten bilden, wird als in situ – Rekristallisation bezeichnet. Da
keine Orientierungsänderungen über größere Volumen auftreten,
bleibt die vor der Rekristallisation vorhandene Orientierungsverteilung (z.B. Walztextur) erhalten.

287

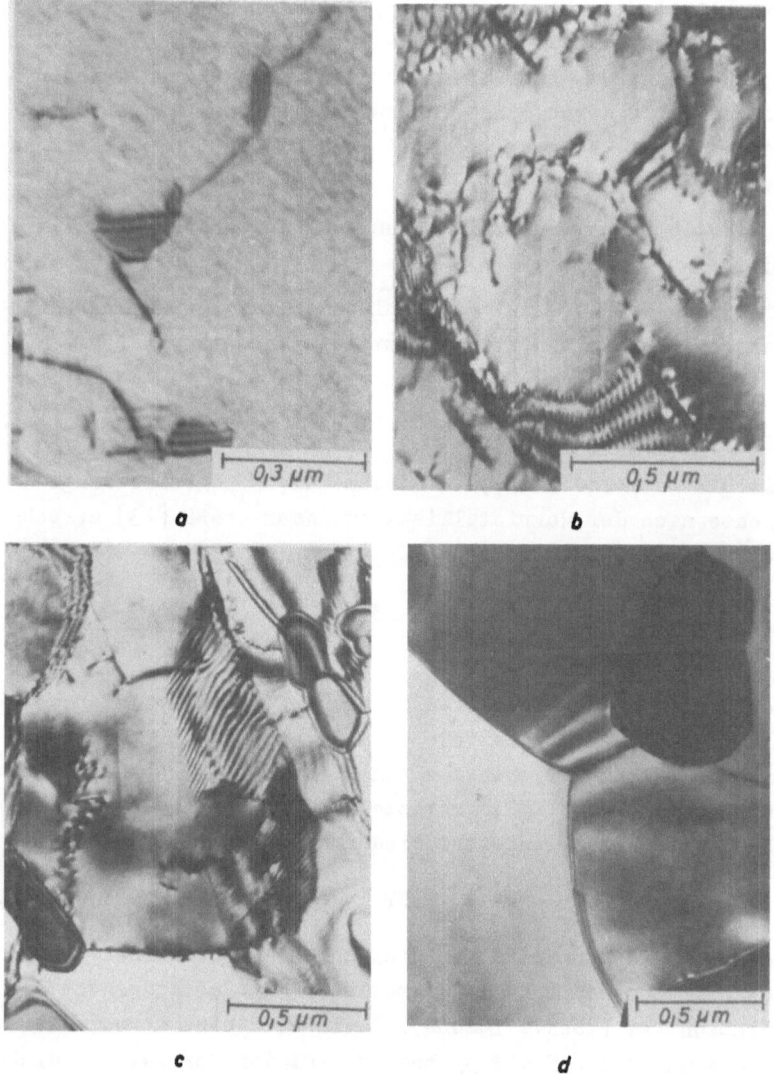

a

b

c

d

Abb.5a-d. In situ-Rekristallisation in Al-5 Gew% Cu Legierung (nach U. KÖSTER). a) Versetzungen werden durch Bildung von Θ'-Teilchen festgehalten. Al + 3 Gew% Cu, 1.5 h 200°C. b) Versetzungsnetze werden durch Θ'Teilchen festgehalten Al + 3 Gew% Cu, 25% verformt, isochron auf 370°C aufgeheizt. c) Nach Umwandlung $\Theta' \rightarrow \Theta$ werden einzelne Subkornknoten festgehalten, während andere durch Rotation (Erniedrigung der Versetzungsdichte in der Grenze) und Wanderung ausheilen können. Al + 5 Gew% Cu, 90% verformt, 170h 240°C.d) Gefüge nach in situ-Rekristallisation (große Θ-Teilchen). Al + 5 Gew% Cu, 90% verformt, 170 h 240°C.

Bei Blasen, die Gase enthalten, die sich im Grundgitter nicht
lösen, besteht keine Möglichkeit zur Umlösung. Dies gilt z.B.
für He, das bei Betrahlung entsteht. Die Blasen wachsen in
diesem Falle durch Wanderung und Koaleszenz [31]. Dieser
sehr langsame Vorgang bestimmt dann die Geschwindigkeit der
Rekristallisation.

4.1.2.4 Diskontinuierliche Ausscheidung und Rekristallisation

Die Größe $p = - \frac{\delta G}{\delta V}$ ist die treibende Kraft der Korngrenzenbewegung. Für einen Kristall mit der Versetzungsdichte N_o
beträgt sie

$$+ p_V = (N_o - N_1) b^2 / \mu \quad [\text{in dyn} \cdot \text{cm}^{-2} \text{ oder erg} \cdot \text{cm}^{-3}] \quad (8)$$

mit b = Burgersvektor, μ = Schubmodul, N_1 = Versetzungsdichte nach der Rekristallisation. Nach ZENER [23] errechnet sich eine rücktreibende Kraft zu

$$- p_T = \frac{3 \, f \, \gamma_{KG}}{2 \, r} \quad (9)$$

falls sich stabile Teilchen im Grundgitter mit Gleichgewichtskonzentration befinden. Falls sich in der Rekristallisationsfront Teilchen aus einem übersättigten Mischkristall
ausscheiden, oder falls metastabile Teilchen in stabile um-
gewandelt werden, entsteht wiederum eine treibende Kraft:

$$+ p_D = RT \ln c_o / c_1 \quad (10)$$

wobei c_1 die Konzentration nach Durchlaufen der Rekristallisationsfront und c_o die des metastabilen Zustandes vor der
Reaktion ist (ideale Lösung). Falls sich eine Korngrenze
allein unter der Kraft p_D bewegt, ist der Vorgang als diskontinuierliche Ausscheidung bekannt [30]. Da anzunehmen ist, daß
p_V und p_D additiv sind, ist in verformten und übersättigten
Kristallen eine diskontinuierliche Reaktion möglich, bei der
gleichzeitig die Versetzungen ausheilen und die Atome sich
ausscheiden. Die Geschwindigkeit ist bestimmt durch den
effektiven Wert von $p_V + p_D$, und die Bedingung für die Bewegung wäre $p_V + p_D > p_T$ falls nicht noch durch Segregation
an der Korngrenze (die hier nicht behandelt werden soll)

eine weitere rücktreibende Kraft - p_s ausgeübt wird:

$$p_V + p_D \rangle p_T + p_s \qquad (11)$$

Diskontinuierliche Ausscheidung allein kann zu völliger Rekristallisation führen, besonders dann wenn die durch Volumenänderung bei der Ausscheidung hervorgerufenen Spannungen eine häufige Richtungsänderung der Reaktionsfront bewirkt. Die Kristalle zerfallen dann in Zellen mit regelloser Orientierung.

Eine Erhöhung der Versetzungsdichte zusätzlich zur Übersättigung bewirkt jedoch nicht einfach eine zusätzliche Kraft
p_V. Diese Versetzungen können nämlich den Übersättigungszu-
stand vor der Reaktionsfront durch Bereitstellen von Keim-
bildungsorten niedriger Energie ändern. Dadurch wird einmal
p_D durch Erniedrigung von c_o verkleinert, p_T aber durch die
an Versetzungen gebildeten Teilchen vergrößert. Aus diesem
Grunde ist es möglich, daß bei zunehmendem Verformungsgrad
eine kombinierte diskontinuierliche Ausscheidungs- und Rekristallisationsreaktion in kontinuierliche Reaktionen, i.e.
individuelle Ausscheidung an Versetzungen und in situ-Rekristallisation, übergeht. Beide Mikromechanismen sind in
Abb.6 an Ni-Be-Legierungen zu sehen. Um das Zusammenwirken
der beiden diskontinuierlichen Reaktionen zu verstehen, muß
demnach die Wirkung der Gitterbaufehler auf die Keimbildung
genau bekannt sein. Die Bedingungen vor der Reaktionsfront
sind immer zeitabhängig. Bei Ausscheidung einer stabilen
Phase wird mit zunehmender Zeit t die treibende Kraft p_D
ab- und die Hinderniskraft p_T zunehmen:

$$p_V^o + (p_D^o - \Delta p_D(t)) + (p_T^o + \Delta p(t)) = p , \qquad (12)$$

bis die Reaktionsfront unbeweglich wird: $p = o$.

4.1.2.5 Zusammenhang zwischen Übersättigung, Versetzungsdichte, Temperatur und Rekristallisationsverhalten

In Kap. 4.1.2.1 wurde die Änderung der Reihenfolge von Aus-
scheidungs- und Rekristallisationsreaktion besprochen
(Abb.2). Dieses Schema muß im Bereich $T \langle T_{II}$ ergänzt werden, in dem die Teilchen und Versetzungen einander beein-

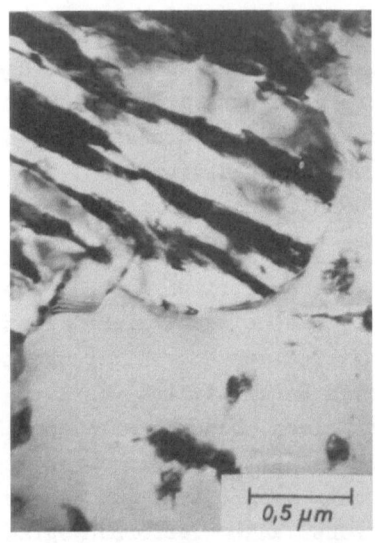 

a   *b*

Abb.6 a und b. Diskontinuierliche Ausscheidung in verformter
                Ni-Be-Legierung

a) Bewegung der Reaktionsfront der diskontinuierlichen Ausscheidung in nur teilweise an Versetzungen entmischten Kristallen Ni + 7,3 At% Be, bei 600°C 15 min angelassen
b) Die Reaktionsfront bewegt sich nicht, da der Mischkristall
an Stellen hoher Versetzungsdichte bereits entmischt ist.
Ni + 7,3 At% Be, 10 % verformt, bei 700°C 15 min angelassen

flussen. Für eine Legierung der Konzentration c und mit der
Versetzungsdichte N (Abb.7) steigt der Volumenanteil der
sich an Versetzungen ausscheidenden Teilchen von $f = 0$ bei
T_{II} mit abnehmender Temperatur an. Unterhalb einer bestimmten Temperatur T_{III} findet keine diskontinuierliche Reaktion
mehr statt. Wir können somit vier Temperaturbereiche mit
verschiedenem Rekristallisationsverhalten erwarten. Zusätzlich zu den in Kap. 4.1.2.1 erwähnten:

3a) $T_{II} > T > T_{III}$: durch Teilchen beeinflußte diskontinu-
                            ierliche Rekristallisation
3b) $T < T_{III}$: in situ-Rekristallisation.

Das gilt jedoch nur dann, wenn sich an den Versetzungen die
stabile Phase ausscheidet. Scheidet sich eine metastabile

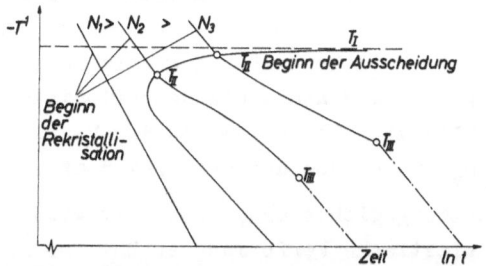

Abb.7. Einfluß von Versetzungsdichte (oder Verformungsgrad) auf die Übergangstemperatur T_{II}, für die
Voraussetzung, daß die Versetzungen den Beginn
der Ausscheidung nicht beeinflussen (schematisch)

Phase aus, so kann infolge der mit der Umwandlung in die
stabile Phase verbundenen Energie die diskontinuierliche Reaktion noch bei tieferen Temperaturen auftreten.

Durch die Erhöhung der Versetzungsdichte N (bei const c_o)
des Mischkristalls werden die Zeiten bis zum Beginn der
Kristallisation t_R und t_A zu kleineren Werten verschoben
(Kap. 4.1.2.1).

Es folgt daraus, daß die Temperatur T_{II} durch Erhöhung der
Versetzungsdichte zu niedrigeren Werten verschoben wird, da
dann die Rekristallisation früher beginnt. Umgekehrt bewirkt eine Erhöhung der Konzentration des Mischkristalls
eine Erhöhung dieser Temperatur. Die Übergangstemperaturen
sind schematisch für ein Legierungssystem in Abb.8 eingezeichnet worden.

Für die Temperaturabhängigkeit der Zeit, bis zu der ein bestimmtes Stadium der Rekristallisation eingetreten ist (Abb.
2), ergibt sich folgender Zusammenhang: Der Bereich 1 (Rekristallisation im Mischkristall) geht direkt in den Bereich 2 (Rekristallisation vor der Ausscheidung) über, ohne
daß die Rekristallisation mehr als durch Segregation verzögert wird. Im Bereich 3a nimmt die Kraft p_T mit abnehmender Temperatur zu, bis die Reaktionsfront unbeweglich wird.
Die Kurve, die ein bestimmtes Stadium der Rekristallisation
anzeigt, wird unterhalb T_{II} nach längeren Zeiten verschoben. Da die Kraft p_T eine Funktion der Temperatur ist, kann

ein Arrheniusansatz zur Bestimmung einer Aktivierungsenergie
der Rekristallisation nicht mehr angewandt werden. Unterhalb
von T_{III} wird die Teilchenumlösung geschwindigkeitsbestim-
mend. Nach Kap. 4.1.2.3 hängt die Geschwindigkeit der in-
situ-Rekristallisation vor allem von der Versetzungsdichte,
und über D_p, D_v und c_∞ von der Temperatur ab.

Wird die Versetzungsdichte eines übersättigten Mischkristalls durch plastische Verformung erhöht, so stehen die
treibenden Kräfte p_v und p_D am Anfang für die Bewegung der
Reaktionsfronten zur Verfügung (Kap. 4.1.2.4). Mit der An-
laßzeit (bei einer Temperatur, bei der thermisch aktivierte
Prozesse möglich sind) ändert sich sowohl p_v als auch p_D.
Dies wird in Abb.9 (schematisch) gezeigt. Dabei soll ange-

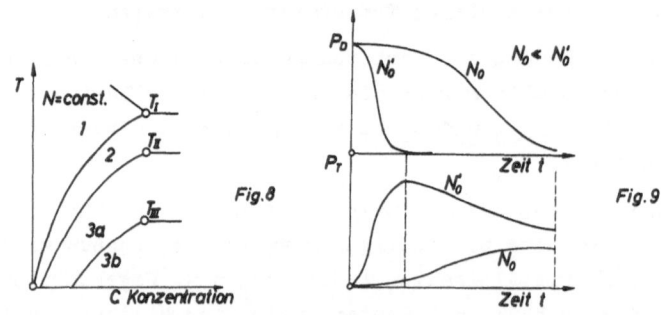

Abb.8. Die Bereiche:

1. Rekristallisation des Mischkristalls

2. Rekristallisation mit anschließender Ausscheidung

3a. Rekristallisation durch Bewegung von Reaktionsfronten bei
gleichzeitiger Ausscheidung

3b. In situ-Rekristallisation in einem Legierungssystem mit
zu niedrigen Temperaturen abnehmender Löslichkeit für konstanten Verformungsgrad,
(schematisch)

Abb.9. Die treibenden p_D und rücktreibenden p_r Kräfte auf
eine Reaktionsfront D, wenn sich der Ausscheidungsgrad vor
dieser Front abhängig von der Versetzungsdichte N_o (bei der
Zeit t = o) ändert (schematisch)

nommen werden, daß zunehmende Zahl der Versetzungen $N_o \leftarrow N_o'$
die Ausscheidung beschleunigt, i.e. p_D erniedrigt, daß aber
p_v mit der Alterungszeit konstant bleibt, da die Versetzun-

gen durch die Teilchen festgehalten werden. Da sowohl p_V als auch p_D aber mit umgekehrten Vorzeichen von der Versetzungs- dichte abhängen, gibt es eine Versetzungsdichte (oder Ver- formungsgrad), bei der $p_V + p_D - p_T$ den größtmöglichen Wert besitzt. Daraus folgt, daß (falls sich eine stabile Phase an den Versetzungen ausscheidet) es eine mittlere Versetzungsdichte gibt, bei der die diskontinuierliche Re- aktion auftritt, während die starke Beschleunigung der Aus- scheidung bei höheren Versetzungsdichten zu Unbeweglichkeit der Reaktionsfront führt.

4.1.2.6 Beeinflussung der Rekristallisationstexturen

Aus dem besprochenen Rekristallisationsverhalten lassen sich einige Schlüsse auf die zu erwartenden Texturen ziehen. Lei- der fehlt uns aber noch grundsätzliche Information über den Einfluß von Teilchen und der Größe der treibenden Kraft auf die Anisotropie der Korngrenzenbewegung.

Die Verhältnisse sind einfach in den Bereichen 2 und 3b. Im Bereich 2 ist die gleiche Rekristallisationstextur zu er- warten wie für den stabilen Mischkristall (Bereich 1). Die anschließende kontinuierliche Ausscheidung der Teilchen kann die Textur nicht mehr beeinflussen.

Bei in situ Rekristallisation sollte sich die makroskopisch gemessene (mit Röntgenstrahlen) Textur kaum von der Aus- gangstextur unterscheiden, da keine Umorientierung großer Volumina stattfinden kann. Mikroskopisch unterscheiden sich beide Zustände dadurch, daß kontinuierliche Orientierungs- änderungen die durch Versetzungen und elastische Spannungen hervorgerufen wurden, durch diskontinuierliche (i.e. Korn- grenzen) in fehlerfreien Kristalliten ersetzt wurden.

Es bleibt der Bereich 3a, in dem die Reaktionsfront mit den Teilchen in Wechselwirkung ist. Es sind drei Möglichkeiten für die Ausbildung der Textur denkbar. Es entsteht:

1) die gleiche Textur wie im Mischkristall (Bereich 1 und 2)

2) eine regellose Orientierungsverteilung

3) eine neue Rekristallisationstextur[+)].

Der Fall 1 ist zu erwarten, wenn die Rekristallisationskeime
große Strecken zur Wachstumsauslese zurücklegen könnten,i.e.
wenn der Volumenanteil der Teilchen mit der Zeit nicht stark
zunimmt. Der Fall 2 sollte auftreten, wenn regellos gebildete Rekristallisationskeime keine Möglichkeit zur Wachstumsauslese haben oder isotrop wachsen. Über die Voraussetzung
zur Isotropie ist nicht viel bekannt. Die Anisotropie soll
mit zunehmender Reinheit des Kristalls am stärksten sein[32]
und durch Segregation abnehmen. Es ist möglich, aber noch
nicht eindeutig nachgewiesen, daß Wechselwirkung mit dispergierten Teilchen (Kap. 4.1.2.2) und/oder eine hohe treibende Kraft p zu starker Isotropie führt. Ein Hinweis darauf
ist das Abschwächen oder Verschwinden der Textur unter diesen Umständen.

Das Entstehen einer neuen Textur bedingt eine andere Wachstumsauslese oder Keimbildungsmechanismus als im homogenen
Mischkristall. Da im Bereich 3a die Rekristallisation in
einem entmischten Grundgitter erfolgt, kann z.B. durch Änderung der Stapelfehlerenergie eine Änderung der Zwillings-
häufigkeit und damit der Rekristallisationstextur verbunden
sein.

Natürlich treten im Temperaturbereich um T_{II} und T_{III}
Mischtexturen auf, wenn in verschiedenen Teilen der Proben
verschiedene Reaktionen ablaufen. Diese sind aber keine neue
Rekristallisationstextur.

Dieser allgemeinen Behandlung sollen Beispiele folgen, die
die wichtigsten Erscheinungen an Hand von einfachen Legierungen zeigen.

4.1.3 Beispiele

4.1.3.1 Dispersionen nicht-verformbarer Teilchen in Kupfer

Derartige Dispersionen von Kristallarten können pulverme-

[+)] Falls sich nur Grenzen von Körnern der Walztextur ver-
schieben, kann damit eine Erhaltung oder Verschärfung
der Lagen der Walztextur verbunden sein. Dies ist besonders in der Nähe des Übergangs von Bereich 3a/3b zu erwarten (Abb.8).

tallurgisch (B_4C) [19] oder durch innere Oxidation von Mischkristallen (SiO_2) [18], (Al_2O_3) [33] oder durch Ausscheidung ($Al_2Cu = \theta$) [8] hergestellt werden.

An Cu-SiO_2 Legierungen ist eindeutig nachgewiesen worden, daß sowohl Beschleunigung als auch Verzögerung der Rekristallisation auftreten kann, je nachdem welcher Teilchenabstand vorhanden ist. Für $f \approx 2\%$, 30% plastischer Verformung durch Walzen, tritt eine meßbare Beschleunigung oberhalb eines Teilchenabstandes von $\approx 10^4$ Å auf [18]. Die starke Verzögerung der Rekristallisation ist durch Festhalten der Subkorngrenzen durch die Teilchen zu erklären. Allerdings erfolgt die Umlösung in den genannten Legierungen infolge der extrem niedrigen Löslichkeiten c_∞ so langsam, daß eine durch Teilchenwachstum kontrollierte in situ-Rekristallisation praktisch nicht auftreten kann (Kap.4.1.2.3).

Der Mechanismus der Beschleunigung der Rekristallisation (Abb.10) wurde an einer groben Dispersion von B_4C in Cu untersucht [19]. Bei einem Volumenanteil von $f = 0,04$ Vol%, und einem mittleren Teilchenradius $\bar{r} = 2 \cdot 10^4$ Å ergab sich ein Teilchenabstand $\bar{d} = 4 \cdot 10^5$ Å. Die Rekristallisation in stark verformten Proben (90%) beginnt an allen Grenzflächen

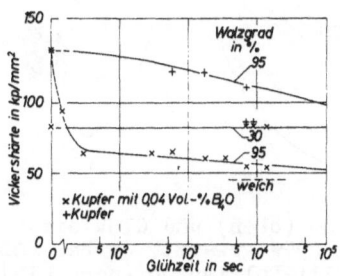

Abb.10. Beschleunigung der Rekristallisation durch in Kupfer dispergierte B_4C-Teilchen bei hohen Verformungsgraden (nach [19])

zwischen den unverformten Teilchen und dem Kupfer [11,12]. Durch das Umfließen der Teilchen ist dort die Verformung isotroper, die Versetzungsdichte höher als in einer Entfernung von $> 2\ \bar{r}$ von den Teilchen. Als erstes Stadium der

Rekristallisation bilden sich eine große Zahl von Keimen,
deren Abmessungen r_K sehr viel kleiner sind als der Teilchendurchmesser ($r_K < 10^3$ Å). Ihre beobachtete Zahl liegt
zwischen 10 und $5 \cdot 10^2$ je nach Größe des Teilchens. Bei
weiterem Wachstum sind davon bald nur noch wenige übrigge-
blieben, die solange weiterwachsen, bis sie auf Rekristallisationsfronten stoßen, die von anderen Teilchen ausgingen.
Es stellt sich eine Korngröße ein die demselben Teilchenabstand entspricht und die sich mit der Zeit nur noch langsam
ändert.

Die Untersuchung der primär gebildeten Keime mit Elektronen-
beugung zeigt, daß keine bevorzugte Orientierung auftritt.
Im Gegensatz dazu ist nach Beendigung des Wachstums eine
stark ausgeprägte Würfeltextur vorhanden. Diese Polfigur
wird in Abb.11 mit der von gleich behandelten reinem Kupfer

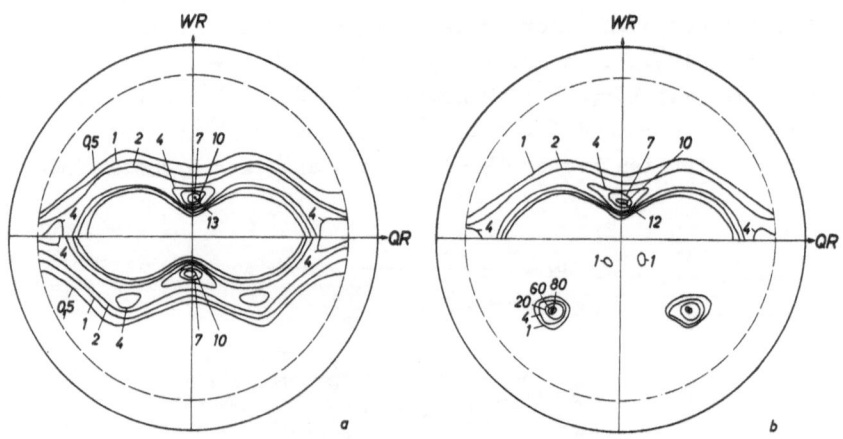

Abb.11a und b. Walz- (oben) und Glühtextur (unten) von gleich
behandeltem Kupfer (95% gewalzt, 4h bei 200°C angelassen).
{111}-Polfiguren (nach [19])

a) reines Kupfer   b) Kupfer mit 0,04 Vol% B_4C

verglichen bei dem die Walztextur noch vorhanden ist. Auch
wenn nach längeren Anlaßzeiten eine vollständige Rekristallisation erfolgen kann, ist die Textur nicht so ausgeprägt.
Das deutet darauf hin, daß nach dem Keimbildungsprozeß an
Teilchen günstige Voraussetzungen für Wachstumsauslese bestehen.

4.1.3.2 Reihenfolge der Reaktionen in Ni-Al und Ni-Cr-Al

Untersuchungen über das Rekristallisationsverhalten von übersättigten Ni-Al [34] und Ni-Cr-Al [16] Legierungen ergaben,
daß durch geeignete Variation von Konzentration, Verformungsgrad und Auslagerungstemperatur die Reihenfolge von
Ausscheidung und Rekristallisation geändert werden kann.

Das Ausscheidungsverhalten dieser Legierungen ist aus früheren Arbeiten bekannt [35,36]. Aus dem übersättigten γ-Misch-
kristall scheidet sich die geordnete γ'-Phase (Ni_3Al) völlig
kohärent im flächenzentrierten Gitter aus. Die Differenz der
Gitterkonstanten beträgt in der Ni-Al-Legierung etwa 0,4 %,
in Ni-Cr-Al < 0,05%. Außer der Verminderung der Kohärenz-
spannungen bewirkt das Chrom noch eine Abnahme der Löslichkeit von Aluminium in Nickel.

Bei einem vorgegebenen Verformungsgrad ε wird, ausgehend
vom homogenen Mischkristall, mit zunehmender Übersättigung
der Beginn der Rekristallisation verzögert, der Beginn der
Ausscheidung beschleunigt. Bei einer Übersättigung δ ≈ 10%
beginnen beide Prozesse gleichzeitig. Für drei Legierungen
mit verschiedener Übersättigung ergaben sich folgende charakteristische Fälle:

1) δ = 5% (Ni - 9,5 At% Al) t_R < t_A . Die Rekristallisation
   ist vor Beginn der Ausscheidung abgeschlossen, sie wird
   nicht durch Teilchen beeinflußt T_I > T > T_{II}. Die aus
   der Temperaturabhängigkeit des Rekristallisationsbeginns
   bestimmte Aktivierungsenergie betrug 2,4 eV, der gleiche
   Wert wurde auch in einer homogenen Legierung gefunden.

2) δ = 10% (Ni - 12,8 At% Al) t_R ≈ t_A. Die Rekristallisationskeime bilden sich bei höheren Temperaturen T > T_{II}
   vor, bei T < T_{II} nach Beginn der Ausscheidung. Im zweiten Fall wird die zur Bildung der Korngrenzen notwendige Versetzungsumordnung durch Teilchen behindert,
   die Teilchendichte reicht aber nicht aus, um eine Bewegung der Korngrenzen zu verhindern. Die Rekristallisation erfolgt diskontinuierlich. Aus der Temperaturabhängigkeit ergibt sich eine scheinbare Aktivierungsenergie von 4,1 eV.Dieser hohe Wert wird durch eine Abnahme der effektiven treibenden Kraft für die Versetzungsumordnung (d.h. einer Vergrößerung des Präex-
   ponentialfaktors in der Arrheniusgleichung) vorgetäuscht.

3) δ = 15% (Ni - 18,5 At% Cr - 7,5 At% Al) t_R > t_A, T < T_{III}.
   Teilchen hoher Dichte behindern die Bildung von Korngrenzen und verhindern deren Bewegung. Es findet eine Rekristallisation in situ statt. (Kap. 4.1.2.2).

Die Stapelfehlerenergie von Ni nimmt durch Zulegieren von
Aluminium im Bereich bis zu 13 At% nur geringfügig ab [37].
Nach Kap. 4.1.2.6 ist demnach als Rekristallisationstextur
im Fall 1 die Würfellage, im Fall 2 ein Gemisch aus Würfellage und Walztextur und im Fall 3 die völlige Erhaltung
der Walztextur zu erwarten.

Segregation von Aluminiumatomen in der Korngrenze führt
jedoch zu einer isotropen Beweglichkeit und damit zur Unter-
drückung der Wachstumsauslese. Die Polfiguren zeigen daher
eine starke Tendenz zur regellosen Orientierungsverteilung
in den Fällen 1. und 2.. Im Fall 3 tritt die zu erwartende
Erhaltung der Walztextur auf (Abb.12). Ein Vergleich mit
anderen Legierungssystemen (z.B. Ni-Be und Fe-Cu) zeigt,
daß im allgemeinen eine wesentlich geringere Übersättigung
von $\approx$ 5% ausreicht, um bei der gleichen Verformungsenergie
und der gleichen Teilchengröße eine Bewegung der Korngrenze
zu verhindern. Zur Deutung dieses Phänomens muß zusätzlich
zu den in Kap. 4.1.2.2 und 4.1.2.4 diskutierten Möglichkeiten der Wechselwirkung zwischen Korngrenzen und ausgeschiedenen Teilchen eine weitere betrachtet werden.

1. Teilchen verhalten sich gegenüber der Korngrenze stabil
   (Kap.4.1.2.2).

$$p_{eff} \; (1) = p_V - p_T \quad - (\frac{\delta G}{\delta V}) = p_V \qquad (13)$$

2. Die Korngröße bewirkt eine Umlösung der Teilchen in eine
   Anordnung geringerer Oberflächenenergie, z.B. größere
   Teilchen ($\Delta r > o$) oder Lamellen (Kap.4.1.2.4).

$$p_{eff} \; (2) = p_V + p_O \quad - (\frac{\delta G}{\delta V}) > p_V \qquad (14)$$

3. Die Teilchen werden in der Korngrenze aufgelöst und scheiden sich hinter der Korngrenze erneut aus. ($\Delta r < o$)

$$p_{eff} \; (3) = p_V - p_O \quad - (\frac{\delta G}{\delta V}) < p_V \qquad (15)$$

Der erste Fall tritt am häufigsten auf. Dagegen kann ein
Auflösen oder Umlösen der Teilchen in solchen Legierungssystemen erwartet werden, die eine geringe Lösungswärme
ΔH und eine geringe Energiedifferenz zwischen Matrix und
ausgeschiedener Phase aufweisen. (z.B. Ni-Al: $\Delta H \approx 2$ Kcal/Mol)

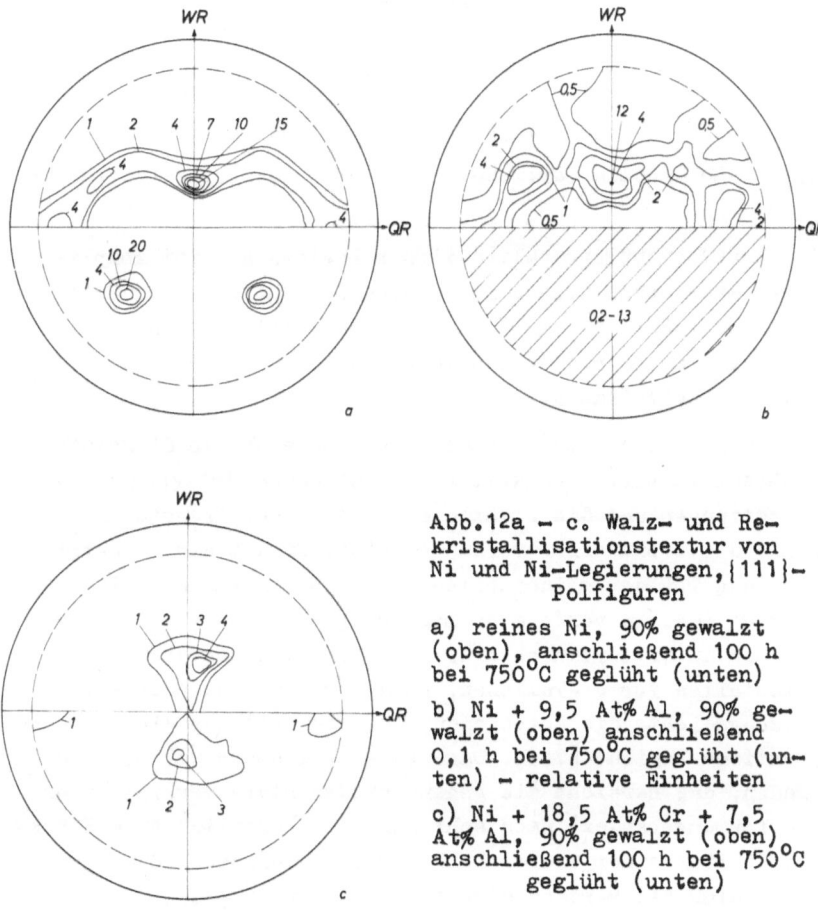

Abb.12a – c. Walz– und Re-
kristallisationstextur von
Ni und Ni-Legierungen, {111}-
Polfiguren

a) reines Ni, 90% gewalzt
(oben), anschließend 100 h
bei 750°C geglüht (unten)

b) Ni + 9,5 At% Al, 90% gewalzt (oben) anschließend
0,1 h bei 750°C geglüht (unten) – relative Einheiten

c) Ni + 18,5 At% Cr + 7,5
At% Al, 90% gewalzt (oben)
anschließend 100 h bei 750°C
geglüht (unten)

Eine Umlösung (Fall 2) ist dabei energetisch günstiger als
ein Auflösen (Fall 3). In Ni-Al und Ni-Cr-Al wird ein Umlösen nur in schwach verformten Legierungen (geringe Korn-
grenzengeschwindigkeit) beobachtet [36]. Widerstandsmessun-
gen und elektronenmikroskopische Untersuchungen in stark verformten Ni-Al-Legierungen [31] deuten darauf hin, daß bei
hoher Geschwindigkeit der Korngrenzen ein Auflösen der Teilchen stattfindet. Die Geschwindigkeit ist zu hoch, um die
diffusionsbegrenzte Umlösung zu ermöglichen. Bei einer weiteren Erhöhung der treibenden Kraft sollte für die gleiche

Teilchendispersion der Fall 1 auftreten. Die für die drei
Prozesse nötige Diffusionszeit verhält sich

$$t(1) = o < t(3) < t(2) \tag{16}$$

Hier liegt also ein Fall vor, in dem die Pinnkraft der Teilchen von der Geschwindigkeit der Korngrenze abhängt.

4.1.3.3 Diskontinuierliche und in situ-Rekristallisation in
Al-Cu

Das Ausscheidungsverhalten dieser Legierungen ist kompliziert aber exemplarisch für den häufig auftretenden Fall,
daß sich außer der stabilen mehrere metastabile Phasen ausscheiden. Diese wiederum bevorzugen verschiedene Defekte als
Orte der Keimbildung.

In unverformten Mischkristallen bilden sich die θ"-Phasen
kohärent im perfekten Gitter, die θ'-Phase teilkohärent an
Versetzungen und die stabile θ-Phase an Korngrenzen. Falls
keine Gitterbaufehler vorhanden sind, findet ein direkter
Übergang von der metastabilen zur stabilen Phase statt:
θ" → θ' → θ. In verformten Legierungen dominiert die Keimbildung an Gitterbaufehlern. Alle Versetzungen dienen als
Keimstellen für θ'-Teilchen, so daß mit zunehmender Versetzungsdichte die Ausscheidung mit der Bildung dieser Phase
praktisch beginnt [7,9]. Dazu kommt bei hohen Verformungsgraden, daß Bereiche die gegeneinander stark verkippt sind
(Verformungsbänder) nach sehr kurzen Anlaßzeiten eine Struktur ähnlich einer Großwinkelkorngrenze annehmen können, an
denen sich die θ-Phase direkt bilden kann [7,38,39].

Das Ausheilen der Versetzungen kann wie in Kap. 4.1.2.2
und 4.1.2.3 beschrieben, kontinuierlich und diskontinuierlich
erfolgen. Der Ablauf der in situ Rekristallisation soll an
Hand von Abb. 5 beschrieben werden: Die durch die Verformung
erzeugten Versetzungen werden durch Aufspaltung zur Bildung
von θ'-Teilchen sehr effektiv festgehalten. Ihre Dichte ändert sich nur durch Auflösen der θ'-Teilchen. Bei kleiner
Ausgangsversetzungsdichte finden sich einzelne Versetzungen,
bei großer Versetzungsnetze zwischen den Teilchen (Abb.5a+b).
Der Umlösung der θ'-Teilchen überlagert sich die Umwandlung
der größten Teilchen in die stabile θ-Phase. Mit dieser

Phase ist ein Al-Mischkristall mit geringer Cu-Konzentration
im Gleichgewicht. Die Versetzungsnetze werden deshalb kaum
noch durch Segregation, sondern nur durch die θ - Teilchen
festgehalten (Abb. 5c). Wenn ein Knoten durch Umlösung von
seinem θ - Teilchen befreit ist, können die angrenzenden
Subkorngrenzen schnell ausheilen. Die Subkorngröße und die
Winkel zwischen den Körnern nimmt zu, bis ein Gefüge wie
in Abb. 5d entsteht, in dem Korngrenzen von 1° - 5° zu sehen
sind. Falls die Zahl der Versetzungen hinreichend groß (oder
die Übersättigung hinreichend klein) ist für die Bildung von
Rekristallisationsfronten, können diese sich unter drei verschiedenen Bedingungen bewegen (Abb. 13):

Abb. 13a: Vor der Rekristallisationsfront befinden sich
θ'-Teilchen, Versetzungsnetze, sowie wenige an besonderen Stellen (Knoten von Versetzungsnetzen) gebildeten θ - Teilchen. Die wirksame treibende Kraft
p setzt sich zusammen aus (Kap. 4.1.2.4):

$$+ p_V + p_{\theta' \rightarrow \theta} - p_\theta = p \qquad (17)$$

Die Umwandlung θ'→ θ fördert, die schon vorhandenen θ - Teilchen hemmen die Bewegung der Front.

Abb. 13b: Es sind Versetzungsnetze und θ - Teilchen vorhanden.
Es handelt sich um reine Rekristallisation ohne
überlagerte diskontinuierliche Ausscheidung.

$$+ p_V - p_\theta = p \qquad (18)$$

Abb. 13c: Eine Korngrenze bewegt sich durch eine Dispersion
Al-θ ohne Gitterbaufehler. Die treibende Kraft ist
lediglich die Erniedrigung der Korngrenzenenergie
= Kornwachstum.

Eine Vorhersage, unter welchen Bedingungen welcher Prozess
auftritt, ist aus den in Kap. 4.1.2 besprochenen allgemeinen
Prinzipien möglich.

Mit zunehmender Übersättigung ist bei konstanter Versetzungsdichte zunehmende Behinderung der Versetzungsumordnung zu
Rekristallisationskeimen und damit Auftreten von in situ-
Rekristallisation zu erwarten (Abb.5). Eine zunehmende Versetzungsdichte führt sowohl zur Bildung von Rekristallisa-

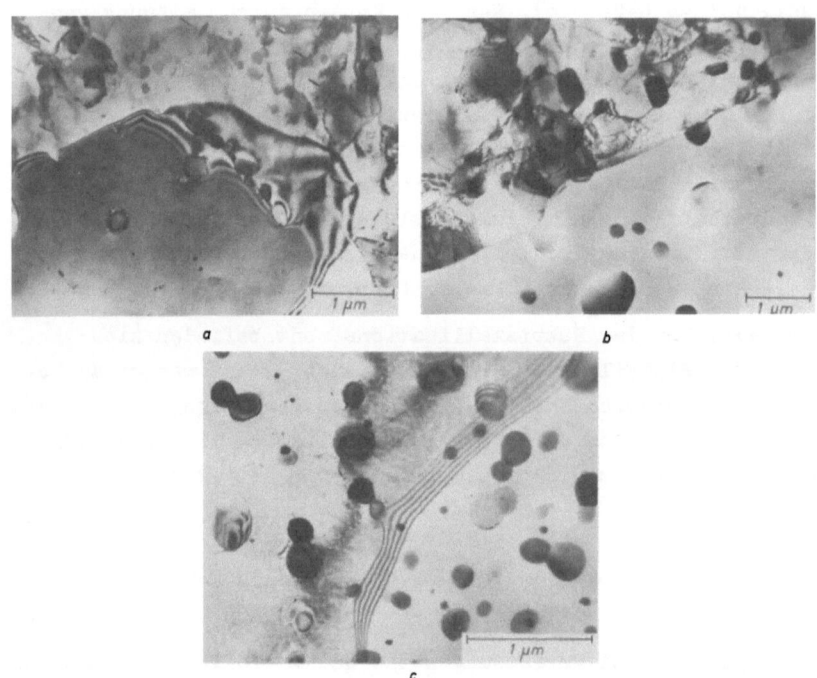

Abb. 13a - c. Diskontinuierliche Rekristallisation in Al-Cu-
Legierungen

a) Umwandlung θ'→ θ und Auflösung der Versetzungsnetze durch
die Reaktionsfront. Al + 2 Gew% Cu, 90% verformt, isochron
auf 300°C aufgeheizt
b) Auflösung der Versetzungsnetze und Wechselwirkung mit
vor der Front ausgeschiedenen θ - Teilchen. Al + 5 Gew% Cu,
    50% verformt, isochron auf 400°C aufgeheizt
c) Wechselwirkung mit θ - Teilchen beim Kornwachstum. Al +
5 Gew% Cu, 90 % verformt, isochron auf 350°C aufgeheizt

tionskeimen als auch zu Keimstellen für die stabile θ - Phase.

Das führt dazu, daß die Reaktion von Abb.13a nur in einem

mittleren Bereich von Versetzungsdichten auftreten kann, in

dem sich einige Rekristallisationsfronten, aber nicht zu

viele θ - Teilchen vor der Front bilden können. Diese Reak-

tion tritt deshalb bei höheren Verformungsgraden und hoher

Übersättigung nicht mehr auf. Die Reaktion Abb.13b ist zu
erwarten, wenn bei hohen Versetzungsdichten und geringer
Übersättigung (i.e. Teilchendichte) die Bedingung $p_v > p_\Theta$
erfüllt ist.

Die Texturen, die in verschiedenartig wärmebehandelten Al-
Cu-Legierungen auftreten, können eindeutig dem mikroskopi-
schen Ablauf der Rekristallisation zugeordnet werden [40]:
In situ Rekristallisation führt zur Beibehaltung der Walztextur (Abb.14a), die diskontinuierliche Reaktion zur Würfellage als Rekristallisationstextur (Abb.13a, b, 14b), falls
die Reaktion vollständig nach diesem Mechanismus ablaufen
konnte. Das ist jedoch bei hohen Cu-Gehalten und hohen Verformungsgraden nicht möglich. Es treten dann gemischte Texturen (Walz- + Würfellage) auf, die kennzeichnend für Glühtexturen vieler ausscheidungsfähiger Al-Legierungen sind
(Abb.14c).

4.1.3.4 Diskontinuierliche Ausscheidung und Rekristallisation in Cu-Co und Ni-Be

In unverformten Ni-Be-Legierungen tritt in einem begrenzten Temperatur-Konzentrationsbereich eine diskontinuierliche
Ausscheidung auf [36]. Die sich als Reaktionsfront bewegen-
den Korngrenzen bewirken eine lamellenförmige Ausscheidung
der Gleichgewichtsphase NiBe oder eine Umwandlung der bereits ausgeschiedenen metastabilen Phase in die Gleichgewichtsphase. Die Reaktion kommt zum Stillstand, wenn sich
vor der Korngrenze Teilchen der stabilen Phase bilden können.

In unverformten Cu-Co-Legierungen scheidet sich aus dem
übersättigten Mischkristall Kobalt mit etwa 10 At% Kupfer bei
$T > 340^\circ C$ als kfz Mischkristall in Form kleiner kohärenter
Teilchen aus. Die Teilchen sind statistisch verteilt. Durch
eine Verformung von 5 bis 20% kann eine diskontinuierliche
Ausscheidung hervorgerufen werden. Dabei entsteht hinter der
Reaktionsfront eine kettenförmige Anordnung der Teilchen [34].
Die Differenz der Gitterkonstanten von 0,4% ist zu gering,
um eine Änderung der Wachstumsrichtung infolge der Verspannung zu induzieren (Abb. 15b). Die Bewegung der Korngrenzen
erfolgt isotrop. Geht man davon aus, daß jedes Korn mit der

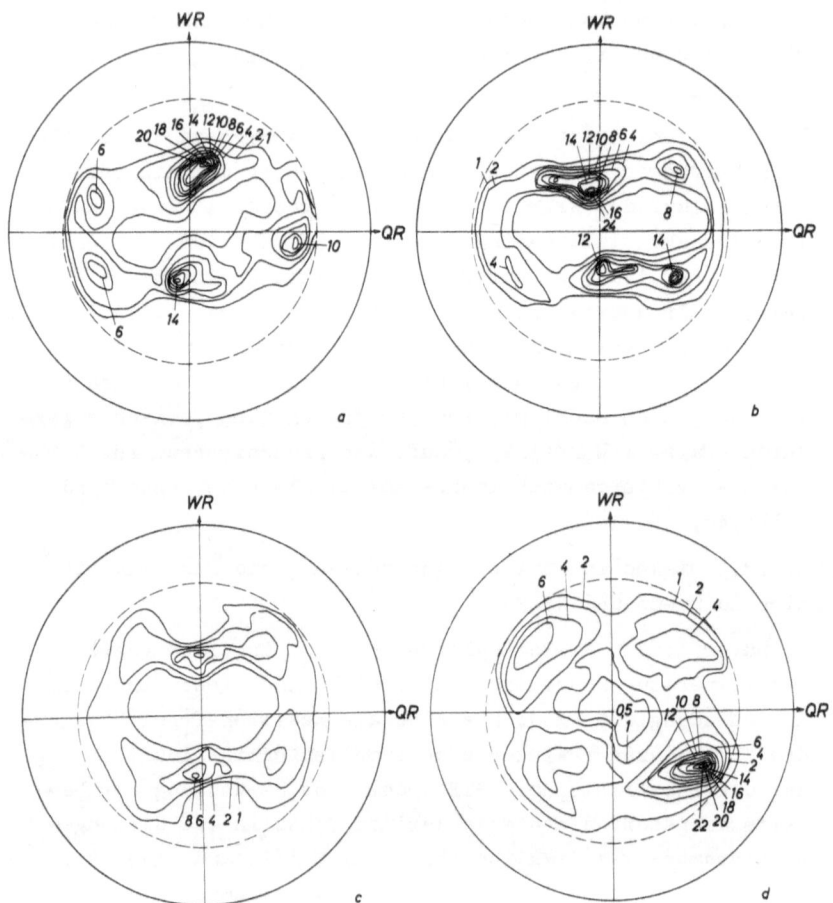

Abb.14 a – d. Glühtexturen der Al–Cu–Legierungen, {111}–Polfiguren – relative Einheiten

a) Walztextur von Al + 2 Gew% Cu, 90% bei 20°C gewalzt
b) Beibehaltung dieser Textur nach dem Weichglühen in Al + 5
Gew% Cu, 90% gewalzt, 140h bei 280° geglüht (in situ Rekristallisation)
c) Mischtextur aus Walztextur und Würfeltextur, Al + 2 Gew%
Cu 90% verformt, isochron auf 310° aufgeheizt
d) Würfeltextur, die Rekristallisation erfolgte diskontinuierlich, Al + 2 Gew% Cu, 90% gewalzt, 0,5 h bei 320°C geglüht

gleichen Wahrscheinlichkeit wachsen oder schrumpfen kann, so
ist im Mittel keine Änderung der Orientierungsverteilung zu
erwarten. Der Laufweg der Korngrenzen ist zu gering, um eine

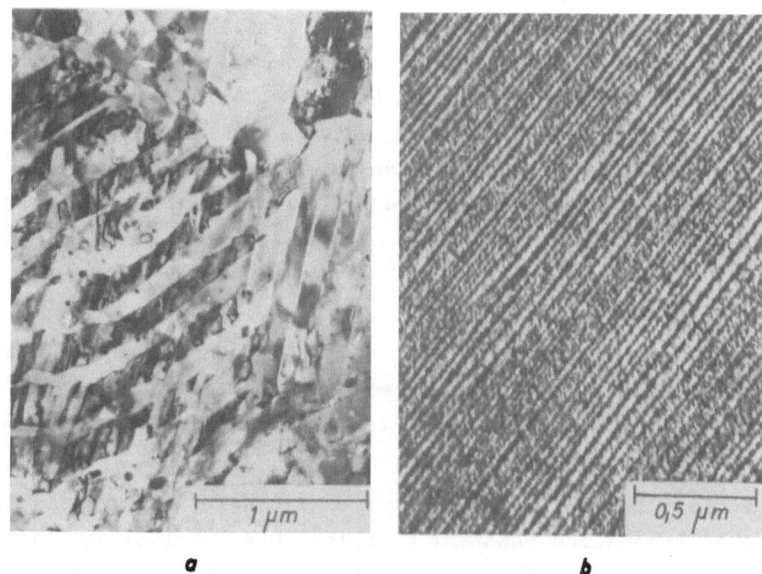

a   b

Abb. 15a und b. Diskontinuierliche Ausscheidung in Ni-Be und
Cu-Co-Legierungen

a) Änderung der Wachstumsrichtung der Reaktionsfront führt
zur Aufteilung in kleine Gebiete regelloser Orientierungsverteilung (Korngröße nach dem Walzen $\approx$ 150 μ). Ni + 7,3 At%
Be, 50% gewalzt, 1h 600°C
b) Keine Änderung der Wachstumsrichtung in Cu + 2,2 At% Co,
10% gewalzt, 10h 550°C

Erhaltung der Ausgangstextur nachweisen zu können.

Das Rekristallisations- und Ausscheidungsverhalten einer
Ni-7,3 At% Be-Legierung ist in Abhängigkeit vom Verformungsgrad untersucht worden [34]. Eine Verformung bewirkt eine
Beschleunigung der diskontinuierlichen Ausscheidung durch
eine Erhöhung der treibenden Kraft um p_v (Kap. 4.1.2.4) [41]
und eine Beschleunigung bei der Bildung der Gleichgewichtsphase, die an Versetzungen eine geringere Keimbildungsener-
gie erfordert.

Bei geringen Verformungsgraden ($<$ 50%) überwiegt der Ein-
flüß auf die diskontinuierliche Ausscheidung; bevor sich die
Gleichgewichtsphase an Versetzungen bilden und dadurch einen

Stillstand der diskontinuierlichen Reaktion verursachen
kann, hat diese bereits 70 - 100% des Volumens erfaßt.

Bei hohen Verformungsgraden (> 60%) bewirkt die hohe Versetzungsdichte zusätzlich zur Verminderung der Keimbildungs-
energie durch erhöhte Diffusion entlang von Versetzungska-
nälen eine weitere Beschleunigung bei der Bildung der stabilen Phase. Sie scheidet sich aus, bevor die diskontinuierliche Reaktion beginnen kann. Die diskontinuierliche Ausscheidung wird verhindert, die Rekristallisation erfolgt in
situ.

In einer 50% verformten Probe erfaßt die diskontinuierliche
Ausscheidung das gesamte Volumen, während in einer 90% verformten Probe die diskontinuierliche Ausscheidung auf Kosten einer in situ Rekristallisation völlig unterdrückt werden konnte.

Für diese beiden Grenzfälle wurden die Walz- und Rekristallisationstexturen bestimmt (Abb.16). Die Rekristallisationstextur der 50% verformten Proben zeigt eine statistische

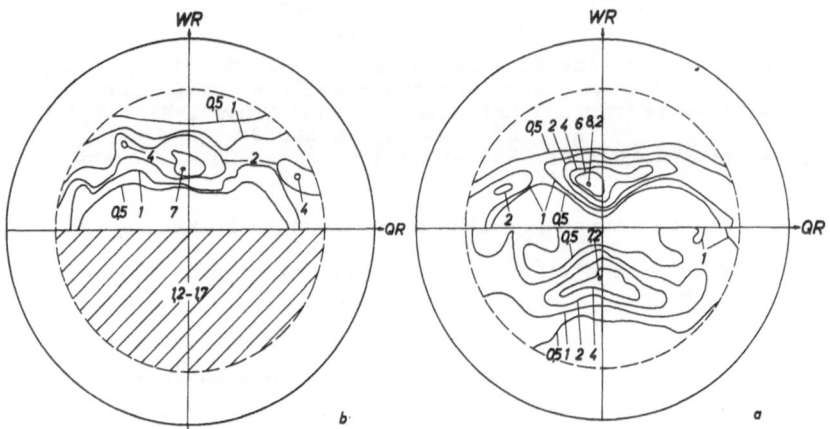

Abb.16. Walz- und Rekristallisationstexturen einer Ni + 7,3
    At% Be-Legierung, {111}-Polfigur, relative Einheiten
a) 90% gewalzt (oben), anschließend 200 h bei 500°C + 16 h
        bei 550° geglüht (unten)
b) 50% gewalzt (oben), anschließend 26 h bei 600°C geglüht
        (unten)

Orientierungsverteilung der Körner. Die Volumenänderung von
ca. 3% bei der diskontinuierlichen Ausscheidung führt zur
Aufteilung der Kristalle in kleine Bereiche regelloser Orientierung (Kap. 4.1.2.4). Abb. 15a zeigt diese Aufteilung. Bei
der Rekristallisation der 90% verformten Probe blieb die
Walztextur erhalten, was im Einklang mit den Ergebnissen
aus Kap. 4.1.2.6 steht.

Es ist zu erwarten, daß in 50 - 60% verformten Proben diskontinuierliche Ausscheidung und in situ Rekristallisation
nebeneinander auftreten und zu einer Überlagerung der beiden Texturen führen.

4.1.3.5 Ausscheidung vor und nach plastischer Verformung in
Al-Fe-Legierungen

Die maximale Löslichkeit von Eisen in Aluminium beträgt
0,026 At% bei 655°C. Dieses Element ist als Verunreinigung
im technischen Al immer vorhanden. Es hat einen großen Einfluß auf dessen Rekristallisationsverhalten [3-5, 42-44].
Folgende Erscheinungen sind zu deuten:

1. Sowohl Eisenatome in Lösung als auch in feiner Dispersion führen nach Verformung zu einer Verzögerung der Rekristallisation, während grob verteilte Teilchen eine
   Beschleunigung bewirken (Abb.17)[45].
2. Während in reinstem Al die Würfellage als Rekristallisationstextur (Abb.18a) auftritt, bewirkt ein Zusatz von
   sehr kleinen Mengen von Fe eine Beibehaltung der Walztextur (Abb. 18b) [42].
3. Durch Zulegieren von Elementen, die die Löslichkeit des
   Eisens in Al durch Bildung stabiler intermetallischer
   Phasen verringern, wird die Würfeltextur wieder hergestellt (Abb. 18c) [42,46].
4. Mit zunehmendem Verformungsgrad nimmt die Tendenz zur
   Beibehaltung der Walztextur zu [3].

Es ist nachgewiesen worden, daß diskontinuierliche Rekristallisation für die Entstehung der Würfellage, in situ Rekristallisation für die Erhaltung der Walztextur verantwortlich
sind [3]. Über den Ausscheidungsvorgang sind einige mikroskopische Untersuchungen gemacht worden [45,3,40,47]. Die Aus-

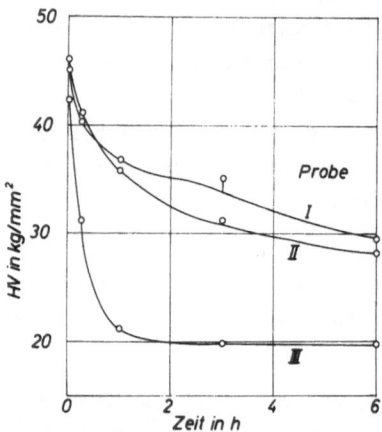

Abb.17. Einfluß des Ausscheidungszustandes auf den
Ablauf des Weichglühens einer Al + 0,04 Gew% Fe-Legie-
rung; nach 96,5% Verformung durch Walzen bei 20°C
wurde bei 300°C ausgelagert. Vor der Verformung hatte
         das Eisen folgende Verteilung:
Probe I   : homogener Mischkristall
Probe II : 20% des Fe ausgeschieden, $1.7 \cdot 10^{14}$ Teil-
              chen/cm³
Probe III : 85% des Fe ausgeschieden, $3 \cdot 10^{12}$ Teil-
        chen/cm³ (nach MIKI und WARLIMONT [45])

scheidung erfolgt als Al_3Fe (oder als eine noch unbekannte
metastabile aber nicht-kohärente Phase), aber mit meßbarer
Geschwindigkeit erst bei für Al-Legierungen hoher Temperatur
(> 300°C). Das kann durch einen geringen Diffusionskoeffizienten von Fe [45], oder aber durch eine hohe Aktivierungsenergie der Keimbildung von Al_3Fe erklärt werden. In verformten homogenen Mischkristallen sind folgende Anlaßstadien zu
erwarten:

a) Umordnung der Versetzungen zu Netzen, Segregation von Fe-
   Atomen an den Versetzungen, dadurch verzögernd der Umordnung.

b) Keimbildung an bevorzugten Stellen, offensichtlich nicht
   an den Versetzungslinien, sondern nur an einigen Knoten
   und dichten Netzen (Abb.19).

c) Wachstum der Subkörner ist bestimmt durch Umlösung der
   Teilchen, wie in Kap. 4.1.2.3 und 4.1.3.3 besprochen.

Entsprechende Stadien zu b) und c) sind zu erwarten, wenn

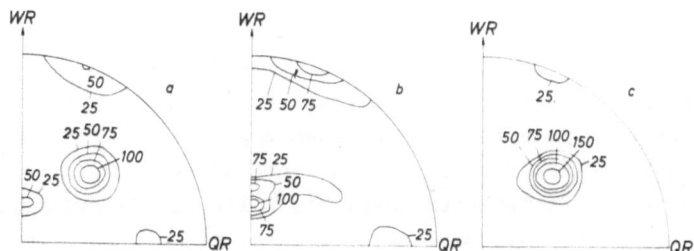

Abb. 18 a – c. Beeinflussung der Glühtextur von Aluminium durch Eisen, alle Legierungen wurden 90% verformt und 7h bei 320°C geglüht. {111}-Polfiguren, relative Einheiten (nach BUNK [42])

a) Reinstaluminium 99,99% Al, Würfellage
b) Al + 0,1 Gew% Fe, Walztextur weitgehend erhalten
c) Al + 0,1 Gew% Fe + 0,1 Gew% Be, Würfellage

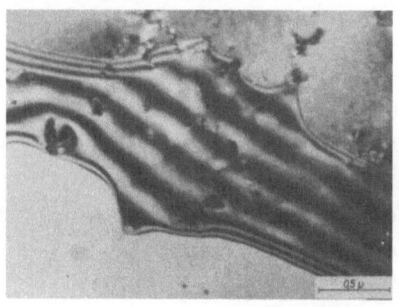

Abb. 19. Gefüge einer Al + 0,042 Gew.% Fe Legierung, 95% gewalzt, 2 min bei 440°C geglüht

die Teilchen fein verteilt ausgeschieden sind. Die Versetzungen ordnen sich zu Netzen an diesen Teilchen um, die dann deren Weiterwachsen bestimmen (Abb. 18).

Falls kein Eisen vorhanden ist, oder falls es in Form von grob verteilten Teilchen vorliegt, wird die Umordnung der Versetzungen durch Segregation oder kleine Teilchen nicht gehindert. Es können sich Keime für diskontinuierliche Rekristallisation bilden, die im reinen, oder durch vorherige

Ausscheidung gereinigten Al günstige Bedingungen für Wachstumsauslese vorfinden.

Der Übergang zu in situ Rekristallisation bei hohen Verformungsgraden kann ähnlich wie das Verhalten der Legierungen
mit diskontinuierlicher Ausscheidung erklärt werden (Kap.
4.1.2.4 und 4.1.3.5). Bei mittleren Verformungsgraden sind
Versetzungen vorhanden, an die Eisen segregiert, aber noch
wenige Keimstellen für Al_3Fe, von denen aus eine grobe Verteilung von Al_3Fe Teilchen entsteht. Falls ein Rekristalli-
sationskeim entsteht, so findet er nicht nur die Energie
der Versetzungen, sondern auch die in den Segregationszonen
oder in Lösung befindlichen Eisenatome als treibende Kraft.
$p = p_v + p_{(Fe)}$. Ist der Verformungsgrad hoch, so ist die
Keimdichte durch die größere Versetzungsdichte höher. Infolge der größeren Zahl von Keimstellen für Al_3Fe ist aber
die Entmischung weiter fortgeschritten und dabei vor der
Rekristallisationsfront Teilchen in kleiner Dispersion entstanden, die deren Bewegung hemmen.

$$p = p_v - p_{(Al_3Fe)} \tag{19}$$

Literatur

1. D. Merz und G. Wassermann: Z. Metallkde. 56(1965)516.
2. I. Miki und H. Warlimont: Z. Metallkde. 59(1968)408.
3. J. Grewen und M.v. Heimendahl: Z. Metallkde. 59(1968)205.
4. W. Bunk: Z. Metallkde. 57(1966)345.
5. W. Normann: Z. Metallkde. 58(1967)151.
6. G. Masing, K. Lücke und P. Nölting: Z. Metallkde. 47(1956)64.
7. T.J. Koppenaal und M.E. Fine: Trans.Met.Soc.AIME 221(1961)1178.
8. R.D. Doherty und J.W. Martin: J. Inst.Metals. 31(1963)332.
9. U. Köster und E. Hornbogen: Z. Metallkde. 59(1968)792.
10. W.C. Leslie: Trans.Met.Soc. AIME 221(1961)752.
11. W. C. Leslie, J.T. Michalak und F.W. Aul in: Iron and Its Dilute Solid Solutions, Interscience 1963, S.119.
12. W.C. Leslie, J. T. Michalak und R.J. Sober: Trans. Amer. Soc.Metals 58(1965)673.
13. G. Venturello, C. Antonione und F. Bonaccorso: Trans.Met. Soc.AIME 227(1963)1433.
14. H. Hu: Vortrag auf der DGM Tagung in Hannover 1968.

15. E. Hornbogen und H. Kreye: Z. Metallkde. 57(1966)122.

16. F. Haeßner, E. Hornbogen und M. Mukherjee: Z. Metallkde. 57(1966)270.

17. A.K. Chakraborty und E. Hornbogen: Z. Metallkde. 58(1967) 46.

18. F.J. Humphreys und J.M. Martin: Acta Met. 14(1966)775.

19. F. Haeßner, E. Hornbogen und M. Mukherjee: Z.Metallkde. 57(1966)171.

20. J. Grewen und W. Leo: Z. Metallkde. 59(1968)770.

21. J. Grewen und B. Scholz: Metall 22(1968)1119.

22. E. Hornbogen: Z. Metallkde. 56(1965)133.

23. C. Zener, zitiert in: C.S. Smith: Trans. AIME 175(1948)15.

24. M.T. Ashby und J. Lewis: Harvard Univ. Techn. Rep. 547 (1967).

25. M.T. Ashby und J.G. Palmer: Acta Met. 15(1967)420.

26. C. Wagner: Z. Elektrochem. 95(1961)581.

27. A.J. Ardell und R.B. Nicholson: Acta Met. 14(1966)1295.

28. H. Kreye: Priv. Mitteilung

29. J.C.M. Li: Recrystallization, Grain Growth and Textures H. Margolin, Hersg., Amer. Soc. Metals, Metals Park, Ohio 1966, S. 45.

30. C.S. Smith: Trans.Amer.Soc.Metals 45(1953)564.

31. C.A. Johnson: Private Mitteilung.

32. H. Hu: Bisher unveröffentl. Ergebnisse über reines Al.

33. J.L. Brimhall, M.J. Klein und R.A. Huggins: Acta Met. 14(1966)459.

34. H. Kreye: Dissertation, Stuttgart 1968.

35. R.O. Williams: Trans.Met.Soc. AIME 215(1959)1026.

36. E. Hornbogen und M. Roth: Z. Metallkde. 58(1967)842.

37. B.E.D. Beeston und L.K. France: J. Inst. Metals 96(1968) 105.

38. E. Hornbogen: Aluminium 43(1967)166.

39. M.v. Heimendahl: Acta Met. 15(1967)417.

40. H. Ahlborn, E. Hornbogen und U. Köster: Unveröffentlichte Ergebnisse.

41. H. Böhm, M. v. Heimendahl und G. Vierling: Z. Metallkde. 52(1961)746.

42. W. Bunk: Z. Metallkde. 56(1965)645.

43. J. Grewen: Z. Metallkde. 57(1966)418.

44. J. Grewen: Z. Metallkde. 59(1968)236.

45. I. Miki und H. Warlimont: Z. Metallkde. 59(1968)254.

46. H. Hug: Metall 14(1960)885.

47. K. Holm: Diplomarbeit, Göttingen 1969.

4.2 Formation of Recrystallized Grains in Heavily Rolled Sheet — Influence of Chemical Driving Force on the Formation of Recrystallized Grains —
Entstehung von rekristallisierten Körnern in stark gewalztem Blech — Einfluß der chemischen treibenden Kraft auf die Bildung rekristallisierter Körner —

by S. Horiuchi and I. Gokyu[+)]

Zusammenfassung

In einer Al-Legierung mit 11,3% Zn, die um 95% bei -70°C gewalzt und dann für kurze Zeit (< 1 Std.) bei Raumtemperatur gelagert worden war, wurden rekristallisierte Körner (< 2μⵁ) beobachtet. Diese Körner wuchsen bei weiterer Lagerung bei Raumtemperatur sehr langsam. Jedes Korn bildete mit der es umgebenden verformten Matrix eine Großwinkelkorngrenze. In den Körnern schieden sich Stäbchen der ß-Phase (Zn) aus, deren Längsrichtung senkrecht zur Korngrenze stand.

Diese Beobachtungen werden aufbauend auf der Gleichung für die Korngrenzenwanderung V = MP diskutiert, in der V die Geschwindigkeit, M die Beweglichkeit und P die treibende Kraft sind. Dabei ergibt sich, daß P einen neuen Typ von treibender Kraft, Pc, enthält. Tritt eine Ausscheidung an der wandernden Korngrenze auf, so verringert sich die freie chemische Energie im rekristallisierten Bereich. Diese Herabsetzung der freien Energie verursacht die treibende Kraft der Korngrenzenwanderung Pc. Pc wurde berechnet und liegt in der Größenordnung von 10^8dyn/cm^2.

4.2.1 Introduction

Under the assumption that the grain is a sphere of radius R, the migration velocity V of a grain boundary is given by the following equations,

$$V = \frac{dR}{dt} = MP \qquad (1)$$

$$P = Pe + Pf + Pr \qquad (2)$$

[+)] Dept. of Metallurgy, Fac. of Engineering, University of Tokyo, Hongo, Bunkyo-ku, Tokyo, Japan

where P is the driving force for boundary migration: Pe due
to the strain energy, Pf due to the interaction of solute
atoms or of precipitates with boundary and Pr due to the
energy of boundary. M is the mobility.

During deformation of an alloy, in which GP zones or
clusters exist, they are broken into pieces and after the
deformation of high reduction solute atoms are redistributed
in the matrix nearly at random. If precipitation could occur
at a moving boundary, during annealing immediately after
rolling, the chemical free energy would reduce in the region
passed by the boundary. This reduction in the chemical free
energy induces the driving force Pc, called here the
chemical driving force. Then, eq.(2) should be rewritten as
follows:

$$P = Pe + Pf + Pr + Pc \qquad (3)$$

In the present study the values of Pe, Pf, Pr, Pc and M can
be obtained by direct measurements or calculations. On the
basis of these values the kinetics for the process, in which
recrystallization nuclei grow, is discussed in terms of eq.(1).

4.2.2 Experimental Procedure

The chemical composition of Al-Zn alloy used is shown in
Table 1. This alloy was quenched into water ($0^{o}C$) after
solution-treatment at $500^{o}C$, left at room temperature for 1
day and rolled 95 % at $- 70^{o}C$ (dry ice - ethanol). The final
thickness of the rolled sheet was 0.2 mm. The sheet was
annealed at $20^{o}C$. A part of the sheet was further annealed
at $200^{o}C$ to $250^{o}C$. Specimens were thinned by electropolishing
and observed in an electron microscope (JEM150).

Table 1. Chemical composition of the specimen

component	Zn	Fe	Si	Cu	Pb	Al
wt %	11.27	0.013	0.005	0.0008	tr.	bal.

4.2.3 Experimental Results

A structure in the specimen annealed at $20^{o}C$ for about 1 hr
is shown in Fig. 1. Dislocations of high density form complex

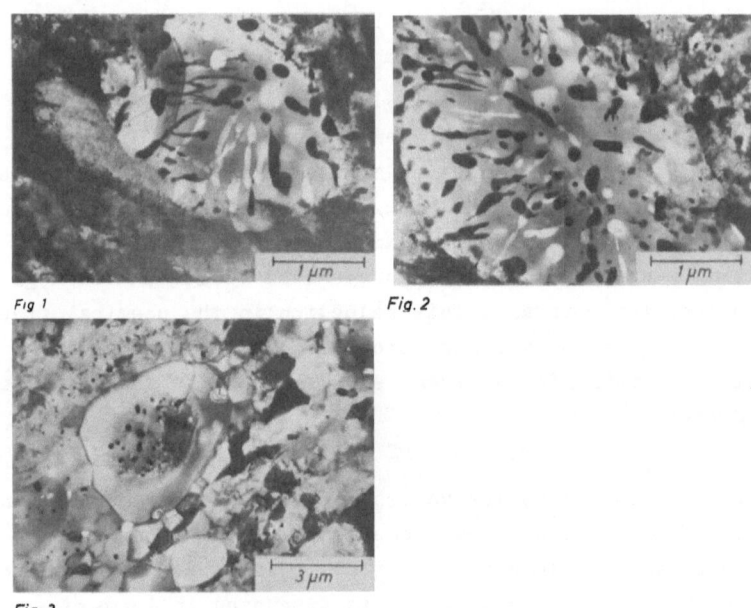

Fig 1

Fig. 2

Fig. 3

Figs. 1 – 3. Transmission electron micrographs of an Al–11,3% Zn alloy, quenched into water after solution-treatment, aged for 1 day at room temperature, rolled 95 % at –70° and finally annealed for 1 hr at 20°C (Fig.1) and for 30 days at 20°C (Fig.2). The specimen in Fig.3 was annealed for 1 hr at 20°C and then for 30 min at 200°C. The micrograph shows a recrystallized grain and the polygonized matrix

arrangements in the deformed matrix. The GP zones, which existed before rolling [1], cannot be observed, since they were destroyed during rolling. Recrystallized grains smaller than 2 μ in diameter appear in the matrix. Their dislocation density is very low, indicating that the grains formed after rolling. Precipitates of β phase are observed as well in the matrix as in the grains. The precipitates in the latter are elongated in the direction normal to the boundary; they grow in the same direction as the boundary migrates. In most cases, a large angle boundary is formed between a grain and the surrounding matrix.

On further annealing at 20°C grains grow very slowly. Fig.2

shows a structure in the specimen annealed at 20°C for 30 days. The density and arrangement of the dislocations in the matrix and the dimensions and the arrangement of the precipitates as well in grains as in the matrix are very much similar to those in the specimen annealed for 1 hr (Fig.2). The average diamter of large grains is 4μ and hence the average velocity of boundary migration is 3.8×10^{-10}cm/sec.

A structure of a specimen annealed at 20°C for 1 hr and further at 200°C for 30 min is shown in Fig.3. Polygonization has occurred in the matrix. Polygons are uniformly small ($< 1.5\mu$) and their orientations are nearly the same as those before annealing at 200°C. The grains become larger and consist of the shell of about 1.5μ in thickness and the core smaller than 2μ in diamter. The latter corresponds to the grains formed at 20°C and contains precipitates. The average growth velocity of the shell is 8.3×10^{-8}cm/sec.

At the initial stage of annealing at 250°C, an average growth velocity is 3.7×10^{-7}cm/sec.

4.2.4 Discussion

The magnitude ΔF of reduction in chemical free energy by precipitation at a moving boundary is assumed to be equal to that by precipitation from super-saturated solid solution, which was obtained by quenching after solution treatment. Then [2]:

$$\Delta F = ZVe[Co(1-Co)-C_i(1-C_i)] + kT[ColnCo-C_ilnC_i + (1-Co)ln(1-Co)-(1-C_i)ln(1-C_i)] \qquad (4)$$

where the solute concentration in the matrix changes from Co to C_i by the precipitation, Z is a coordination number, k Boltzmann's const., T absolute temperature and Ve an exchange interaction energy. From the phase diagram, Ve = 1.1×10^{-2}eV. By substituting Co = 4.8×10^{-2} and $C_i = 1.0 \times 10^{-2}$ into eq.(4),

$$\Delta F = 1.4 \times 10^{-3} \text{ eV/atom}$$

The change ΔEc in chemical free energy per unit volume by precipitation is

$$\Delta Ec = (4/a^3) F \qquad (5)$$

where a is the lattice constant. Finally Pc is given by

$$Pc = -\frac{1}{4\pi R^2} \frac{d}{dR} \left[\frac{4}{3}\pi R^3 (-\Delta Ec)\right] = \Delta Ec \qquad (6)$$

$$= 1.3 \times 10^8 \text{ dynes/cm}^2$$

When the precipitation can easily occur at a moving boundary, we have only to take into account the jump of atoms across the boundary. Hence after LÜCKE and STÜWE [3] M in eq.(1) can be written by the following equations:

$$M = D \cdot \exp(-\frac{Q}{kT}) \qquad (7)$$

$$D = \frac{\alpha b \gamma}{NkT} \qquad (8)$$

$D=7.3 \times 10^{-5}$ for $b=2.9 \times 10^{-8}$ cm, $N=6.0 \times 10^{22}/\text{cm}^3$ and $\alpha=1$. The orientation dependence of D has not yet been formulated. It is assumed to be sinusoidal:

$$D = \frac{\alpha b \gamma}{NkT} \frac{\sin(\theta/2)}{\sin(\pi/12)} \qquad (8')$$

Here θ is an orientation difference between the grain and its neighboring matrix. Substitution of values, V_1 at $T_1 = 200°C$, V_2 at $T_2 = 250°C$ and $\theta=\frac{\pi}{6}$, into eq.(1) leads to:

$$Q = \frac{k}{(1/T_1 - 1/T_2)} \ln (\frac{V_2}{V_1}) = 0.6 \text{ eV}$$

where the driving forces are assumed to be equal at 200°C and 250°C.

Pr is given by the following equation:

$$Pr = -\frac{1}{4\pi R^2} \frac{d}{dR}(4\pi R^2 \sigma) = -\frac{2\sigma}{R} \qquad (9)$$

where σ is the boundary energy. σ is conventionally written as

$$\sigma = A\theta(B-\ln\theta) \qquad (10)$$

where A and B are constants and assumed here to be 1.0×10^3 and $1+\ln\frac{\pi}{6}$, respectively.

After prolonged annealing at 20°C, Pc $\cong$ 0. Then, from the result obtained with regard to migration velocity:

$$P = Pe+Pf+Pr = 5.2 \times 10^4 \quad dynes/cm^2$$

Since $Pr=-1.0 \times 10^7$ dynes/cm^2 for $\theta=\pi/6$ and $R=1.0 \times 10^{-4}$,

$$Pe+Pf = 1.0 \times 10^7 \quad dynes/cm^2$$

At the initial stage of annealing, on the other hand, $P=Pe+Pf+Pr+Pc=1.4 \times 10^8$ dynes/cm^2. From the equation, $V = 0$, the critical radii Rc are calculated to be 7.4×10^{-6} cm for the migration of large angle boundary $(\theta=\frac{\pi}{6})$ and 1.1×10^{-6} cm for the migration of small angle boundary $(\theta=\frac{\pi}{180})$.

The kinetics for the growth of nuclei is discussed in terms of eq.(1) under the assumption that Pe+Pf, Pc and Q are constant, which applies for the initial stage of annealing.

4.2.4.1 Kinetics for the growth where angle remains constant during boundary migration

4.2.4.1.1 For the large angle boundary $(\theta=\pi/6)$. Eq.(1) can be written as

$$\frac{dR}{dt} = G + \frac{H}{R} \tag{1'}$$

where $G=M(Pe+Pf+Pc)$, $H=-M\sigma$ and $M=7.3 \times 10^{-5} \cdot \exp(-\frac{Q}{kT})$. Solving eq.(1') analytically:

$$\frac{1}{G}R - \frac{H}{G^2} \ln\left|R + \frac{H}{G}\right| = t + C' \tag{11}$$

Giving the initial conditions, $t=0$ and $R=Rc=7.4 \times 10^{-6}$, $C'= -42$ for $T=20°C$. From eq.(11), the relation of R vs. t is shown as the curve 1 in Fig.4. The nucleus can grow very fast.

4.2.4.1.2 For the small angle boundary $(\theta=\pi/180)$. From eq.(7), $M=2.5 \times 10^{-6} \cdot \exp(-\frac{Q}{kT})$. Solving eq.(1) with the initial conditions, $t=0$ and $R=Rc=1.1 \times 10^{-6}$, the relation of R vs.t can be expressed as the curve 2 in Fig.4. The nucleus grows very slowly.

4.2.4.2 Kinetics for the growth where angle gradually increases

It is assumed that in a range, $R<Rs$, the angle increases as a linear function of R as $\theta=\frac{\theta s}{Rs} \cdot R$, while in another range, $R \geq Rs$, $\theta=\theta s=\pi/6$. Then from the equation, $V = 0$,

318

$$Rc = \frac{Rs}{\theta s} \cdot exp(B - \frac{P-Pr}{2A} \cdot \frac{Rs}{\theta s}) \qquad (12)$$

On the basis of eq.(12) the relation of Rs vs. Rc is shown in Fig.5. The smaller Rs the larger Rc. The nucleus of radius R, which is larger than Rc, can grow. Eq.(1) can be solved numerically. For example, the relation of R vs. t is shown as the curve 3 in Fig.4 for the case of Rs=1.0x10^{-5} and Rc=6.9x10^{-6}. At point 1 in Fig.5, the nucleus grows from the beginning with θ=π/6. Its growth velocity rapidly reduces with increase of Rs.

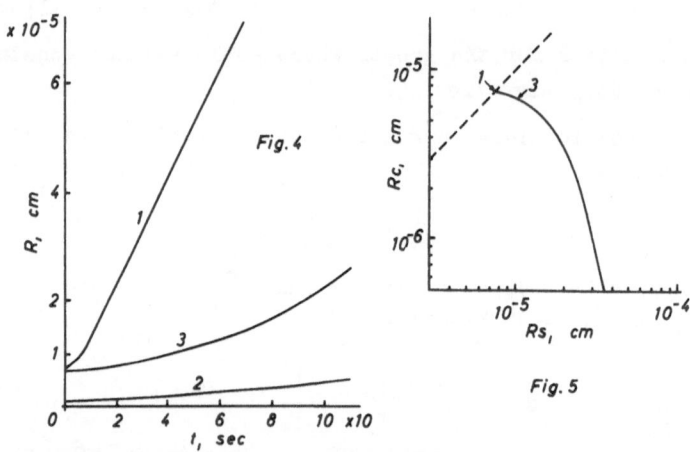

Fig.4. Growth velocity of recrystallized grain in heavily rolled sheet of Al-11,3% Zn.Angle remains constant during boundary migration at curve 1 and 2, while angle increases at curve 3.

Curve 1: $\theta = \frac{\pi}{6}$ (const.);   curve 2: $\theta = \frac{\pi}{180}$ (const.);
curve 3: $\theta = \frac{\pi}{180} \rightarrow \frac{\pi}{6}$

Fig.5. Critical radius Rc of recrystallization nucleus. Angle linearly increases from $\frac{\theta s}{Rs} \cdot Rc$ to θs $(= \frac{\pi}{6})$ during boundary migration

References

1. A. Kelly and R.B. Nicholson in: Chalmers' Progr.Mat. Science 10(1961)205.

2. H.K. Hardy and T.J. Heal in: Chalmers' Progr. Met.Physics 5(1954)146.

3. K. Lücke and H.-P. Stüwe in: Recovery and Recrystallization of Metals, L. Himmel, Ed., Interscience, New York 1963, p.171.

4.3 A Study of the Recrystallisation Behaviour of Iron Containing Coarse Second Phase Particles, with Special Reference to the Formation of Annealing Textures

Untersuchung des Rekristallisationsverhaltens von Eisen mit groben Teilchen einer zweiten Phase unter besonderer Berücksichtigung der Entstehung von Glühtexturen

by D. T. Gawne and G. T. Higgins[+]

Zusammenfassung

Dieser Beitrag behandelt elektronenmikroskopische und röntgenografische Untersuchungen der Rekristallisation und der Texturbildung in einer Eisenlegierung mit 0,4% Kohlenstoff, die kugelförmige Karbidteilchen enthielt. Es zeigte sich, daß sich die Rekristallisationskeime zu Beginn der Rekristallisation ausschließlich an den Karbidteilchen bilden. Eine Untersuchung des um die Karbide liegenden, verformten Subgefüges ergab, daß es von der übrigen Matrix abweicht.

Die stimulierende Wirkung der Karbidteilchen verändert im Vergleich zu reinem Eisen den Typ der bei der Rekristallisation gebildeten Orientierungen und beeinflußt damit die Textur. Der bemerkenswerteste Befund ist das Auftreten einer starken {100}⟨001⟩ Orientierung.

4.3.1 Introduction

There is now considerable evidence to support the view that coarse second phase particles accelerate recrystallisation both in iron [1-3] and in non-ferrous metals [4-7]. However, the bulk of this work is of a purely qualitative nature.

In the programme of work at the University of Liverpool the object has been to establish the degree of association between carbide particles and recrystallising grains in an iron -0,4% carbon alloy and to examine the influence of such particles upon the orientations of the recrystallising grains and therefore upon the final recrystallisation texture.

[+] Department of Metallurgy, University of Liverpool, Liverpool, England

Table 1

Alloy	Analysis wt.%										
	C	Si	Cr	Mo	V	S	P	Al	Mn	O	P
0.4%C Steel	0.4	<0.1	<0.01	<0.01	<0.01	0.005	Not detected	0.008	<0.1	Not determined	
Pure Iron	0.004	0.006	Not determined			0.003	0.002	Not determined	0.006	0.006	0.001

Table 2. Proportions of various orientations in 'As Rolled' material and of re-crystallised grains after annealing

Conditions of Material	Number of Observations	Planes in Surface of the Foil (Percentage)				
		{111}	{100}	{112}+{113}	{110}+{331}	{210}
0.4%C Steel As Rolled	82	49.4	22.6	14.6	9.8	1.2
(5% Recrystallised)	104	51	33.7	9.6	3.8	1.9
(17% Recrystallised)	159	62.3	27.0	3.8	6.3	0.6
(50% Recrystallised)	124	59.6	17.7	15.3	7.2	0
(100% Recrystallised)	68	55	16	22	6	1
Pure Iron As Rolled	105	48	19	19	6	0
2% Recrystallised	28	61	7	18	7	0

4.3.2 Experimental Techniques

The analysis of the steel used in this investigation is given
in Table 1. Results for a pure iron cold rolled 65%, from
another source [8], are introduced into this paper for
comparison purposes and the analysis of this material is also
included in Table 1.

To produce a uniform dispersion of spheroidal cementite
particles the 0.4% carbon alloy was quenched, tempered, cold
rolled and annealed prior to a final cold rolling reduction
of 65 %.

All the high temperature annealing treatments were carried
out in an atmosphere of argon and the recrystallisation
annealing treatments in a salt bath.

Optical microscopy proved to be unsuitable for the examination
of the recrystallisation process because of the fine grain
size and the poorly delineated grain structure arising from
the preferential etching of the carbides. Consequently
transmission electron microscopy has been used as the chief
tool combined with selected area electron diffraction for
detailed orientation determinations.

In addition to the electron microscopy parallel
determinations of texture variations were achieved using a
conventional Schultz goniometer in both reflection and
transmission.

4.3.3 Results and Discussion

The processing treatments employed resulted in a particle
size of 0.60 microns with a spacing of 2 microns.

The texture of the material after cold rolling 65% is
illustrated in Fig.1, information relating to the orientations
present are also to be found in Table 2 and Fig.2 and 3 from
direct electron diffraction data in the electron microscope
and in Table 3 from inverse pole figure information.

These results suggest that the deformation texture may be
adequately represented by a combination of the following
ideal components –

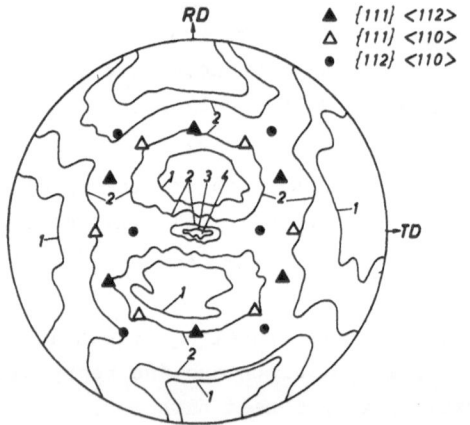

Fig.1. {200} pole figure for the 0,4% carbon alloy
cold rolled 65%

{111}⟨110⟩ – {111}⟨112⟩ spread; {100}⟨110⟩; {112}⟨110⟩

Recrystallisation was found to occur readily at 480°C and
the rate of recrystallisation at this temperature is
compared with that for pure iron at 500°C in Fig.4. It is
notable that not only is the rate of recrystallisation
greater in the steel containing spheroidal carbide particles
but the process goes to completion, whereas in the pure iron
it does not, even at 500°C. In pure iron this effect has been
attributed to the sluggish recrystallisation of those grains
possessing low stored energies [8,9]. If this is so then
clearly these variations are not as important in the alloy
containing particles.

Though this is clear evidence of a qualitative effect, it
is of interest to know the extent to which particle stimulated
nucleation occurs. Clearly sectioning must result in some
recrystallised grains associated with carbides in three
dimensions appearing unassociated in a thin foil. To account
for this an analysis has been carried out [10] which shows
that at the 5% stage, where the mean recrystallised grain
size is 0.96 microns, even if all the recrystallised grains

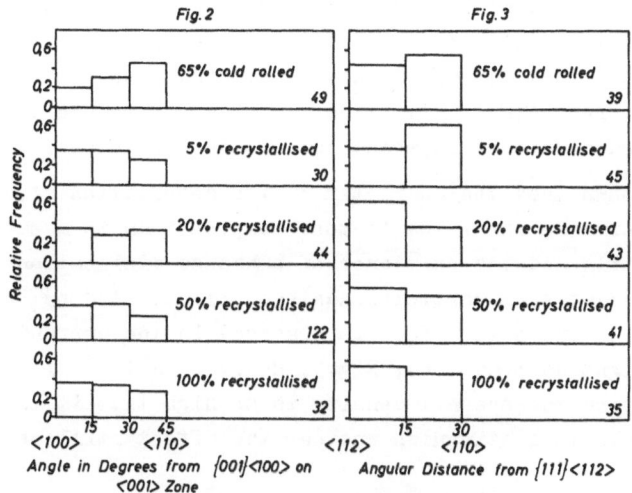

Fig.2. 0,4% C-Fe proportions of {100} components at various stages of recrystallisation

Fig.3. Proportion of {111} components at various stages of recrystallisation in Fe-0,4% carbon alloy

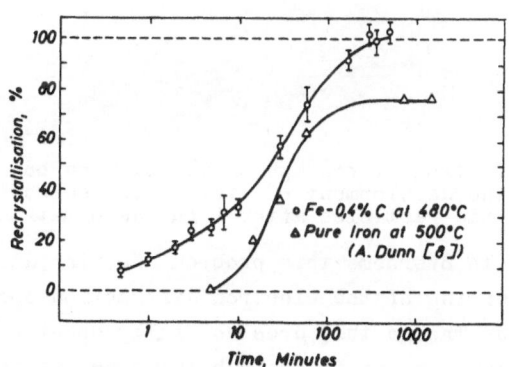

Fig.4. Recrystallisation curves for the Fe-0,4% carbon alloy and pure iron determined from hardness measurments

are associated with carbides only 69% will appear to be so in a thin foil.

With this analysis in mind a series of random observations

of recrystallisation grains orientations was carried out
upon material annealed to give 5% recrystallisation. Of a
total of 167 grains examined 117 were associated with carbide
particles, that is 70%. These results indicate that at this
stage of recrystallisation the new grains form exclusively
in the substructure adjacent to carbide particles.

It is known that the character of the deformation around
non-deformable particles is more complex than in a particle
free matrix and the existence of enhanced dislocation
densities at such particles has been illustrated by LESLIE
et al [1]. The same effect is observed in the current material
after light deformations, Fig.5. However, after 65%
deformation the overall density is so high that it is
impossible to distinguish whether the effect still persists.

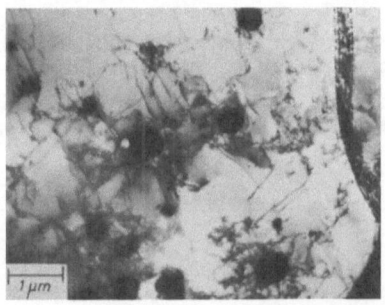

Fig.5. Electron micrograph of the 0,4% carbon steel
showing the development of dislocation networks around
carbide particles after light deformation

In an effort to overcome this problem misorientations as
revealed by arcing of the electron diffraction spots have
been measured. Though this does not fully specify the
misorientations present in the substructure, it does provide
a meaningful qualitative indication. The diffraction aperture
used was fixed at 1 x 1 microns and the results for the areas
adjacent to and removed from carbides in {111} oriented
grains are given in Fig.6. The results in this figure for the
misorientations away from carbides have been corrected to
take into account the fact that for the particular circumstances in question 35% of the areas would be influenced by

325

carbides either above or below the areas examined but not
seen because of sectioning. In Fig. 7 corresponding data for
the misorientations in {100} oriented grains is given.

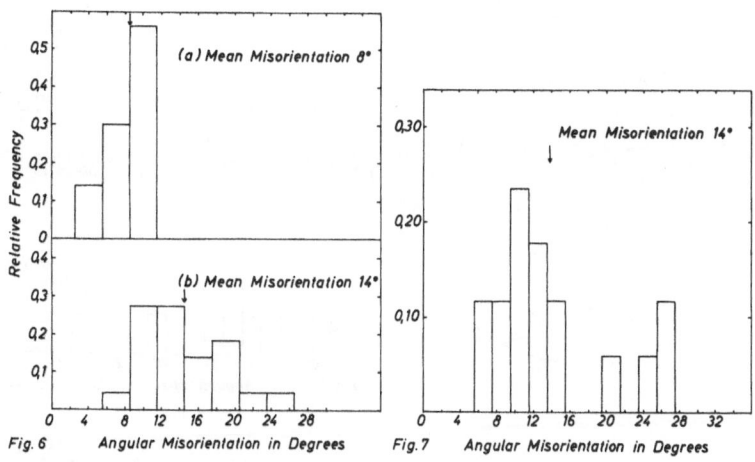

Fig. 6

Fig. 7

Fig.6. Misorientation for {111} oriented grains.
a) removed from particles, b) adjacent to particles

Fig.7. Distribution of misorientations adjacent to particles
in {100} oriented grains

It is clear therefore that in the vicinity of the carbides
there are no differences in the misorientations on the basis
of grain orientation. This is to be contrasted with
observations upon pure iron and a number of iron alloys [8,9].
The results in Fig.8 also illustrate that away from carbides
the differences in misorientation on the basis of grain
orientation reappear. It should be noted that these results
correspond to a diffraction aperture of 4 x 4 microns and
hence are greater than those in Figs. 6 and 7. Observations
by DUNN [8] on pure iron using such an aperture indicate
misorientations of 8°, 5° and 6° for {111}, {142} + {113}
and {100} oriented grains respectively. In the case of a 2%
copper steel treated to give particles 500 Å in size we have
also observed corresponding misorientations of 10°, 6° and
4°. Thus, though there are differences in misorientation
according to grain orientation in these areas the general
level of misorientation is very much higher as a consequence
of the coarse carbide particles.

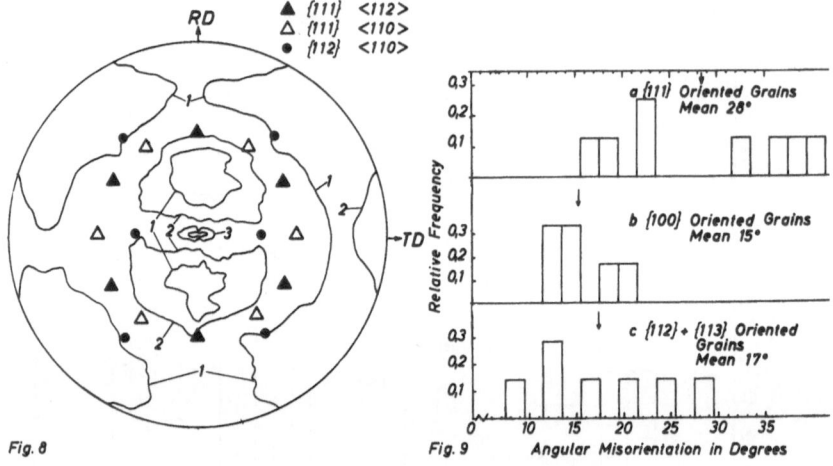

Fig. 8

Fig. 9   Angular Misorientation in Degrees

Fig.8. Misorientations in 0,4% carbon alloy cold rolled 65%.
Diffraction aperture 4 μ x 4 μ

Fig.9. {200} pole figure for the 0,4% carbon alloy recrystallised 100%

The existence of high misorientations adjacent to carbide
particles explains why there is such a high degree of
association between carbides and recrystallising grains for
the development of subgrains with highly misoriented
boundaries, required for the initiation of recrystallisation,
is more readily achieved in these regions. In addition the
generally high misorientation in the substructure suggests
that the dislocation density is higher in the presence of
the particles and therefore that the driving force for
recrystallisation is higher than would be the case in pure
iron strained to the same nominal level.

The absence of differences in misorientation and therefore
nucleation tendency adjacent to particles with respect to
grain orientation has a direct bearing upon the orientations
of the recrystallised grains that are developed. This is
evident from the spectrum of orientations observed in the
electron microscope, Table 2, when compared with those
reported by DUNN [8] for pure iron.

The main feature of these results is the absence of any gross change in the proportions of the various orientations in the sheet surface during recrystallisation of the 0.4% carbon steel, Fig.9. This is to be contrasted with the pure iron results where the percentage of the {100} oriented grains drops sharply. The same effect is shown by the inverse pole figure data in Table 3. It is concluded that the carbide particles stimulate nucleation in all orientations, thus destroying the differentiation that normally occurs due to the differences in the ease of nucleation in the different components. A further consequence is the close similarity between the cold rolled and recrystallised textures on the basis of sheet surface orientations.

Table 2 indicates that in detail there is a tendency for the relative importance of the {100} and {111} oriented grains in the recrystallised structure to move in favour of the latter as recrystallisation proceeds. The results on pure iron provide an explanation for this drop, for once the carbide stimulated nucleation has been exhausted the {100} oriented grains will be slow to provide further nuclei whereas in the {111} grains of the deformation structure nucleation may continue in regions removed from carbides because of the higher stored energy in this component.

These comments infer that the sheet surface orientations of the recrystallising grains are the same as the deformed matrix in which they develop. Observations of these relationships indicated that 88% of the recrystallised grains examined were growing into a matrix of similar orientation and the above assumption is therefore justified. This is also borne out by observations upon the 100% recrystallised material where there was a marked tendency for grains to be surrounded by grains of like orientation.

Although there is no marked change in the intensities of grain orientations lying in the sheet surface on recrystallisation, Figs. 1, 9 and 2 shows that there is a change in the inclination of the {100} component with respect to the rolling direction from {100}⟨110⟩ to {100}⟨001⟩ on recrystallisation. Reference to Fig.7 shows that the average misorientation

Table 3. Intensities of reflections from the sheet surface

Condition of Material	Intensity of Reflection x Random			
	{111}	{200}	{112}	{110}
0.4%C Steel As Rolled	2.6	2.5	2.1	0.2
3 mins.480°C (20% Recrystallised)	2.6	2.9	2.0	0.2
24 mins.480°C (50% Recrystallised)	2.8	3.2	1.9	0.2
83 mins.480°C (80% Recrystallised)	2.7	2.7	1.8	0.2
5 hours, 480°C (100% Recrystallised)	2.7	2.5	1.9	0.2
Pure Iron As Rolled	2.8	2.3	2.3	0.2
24 hours 600°C (100% Recrystallised)	3.6	0.8	1.7	0.2

adjacent to carbides in {100} grains is $14°$. It is assumed that the most likely nucleus is that subgrain with the maximum misorientation relative to the adjacent matrix, then each grain in the deformed material may be considered to produce recrystallised grains misoriented from itself by $14°$. Consequently, if each block in the histogram for the cold rolled material in Fig. 2 is moved $14°$ to each side of its mid position the histogram of the recrystallised material is obtained.

The small texture change in the {111} component, Fig.3, may be rationalised in the same way.

References

1. W.C. Leslie, J.T. Michalak and F.W. Aul in: Iron and its Dilute Solid Solutions AIME ,1963, p. 119.

2. A.T. English and W.A. Backofen: Trans.Met.Soc. AIME 230(1964)396.

3. C. Antione et al: Trans.Met.Soc.AIME 230(1964)700.

4. P. Cotterill and P. Mould: J. Materials Sci. 2(1967)241.

5. R.D. Doherty and J.W. Martin: Trans.Amer.Soc.Metals 57(1964)874.

6. J.W. Martin: Metallurgia, 55(1957)161.

7. F. Haeßner, E. Hornbogen and M.Mukherjee: Z. Metallkde. 57(1966)171.

8. A. Dunn: Ph. D. Thesis, University of Liverpool 1968.

9. I. Dillamore, C. Smith and T. Watson: Metal Sci. J., 1(1967)49.

10. D.T. Gawne and G.T. Higgins: To be published.

4.4 Some Aspects of the Recovery and Recrystallization Behaviour of an Iron-0.02% Carbon Alloy and a Commercial Rimming Steel

Einige Gesichtspunkte zum Erholungs- und Rekristallisationsverhalten einer Eisen-Kohlenstofflegierung mit 0,02% C und eines unberuhigten Stahls

by C. Dasarathy[+)]

Zusammenfassung

Das Erholungs- und Rekristallisationsverhalten einer Legierung aus Eisen mit 0,02% C, die feine Karbide enthielt sowie eines unberuhigten Stahls (0,06% C) mit groben Karbiden wurden untersucht. Sowohl die Erholungs- als auch die Rekristallisationsgeschwindigkeit waren beim unberuhigten Stahl größer als bei der Legierung. Erholungs-Vorglühungen beschleunigten die Rekristallisation im unberuhigten Stahl, während sie bei der Legierung rekristallionsverzögernd wirkten.

Die Verformungstexturen beider Werkstoffe waren ähnlich, mit Ausnahme eines etwas höheren Anteils an (001) bei der Legierung. Die Rekristallisationstexturen beider Materialien zeigten gewisse Unterschiede. Aufbauend auf bestehenden Theorien der Rekristallisationstexturen einschließlich der Begünstigung der Keimbildung durch Teilchen werden die Ergebnisse diskutiert.

This paper reports the results of a study of some effects due to second phase particles and prior recovery treatments on the recovery and recrystallization behaviour of a laboratory prepared iron-0.02%C alloy, containing fine carbides and a commercial rimming steel of 0.06% carbon containing coarse carbides and oxides.

4.4.1 Experimental Techniques

The hot rolled materials were cold reduced by 65% prior to annealing in salt baths at 600°C and 700°C with and without prior recovery treatments of 15 minutes at 300°C, 400°C or

[+)] R.T.B. Research Laboratories, British Steel Corporation (South Wales Group), Aylesbury, Bucks, England

500°C. After recovery treatments, specimens were rapidly
up-quenched to the annealing temperature or quenched to room
temperature.

The progress of recovery and recrystallization was studied
qualitatively by X-ray line broadening using the sharpness
factor [1] on the (310) reflection. Recrystallization was
also followed by light microscopy, the as-recrystallized
grain sizes being estimated by the mean-lineal intercept
method. The size of the particles was evaluated as described
in the literature [2]. Recrystallization textures were
obtained on specimens that were thinned to half thickness
from one side only using the inverse pole figure method with
the (100), (211), (111), (332) and (110) reflections.

4.4.2 Experimental Results

Particle size analysis in the hot rolled materials showed
the particles to be finer (0.25/u) in the alloy than in the
steel (1.2/u). They were situated predominantly at the grain
boundaries in the alloy, but distributed at random in the
steel.

The X-ray line broadening results (Fig.1) on specimens held
at temperature for 15 minutes show that the first major
change in broadening occurs around 400°C in both materials.
Above 400°C, recovery proceeds faster in the steel than in
the alloy. X-ray measurements and metallographic examination
showed that recrystallization in specimens annealed at 600
or 700°C without a prior hold was faster in the steel than
in the alloy. Recovery treatments accelerated recrystallization in the steel whereas they retarded recrystallization
in the alloy (Figs. 2 and 3). The just recrystallized grain
size in the steel was in general slightly finer than in the
alloy (Table 1). Prior recovery tended to decrease the grain
size in the steel but it had no detectable effect in the
alloy.

The recrystallization textures of the steel (Fig.4a) show
that a) recrystallization at 700°C with no prior treatment
causes a general drop in all components other than (110),
which has increased by a factor of about six, relative to

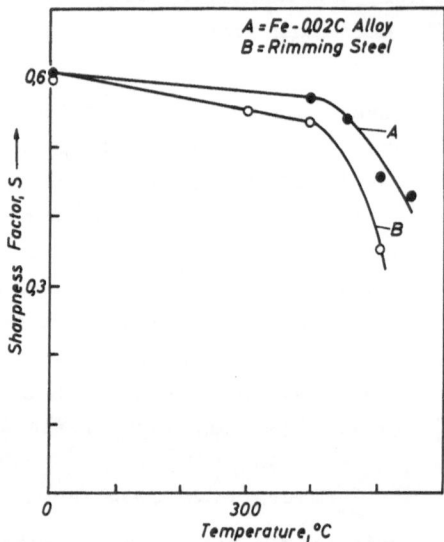

Fig.1. Change in sharpness factor with temperature

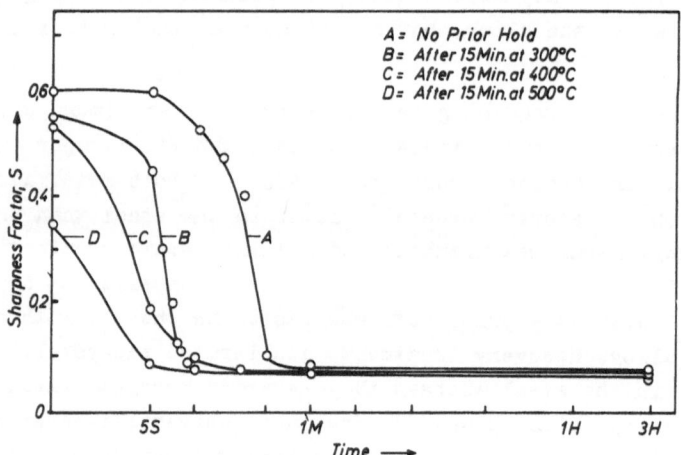

Fig.2. Change in sharpness factor with time for the steel annealed at 700°C

the deformation texture, b) on annealing at 700°C, prior treatment produces only certain small changes in texture, c) with no prior treatment, the texture at 600°C is slightly different from that at 700°C, the changes being in the (211), (111) and (332) components. However, with a prior treatment

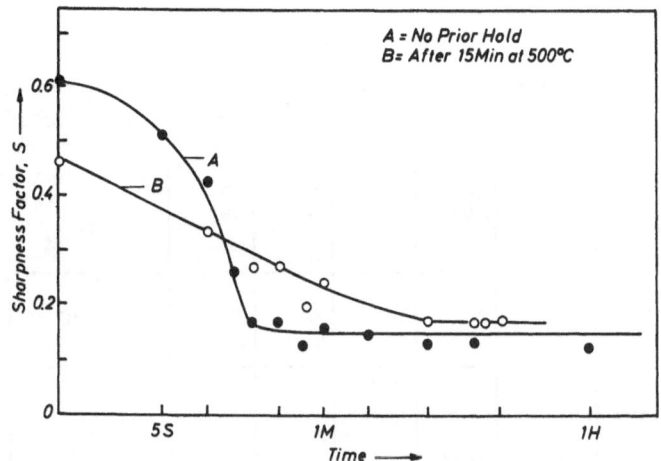

Fig.3. Change in sharpness factor with time for the
alloy annealed at 700°C

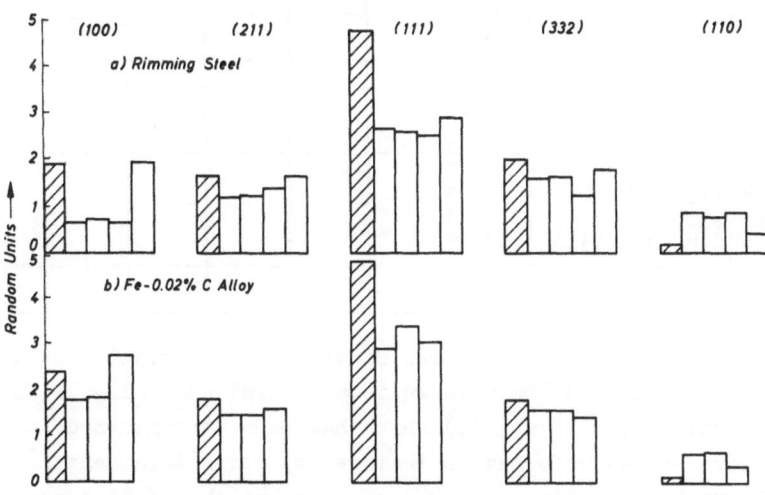

Fig.4. Recrystallization textures of specimens annealed at
600°C and 700°C with and without the prior recovery treatment
Key to each histogram: Column1 = deformation texture (shown
hatched), column2 = annealed at 700°C with no prior treatment,
column 3 = annealed at 700°C with a prior treatment, column
4 = annealed at 600°C with no prior treatment, column 5 =
annealed at 600°C with a prior treatment

before annealing at 600°C, there is a general strengthening
of all components other than (110), which has been weakened.

334

Table 1. Recrystallization times and grain sizes

		No prior hold		15 min at 300°C		15 min at 400°C		15 min at 500°C	
		600°C	700°C	600°C	700°C	600°C	700°C	600°C	700°C
Rimming Steel	Start of recrystall.	30 sec	<20 sec	<15 sec	5 sec	15 sec	5 sec	just started after 15 min at 500°C <2%	
	Completion of recrystall.	5 min	30 sec	1 min	10 sec	1 min	10 sec	30 sec	5 sec
	Just re-crystallized grain size width μ	9.9	10.4					9.5	10.0
	thickness μ	7.9	9.2					6.7	7.4
Fe-0.02% C Alloy	Start of recrystall.	5 min	5 sec	-	-	-	-	1 hr	20 sec
	Completion of recrystall.	>90% 8 hrs	>90% 1min	-	-	-	-	~50% 8 hrs	>90% 10min
	Just re-crystallized grain size width μ	11.3	11.9						11.3
	thickness μ	7.4	8.4						8.2

The recrystallization textures of the alloy (Fig.4b) show that
a) on annealing at 700°C there is a general weakening of all
components other than (110) which has been strengthened,
relative to the deformation texture, b) prior treatment
followed by annealing at 700°C produces no change in texture
apart from a slight enhancement of the (111) component, c) in
the absence of prior treatment, annealing at 600°C enhances
the (100) component to a level higher than that in the
deformation texture whilst the (110) component is weakened,
relative to the texture at 700°C, d) the effect of the prior
treatment on the material annealed at 600°C, could not be

studied since the specimen had recrystallized to only about
50%, after 8 hours (the longest time used) at temperature.

4.4.3 Discussion

It is clear from this investigation that certain differences
in the recovery and recrystallization characteristics exist
between the two materials annealed with and without the prior
treatments. These differences may arise from initial
differences in grain size, deformation textures and the size
of the particles, in the two materials. However, the initial
grain sizes were found to be essentially similar and the
deformation textures are also almost the same except that the
iron—carbon alloy has a higher proportion of the (100)
component. This difference in a component known [3] to recrystallize with difficulty may partially account for the
delayed recrystallization in the alloy. However, it is
considered that the main reason for the differences in the
recovery and recrystallization behaviour is due to the size
and distribution of the particles in the two materials.
Previous work by DOHERTY and MARTIN [4] has shown that when
second phase particles are present in a material, recrystallization was increasingly retarded with increasing fineness
of the particles. They suggest that for easy nucleation, the
nuclei should be above a critical size before they impinge
upon surrounding particles. GAWNE and HIGGINS [5] have found
that easy nucleation may occur in the regions around coarse
carbides where the deformation structure is disturbed locally
by the presence of the carbide [6].

In the present case it is considered that the difference in
the rate of recovery between the two materials is caused by
the retarding effect of the fine carbides in the alloy and
the accelerating effect of the highly distorted regions
around the coarse particles in the rimming steel. When recrystallization takes place in the rimming steel without a
prior recovery treatment, it is considered that nucleation
of recrystallization occurs predominantly at the coarse
particles. With a recovery treatment, additional nuclei may
form at the cold worked grain boundaries, which on heating

to the higher temperature, immediately recrystallize. The effect is to produce a fine grain size and an acceleration of the recrystallization rate. In the alloy, recovery must also lead to the formation of recrystallization nuclei but its dominating effect is considered to be one of the dissipation of stored energy which subsequently causes a significant decrease in the recrystallization rate. The difference in recrystallization rate between the two materials when no recovery treatment was given is considered to be due to the regions of high stored energy around the coarse particles in the rimming steel. In contrast, the recrystallization of the alloy particularly at 600°C is likely to be impeded by the fine carbides. The very much greater increase in the relative rate of recrystallization on going from 600 to 700°C in the alloy compared with the rimming steel is clearly due to the almost complete dissolution of the carbides in the alloy, on recrystallization at 700°C.

The differences in recrystallization texture between the various samples were relatively small. Nevertheless, they can be interpreted at least in part on the same basis as the differences in recrystallization behaviour. During this discussion it is assumed that the change in stored energy of deformation with orientation in the present materials is qualitatively the same as in cold worked high purity iron[7], i.e. stored energy increases in the order (100), (211), (111), (332), (110). On this basis, the (110) should be the component which recrystallizes most easily and the (100) the one which recrystallizes with most difficulty, except when recrystallization is influenced by large second phase particles. If it is accepted that the grain boundary nucleation postulated for the steel treated at 500°C prior to annealing at 600°C gave predominantly (100) orientation, then the difference in texture for this material annealed with and without the prior hold may be understood. Presumably the effect of the hold at 500°C is not large enough to show itself when recrystallization occurs at 700°C. In the alloy, it is considered that recovery prior to recrystallization tends to promote the formation of the high energy orientations

(111) and (110). On this basis, recrystallization at $700^{\circ}C$
with a prior hold gives a higher (111) and recrystallization
at $600^{\circ}C$ without a prior hold gives a lower (110) and (111).
The reason is that, inspite of the fact that recrystallization
at the lower temperature normally gives the higher amount of
recovery, in this case, recrystallization at the higher
temperature gives the higher amount of recovery because of
the almost complete dissolution of the fine carbide
particles. The higher proportion of (100) orientation in
samples annealed at $600^{\circ}C$ compared with those annealed at
$700^{\circ}C$ may be due to the influence of grain boundary
nucleation as was postulated in the steel. The texture of the
sample annealed at $600^{\circ}C$ with a prior hold is not considered
here because it did not completely recrystallize.

It might be expected that in general the proportion of (100)
in the steel should be higher than in the alloy due to the
influence of particle aided nucleation favouring the lower
energy orientations. However, the present results showed the
reverse of this. This result may be partially due to the
high proportion of (100) in the alloy deformation texture but
the main reason for this effect is not yet clear. It may be
that, the (100) component in the steel increased during the
initial stages of recrystallization but was later consumed
before recrystallization was complete as was observed in a
high carbon–iron alloy by GAWNE and HIGGINS [5]. The
consumption of (100) in the alloy may not have occurred to
the same extent, presumably due to the influence of the
fine carbide particles. However, a partial consumption of
this component at $700^{\circ}C$ when the particles have dissolved
can account for its low proportion in the alloy.

4.4.4 Conclusions

1. Material containing coarse particles recrystallizes more
rapidly than that containing fine particles. 2. Prior
recovery treatment accelerates recrystallization in material
containing coarse particles, whereas it retards recrystallization in material containing fine particles. 3. The size
and distribution of particles influence recrystallization

textures. 4. Prior recovery treatment may cause significant differences in the recrystallization textures. 5. Treatments giving a high proportion of (100) give a low proportion of (110).

Acknowledgements

The author wishes to thank Mr. P. Newman and Mr. W.C. Lake for carrying out most of the experiments, Dr. P. Moore and Mr. R.C. Hudd for many helpful discussions and the Directors of Richard, Thomas and Baldwins Ltd. for permission to publish this paper.

References

1. H. Hu in: Recovery and Recrystallization of Metals, L. Himmel, Ed., Interscience,New York 1963, P. 318.

2. R.B. Shaw, L.A. Shepard, L.A. Starr and J.E. Dorn: Trans. Amer.Soc.Metals 45(1953)249.

3. C. Dasarathy and R.C. Hudd: To be published.

4. R.D. Doherty and J.W. Martin: J.Inst.Metals 91(1962/63)332.

5. D. Gawne and G.T. Higgins: Private communication.

6. D. Gawne and G.T. Higgins: This volume, chapter 4.3.

7. I.L. Dillamore, C.J.E. Smith and T.W. Watson: Met.Sci.J. 1(1967)49.

4.5 Texture Development during Grain Growth in Low Carbon Steel Sheet
Texturbildung beim Kornwachstum in Stahlblech mit niedrigem Kohlenstoffgehalt

by W.B. Hutchinson, C.J.E. Smith, T.W. Watson and I.L. Dilla-more[+)]

Zusammenfassung

Die Entstehung von Vorzugsorientierungen beim Kornwachstum in kaltgewalzten und isotherm geglühten Stählen mit niedrigem Kohlenstoffgehalt wurde untersucht. Wenn unmittelbar nach der primären Rekristallisation eine zweite Phase in feiner Dispersion vorlag, entstanden starke Texturen. Die Beseitigung der Teilchen der zweiten Phase durch Entkohlung führte zu einem Kornwachstum, das eine Zunahme der {111} und eine Abnahme der {100}-Orientierung zur Folge hatte. Die Ursachen dieses Verhaltens werden mit Hilfe der bekannten Mechanismen für Rekristallisation und Kornwachstum diskutiert. Über experimentelle Ergebnisse, die für diese Mechanismen sprechen, wird berichtet.

Vom wirtschaftlichen Standpunkt aus gesehen ist dieser Prozess insofern interessant, als das so behandelte Material einen hohen plastischen Anisotropiekoeffizienten zeigt ($\bar{R} \sim 2,0$).

It is now well known, following the work of LANKFORD et al [1], that drawability of sheet is governed largely by plastic anisotropy behaviour, and this is known to originate in texture. WHITELEY and WISE [2] have demonstrated for steel sheet that good drawability (high R-values) results from high densities of {111} planes, and low densities of {100} planes parallel to the sheet. The work described here is part of a general programme aimed at understanding the formation of annealing textures, and their relationship with mechanical properties.

[+)]Birmingham University, Birmingham, England

It has been observed in low carbon steel sheet that during
grain growth following cold rolling and recrystallisation,
the volume of metal oriented with {111} parallel to the sheet
usually increases while that with {100} decreases. The
degree of texture development depends on a number of factors.
To obtain strong textures during grain growth it is
necessary that the carbon content exceeds the solid
solubility limit, and that the carbides are finely dispersed
throughout the matrix. A cold rolling reduction of between
70% and 85% is most suitable, and annealing should be done
at a temperature above the eutectoid transformation so that
the carbides are replaced by islands of austenite. To promote
grain growth, the particles must be removed during annealing,
and so a decarburising atmosphere is necessary.

Figs. 1 and 2 show pole intensity data as a function of grain
size for a rimming steel annealed at 780°C after cold rolling
84%. Before cold rolling, the steel had been heat treated to
form fine or coarse dispersions of pearlite (average pearlite

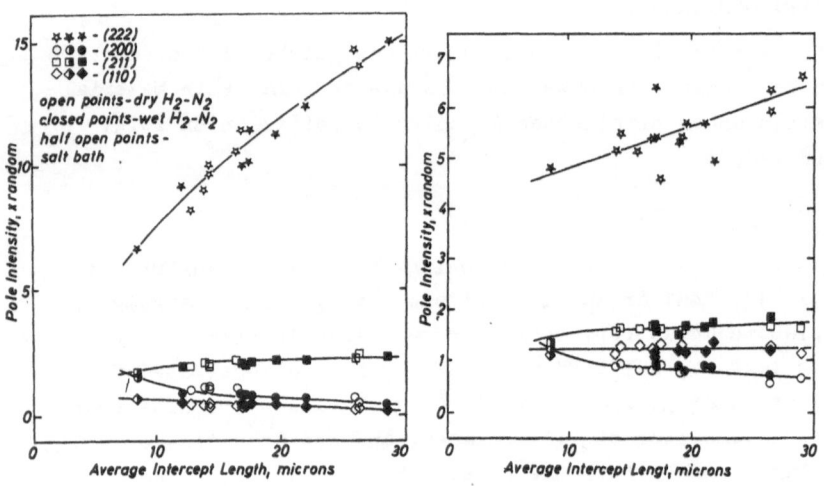

Fig.1. Relationship between
pole intensity and grain size
for rimming steel with fine
2nd.phase dispersion annealed
at 780°C

Fig.2. Relationship between
pole intensity and grain size
for rimming steel with coarse
2nd.phase dispersion annealed
at 780°C

nodule diameters 3.5 μ or 25 μ). Qualitatively, the results
are similar in both cases. The {222} and {112} intensities
increased during grain growth while the {200}, {110}, and
also the {310} decreased. However, the magnitude of the {222}
intensity became much greater in the samples which had been
previously treated to contain a fine second phase dispersion.
Within the range of grain growth studied there was
apparently no effect of decarburising rate on texture,
although this did affect the rate of grain growth. Fig.3
shows a {200} pole figure for another rimming steel with a
fine second phase dispersion, just recrystallised after
rolling 80%. Further heating for 64 hours at 780°C in a
decarburising atmosphere caused a change in texture to that
shown in Fig.4. There is a strong increase in the fibre
component with ⟨111⟩ perpendicular to the sheet. Peaks occur
corresponding to orientations {554}⟨225⟩. Components with
{100} planes parallel to the sheet always decreased during
grain growth but the absolute value of the {200} intensity
was lowest following reductions between 70% and 80%.

Fig.5 shows R-values for rimming steel rolled to different

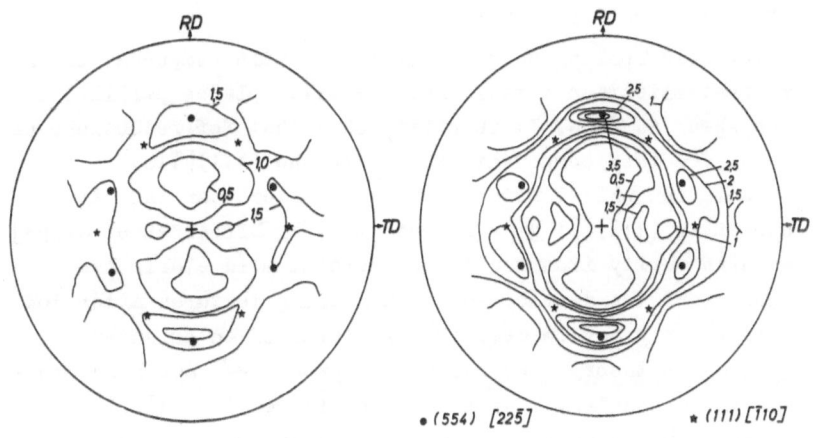

• (554) [225]   * (111) [110]

Fig.3. {200} pole figure for   Fig.4. {200} pole figure for
rimming steel with fine second   rimming steel with fine second
phase dispersion cold rolled   phase dispersion cold rolled
  80%; as recrystallized   80%; annealed at 780°C 64
                                            hours

reductions and annealed for 64 hours in a decarburising atmosphere. For prior reductions of 70% to 80%, the average R-value is approximately 2.0.

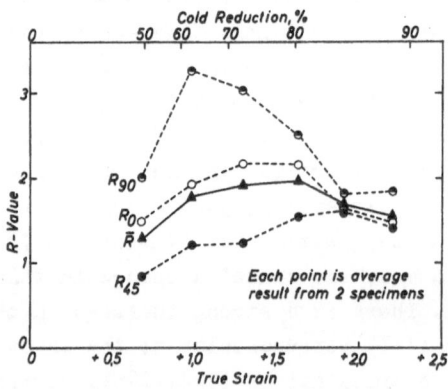

Fig.5. Relationship between R-value and cold-reduction for specimens annealed 64 hrs. in dry H_2-N_2

In order to understand the processes occurring during grain growth, it is necessary to look first to the primary recrystallisation process.

After cold rolling and annealing, the main component of the recrystallisation texture has the {111} planes parallel to the sheet surface. It is often found that for reductions of up to 70% this component is near to the (111)[$\bar{1}$10] orientation whilst for reductions greater than this it is closer to (111)[$\bar{2}$11]. A recent study by DILLAMORE et al.[3] on high purity iron and two commercial mild steels has shown that the development of annealing textures after 70% cold rolling may be described in terms of an oriented nucleation theory. Two nucleation processes, subgrain growth and grain boundary bowing were considered. The driving energy for nucleation by subgrain growth was estimated as a function of the grain orientation using electron microscopy and was found to increase from the (001)[$\bar{1}$10] to the (110) [$\bar{1}$10] orientation. It was agreed that the most favoured orientation would be near to (111)[$\bar{1}$10] since it has a high driving energy for recrystallisation and is also strongly

represented in the deformation texture. It has been shown
by SMITH and MORRIS [4], however, that at higher deformations
the (111)[$\bar{1}$10] is less favoured and it is proposed that some
additional mechanism must operate which allows the (111)[$\bar{2}$11]
component to develop in preference to the (111)[$\bar{1}$10]
orientation.

Most of the studies which have been made on the development
of preferred orientation during cold rolling are agreed that
at low deformations a component near to (111)[$\bar{2}$11] is
prominant whilst at high deformations the (111)[$\bar{1}$10]
orientation is important. There is evidence from texture and
electron microscope studies that as the deformation is
increased the (111)[$\bar{2}$11] component rotates about the [111]
direction to either (111)[$\bar{1}$10] or (111)[$\bar{1}$01]. Since
rotations in either sense are equivalent, a particular grain
with the (111)[$\bar{2}$11] orientation may rotate towards either
(111)[$\bar{1}$10] or (111)[$\bar{1}$01]. Alternatively both rotations may
occur by the formation of a transition band across the grain
in question, and its simplest form will consist of low angle
boundaries made up of edge dislocations. Clearly material in
the centre of the band will have an orientation near to (111)
[$\bar{2}$11] so that nuclei forming in the band will have this
orientation.

The actual details of nucleation within a transition band
have been extensively examined by HU [5] and WALTER [6] so
that they need not be considered further here. The important
feature of these bands, however, is that they allow a potential
nucleus to very quickly attain a high angle boundary. The
rate of growth, G, of a recrystallising grain is governed
both by the mobility M and the driving energy E as follows,
$$G = M \cdot E. \tag{1}$$
At high deformations, the value of E for grains nucleated
either within the grain or within the transition band itself
will be approximately the same, but the value of M will differ,
being greatest for grains which form in the centre of a
transition band. As a result these grains, which will have an
orientation near to (111)[$\bar{2}$11], will be able to grow at a
much faster rate than those in the matrix and will therefore

dominate the recrystallisation process.

In addition to subgrain growth, it was stated that re-crystallised grains might be nucleated by grain boundary bowing. An anlysis of this made by DILLAMORE et al. [3] has shown that this is only probable between grains with the (001)[$\bar{1}$10] and (110)[$\bar{1}$10] orientations. Since there is very little of the (110)[$\bar{1}$10] component present at high deformations, this process will not be particularly important and may therefore be neglected.

So far the presence of second phase particles has been ignored, but in the present work their main role will be to enhance nucleation. Since they are uniformally distributed between all orientations, the over riding factors in determining which orientations will prosper will still be the anisotropy of stored energy and the boundary mobility.

As a result of the nature of the recrystallisation process, the structure of the metal in the just recrystallised condition consists of a distribution of grain sizes in which the large grains are oriented with {111} planes parallel to the rolling plane. Experimental evidence to confirm this has been found by examining magnetic closure domain patterns (Bitter patterns) on polished sections from annealed specimens. These patterns are orientation dependent, and large grains almost invariably show patterns which are characteristic of {111} planes.

During grain growth, there is always a tendency for large grains to grow at the expense of small ones, and so the volume of material in the {111} orientation would be expected to increase. In order to understand why second phase particles are so important to the development of texture during grain growth, we shall refer to the theory of grain growth proposed by HILLERT [7]. The equation for the growth rate of a grain of diamter D may be written

$$\frac{dD}{dt} = K \left(\frac{1}{D_{cr}} - \frac{1}{D} \pm Z \right) \tag{2}$$

D_{cr} is a characteristic grain size which increases

continuously during grain growth; grains for which $D > D_{cr}$ grow, while those for which $D < D_{cr}$ shrink. The effect of second phase particles is described by the term Z which is a reaction term which cannot exceed the value Z. It always acts to decrease the absolute value of $(\frac{dD}{dt})$.

In the absence of effective second phase particles (coarse second phase dispersion, $Z = 0$) Hillert calculates a steady state distribution of relative grain sizes, i.e. P $(\frac{D}{D_{cr}})$ against $(\frac{D}{D_{cr}})$, and finds the maximum grain size to be $2D_{cr}$. Also, D_{cr} is then approximately equal to the arithmetic average $\bar{D}$. The grain size distribution which exists at the end of primary recrystallisation will not generally be the same as the steady state distribution calculated by Hillert. In general, the initial relative grain size distribution will be wider than the steady state distribution, but in the absence of second phase particles, it will contract to this form during grain growth. Thus, although the large grains increase their absolute size, they do so slowly, and decrease their relative size. The possibility of growth accidents occuring is increased, and these may reduce the size advantage of the large grains, or create new large grains. In this situation, only a modest increase in the {111} component may be expected.

When effective second phase particles are present (as is the case for the steels with the fine austenite dispersions) grains of around the average size are completely inhibited from growing. Only much larger grains can grow, thus increasing further their original size advantage, and widening the relative grain size distribution. Eventually, all grain growth will slow down and stop in the presence of the inhibiting particles. If, however, the particles are progressively removed, by decarburising in the present case, grain growth may continue. The {111} oriented grains which now have increased their size advantages continue to grow selectively during further grain growth and this causes a strong increase in the {111} component.

Grain size distributions have been measured on a series of specimens which contained fine dispersions of austenite during annealing at 780° following 85% cold rolling. The grain size distributions measured on the plane of polish were converted to true spatial grain size distributions, and these were then expressed as relative size distributions $P(\frac{D}{\bar{D}})$ against $(\frac{D}{\bar{D}})$. Fig.6 shows the distribution immediately after recrystallisation, and Fig.7 after one hour at 780°C. The dashed curves show Hillert's solution in the absence of second phase particles. The predicted widening of the relative grain size distribution during grain growth is confirmed. During grain growth on further annealing, the distribution remained similar to that in Fig.7, and there was an accompanying strong increase in the {111} texture components.

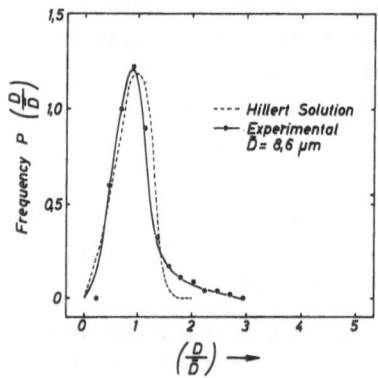
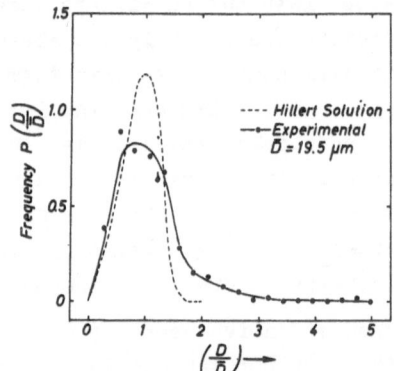

Fig.6. Relative grain size distribution after recrystallization. Rimming steel with fine second phase dispersion, cold rolled 85 %

Fig.7. Relative grain size distribution after annealing for 1 hour at 780°C. Rimming steel with fine second phase dispersion, cold rolled 85 %

Conclusions

During grain growth in steel following cold rolling and recrystallisation, strong textures suited to deep drawing operations may develop. A fine dispersion of second phase particles must be present initially and removed during annealing. Average strain ratio values of 2.0 or higher may result in steel sheet treated in this way.

347

Acknowledgments

The authors are grateful to Richard Thomas and Baldwin Ltd.,
for material and financial assistance.

References

1. W.T. Lankford, S.C. Snyder and J.A. Bauscher: Trans.Amer.
   Soc.Metals 42(1950)1197.

2. R.L. Whiteley and D.E. Wise: Flat Rolled Products III,
   E.A. Earhart, Ed., Amer.Soc.Metals,Chicago 1962, p.47.

3. I.L. Dillamore, C.J.E. Smith and T.W. Watson: Met.Sci. J.
   1(1967)49.

4. C.J.E. Smith and P.L. Morris: To be published.

5. H. Hu in: Recovery and Recrystallisation of Metals, L.
   Himmel, Ed.,Interscience, New York 1963.

6. J.L. Walter: Ibid, in discussion.

7. M. Hillert: Acta Met. 13(1965)227.

4.6 Influence of the Hot Rolling Parameters, Annealing Cycle and Atmosphere on the Textures of Low Carbon Rimmed and Aluminium Killed Steel Sheets

Einfluß des Warmwalzens, der Glühungen und der Glühatmosphäre auf die Texturen von Blechen aus unberuhigtem und Al-beruhigtem Stahl mit niedrigem Kohlenstoffgehalt

by G.Pomey[+], T.Sakamoto[++], M.Grumbach[+] and Ph.Charpentier[+]

Zusammenfassung

Es wurden die Warmwalztexturen in verschiedenen Schichten von Tiefziehblechen aus unberuhigtem und Al-beruhigtem Stahl mit niedrigem Kohlenstoffgehalt in Abhängigkeit von einem großen Temperaturbereich für das Walzen und Haspeln untersucht. Die Texturen waren sehr inhomogen. In der Mitte der Bleche traten zwei verschiedene Texturen auf, die einen Zusammenhang mit dem Parameter Walzendtemperatur-Haspeltemperatur zeigten. Für Bleche aus beiden Stahlsorten, die auf herkömmliche Weise kaltgewalzt worden waren, wurde der Einfluß der Glühbedingungen eingehend untersucht.

Textures of low carbon steel sheets have been studied extensively for many years and since it has been shown that preferred orientations are related to deep drawability, the influence of the production parameters upon textures has been investigated by various authors. However, we thought that a more systematic study of the influence of the hot rolling parameters and annealing conditions on the textures of commercial low carbon steel sheets would be useful[1].

4.6.1 Hot Rolling Textures

Eighteen samples, taken after coiling, from 2.5 mm thick hot rolled sheets of commercial low carbon (0.02 to 0.09 weight per cent of carbon) rimmed and aluminium killed steel were examined. The finishing and coiling temperatures ranged from $830^\circ C$ to $905^\circ C$ and from $520^\circ C$ to $725^\circ C$ respectively.

[+] Institut de Recherches de la Sidérurgie Française, 78-Saint-Germain-en-Laye – France
[++] Central Research Lab., Fuji Iron and Steel Co., Sagamihara, Japan

Conventional {110} pole figures were determined at the
surface and in the center of each sample by the Schulz's
reflexion method using a Siemens texture goniometer and
Molybdenum K α radiation.

4.6.1.1 Results

The center textures can be classified into two types. Type I
(Fig.1) is a partial ⟨110⟩ fiber texture with its axis
parallel to the rolling direction. Examination of the partial
pole figure obtained through the Schulz's reflexion method
indicates that the orientation {100}⟨110⟩ is at the center
of the spread which extends to {112}⟨110⟩. Type II (Fig.2)
can be described as follows: it is deduced from type I
through a reinforcement of the {100}⟨110⟩ component of the
partial fiber texture. Moreover, a weak {110} component has

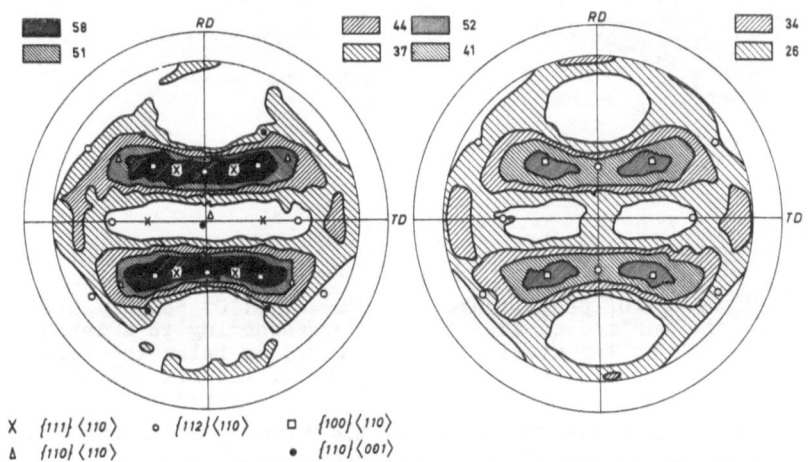

X {111}⟨110⟩   o {112}⟨110⟩   □ {100}⟨110⟩
Δ {110}⟨110⟩   • {110}⟨001⟩

Fig.1. {110} pole figure   Fig.2. {110} pole figure
corresponding to the center   corresponding to the center
of the hot rolled sheet.   of the hot rolled sheet.
Type I texture   Type II texture

appeared as can be seen from the observation of the center
of the pole figures. Those two types of texture were
observed in rimmed as well as in aluminium killed steels.
We found that the type of texture at the center of a given
sheet can be related to the parameter ΔT = Finishing
temperature — coiling temperature. When 205°C < ΔT < 260°C

type II is obtained while type I is present if $\Delta T > 260^{\circ}C$ and $\Delta T < 205^{\circ}C$.

The surface textures are essentially the same for all the samples and can be described as $\{110\} \langle u\,v\,w \rangle$ plus $\{112\}$ $\langle 111 \rangle$ (Fig.3). The evolution of the texture as a function of depth was followed with one sample through chemical thinning and determination of the $\{110\}$ pole figure after each thinning operation. The surface texture extends 0.3 mm under the surface. As we go deeper below the surface, a

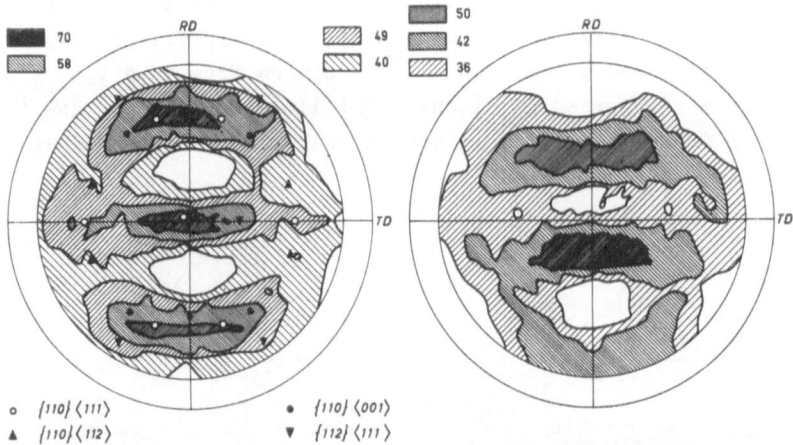

Fig.3. $\{110\}$ pole figure at the surface of the hot rolled steel

Fig.4. $\{110\}$ pole figure corresponding to a depth of 0.5 mm below the surface of the hot rolled sheet

transition zone appears in which the texture is deduced from the surface texture approximately through a rotation about the transverse direction. This transition zone extends from 0.3 mm to 0.75 mm below the surface at which point we reach the center texture. Fig.4 shows the $\{110\}$ pole figure corresponding to a depth of 0.5 mm.

4.6.1.2 Discussion

The coexistence of two different textures at the surface and in the center of a sheet had been noted in 1955 by POMEY and CRUSSARD [2] in low carbon steels, hot rolled in the temperature range $810^{\circ}C$ to $895^{\circ}C$. In 1958, MÖLLER and

STÄBLEIN [3] observed the existence of these two textures in silicon iron sheets. The more recent studies have all recognized this coexistence, not only after hot rolling but also after cold rolling in certain conditions [4]. Concerning the two types of textures found in the center of the hot rolled sheets, it is interesting to note that NAGASHIMA and his co-workers [5] published two {110} pole figures (Figs. 18 and 20 of their article) obtained in the centers of two rimmed steel sheets which had been hot rolled in the following conditions: 1) Finishing temperature 830°C – Coiling temperature 620°C. 2) Finishing temperature 805°C – Coiling temperature 550°C. In the first case (830°C, 620°C) the pole figure corresponds to a texture of what we called type I, whereas in the second case it represents a texture of our type II. These results are consistent with our correlation since the two ΔT's are equal respectively to 210°C and 260°C.

A rotation of the center texture of 30° about the transverse direction gives a close approximation of the surface texture. However, the rotation mechanism does not seem sufficient to rationalize perfectly the relationship between the two textures. It is probable that a recrystallization mechanism operates at the same time.

4.6.2 Recrystallization Textures

We have annealed under various conditions (see below) several samples from four different commercial cold-rolled coils of approximately 0.8 mm thick sheet corresponding to the following reductions in thickness: UK 68 %, SK 60 %, UR 64 %, SR 55 %. Two coils were of low carbon rimmed steel (R), the other two of low carbon aluminium killed steel (K).

4.6.2.1 Cold Rolling Texture of the Sheets

The cold rolling textures which were determined for the four sheets are very similar. The corresponding {110} pole figure determined by using the composite sampling technique described by ELIAS and HECKLER [6] (their Fig.3) is shown in Fig.5. From the analysis of this pole figure it appears reasonable to describe the cold rolling texture as a

superposition of two partial ⟨110⟩ fiber textures which we
call ⟨110⟩ // RD and ⟨110⟩ // 60° RD. One of the fiber axes
is the rolling direction (⟨110⟩ // RD). The other is
obtained by a 60° rotation of the rolling direction about the
transverse direction (⟨110⟩ // 60° RD). There was no
difference between the textures at the surface and in the
center of the four cold rolled sheets.

4.6.2.2 Recrystallization Textures of Rimmed Steels

4.6.2.2.1 Annealing treatments in argon. In a first step,
nine annealing cycles were conducted in argon to study the
influence of the heating rate (100°C per hour and 800°C per
hour), soaking time (from 1 to 24 hours) and temperature
(from 600°C to 800°C) on the textures formed. After each
annealing cycle {110} pole figures were determined. A careful
examination of the pole figures leads to the following
conclusions: 1) When the annealing temperature is low (600°C)
and the soaking time short (2 hours), the texture is very
similar to the cold rolling texture. Recovery has occured as
shown by the hardness values, but recrystallization has not
been completed. 2) As the soaking time and/or temperature
increase, we can observe an evolution of the textures towards
what can be considered as the typical recrystallization
texture of rimmed steel sheets and which will be described
later. The nature of the texture obtained is not affected by
the heating rate when it varies in the limits of our
experiments.

In a second step, we have performed a series of annealing
treatments at 700°C with rimmed steel to follow the evolution
of the texture as well as of r and n values as a function of
time. Table 1 describes these cycles together with the values
of r_m and n_m obtained.

As the soaking time increases the values of r_m and n_m rise,
but there is no clear difference between the pole figures
obtained. After the eight hours anneal the {110} pole figure
is essentially the same as after the 96 hours anneal. Since
there is an increase in r_m as the soaking time gets longer,
there must be a slight evolution in the textures, but our
description by a single pole figure is probably not precise

Table 1

Average heating rate °C per hour	Soaking temperature	Soaking time in hours	n_m [+]	r_m [++]
4000	700°C	8 h	0,250	1,123
4000	700°C	19 h 30	0,257	1,136
4000	700°C	24 h	0,255	1,043
4000	700°C	48 h	0.274	1,160
4000	700°C	96 h	0,274	1,244

[+]
$$n_m = \frac{n_0 + 2\,n_{45} + n_{90}}{4}$$

[++]
$$r_m = \frac{r_0 + 2\,r_{45} + r_{90}}{4}$$

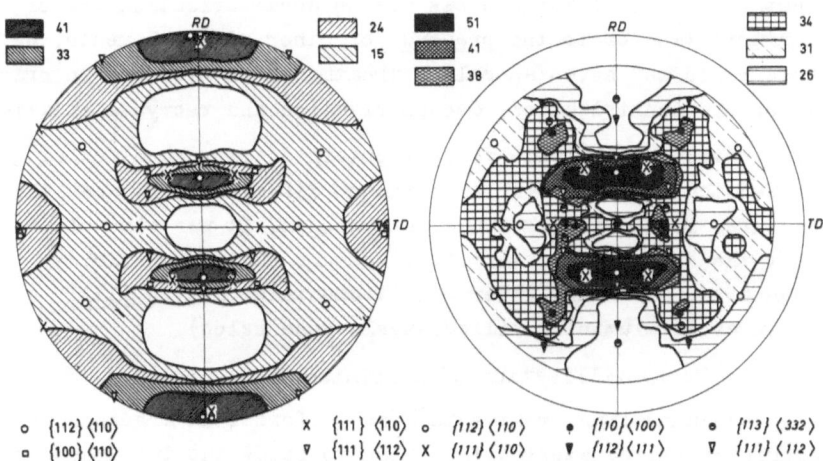

	41		24		51		34
	33		15		41		31
					38		26

o {112}⟨110⟩   x {111}⟨110⟩   o {112}⟨110⟩   • {110}⟨100⟩   • {113}⟨332⟩
□ {100}⟨110⟩   ▽ {111}⟨112⟩   x {111}⟨110⟩   ▼ {112}⟨111⟩   ▽ {111}⟨112⟩

Fig.5. {110} pole figure in the cold rolled condition

Fig.6. {110} pole figure representing the recrystallization texture of rimmed steel

enough to allow an observation of this evolution. In Fig.6, we show the {110} pole figure corresponding to the 96[h] anneal. This pole figure is characteristic of the recrystallization texture of rimmed steel which can be described as follows:

Description of the recrystallization texture: 1) The peaks of the cold rolling pole figure are maintained. 2) A Goss component, {110} ⟨110⟩, has appeared together with {112} ⟨111⟩ orientations. The sum of these new components can be described approximately as a limited fiber texture in which the ⟨110⟩ axis is parallel to the transverse direction. 3) The third characteristic of this texture is the disappearance of the {112}⟨110⟩ component of the cold rolling texture.

4.6.2.2.2 Annealing treatments in hydrogen. In recent years, it has been reported that the so called {111} texture obtained industrially with aluminium killed steels had also been obtained with rimmed steels which had been decarburized and denitrided in order to avoid their strain ageing and to which, moreover, some minor elements had been added in order to keep suitable mechanical properties. In order to determine whether the {111} texture was due to decarburization and/or denitridation or to the presence of other added elements, we have performed seven annealing treatments in hydrogen in order to obtain denitridation, decarburization and recrystallization.

Examination of the pole figures obtained after each anneal showed that there are no significant differences between the textures formed in argon and in hydrogen. It was also clear that the obtention of the {111} texture was not possible even after annealing under conditions which led to {111} textures in aluminium killed steels (see below).

4.6.2.3 Recrystallization of Aluminium Killed Steels

Eighteen annealing treatments were performed in argon on the aluminium killed steel with a view to study the influence on the textures formed of the heating rate (100°C per hour and 800°C per hour), soaking time and temperature, existence of a plateau below the recrystallization temperature at '550°C. Following each annealing treatment, grain shape was observed and {110} pole figures were determined by the Schulz's reflexion method.

Results 1) When the heating rate is equal to 800°C per hour, the grains are equiaxed after recrystallization and the texture can be described as the texture of rimmed steels without the Goss component. Therefore, we cannot speak any more of a ⟨110⟩ fiber texture with the ⟨110⟩ axis parallel to the transverse direction.

2) When the heating rate is equal to 100°C, the grains are pancake-like after recrystallization and the texture can be described as a {111} type texture.

3) The existence of a plateau at 550°C before the final anneal leads to a very well developed {111} type texture (Fig.7) and to pancake-like grains. In this connection it must be remembered that a {111} type texture actually consists of two components: the main component corresponds to the highest intensity regions of the cold rolling texture, the secondary component is {111}⟨u v w⟩.

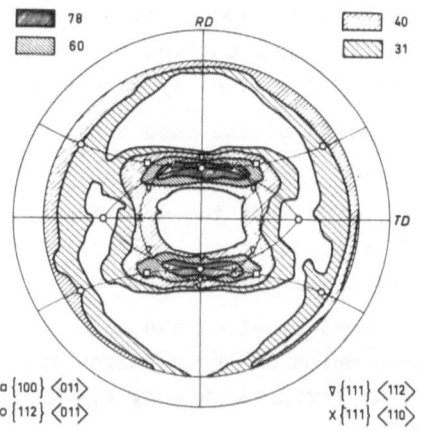

Fig.7. {110} pole figure of the {111} type texture obtained after recrystallization of aluminium killed steel

4.6.2.4 Discussion

Our results concerning the recrystallization textures of aluminium killed steels are in general agreement with those published in the literature. They are consistent with the hypothesis of a precipitation or of a pre-precipitation of NAl before recrystallization.

In the case of rimmed steels, the observed existence of the highest intensity regions of the cold rolling {110} pole figure in the recrystallization {110} pole figure has been reported several times but, to our knowledge, the development of a partial ⟨110⟩ fiber texture with the ⟨110⟩ axis parallel to the transverse direction has not been reported as yet. The fact that the textures obtained with rimmed steels are not very different from those obtained with aluminium killed steels after a high heating rate is consistent with the results of AKAMATSU and SAKAMOTO, who found that the textures of rimmed and killed steels can be the same [7].

4.6.3 Conclusions

1) The center textures of hot rolled sheets of commercial low carbon rimmed and aluminium killed steel processed under normal conditions can be classified into two types, the appearance of which is related to the parameter ΔT = Finishing temperature minus Coiling temperature. These two types are ⟨110⟩ // RD and ⟨110⟩ // RD + {100}⟨110⟩.

2) The surface textures of the sheets are all the same, namely {110}⟨u v w⟩ + {112}⟨111⟩.

3) As a first approximation, the surface texture of a given sheet is deduced from the center texture through a $30°$ rotation about the transverse direction.

4) Annealing of rimmed steel sheets cold rolled 55 % and 64 % leads to a texture which is the superposition of three partial ⟨110⟩ fiber textures, namely ⟨110⟩ // RD, ⟨110⟩ // $60°$RD, ⟨110⟩ // TD.

5) Aluminium killed steel sheets, cold rolled 60 and 68 % have a well defined {111} type texture and pancake-like grains only when annealed under suitable conditions.

References

1. T.Sakamoto, G.Pomey and M.Grumbach:Rapports Extérieurs IRSID 239 (mai 1967) et 260 (août 1967).
2. G.Pomey and C.Crussard:Rev.Mét. 52(1955)401.
3. H.Möller and H.Stäblein:Arch.Eisenhüttenw. 29(1958)377.
4. C.A.Stickels:Trans.Met.Soc.AIME 239(1967)1857.
5. H.Takechi, H.Kato and S.Nagashima:Trans.Met.Soc.AIME 242(1968)56.
6. J.A.Elias and A.J.Heckler:Trans.Met.Soc.AIME 239(1967)1237.
7. T.Akamutsu and T.Sakamoto:Trans.Jap.Inst.Metals 7(1966)81.

4.7 Einfluß von dispergiertem α-Al$_2$O$_3$ auf die Verformungs- und Rekristallisationstextur von Kupferdrähten
Influence of dispered α-Al$_2$O$_3$ on the Deformation- and Recrystallization-Texture of Copper Wires

von Ch. Liesner und G. Wassermann[+)]

Abstract

The results show that especially small amounts of fine particles of 0,05 and 0,3 μm influence the development of deformation textures very strongly. The deformation textures are retained even by annealing at 1000°C.

Die Verformung von Metallkristallen wird durch nicht verformbare Zusätze einer zweiten Phase wesentlich verändert. Diese an einzelnen Beispielen [1] seit langem bekannte Erscheinung ist von MERZ und WASSERMANN [2] systematisch untersucht worden. Es ergab sich, daß in Stangen und Drähten mit einer Matrix aus Aluminium, Kupfer oder Eisen [3] die Verformungstexturen durch dispergierte Teilchen aus Stoffen, deren Fließgrenzen über der der Matrix liegen oder die überhaupt nicht plastisch deformierbar sind, mit zunehmendem Mengenanteil immer mehr verschlechtert wurden.

4.7.1 Versuchsdurchführung

Stangen und Drähte aus Kupfer mit dispergiertem α-Al$_2$O$_3$ (Korund) wurden pulvermetallurgisch durch Strangpressen und Kaltziehen hergestellt.

4.7.2 Strangpreßtexturen

Beim Strangpressen von zusatzfreiem Material entsteht die bekannte ⟨211⟩-Fasertextur des Kupfers. Abb. 1a gibt eine Übersicht der Preßtexturen für Stangen mit Oxidzusätzen in Abhängigkeit von Teilchendurchmesser und zugesetzter Menge. Die ⟨211⟩-Textur des zusatzfreien Kupfers wird nicht verändert, solange der Teilchendurchmesser groß und der Anteil an 2.

[+)] Institut für Metallkunde und Metallphysik der Technischen Universität Clausthal, Clausthal, Bundesrepublik Deutschland

Phase gering ist. Kleine Teilchen verändern dagegen die Textur schon bei sehr geringer Menge. Es entsteht dann beim Pressen eine äußerst scharfe ⟨100⟩-Fasertextur. Generell läßt sich sagen, daß auch hier die Texturausbildung nach ⟨211⟩ wie auch nach ⟨100⟩ umso stärker behindert wird, je größer der Teilchendurchmesser und die zugefügten Mengen sind [2,3].

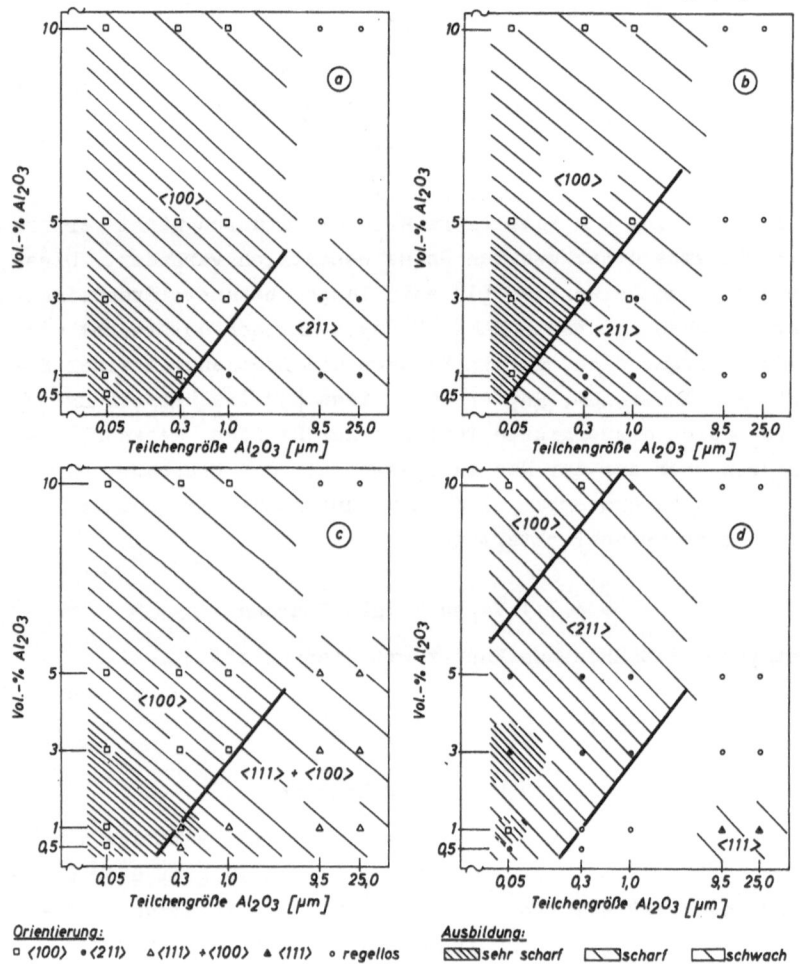

Abb.1a - d: Einfluß von Teilchengröße und -menge des Korunds auf die Texturausbildung von Kupfer. Auswertung aus Drehkristallaufnahmen aus der Stangenmitte.

a) Stranggepreßt, Preßtemperatur 1000°C, Warmverformungsgrad 98%. - b) Wie a) zusätzlich 2h/1000°C geglüht. - c) Wie a), um 90% (1-5 Vol.%) bzw. 60% (10 Vol.% Korund) kalt nachgezogen. - d) Wie c) zusätzlich 10'/1000°C geglüht

Abb.1b zeigt die Texturausbildung nach einer an das Strangpressen anschließenden Glühung von 2 Std. bei $1000^{\circ}C$.

Gegenüber Abb.1a ist für geringe Zusatzmengen eine Verschiebung der Bereichsgrenze zu kleineren Teilchen erfolgt.

4.7.3 Drahttexturen

Wie Abb.1c erkennen läßt, wird durch Nachziehen um 60 bzw.
90% (ohne $1000^{\circ}C$-Zwischenglühung) in den Preßstangen mit
$\langle 211 \rangle$-Textur die normale Ziehtextur nach $\langle 111 \rangle + \langle 100 \rangle$ hervorgerufen. Die Abnahme der Texturschärfe mit zunehmenden
Teilchendurchmessern und -mengen bis zur Regellosigkeit tritt
auch hier auf. Die $\langle 100 \rangle$-Preßtextur bleibt dagegen erhalten,
sie wird durch das Nachziehen noch deutlich verschärft.

Abb.1d zeigt die Ergebnisse nach einer auf das Ziehen folgenden Glühung von 10 Min. bei $1000^{\circ}C$. Die Drähte mit der dop-
pelten Fasertextur (Abb.1c) sind (mit 2 Ausnahmen) regellos
rekristallisiert. Die Proben mit der reinen $\langle 100 \rangle$-Ziehtextur
zeigen nun eine $\langle 211 \rangle$-Rekristallisationstextur. Nur die Probe mit 1 Vol.-% Korund von $0,05\,\mu m$ hat auch nach der $1000^{\circ}C$-
Glühung eine äußerst scharfe $\langle 100 \rangle$-Textur, ferner gilt dies
für die beiden Proben mit 10 Vol.-% mit 0,05 und $0,3\,\mu m$ Teilchengröße.

4.7.4 Diskussion der Ergebnisse

Die $\langle 211 \rangle$-Textur des zusatzfreien und des mit grobdispersen
Teilchen durchsetzten Kupfers ist eine Rekristallisationstextur [4], während die $\langle 100 \rangle$-Strangpreßtextur des Kupfers mit
vorwiegend feinen Teilchen wahrscheinlich eine bei der Warm-
verformung gebildete Verformungstextur ist, in der die fein-
dispergierten Teilchen bei geeignetem Verhältnis von Teilchengröße und Teilchenmenge eine Wanderung von Großwinkelkorngrenzen (Rekristallisation) verhindern. Eine Nachglühung
bei $1000^{\circ}C$ verschiebt die Grenze des $\langle 211 \rangle$-Bereiches allerdings noch ein wenig.

Beim Nachziehen bleibt die reine $\langle 100 \rangle$-Orientierung erhalten
und wird sogar noch verschärft. Offensichtlich können die
dispergierten Korundteilchen die durch das Pressen bereits

hoch entwickelte Verformungstextur nicht mehr nachteilig be-
einflussen. Die Entwicklung von $\langle 111 \rangle$ + $\langle 100 \rangle$ aus $\langle 211 \rangle$ entspricht den Erwartungen [1]. Da sich hier die Verformungs-
texturen erst entwickeln müssen, tritt der behindernde Einfluß des Korunds deutlich in Erscheinung. Bei der Glühung
der nachgezogenen Drähte bei 1000°C entspricht der Übergang
der Doppelfasertextur zu regelloser Orientierung bzw. von
$\langle 100 \rangle$ in die $\langle 211 \rangle$-Textur dem von Kupferdrähten bekannten
Verhalten [1,4]. Der Einfluß der Korundteilchen kommt aller-
dings in einigen Fällen in abweichenden Ergebnissen zum Ausdruck. Bemerkenswert ist vor allem die trotz der Glühung
scharf erhaltene $\langle 100 \rangle$-Textur für 1 Vol.-% 0,05 μm Al_2O_3.
Die in der entsprechenden Fasertextur vorhandenen, dünn aus-
gezogenen Kristalle lassen eine Veränderung durch Rekristall-
lisation nicht erkennen. Trotzdem ist nicht vorstellbar, daß
solche Kristalle noch kalt verfestigt sind. Sie werden durch
einen nicht festgestellten Mechanismus (z.B. Rekristallisation in situ, örtliche Neuorientierung) entfestigt sein.

Literatur

1. G. Wassermann und J. Grewen: Texturen metallischer Werkstoffe, Springer, Berlin 1962.
2. D. Merz und G. Wassermann: Z. Metallkde. 56(1965)516.
3. Ch. Liesner und G. Wassermann: Unveröffentlichte Ergebnisse.
4. J. Grewen: Z. Metallkde. 57(1966)581.

4.8 Über den Zusammenhang zwischen Zipfelhöhe und Zustands-
schaubild bei Aluminium
Connection between Ear-Height and Constitution Diagram of
Aluminium Alloys

von W. Normann[+)]

Abstract

Starting from a discussion given at the symposium on "earing
behaviour and texture", earing results from a number of
papers were examined concerning their connection with the
constitution diagram Al-Fe-Si. These results obtained from
materials annealed and water quenched prior to last cold
rolling were included in this examination, because these
cases represent nearly the same definite state of quenching.
Hot rolling immediately after annealing treatment should
gave the same conditions. It·was found that material
possessing an equlibrium state near the phase boundary
between the phase regions $Al + Al_3Fe$ and $Al + Al_3Fe + \alpha FeSi$
showed highest $45°$-ears after last annealing.

Anläßlich des Symposiums über Zipfelbildung und Textur in
Frankfurt 1965 wurde im Anschluß an einen Vortrag von SIE-
BEL [1] der Versuch gemacht, die dargebotenen Versuchsergebnisse in einen Zusammenhang mit dem Zustandsschaubild
Al-Fe-Si [2] zu bringen. Dieser damals entwickelte Zusammenhang wurde inzwischen anhand weiterer veröffentlichter Arbeiten überprüft und die Ergebnisse zusammenfassend dargestellt.

Als Ausgangspunkt sei nochmals die seinerzeit gezeigte Abb.1
dargestellt. Über der Barren- und Platten-Glühtemperatur ist
die Zipfelhöhe weichgeglühter Bleche aufgetragen. Auf den
Temperaturachsen sind außerdem, in Abkürzungen, die Phasenräume aus dem Zustandsschaubild Al-Fe-Si angedeutet, die bei
den jeweiligen Fe- und Si-Gehalten im Gleichgewicht vorliegen. Die fünf verschiedenen Kurven unterscheiden sich durch
verschiedene Fe- und Si- Gehalte und/oder durch unterschiedliche technologische Verfahrensgänge. Diese (rechts in Abb.1

[+)] Julius und August Erbslöh, Wuppertal-Barmen, Bundesrepublik Deutschland

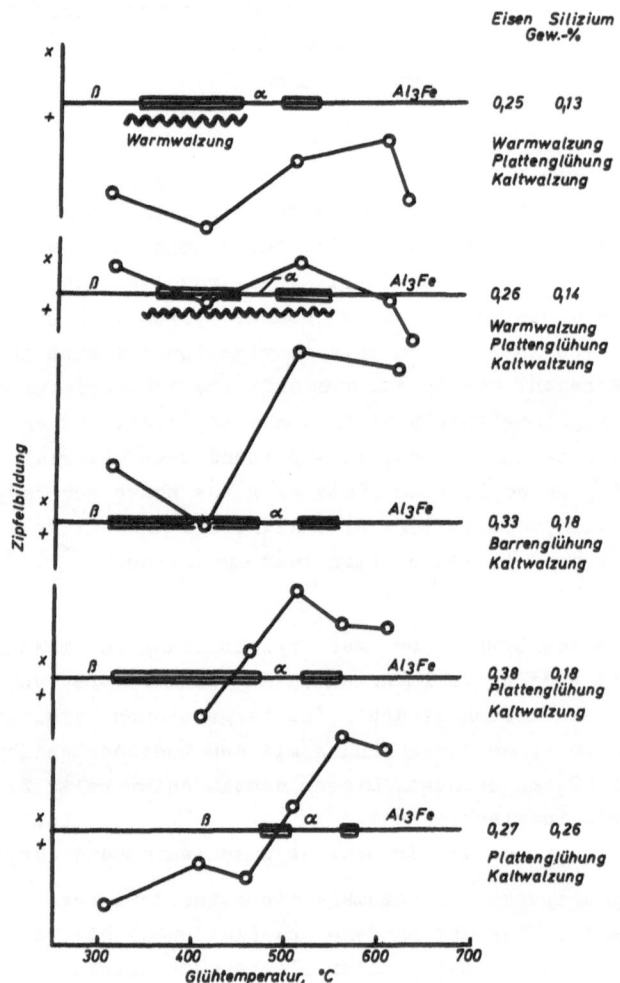

Abb.1. Zipfelwerte nach SIEBEL [1] aufgetragen über
der Glühtemperatur

angeschrieben) bedingen jeweils Allgemeinverschiebungen der
Zipfelwerte zu höheren bzw. niederen Werten. So sind z.B.
alle Zipfelwerte der obersten Kurve unter Null (bzw. 90°),
während demgegenüber alle Zipfelwerte der mittleren Kurve

über Null (45°) liegen.

Es wurde seinerzeit darauf hingewiesen, daß allen Kurven ein Maximum in der Nähe der Phasenraumgrenze Al-Al$_3$Fe – Al+Al$_3$Fe +α-FeSi gemeinsam ist. Das Vorhandensein dieses Maximums ist an diesem Bild deutlich sichtbar, um es jedoch mit den Werten anderer Arbeiten graphisch vergleichen zu können, wurden mit diesen sowie allen anderen Werten zwei Schritte zu einer Normierung durchgeführt:

1. Die Zipfelwerte werden nicht mehr über der Temperaturachse aufgetragen, sondern auf einer Achse, auf der nur die Phasen einheitlich aufgetragen sind. Aus Abb.1 wird dann Abb.2.

2. Alle Kurven, die aus verschiedenen technologischen Arbeitsverfahren stammen, werden gegeneinander, mangels absoluter Werte beliebig, parallel der Zipfelachse verschoben, bis die Gesamtstreuung der Punkte minimal wird. Aus Abb.2 wird Abb.3. Man sieht, man kann fast alle Punkte in einem Band mit einem deutlichen Maximum unterbringen.

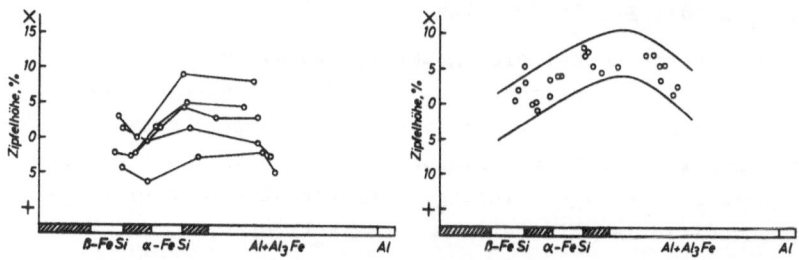

Abb.2. Zipfelwerte nach SIEBEL [1] (Abb.6, 9, 10, 13 und 14) aufgetragen über den Phasenräumen

Abb.3. Zipfelwerte nach SIEBEL [1] aufgetragen über den Phasenräumen und verschoben zu geringster Streuung

Dasselbe Verfahren wurde nun auf eine Reihe von Kurven aus anderen Veröffentlichungen angewandt, um deren Übereinstimmung mit den Ergebnissen nach SIEBEL [1] zu überprüfen. Um solche Werte benutzen zu können, müssen immer drei Voraussetzungen erfüllt sein:

1. Wegen der evtl. notwendigen Parallelverschiebung zur Zipfelachse muß stets eine Kurve aus mindestens 2 Werten vorliegen, deren Werte von Versuchen mit gleichen technologischen Arbeitsverfahren stammen.

2. Die Eisen- und Siliziumgehalte müssen bekannt sein und

3. es muß jeder Wert eindeutig einer bestimmten Temperatur
zuzuordnen sein.

Dabei spielt diejenige Glühbehandlung die entscheidende Rolle, die vor dem letzten Kaltwalzen des Materials stattgefunden hat, während die Weichglühtemperatur des gewalzten Materials in der Praxis offensichtlich keinen wesentlichen
Einfluß hat. Es muß also vor dem letzten Kaltwalzen eine
vielstündige Glühbehandlung mit anschließendem Abschrecken
stattgefunden haben, von der man hoffen kann, daß durch sie
der zu dieser Temperatur gehörige Gleichgewichtszustand eingestellt worden ist. Ob diese Hoffnung immer erfüllt wird,
ist jedoch leider fraglich, besonders im Phasenraum $Al+Al_3Fe$.
Erst seit kurzem liegen eigene Versuchsergebnisse vor, die
besagen, daß im Raume $Al+Al_3Fe$ bei $350^{\circ}C$ auch nach 600stün-
diger Glühung die Leitwertänderung noch nicht abgeschlossen
ist. Zumindest in diesem Phasenraum ist also die Gleichge-
wichtseinstellung, d.h. die vollständige Ausscheidung von
Al_3Fe, in der Praxis fraglich.

Erfreulicherweise hat sich gezeigt, daß man dem Glühen mit
anschließendem Abschrecken auch ein Warmwalzen gleichsetzen
kann, wenn der Barren vorher viele Stunden bei Warmwalztem-
peratur geglüht und anschließend sofort gewalzt wurde. Dies
Ergebnis ist deswegen wichtig, weil für dieses Arbeitsverfahren bei weitem die meisten Werte vorliegen.

In Abb.4 sind alle untersuchten Werte zusammengezeichnet. Im
wesentlichen stammen sie von Al 99,9 bis 99,0,davon viele
mit natürlichem Fe:Si-Verhältnis, also etwa 2:1. Reihenun-
tersuchungen mit Veränderung des Fe:Si-Verhältnisses gehen
meist nicht bis zum Extremen. Die Temperaturen liegen überwiegend zwischen 400 und $600^{\circ}C$, die Kaltwalzgrade zwischen
85 und 95 bis 99 %.

Man sieht deutlich, wie sich die Mehrzahl der Werte in das
eingezeichnete Band fügt, das ein ausgeprägtes Maximum zeigt.
Geht man in dieser Darstellung von links nach rechts und verändert dabei jeweils einen der drei angegebenen Parameter,
dann durchläuft man diese Zipfelkurve. Bei mittleren Temperaturen und natürlichem Fe:Si-Verhältnis hat man ganz links
etwa Al 98,5 und das Zipfelmaximum liegt bei 99,8. Bei noch

reinerem Material erhält man zunehmend mehr 90°-Zipfel. Bei
kleinem Fe:Si-Verhältnis hat man auf der linken Hälfte be-
kanntlich geringe Zipfelhöhen, eine Erhöhung von Fe:Si führt
zu einer Zipfelerhöhung, das Maximum überschreitet man in der
Praxis kaum. Mit steigender (Walz-) Temperatur nehmen zunächst
die Zipfel zu und nach Überschreiten des Maximums wieder ab,
wie die Versuche von SIEBEL [1] und auch anderen zeigten.

Wie stark die Zipfelwerte von der Verfahrenstechnik abhängen zeigt Abb.5, in der graphisch die an den Kurven vorge-
nommenen Parallelverschiebungen aufgetragen sind. Mit Ausnahme einer geschlossenen Gruppe von 48 Werten, deren starke
Abweichung unerklärlich ist, und den Versuchen mit Warmwalzen und Plattenglühung sind die Abweichungen aller anderen
Werte doch sehr gering und überschreiten die Streubreite des
Bandes nicht.

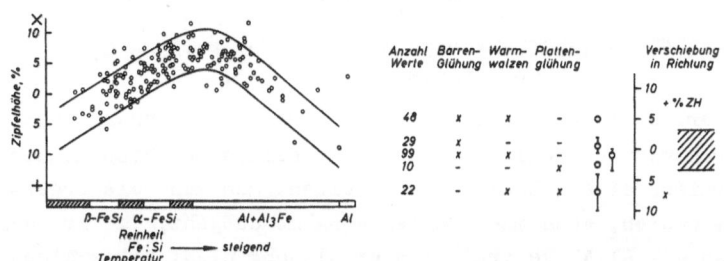

Abb.4. Zipfelwerte verschiedener Autoren aufgetragen über
den Phasenräumen und verschoben zu geringster Streuung

Abb.5. Notwendige Verschiebung der Zipfelwerte um ein Band
geringster Streuung zu erhalten

Zur Erläuterung der vorgetragenen Ergebnisse wird in Abb.6
ein Schnitt für gleichbleibende Gesamtverunreinigung gezeigt.
Die Ergebnisse besagen, daß entlang der Phasenraumgrenze
$Al+Al_3Fe$ – $Al+Al_3Fe+\alpha$-FeSi ein Maximum an 45°-Zipfeln auftritt. Entfernt man sich von dieser Grenze zu niedrigeren
Fe:Si-Verhältnissen oder zu tieferen Temperaturen, dann
nehmen die 45°-Zipfel ab, ebenso, wenn man von der Grenze
aus zu höheren Temperaturen oder höheren Fe:Si-Verhältnissen
übergeht.

In der gezeigten Darstellungsart wurde eine große Zahl be-

366

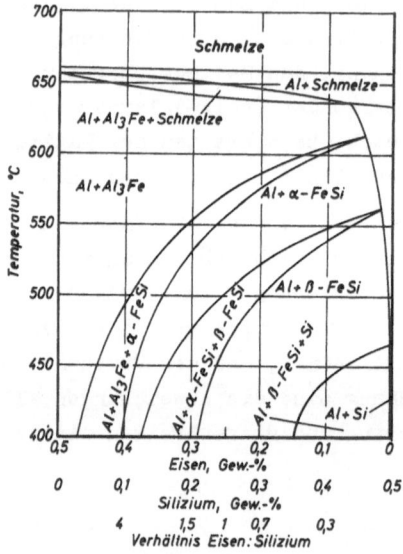

Abb.6. Zustandsschaubild Al-Fe-Si für Fe+Si = 0,5
Gew% (nach PHILLIPS [2])

kannter, aber verstreuter Einzelergebnisse zusammengefaßt.
Dadurch wird es möglich, manche anscheinend widersprüchlichen
Ergebnisse einheitlich zu beschreiben. Man kann die Ergebnisse deuten, wenn man von der Annahme ausgeht, daß im Zwei-
phasenraum Al+Al$_3$Fe trotz langer Glühung nicht alles Eisen
als Al$_3$Fe ausgeschieden ist, sondern teilweise noch über-
sättigt vorliegt. Diese Übersättigung muß dann an der Phasenraumgrenze ein Maximum erreichen. Nach Überschreiten der
Grenze wird die Übersättigung wegen der günstigen Ausschei-
dungsbedingungen durch Bildung ternärer Phasen mehr und mehr
abgebaut.

Es war das Ziel dieser Darstellung, die seinerzeit von BUNK
und ESSLINGER [3] angenommene Bedeutung der Phasenraumgrenze
Al+Al$_3$Fe – Al+Al$_3$Fe+α-FeSi bezüglich des Zipfelverhaltens von
Reinaluminiumblechen herauszustellen und den Gedanken des
Einflusses des Fe:Si-Verhältnisses auf die Zipfel [4] wei-
terzuentwickeln. Um die Abhängigkeit der Zipfel vom dreidimensionalen Al-Fe-Si-Temperatur-Schaubild darzustellen,
wurden die Phasenräume auf einer eindimensionalen Achse

abgebildet. Dabei gehen naturgemäß viele Feinheiten verloren und es wird darum auf eine zweidimensionale Darstellung der Phasenräume verwiesen, die für eine ausführliche Veröffentlichung vorgesehen ist.

Literatur

1. G. Siebel: Z. Metallkde. 57(1966)429.
2. H.W.L. Philips: Inst.Met.Rep.Ser. 25(1959).
3. W. Bunk und P. Eßlinger: Z. Metallkde. 50(1959)278.
4. H. Hug: Aluminium Suisse 7(1957)33.

4.9 Einfluß der Textur auf die Hydridorientierung in Zircaloy-
Hüllrohren
Influence of Texture on the Hydride Orientation in Zircaloy
Tubing

von E. Schelzke[+)]

Abstract

It was possible to show a correlation between texture and
hydride orientation in zircaloy tubing. The platelike
precipitations of zirconium hydride show only a limited
orientation range, which is texture dependent. It
corresponds to the range of habit planes of precipitations
of zirconium hydride and lies in the orientation triangle
of α-zirconium between 5 and 20° to the $[0001]_{\alpha-Zr}$-pole.
Stresses can alter the direction of precipitation only
within this range.

4.9.1 Einleitung

Aus dem gegenwärtigen Stand der Literatur [1-10] zu den an-
stehenden Problemen der Hydridorientierung in dünnwandigen
Rohren aus Zircaloy geht hervor, daß in einigen wichtigen
Punkten keine einheitlichen Ansichten vertreten werden und
die bisher bekannt gewordenen Arbeiten zum Teil widerspre-
chende Ergebnisse aufweisen. Diese Diskrepanzen sind vor
allem bei den Untersuchungen über den Zusammenhang zwischen
Textur und Lage der Zirkonium-Hydridausscheidungen zu ver-
zeichnen.

In der vorliegenden Arbeit [11] wird eine Deutung des Zu-
sammenwirkens von Textur und Spannung auf die Lage der Zir-
konium-Hydridausscheidung in dünnwandigen Zircaloy-Rohren gegeben.

4.9.2 Texturuntersuchungen

Zur Bestimmung der Textur wurden nach dem Zählrohrverfahren

[+)]Metall-Laboratorium der Metallgesellschaft A.-G., Frankfurt (Main), Bundesrepublik Deutschland

die Polfiguren der hexagonalen Basisfläche {0001} und der
Prismenfläche {10T0} des α-Zirkoniums von Rohrproben aus
Zircaloy-2 und Zircaloy-4 aufgenommen. Die Texturbestimmungen ergaben als gemeinsames Merkmal der Verformungstexturen eine [10T0]-Richtung in Längsrichtung (LR) der Rohre;
während der Basispol je nach Verformungsgrad in der Ebene
Radialrichtung-Tangentialrichtung (RR-TR) Kippwinkel
von 0 bis 90° aufwies. In den Abb. 1 a - c sind die
{0002}-Polfiguren von drei ausgewählten Rohrproben

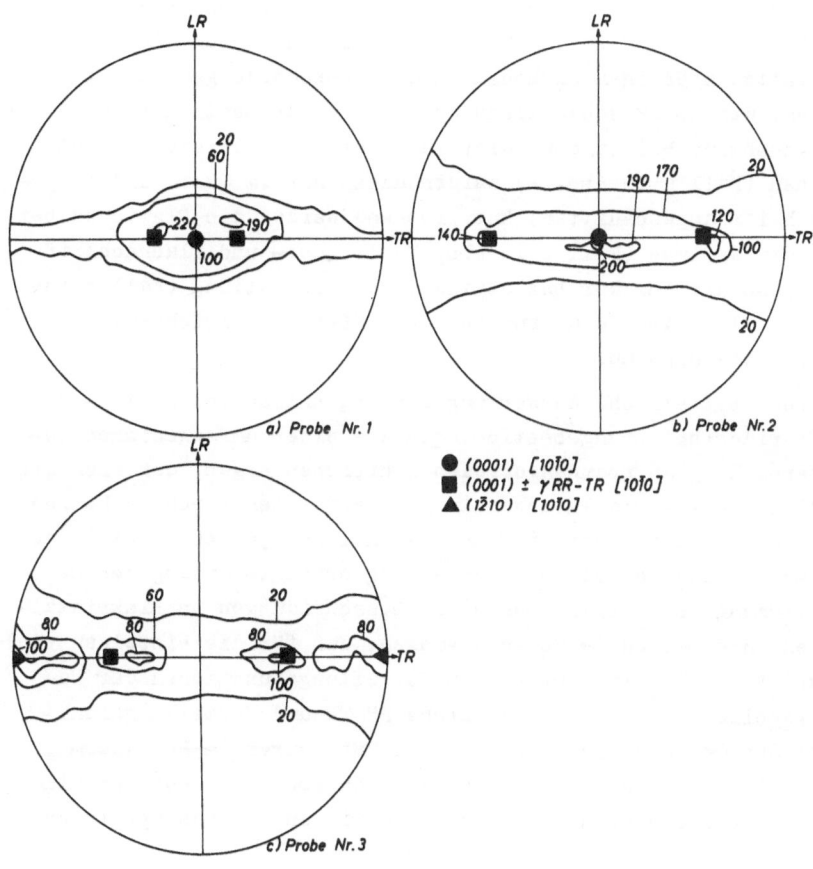

a) Probe Nr. 1   b) Probe Nr. 2

● (0001) [10T0]
■ (0001) ± γ RR-TR [10T0]
▲ (1T10) [10T0]

c) Probe Nr. 3

Abb. 1. {0002}-Polfiguren von kaltgepilgerten Zircaloy-Hüll-
rohren

aus Zircaloy–2 (Nr. 1, 2 und 3) mit sehr unterschiedlicher
Textur wiedergegeben.

4.9.3 Form und Lage der Zirkonium-Hydridausscheidungen

Die Beladung der Proben mit Wasserstoff erfolgte bei $400^{\circ}C$
in einer modifizierten Sievertsapparatur. Die Abkühlungsgeschwindigkeit betrug ca $1^{\circ}C/min$. Zur Sichtbarmachung der
Spuren der ausgeschiedenen Zirkonium-Hydridplättchen wurden
die Proben chemisch geätzt.

Aus dem Verlauf der Ausscheidungsspuren[+) in den drei senkrecht zueinander stehenden Schliffebenen (Ebenen: RR–TR,
RR–LR und TR–LR in Abb.2 konnte gefunden werden, daß die
plättchenförmigen Zirkonium-Hydridausscheidungen in Hüllrohren, wie Abb.2 schematisch darstellt, nur bestimmte Lagen
einnehmen. Bei tangentialer Lage der ausgeschiedenen Plättchen (Fall a in Abb. 2) werden diese nur im Quer- und Längs-
schliff angeschnitten. Entsprechend werden die Plättchen bei
radialer Lage (Fall c in Abb. 2) im Quer- und Flachschliff
angeschnitten. Nur bei statistischer Verteilung (Fall b in
Abb. 2) treten in allen drei Schliffebenen gleichermaßen
Schnittspuren auf.

Eine statistische Auswertung der Ergebnisse von Textur- und
Hydridorientierungsbestimmungen aus einer umfangreichen Untersuchung an spannungsfreien Rohrproben ergab, daß sich die
Zirkonium-Hydridplättchen in kaltverformten Zircaloy-Rohren
in der Nähe der Basisfläche ausscheiden, jedoch nicht in der
Basisfläche selbst. Die Auswertung erfolgte analog der Bestimmung der Habitusebenen für Ausscheidungen in Einkristallen. Hierbei wurde so vorgegangen, daß für die einzelnen vielkristallinen Zircaloy-Proben Verteilungsfunktionen der Ba-
sispolintensitäten in der Ebene RR–TR und Verteilungsfunktionen der Richtung der Ausscheidungsspuren im Rohrquerschnitt ermittelt wurden. Die Verschiebung der Kurvenmaxima
der entsprechenden Verteilungsfunktionen von Basispolinten-

+) Auf eine Wiedergabe der Schliffbilder mußte aus Platz-
mangel verzichtet werden. Hierzu sei auf den in Vorbereitung befindlichen Beitrag [11] verwiesen.

sitäten und Ausscheidungsrichtungen läßt eine Aussage über die häufigsten Winkelabstände zwischen Basisfläche und Habitusebene zu.

Im einzelnen wurde ein Orientierungsbereich für die Habitusebenen der Zirkonium-Hydridausscheidungen ermittelt, der im Orientierungsdreieck des α-Zirkoniums Lagen einnimmt, die einen Winkelabstand von 5° bis 20° zum $[0001]_{\alpha-Zr}$-Pol aufweisen. Für Zircaloy-4 stimmt dieses Ergebnis mit den Angaben von BABYAK [12] weitgehend überein. In einer kürzlich erschienenen Veröffentlichung von WESTLAKE [13] wird im Gegensatz zu früheren Angaben als Ausscheidungsebene in Zircaloy-4 und Zircaloy-2 die $\{10\overline{1}7\}$-Ebene genannt.

4.9.4 Einfluß von Spannungen auf die Hydridorientierung

Durch Verspannen von Rohrproben aus Zircaloy-2 ähnlich Abb.3 fanden LOUTHAN und MARSHALL [3] erstmalig, daß sich unter dem

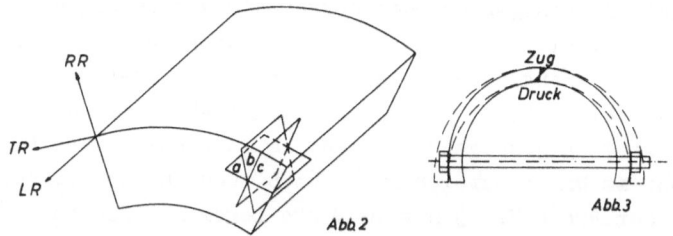

Abb.2. Mögliche Lagen der Hydridplättchen in Rohren aus Zircaloy

Abb.3. Verspannung der Rohrsegmente

Einfluß von Spannungen die Zirkonium-Hydridplättchen parallel zur Druckspannung und senkrecht zur Zugspannung ausscheiden. Im Zusammenhang mit der Existenz bestimmter Winkelbereiche für die Habitusebenen der Zirkonium-Hydridausscheidungen darf angenommen werden, daß sich der Einfluß der Spannungen auf die Lage der Ausscheidungen jedoch nur innerhalb dieser Bereiche bemerkbar macht.

In Abb.4 sind die möglichen Ausscheidungsbereiche der drei sich in der Textur unterscheidenden Proben Nr. 1, 2 und 3 schematisch dargestellt. In diesem Modell sind die aufgrund

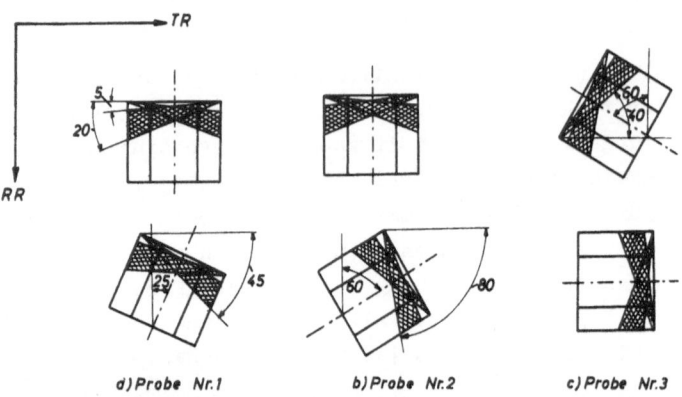

d) Probe Nr.1   b) Probe Nr.2   c) Probe Nr.3

Abb.4. Modell für den Zusammenhang zwischen der Textur
und der Hydridorientierung

der Texturergebnisse bevorzugten Lagen der hexagonalen Elementarzellen, bezogen auf das Koordinatensystem des Rohres, in Seitenansicht im Rohrquerschnitt wiedergegeben. Danach dürfen die Ausscheidungsspuren im Querschliff der Probe Nr.1 nur in dem Winkelbereich zwischen 5° und 45° gegen die Tangentialrichtung geneigt vorkommen, im Querschliff der Probe Nr. 2 nur im Winkelbereich zwischen 5° und 80° und im Querschliff der Probe Nr. 3 nur im Winkelbereich zwischen 40° und 85°.

Diese Voraussagen stimmen mit den experimentellen Ergebnissen gut überein. Abb.5 zeigt Querschliffe der Proben Nr. 1, 2 und 3, die entsprechend Abb.3 verspannt waren und in diesem Zustand hydriert wurden. Es treten nur bei der Probe Nr. 2, aufgrund der sowohl tangential als auch radial liegenden möglichen Habitusebenen, die Ausscheidungsspuren senkrecht zur Zugspannung und parallel zur Druckspannung auf. Dagegen zeigt der Querschliff von Probe Nr. 1 im Zugspannungsgebiet keine radialgerichteten Spuren, sondern nur Spuren mit einer Neigung von 45° gegen die Tangentialrichtung. Analog dazu sind bei der Probe Nr. 3 im Druckspannungsgebiet keine tangential gerichteten Spuren zu erkennen, sondern nur solche mit einer Neigung von 40°.

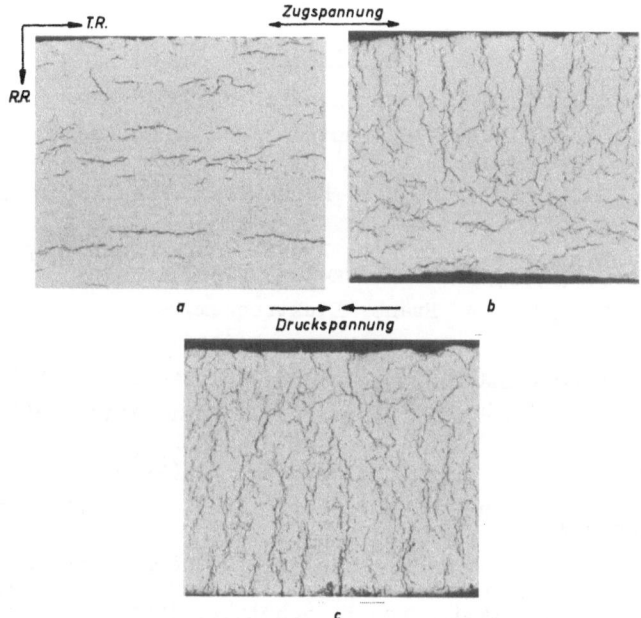

Abb.5. Spuren der Zirkonium-Hydridausscheidungen in den unter Spannungen hydrierten Proben Nr. 1, 2 und 3(V=80)

4.9.5 Diskussion

Als Ursache für das Auftreten einer bestimmten Hydridorientierung in dünnwandigen Rohren aus Zircaloy muß erstens die Existenz eines begrenzten Orientierungsbereiches möglicher Habitusebenen für Zirkonium-Hydridausscheidungen im Zusammenwirken mit der Textur angesehen werden und zweitens die von der Kaltverformung während des Herstellungsverfahrens herrührenden Eigenspannungen.

Die plättchenförmigen Zirkonium-Hydridausscheidungen ordnen sich innerhalb der durch die Textur begrenzten Bereiche bevorzugt auf Habitusebenen an, die der Parallelität zu den Druckspannungen und der senkrechten Lage zu den Zugspannungen am nächsten kommen.

Diese Ergebnisse beschränken sich auf die in kaltverformten, feinkörnigen Blech- und Rohrproben vorwiegend beobachteten transkristallinen Ausscheidungen.

Literatur

1. W. Dürrschnabel: Atomwirtschaft 10(1965)560 u. 682.

2. R.P. Marshall und M.R. Louthan jr.: Trans.Amer.Soc. Metals 56(1963)693.

3. M.R. Louthan jr. und R.P. Marshall: J.Nucl.Mater. 9(1963) 170.

4. P.L. Rittenhouse und M.L. Picklesimer: USAEC-Rep. ORNL-946(1965) und ORNL-TM-1239 (1965).

5. G.W. Parry und W. Evans: Elektrochem.Technol. 4(1966)225.

6. R.P. Marshall: J. Nucl. Mater. 24(1967)34 u. 49.

7. J.J. Kearns und C.R. Woods: J. Nucl. Mater. 20(1966)241.

8. E.D. Hindle und G.F. Slattery: J. Inst. Metals 94(1966) 245.

9. D.G. Westlake: Trans.Met.Soc.AIME 236(1966)1229.

10. D.G. Westlake und E.S. Fisher: Trans.Met.Soc.AIME 224 (1962)254.

11. Auszug aus der von der Fakultät für Bergbau und Hütten-wesen der Technischen Universität Clausthal genehmigten Doktor-Dissertation von E. Schelzke, 1968. Ein ausführ-licher Beitrag befindet sich in Vorbereitung und wird voraussichtlich Anfang 1969 in J.Nucl.Mater.veröffent-licht.

12. W.J. Babyak: Trans.Met.Soc.AIME 239(1967)252.

13. D.C. Westlake: J. Nucl. Mater. 26(Mai 1968)208.

4.10 Texturmessungen an einigen intermediären Phasen, die durch Eindiffusion in Eisen entstanden sind
Texture Determination of some Intermediate Phases, developed by Diffusion into Iron

von H. Kunst[+)]

Abstract

Textures of some intermediate phases developed during technical surface treatments were investigated. Cementite shows no texture, ϵ-nitride a very complicated one and FeB as well as Fe_2B a $\langle 001 \rangle$ fibre texture. The crystal structure and the habit of the phases will be taken into consideration for discussion of the questions why in some cases a texture exists, in others does not, and what is the correlation of the nature and the scatter of the texture with the growth mechanism of the phases.

4.10.1 Einleitung

Die Texturmessungen, über deren Ergebnisse hier berichtet werden soll, dienten zusammen mit anderen Untersuchungen dem Ziel, Aufschluß über den Wachstumsmechanismus einiger intermediärer Phasen zu erhalten, die bei technisch üblichen Behandlungsverfahren für Stahl entstehen.

4.10.2 Meßverfahren

Die Texturmessungen wurden auf röntgenographischem Wege mit einem Siemens-Texturgoniometer durchgeführt. Bestimmt wurde nur der innere Bereich der Polfigur, die Auswertung erfolgte halbquantitativ. Auf die Angabe von Walz- und Querrichtung wurde verzichtet, da diese für das fragliche Problem unwesentlich ist.

4.10.3 Probenmaterial

Als Probenmaterial wurde Elektrolyteisen verwendet, das vor der Diffusionsbehandlung nach zehnmaligem Erhitzen auf $1000^{\circ}C$

[+)] Institut für Härterei-Technik, Bremen, Bundesrepublik Deutschland

und Abschrecken in Wasser sowie einer zweistündigen Glühung
bei 650°C textur- und spannungsfrei vorlag.

4.10.4 Ergebnisse

4.10.4.1 Boride

Bei der Eindiffusion von Bor in Eisen bildet sich – abhängig
vom Borierpotential – entweder eine einphasige Schicht von
Fe_2B oder eine zweiphasige aus FeB (ca. 16 % B) und Fe_2B
(ca. 9 % B) [1]. Das Ergebnis von Texturmessungen an FeB
und Fe_2B zeigen die Abb.1 und 2.

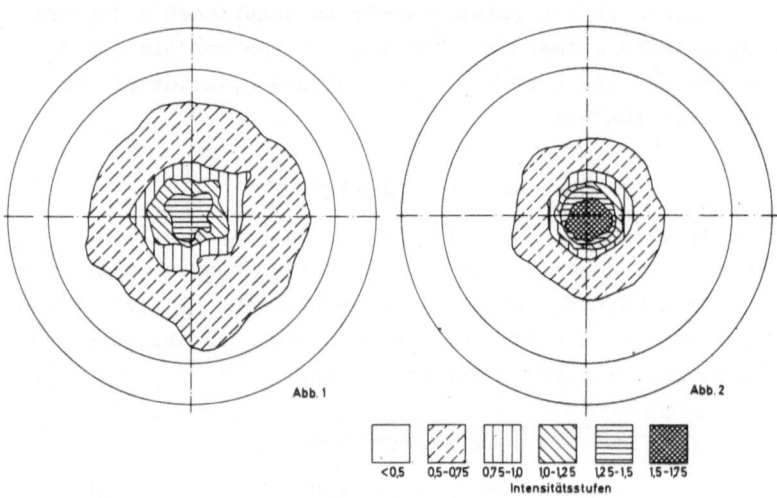

Abb.1. $\{002\}$-Polfigur von FeB; Elektrolyteisen, 3 h 1000°C
in amorphem Bor

Abb.2. $\{002\}$-Polfigur von Fe_2B; Elektrolyteisen, 3 h 1000°C
in 28% Bor, 69% Al_2O_3, 3% BaF_2

Aus der zentralen Belegung der $\{002\}$-Polfiguren (Abb.1 u. 2)
und der ringförmigen Belegung der $\{021\}$- bzw. $\{121\}$-Polfigur
folgt für beide Phasen unmittelbar das Vorliegen einer $\langle 001\rangle$-
Fasertextur relativ geringer Streuung. Demnach ist die Faserachse mit der c-Achse identisch, beide bilden die Normale auf die Oberfläche, d.h. sie liegen in Diffusionsrich-

tung. Die Kristallstrukturen beider Phasen (Abb.3 und 4)
zeigen in c-Richtung Ketten bzw. Reihen von Boratomen recht
geringen Abstandes. Die naheliegende Vermutung, daß die Diffusion des Bors über diese Ketten bzw. Reihen erfolgt, konnte
nach einer anderen Methode bewiesen werden [1]. Die Erleichterung der Diffusion in c-Richtung hat eine starke Anisotropie
zur Folge. Demnach finden bei Überschreiten der Löslichkeits-
grenze, d.h. beim Entstehen der Boridphase, von der Vielzahl

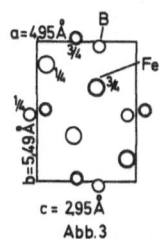
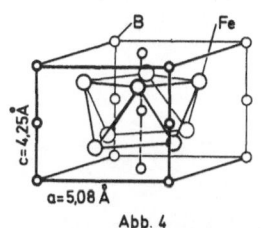

Abb.3. Kristallstruktur FeB   (nach SCHUBERT [3])
Abb.4. Kristallstruktur Fe_2B (nach SCHUBERT [3])

der gebildeten Keime diejenigen die günstigsten Wachstumsbe-
dingungen vor, deren c-Achse in Diffusionsrichtung liegt.
Diese wachsen bevorzugt und führen zur Ausbildung der Textur und zum typischen, spitzzahnigen Aussehen der Boridschicht (Abb.5a).

4.10.4.2 Aluminid

Das Aluminid Fe_2Al_5 entsteht z.B., wenn Eisen in flüssiges
Aluminium getaucht wird. HEUMANN und DITTRICH [2] machten
darauf aufmerksam, daß die entstehende Verbindungsphase eine
Vorzugsrichtung aufweist, verzichteten aber auf quantitative
Texturbestimmungen. Die Polfiguren dieser Phase (Abb.6) zeigen wiederum das Auftreten einer ⟨001⟩-Fasertextur, allerdings mit größerer Streuung. Faserachse, Diffusionsrichtung
und c-Achse fallen demnach zusammen. Die in Abb.7 wiedergegebene Kristallstruktur zeigt eine reihenförmige Anordnung
der Aluminiumatome. Auch hier liegt demnach eine starke Anisotropie der Diffusion vor, deren Folge die Ausbildung einer
Textur und ein charakteristisches, allerdings mehr zungen-
förmiges, Aussehen ist (Abb.5b).

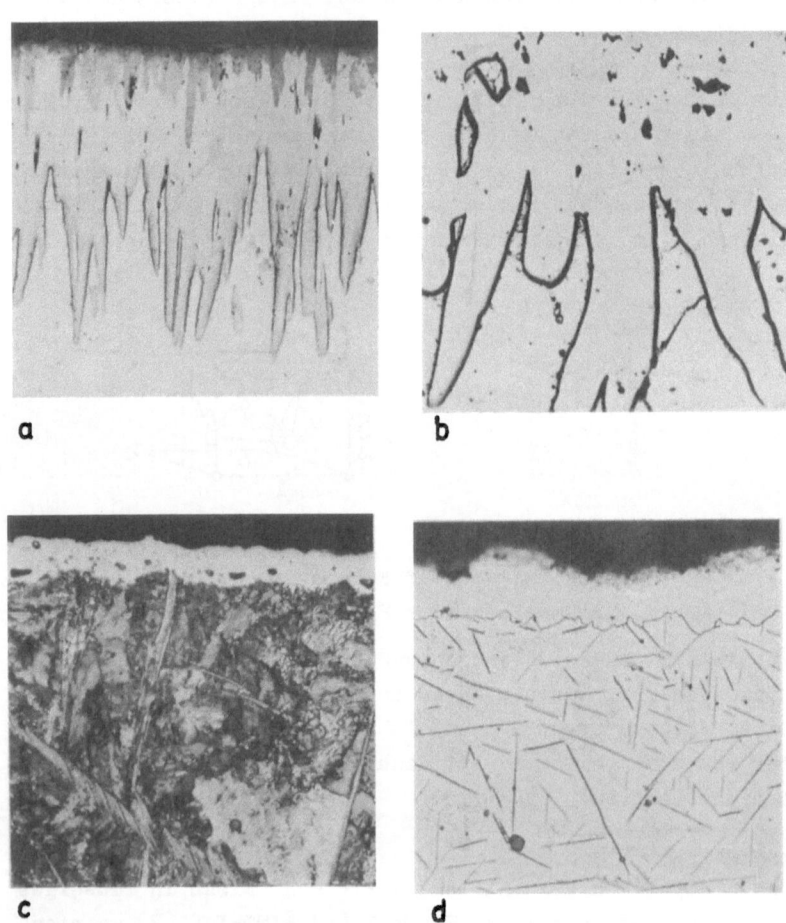

Abb.5a – d. Schliffbilder verschiedener Schichten. a und b:
V = 200, c und d:V = 500

a) Boridschicht (Herstellung s. Abb.1)
b) Aluminidschicht (Herstellung s. Abb. 6)
c) Zementitschicht (16 Mn Cr 5, 75 h 1000°C in aktiviertem
Kohlungspulver)
d) ε-Nitridschicht (Elektrolyteisen, 3 h 570°C in NH$_3$)

4.10.4.3 Zementit

Im Gegensatz zu den bisher behandelten Diffusionsverfahren
ist es bei Zementit unter anderem wegen der hohen Löslichkeit des γ-Eisens für Kohlenstoff schwierig, eine geschlos-
sene Schicht zu erzeugen. Da das auf Elektrolyteisen nicht

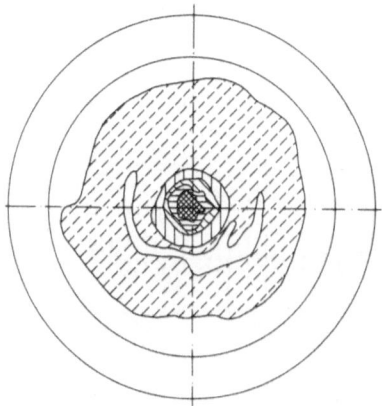

Abb.6. {002}-Polfigur von Fe_2Al_5; Elektrolyteisen, 2 h
780°C in Aluminium

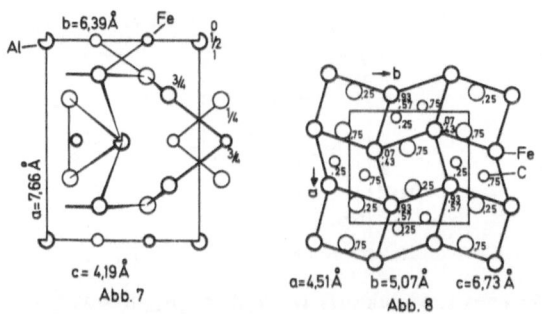

Abb.7. Kristallstruktur Fe_2Al_5 (nach SCHUBERT [3])

Abb.8. Kristallstruktur Fe_3C (nach SCHUBERT [3])

gelang, wurde der Einsatzstahl 16MnCr5 als Probenmaterial
verwendet. Texturmessungen an Zementitschichten zeigten,
daß sie keine Textur aufweisen. Das Kristallgitter (Abb.8)
zeigt zwar eine gewisse Verwandtschaft zum FeB-Gitter, rein
formal besteht eine Anordnung der Kohlenstoffatome in zick-
zackförmigen Ketten, aber die C-C-Abstände sind relativ groß,
auch größer als die Fe-C-Abstände. Infolgedessen findet der
Kohlenstoff nicht die günstigen Diffusionsbedingungen vor wie
die Partner bei den bisherigen Beispielen. Die Folge ist eine
isotrope Diffusion und das Fehlen einer Textur sowie eine
glatt aufliegende Schicht (Abb.5c).

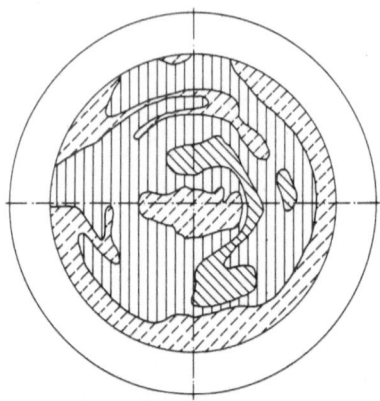

Abb.9. {101}-Polfigur von ε-Fe$_2$N-Fe$_3$N; Elektrolyteisen, 7 h 570° in NaCN-NaCNO

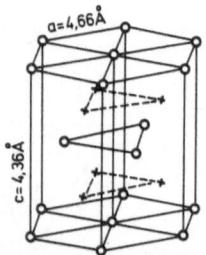

Abb.10. Kristallstruktur ε-Fe$_2$N-Fe$_3$N; Kreuze = mögliche Plätze für N-Atome, c zur besseren Übersicht gestreckt (nach JACK [4])

4.10.4.4 ε-Nitrid.

Die ε-Nitridphase entsteht beim Nitrieren von Stahl im KCN-KCNO-Salzbad oder in Ammoniak. Texturmessungen an dieser Phase ergeben ein sehr verworrenes Bild. Unabhängig vom Diffusionsverfahren, von Temperatur und Dauer ergeben sich Polfiguren (Abb.9), die einer vernünftigen Indizierung nicht zugänglich sind. Um Durchstrahlung der dünnen Nitridschicht und Reste der Walz- oder Glühtextur des Eisens kann es sich nicht handeln, da die Proben vor der Behandlung auf Texturfreiheit geprüft wurden. Auch ein zufälliger Effekt an einzelnen Proben scheidet als Erklärung aus, da die Bestimmung an einer Vielzahl von nach verschiedenen Verfahren und unter verschiedenen Bedingungen nitrierten Proben mit dem gleichen Ergebnis durchgeführt wurden. Selbst wenn man versucht, die Polfiguren als Fasertextur mit beschränkter Drehbarkeit zu interpretieren, ergibt das wegen der Entstehung über einen Diffusionsvorgang keinen Sinn. Das Kristallgitter (Abb.10) gibt ebenfalls keinen Hinweis auf eine Erleichterung der Diffusion in irgendeiner Richtung (wegen des großen Homogenitätsbereiches von 5,7 bis 11,1 % N schreibt man dem ϵ-Nitrid heute die Formel ϵ-Fe$_2$N-Fe$_3$N zu). Auch das Schliffbild liefert keine zusätzlichen Informationen. Zwar sind zuweilen Stengelkörner zu erkennen, normalerweise zeigt die Schicht aber das in Abb. 5d wiedergegebene Aussehen. Die Frage nach der Ursache der Texturausbildung und dem Wachstumsmechanismus des ϵ-Nitrids muß daher zunächst offen bleiben.

4.10.5 Zusammenfassung

Es wurde versucht, Texturmessungen zur Lösung der Frage nach dem Wachstumsmechanismus 5 intermediärer Phasen, die durch Eindiffusion in Eisen entstehen, heranzuziehen. Bei den Boriden FeB und Fe$_2$B sowie beim Aluminid Fe$_2$Al$_5$ wurden ausgeprägte $\langle 001 \rangle$-Fasertexturen gefunden. Mit Hilfe von Kristallstruktur und Schliffbild wird eine Deutung versucht, nach der die Textur durch eine starke, strukturbedingte Anisotropie der Diffusion zustandekommt. Bei Zementit wurde keine Textur gefunden. Auch dafür kann mit Hilfe der Struktur eine Erklärung gegeben werden. Schließlich wurde beim ϵ-Nitrid zwar eine Textur gefunden, aber die Indizierung und Eingruppierung in die bisher entwickelten Vorstellungen bereitet Schwierigkeiten.

Literatur

1. H. Kunst und O. Schaaber: Härterei-Techn.Mitt. 22(1967) 1 u. 275.
2. Th. Heumann und S. Dittrich: Z. Metallkde. 50(1959)617.
3. K. Schubert: Kristallstrukturen zweikomponentiger Phasen, Springer, Berlin 1964.
4. K.H. Jack: Proc.Roy.Soc. Nr. A 195(1948)34.

5 ANISOTROPIE MECHANISCHER EIGENSCHAFTEN, TEXTURVERFESTIGUNG
ANISOTROPY OF MECHANICAL PROPERTIES, TEXTURE STRENGTHENING

5.1 Anisotropie physikalischer und technologischer Eigenschaften von Metallkristallen
Anisotropy of Physical and Technological Properties of Metal Crystals

von E. Schmid[+)]

Abstract

The equilibrium state of a solid is a state of minimum potential energy. The atoms form a regular arrangement, the crystal lattice. As a consequence, the density of the lattice points will vary with direction. This anisotropy is the cause of the anisotropy of the physical and technical properties of crystals. The orientation dependence is qualitatively determined by the crystal class which describes the special combination of the macroscopic elements of symmetry. The symmetry operation implied by the translation through an atomic distance has no influence. Screw axes, rotation axes, glide-reflection and mirror planes, combined with translation or not, all have the same effect. Also the nature of the cohesive forces (e.g. ionic, covalent, metallic and van der Waals binding) is irrelevant as to the number of the parameters needed for the description of the property. The same applies to crystals in which several of the above mentioned bindings are simultaneously effective.

Naturally, the orientation dependence of important physical and technological properties causes an influence of the texture in polycrystals. This is true as well for the properties which are determined by the crystal structure as for those which are influenced greatly by the crystal defects. In numerous cases it is possible to improve metallic properties by a special texture, while in other cases textures must be

---

[+)] II. Physikalisches Institut der Universität Wien, Wien, Österreich

avoided because they lead to failure during application. This
paper is thought to be a concise introduction to the anisotropic behaviour of metal crystals.

Die Richtungsbhängigkeit wichtiger physikalischer und tech-
nologischer Eigenschaften bewirkt naturgemäß einen Einfluß
der Textur auf das Verhalten vielkristalliner technischer
Werkstücke. Dies gilt sowohl für die hauptsächlich durch die
Struktur und nur relativ wenig durch Gitterfehler als auch
für die durch Gitterfehler (Fremdatome, Leerstellen, besetzte
Zwischengitterplätze, Versetzungen) wesentlich mitbestimmten
Eigenschaften. Zu den ersteren gehören die Gitterstruktur,
Elastizität, termische Ausdehnung, Elektronenleitung der Metalle usw., zu den anderen, die SMEKAL strukturempfindlich
nannte, und die man vielleicht klarer fehlstellenempfindlich
nennen sollte, gehören Diffusion, elektrische Leitung in Halbleitern, Lumineszenzerscheinungen, Plastizität usw. In zahl-
reichen Fällen kann durch Ausbildung einer geeigneten Textur
eine entscheidende Verbesserung im Verhalten des Werkstückes
erzielt werden. Auf der anderen Seite können bestimmte Texturen gefährlich werden - sie sind daher nach Möglichkeit zu
vermeiden. Für beide Fälle bringt das Symposium zahlreiche
Beispiele. Dieser Vortrag soll eine gedrängte Übersicht -
unter Hervorhebung einiger wichtiger Beispiele - über das
anisotrope Verhalten der Metallkristalle bringen. Die Beispiele beziehen sich auf das elastische Verhalten, auf verschiedene andere physikalische Eigenschaften und schließlich
auf die Plastizität.

Unmittelbar ergibt sich, daß es unmöglich ist, daß alle Richtungen, die von einem Gitterpunkt in einem Kristall ausgehen,
dieselbe Belegungsdichte mit Gitterpunkten haben. Die Belegungsdichte ist also grundsätzlich anisotrop. Dies führt im
allgemeinen auch zu einer Anisotropie der physikalischen und
technologischen Eigenschaften bei Kristallen. Qualitativ ist
diese Orientierungsabhängigkeit der Eigenschaften durch die
Kristallklasse, die jeweilige Kombination der makroskopischen
Symmetrieelemente, gegeben. Die Deckbewegung Parallelverschiebung wirkt sich für die hier zu beschreibenden Eigenschaften nicht aus: Schraubenachsen und Drehungsachsen, Gleit-

spiegelebene und Spiegelebene unterscheiden sich nicht in
ihrer Wirkung. Auch die Natur der Gitterkräfte - elektrosta-
tische Ionenbindung, Bindung durch lokalisierte Elektronen-
paare, nichtlokalisierte metallische Bindung, schwache v.d.
Waalssche Bindung - spielt für die Zahl der zur Beschreibung
einer Eigenschaft erforderlichen Konstanten keine Rolle. Dies
gilt auch für Kristalle, in denen gleichzeitig mehrere der
erwähnten Bindungskräfte wirksam sind. Tab. 1 gibt eine Übersicht über diese Anzahl in den einzelnen Kristallklassen.
Man erkennt z. B., daß zur Beschreibung des elastischen Verhaltens eines rhombischen Kristalls 9, für das eines kubischen Kristalls 3 voneinander unabhängige Konstanten erforderlich sind. Für zahlreiche physikalische Eigenschaften werden für den Fall niedriger Symmetrie 3 Konstanten, für Kristalle mit einer Hauptachse 2 gebraucht, während bei kubischen Kristallen diese Eigenschaften richtungsunabhängig sind.
Mit aufgenommen in Tab. 1 sind die Ausdrücke zur Berechnung
der Eigenschaft in einer beliebigen Richtung und ihr Mittelwert über alle Richtungen. Für monokline und trikline Kristalle gelten diese Formeln nur, solange die Änderungen der
Achsenwinkel klein bleiben.

5.1.1 Elastische Eigenschaften

In Abb. 1a und b ist für kubische Metalle die Orientierungsabhängigkeit des reziproken Elastizitätsmoduls dargestellt,
in Abb. 2a und b die für hexagonale Metalle und das tetragonale, weiße Zinn. Jeweils ist, um einen unmittelbaren Vergleich der verschiedenen Metalle zu gewinnen, durch den reziproken E-Modul in der Würfelkante bzw. in der Hauptachse
dividiert. Die meisten der untersuchten kub. flz. Metalle
verhalten sich in der Richtungsbhängigkeit des E-Moduls ähnlich: in Richtung der Raumdiagonalen ist der Modul etwa 3mal
so groß (2,23 bis 4,58) wie in der Richtung der Würfelkante.
Erheblich abweichend verhält sich nur Aluminium mit $\frac{E_{111}}{E_{100}}$ =
1,21. Für die im Diamantgitter kristallisierenden
Halbleiter Si und Ge liegen diese Verhältnisse bei 1,45 bzw.
1,51. Von den raumzentrierten Metallen zeigen α-Fe, die bei-
den Alkalimetalle und insbesondere β-Messing (mit $\frac{E_{111}}{E_{100}}$ = 8,3)
eine besonders ausgeprägte Anisotropie. Es sei erwähnt, wo-

385

Tab.1. Zur Anisotropie physikalischer Eigenschaften von Kristallen [1]

| | Zahl der Kristall-klassen | Zahl der zur Kennzeichnung erforderlichen Konstanten | |
		Elasti-zität	therm. Ausdehn.,Leitverm.für Wärme u. Elektrizität,Thomson-Effekt Thermokraft, magnet. Suszeptibilität u.a.
triklin	2	21	3
monoklin	3	13	3
rhomb.	3	9	3

$$\rho_1,\ \rho_2;\quad \rho_{\alpha\beta\gamma} = \rho_1\cos^2\alpha + \rho_2\cos^2\beta + \rho_3\cos^2\gamma$$
$$\bar{\rho} = \tfrac{1}{3}(\rho_1+\rho_2+\rho_3)$$

	Zahl der Kristall-klassen	Elastizität	
hexagon.	2	7	2
	3	6	
	7	5	
tetragon.	3	7	2
	4	6	

$$\rho_{\parallel},\rho_{\perp};\quad \rho_\varphi = \rho_{\perp} + (\rho_{\parallel} - \rho_{\perp})\cos^2\varphi$$
$$\bar{\rho} = \tfrac{1}{3}(\rho_{\parallel} + 2\rho_{\perp})$$

	Zahl der Kristall-klassen	Elastizität	
kubisch	5	3	1
isotrop	2	1	

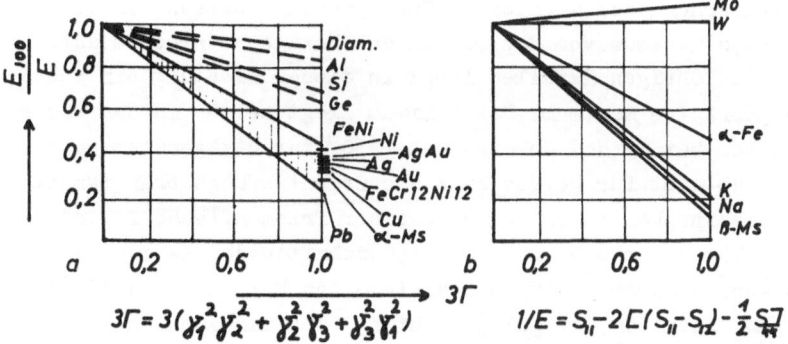

$$3\Gamma = 3(\gamma_1^2\gamma_2^2 + \gamma_2^2\gamma_3^2 + \gamma_3^2\gamma_1^2)$$

$$1/E = S_{11} - 2[(S_{11}-S_{12}) - \tfrac{1}{2}S_{44}]\,\Gamma$$

Abb.1. Anisotropie des Elastizitätsmoduls von kub. Metallen und von Halbleitern (vgl.[2] u.[3];weiterhin für FeCr12Ni12 [4],für Si [5] u. Diamant [6]).a) kub.flz., b) kub.rz.

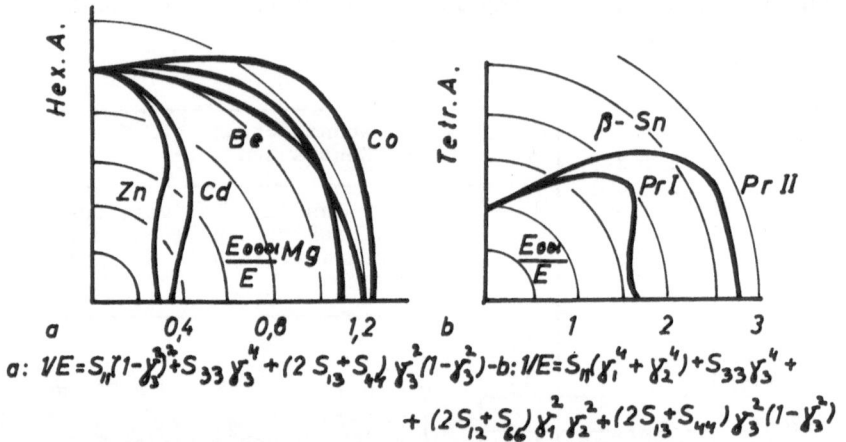

$$a:\ 1/E = S_{11}(1-\gamma_3^2)^2 + S_{33}\gamma_3^4 + (2S_{13} + S_{44})\gamma_3^2(1-\gamma_3^2) - b:\ 1/E = S_{11}(\gamma_1^4 + \gamma_2^4) + S_{33}\gamma_3^4 +$$
$$+ (2S_{12} + S_{66})\gamma_1^2\gamma_2^2 + (2S_{13} + S_{44})\gamma_3^2(1-\gamma_3^2)$$

Abb.2. Anisotropie des Elastizitätsmoduls
hexagonaler Metalle und des β–Zinns (nach
[2] und [3])

a)Hexagonale Metalle

b)β–Zinn

rauf BOAS hinwies, daß dieser Umstand bei der Auswertung von
elektronenmikroskopischen Bildern von Versetzungen unbedingt
zu berücksichtigen ist. Für die hexagonalen Metalle gilt der
Rotationssymmetrie um die hexagonale Achse entsprechend die
in Polarkoordinaten gegebene Darstellung in Abb. 2a. Zn und
Cd mit einem Achsenverhältnis $\frac{c}{a}$ erheblich größer als 1,633
(entsprechend der dichtesten Kugelpackung) weisen ein ausgeprägtes Minimum von E parallel der hexagonalen Achse auf,
bei den übrigen Metallen liegt in dieser Richtung ein schwach
ausgeprägtes Maximum. Das Minimum liegt bei Be in der Basisfläche, bei Mg und Co etwa unter 50° zur hexagonalen Achse.
Abb. 2b gibt für weißes Zinn die Orientierungsabhängigkeit
des reziproken E-Moduls in je einer Prismenfläche I. und II.
Art. Für das rhombische α–U sind Schnitte des E-Modulkörpers
in Abb. 3 dargestellt. Das Minimum des E-Moduls von 15 000
kp/mm^2 liegt in der [010]-Richtung, das Maximum von 29 400
kp/mm^2 in einer Richtung, welche in der (100)-Ebene liegt
und 38° mit der [001]-Richtung einschließt.

Zur Bestimmung der elastischen Konstanten sei kurz angemerkt,

387

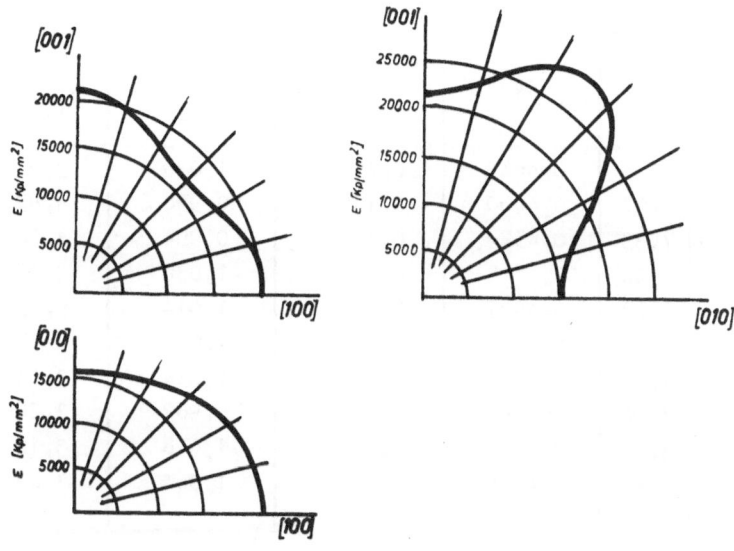

Abb. 3. Anisotropie des Elastizitätsmoduls des rhombischen α-
Urans (nach FISCHER und MC SKIMIN [7], vgl.a. [8])
$s_{11}=0,490_7$, $s_{22}=0,674_3$, $s_{33}=0,479_8$, $s_{44}=0,803_6$, $s_{55}=1,362_0$,
$s_{66}=1,345_3$, $s_{12}=-0,119_4$, $s_{13}=0,008_2$, $s_{23}=-0,262_7 \cdot 10^{-12} cm^2/Dyn$

daß man im allgemeinen heute nicht mehr statische Messungen
oder Ermittlung der Eigenfrequenzen schwingender Proben
anwendet, sondern Ultraschallwellen mit Frequenzen, die weit
über den Eigenfrequenzen liegen. Die Ausbreitungsgeschwin-
digkeit derartiger longitudinaler und transversaler Wellen
liefert den Elastizitäts- und Torsionsmodul der Probe. Bei
Verwendung geeignet orientierter Metallplättchen kubischer
Kristalle (Plättchenebene parallel (101)) gelingt es, mit
e i n e r Probe alle drei elastischen Konstanten zu bestimmen. Dieses Ultraschallverfahren hat zu einer außerordent-
lichen Erweiterung des experimentellen Materials geführt.
Die in Abb. 3 demonstrierten elastischen Konstanten des α-
Urans wurden schon auf diese Weise ermittelt. Weitere Beispiele seien in Tab. 2 kurz beschrieben, welche die elastischen Parameter von Silber und Silbermischkristallen enthält. C_{44} ist der Torsionsmodul eines parallel der Würfel-

Tab.2. Elastische Konstanten von Silber und Silber-Legierungen in 10^{12}dyn/cm^2 (nach BACON u. SMITH [9])

Werk-stoff	At.%	C_{44}	$\dfrac{C_{11}-C_{12}}{2}$	$\dfrac{C_{11}+2C_{12}}{3}$	$\dfrac{2C_{44}}{C_{11}-C_{12}}$
Ag		0,4613	0,1528	1,036	3,03
AgMg	3.07	0,4600	0,1498	0,998	3,09
	7,33	0,4524	0,1462	0,994	3,10
AgZn	2,40	0,4612	0,1484	1,032	3,12
	3,53	0,4577	0,1462	1,014	3,13
AgCd	1,34	0,4611	0,1516	1,026	3,05
	1,92	0,4593	0,1505	1,014	3,04
AgIn	8,36	0,4505	0,1379	0,982	3,26
AgSn	3,17	0,4581	0,1437	1,018	3,18
AgPd	6,22	0,4809	0,1596	1,064	3,03

achse orientierten Stabes, $\frac{1}{3}(C_{11}+2C_{12})$ ist die reziproke kubische Kompressibilität und $\frac{1}{2}(C_{11}-C_{12})$ ist ebenfalls ein Schubwiderstand. Die Zahlen sollen nur zeigen, daß die Änderungen durch Legierung klein sind, die Kristallelastizität also zu den nur wenig fehlstellenabhängigen Eigenschaften gehört. Auf Grund einer von FUCHS [10] angegebenen Theorie können die Änderungen in den Scherungskonstanten auf Grund der elektrostatischen Anziehungskräfte (Metallionen in Elektronengas eingebettet) und der Abstoßung zwischen den Ionen gedeutet werden.

Die geringe Fehlstellenabhängigkeit geht weiter aus Abb. 4 hervor, welche das Verhalten des E-Moduls eines Cu-Kristalls nach steigender n-Bestrahlung zeigt. Nach einer eingestrahlten Dosis von 40 x 10^{11} n/cm^2, die schon fast zur Sättigung führt, ist der Modul nur um etwa 2 % angestiegen. Nach α-Bestrahlung werden deutliche E-Modulabnahmen gefunden. Aus diesen und weiteren Versuchen wird geschlossen, daß der Einbau von Frenkel-Defekten zwei gegenläufige Wirkungen auslöst, eine, die den Modul erhöht (Festlegung der im Metall vorhandenen Versetzungen durch die gebildeten Defekte) und eine, die den Modul erniedrigt (Störungen der Gitterbindungen im

Verzerrungsfeld der Frenkel-Defekte).

Zum Abschluß ist in Abb. 5 ein Beispiel für die Auswirkung
der elastischen Anisotropie des Eisens auf das elastische
Verhalten technischer Werkstücke dargestellt. Insbesondere
im Fall des FeSi-Blechs mit Goss-Textur ((110) ∥ WE, [001] ∥
WR) umfaßt der E-Modul je nach der Richtung im Blech den wei-
ten Bereich von etwa 12 000 bis 28 000 kp/mm^2. Durch Verwendung von geschichteten Werkstücken, deren Elemente in verschiedener Richtung aus einem Blech geeigneter Textur entnommen sind, könnte man zu Federn kommen, deren E-Modul gegenüber Zug- und Biegebeanspruchung verschieden ist, wobei
diese Unterschiede durch Orientierung und Dicke der verwendeten Elemente beeinflußbar sind.

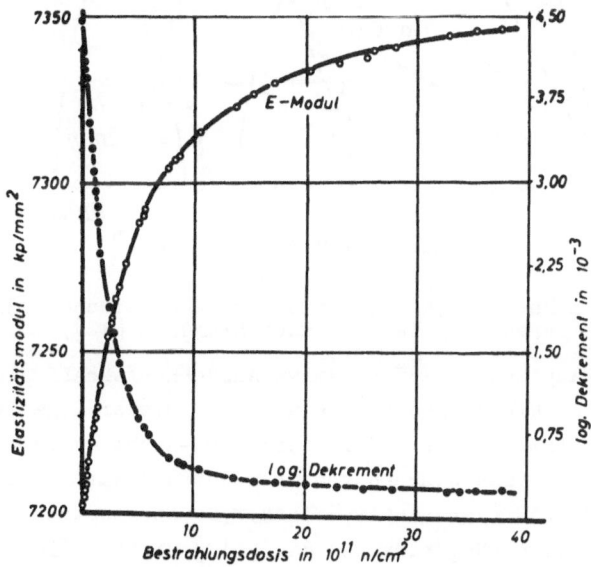

Abb.4. Elastizitätsmodul und Dämpfung eines n-be-
strahlten Cu-Kristalls (nach THOMPSON u.a. [11])

5.1.2 Weitere physikalische Eigenschaften

In Tab. 3 ist für hexagonale, tetragonale und rhombische Kristalle die Anisotropie einiger Eigenschaften zusammengestellt, Eigenschaften, die für kubische Kristalle richtungsunabhängig sind. Besonders hingewiesen sei auf die großen
Unterschiede im Ausdehnungskoeffizienten von Zn und Cd ∥

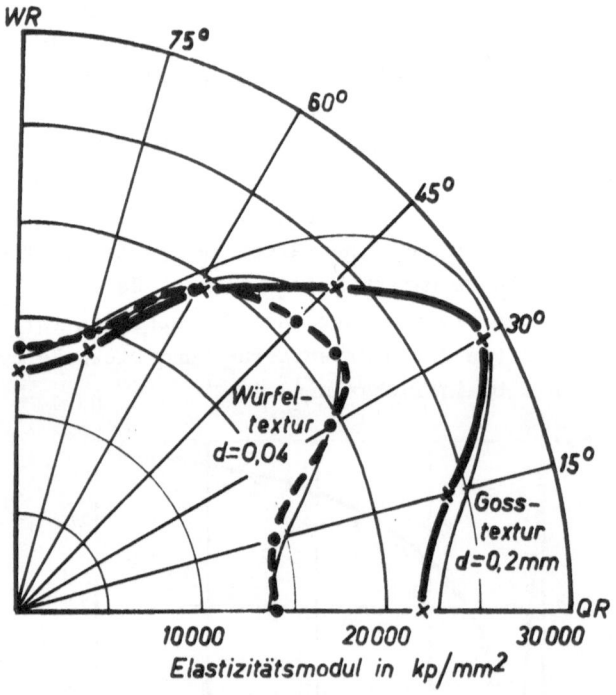

Abb.5. Elastische Anisotropie von Fe-Si-Blechen (3%Si)
mit geregelter Textur (nach STANGLER [12])

und ⊥ zur Hauptachse, auf negative Ausdehnungskoeffizienten
in bestimmten Richtungen bei Te und α-U. Dieser negative Aus-
dehnungskoeffizient des α-U in der [010]-Richtung ist möglicherweise für das Wachsen von Uran durch Reaktorbestrahlung
von Bedeutung. Ein kreiszylindrischer reaktorbestrahlter U-
Kristall, dessen Längsrichtung die [010]-Richtung ist, erfährt durch Bestrahlung in dieser Richtung eine Verlängerung,
sein Querschnitt wird dabei zu einer Ellipse deformiert, wobei in der [100]-Richtung eine Kontraktion erfolgt, während
die [001]-Richtung in ihren Abmessungen erhalten bleibt. Für
die Praxis ergibt sich daraus die Richtlinie, bei Brennstoff-
elementen in der Textur stabförmiger vielkristalliner Proben
den Anteil der [010]-Lage möglichst gering zu halten. Für
die Deutung wird auf innere Spannungen hingewiesen, die
durch die Erwärmung in den durch die Bestrahlung erzeugten

Tab.3. Anisotropie der Metalle: Physikalische Eigenschaften
(vgl.[2]; weiterhin für α-U [13] und β-U [14])

Met.	Gitter-Strukt.	Spez.Widerst.		Therm.Ausdehn.		Thermokr.g.Cu	
		∥	⊥	∥	⊥	∥	⊥
		$\Omega \ mm^2/m$		10^{-6} (zw.20 u.100°)		10^{-6} V/°C	
Mg	hexag.	0,038	0,046	26,4	25,6	1,87	1,66
Zn		0,061	0,058	63,9	14,1	1,32	− 0,50
Cd		0,083	0,068	52,6	21,4	1,60	− 1,74
Se	hexag.	$2,5 \cdot 10^8$	$2 \cdot 10^{10}$				
Te	(trigon.)	280	610	−1,6	27,2		
β-Sn	tetrag.	0,143	0,099	30,5	15,5		
				(zw.662 u.772°)			
β-U				5,6	23,7		
α-U	rhom-bisch			(zw. 20 u.650°)			
				36,7 ∥ [100]			
				−9,3 ∥ [010]			
				34,2 ∥ [001]			

Störbereichen erzeugt werden und in der [010]-Richtung wegen
des negativen Ausdehnungskoeffizienten zu Zugbeanspruchung
führen. Auch eine Anisotropie der Diffusion von Leerstellen
und Zwischengitteratomen wird als Deutung diskutiert [8].

Sowohl Metalle als auch Halbleiter sind Elektronenleiter. Die
verschiedenen Leitungsmechanismen bedingen jedoch, wie schon
erwähnt, weitgehende Unterschiede hinsichtlich des Einflusses von Gitterstörungen. Bei den Metallen ist der spez. Widerstand relativ wenig störungsempfindlich, wenn auch geringe Beimengungen, etwa von P, den Herstellern von Cu-Leitungs-
draht äußerst unangenehm sind. Bei den Halbleitern dagegen
kann der spezifische Widerstand durch eingebrachte Fremdatome um viele Größenordnungen − praktisch fast vom Widerstand des Isolators bis zum Widerstand von Metallen − geändert werden. Der gleiche Unterschied zeigt sich auch hinsichtlich des Einflusses von Bestrahlung. Während bei Metal-

len Korpuskularbestrahlung durch die eingebrachten Frenkel-
Defekte den Widerstand nur um Prozente erhöht (Abb.6), ändert sich das Leitfähigkeitsverhalten von Halbleitern durch
Bestrahlung in außerordentlichem Maße. Neben der Wirkung der
Frenkel-Defekte, deren Komponenten sowohl als Donatoren als
auch als Akzeptoren wirken können, können auch durch Kernum-
wandlungen Donatoren oder Akzeptoren erzeugt und damit zusätzlich eine Änderung der Leitfähigkeit bewirkt werden. Die
Änderung der Zahl der Ladungsträger und ihrer Beweglichkeit
kann durch Messungen des Hall-Effekts unmittelbar nachgewiesen werden. Abb. 7 zeigt, daß durch einen Korpuskularbeschluß
der Leitungsmechanismus verändert, also ein Übergang vom n-
Typ in den p-Typ (oder umgekehrt) herbeigeführt wird. Die

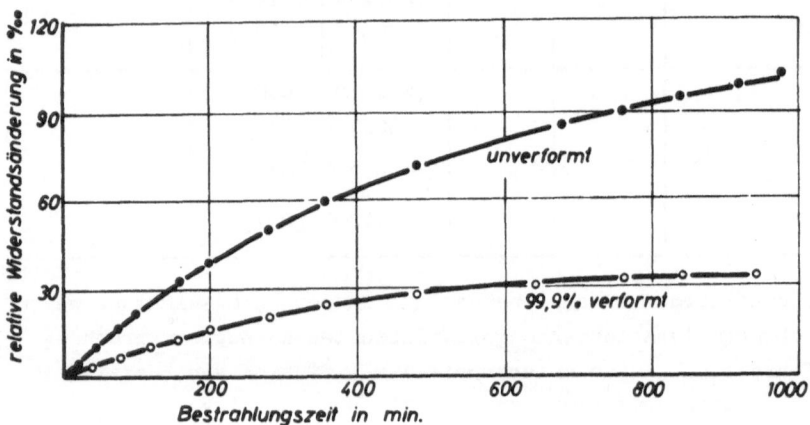

Abb.6. Änderung des Widerstands von Al (99,5) durch α-Bestrahlung bei 78°K (nach HARTMANN [15])

Leitfähigkeit sinkt mit steigender Bestrahlungsdosis, wird
bei einer bestimmten Dosis O und steigt schließlich wieder
an. Der Hall-Koeffizient kehrt bei dieser Dosis sein Vorzeichen um, die ursprüngliche Elektronenleitung geht in Defekt-
elektronenleitung über. Insbesondere im Hinblick auf ihre
Verwendung als Schaltelemente in Raumfahrzeugen ist die Ver-
meidung eines radiation-damage von Halbleitern von erheblicher technischer Bedeutung.

Wenn auch magnetische Probleme nicht in das Programm des

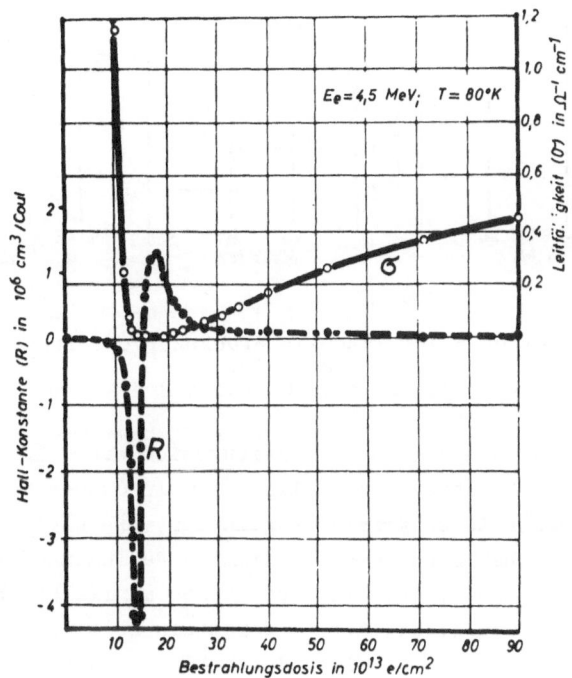

Abb.7. Leitfähigkeit und Hall-Konstante von mit Elektronen bestrahltem InSb (nach AUKERMANN [16])

Symposiums aufgenommen worden sind, so soll doch auf die so wichtigen Feststellungen über die Orientierungsabhängigkeit der Magnetisierungskurve hingewiesen werden. In Abb.8 sind die kennzeichnenden Magnetisierungskurven für Fe-, Ni- und Co-Kristalle dargestellt. Bei Fe wird in Richtung der Würfelkante schon bei kleinen Feldern Sättigung erreicht, die Richtung [111] ist am schwersten magnetisierbar. Bei Ni ist im Gegensatz dazu die Raumdiagonale die Richtung leichtester Magnetisierbarkeit, beim hexagonalen Kobalt ist es die hexagonale Achse.

Für viele physikalische Vorgänge ist die Kenntnis der Oberflächenkräfte von Festkörpern notwendig. Auf verschiedene Weise wird versucht, die Oberflächenspannung zu bestimmen: Winkel in Korngrenzengräben, Gleichgewichtsform kleinster

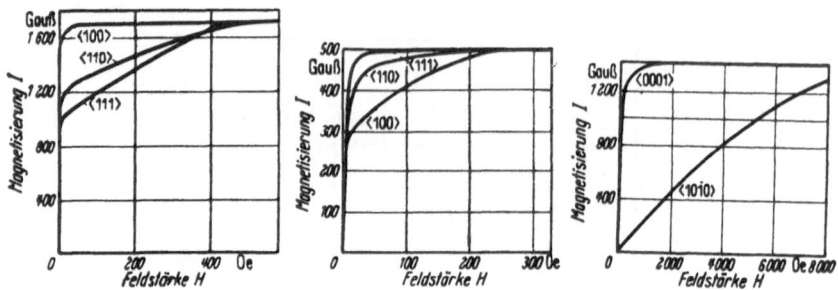

Abb.8. Anisotropie der Magnetisierungskurve ferromagnetischer
Metalle.
Links: Fe, Mitte: Ni, rechts: Co (nach HONDA und KAYA [17,28])

Winkel in Korngrenzengräben, Gleichgewichtsform kleinster
Kristalle, sekundäre Rekristallisation in dünnen Folien, Tem-
pern von geglätteten Kristallkugeln bei verschiedenen Temperaturen, wobei sich unter bestimmten Bedingungen eine Stufenstruktur, zusammengesetzt aus Ebenen minimaler Oberflächenenergie, ausbildet. Abb. 9 zeigt ein Beispiel einer solchen Stufenstruktur, erhalten durch Glühung einer **kub.rz.**
FeSi-Kristallkugel bei Temperaturen zwischen 1200 bis 1100°C
unter gleichzeitiger Gegenwart von 8×10^{-4} Torr Luft. Die
Stufen verlaufen parallel einer Würfelfläche, wie aus der
Vierzähligkeit der Lichtreflexe erkennbar ist. Bei den kub.
flz. Metallen Ni, Au und Ag ergeben sich beim Anlassen bei
Temperaturen nahe dem Schmelzpunkt nur (111)-Terrassen, bei
niedrigeren Temperaturen 1400 bis 1300°C treten (100)-Ter-
rassen hinzu. Diese Beobachtungen stehem mit theoretischen
Überlegungen im Einklang (vgl. [20]). Bei hohen Temperaturen
ergibt die Theorie bei Berücksichtigung nur nächster Nachbarn, daß das Minimum der Oberflächenenergie in der (111)-
Fläche wesentlich ausgeprägter ist als das in der (100)-Flä-
che. Steigt durch Erniedrigung der Temperatur der Temperung
der Ordnungsgrad im Gitter, so nimmt die Mitwirkung zweit-
nächster Nachbarn zu und es ergeben sich nur geringe Unterschiede in den Oberflächenenergien von (111) und (100). Auf
kub.-raumzentrierte FeSi-Kristalle bezieht sich Tab. 4. Sie
zeigt, daß bei Vakuumglühung ($\sim 10^{-6}$Torr) bis zu Glühtempe-

395

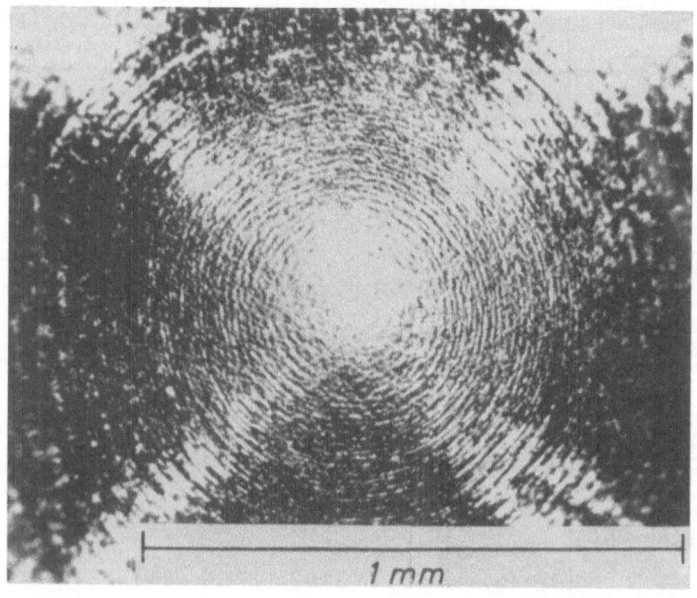

Abb. 9. Stufenstruktur um [100] einer Fe-Si-
kristallkugel. Glühtemperatur 1100 - 1200°C;
8 x 10⁻⁴ Torr Luft nach [19])

raturen von 875°C herunter stets die dichtestbelegte (110)-
Fläche durch kleinste Oberflächenenergie ausgezeichnet ist,
die jeweils angegebenen Winkelbreiten der gebildeten Terras-
sen überragen für (110) weit die zu anderen Flächen gehörigen. Erfolgt die Temperung in Gegenwart geringer Luftmengen,
so ändern sich, wie Tab. 5 zeigt, die Verhältnisse in Bestätigung älterer Versuche [22]. Während bis zu Temperaturen von
1200°C herunter wieder die (110)-Terrassen vorherrschen, treten bei niedrigeren Temperaturen die (100)-Terrassen in den
Vordergrund. Die Sauerstoff-Adsorption führt zu einer Änderung der Reihenfolge der Oberflächenenergie, was wohl mit
unterschiedlichem Adsorptionsvermögen der einzelnen Flächen
zusammenhängt. Mit diesem Übergang des Minimums der Oberflächenenergie auf die Würfelfläche wird die Ausbildung der
technich interessanten "Würfellage" in FeSi-Blechen in Zu-

Tab.4. Reihung der Oberflächenenergien von FeSi bei verschiedenen Temperaturen im Vakuum (nach HAUCK und STANGLER [21])

Temperatur-bereich (°C)	Reihung der Oberflächenenergie γ(hkl) durch Winkelausdehnung					An der Bindung beteil. Nachbarn
I 1500 bis 1170	(110) 20°					1.nächste Nachbarn
II 1160 bis 970	(110) 20°	(111) 11°				
III 960 bis 910	(110) 20°	(100) 3,5°	(111) 1°			2.nächste Nachbarn
IV 900 bis 880	(110) 20°	(100) 10,5°	(111) 38'	(116) 33'		3.nächste Nachbarn
V 870 bis 850	(110) 15°	(100) 5,5°	(111) 34'	(112) 32'	(116) 28'	

Tab. 5. Reihung der Oberflächenenergien von FeSi bei verschiedenen Temperaturen in Gegenwart von Sauerstoff (nach KARNTHALER u. STANGLER [19])

Temperatur-bereich Luftdruck	Reihung der Oberflächenenergie γ(hkl) durch Winkelausdehnung
1500 -1300°C 75x10 Torr	(110)<<(100)
1300 -1200 30	(110)~(100)<<(112)
1200 -1100 8	(100)<(110)<(112)<<(133)
1100 -1000 5	(100)<(112)<(111)<(110)<(133)
1000 - 900 1	(100)<(112)<(111)<(110)<(133)<<(124)~(116)

397

sammenhang gebracht, die sich bei der Sekundärrekristallisation von in geeigneter Weise vorbehandelten Blechen einstellt.

Auch die Lösungsgeschwindigkeit von Metallen in Flüssigkeiten
kann, auch bei kubischen Kristallen, außerordentlich stark
richtungsabhängig sein. Ein Beispiel dafür gibt Abb. 10, die
sich auf die Lösungsgeschwindigkeit von Cu in Essigsäure bezieht. Die Richtungen maximaler Lösungsgeschwindigkeiten
sind keineswegs für das Metall allein charakteristisch; sie
hängen wesentlich auch vom Lösungsmittel ab.

Abb. 10. Lösungsgeschwindigkeit von Kupfer in Essigsäure (nach GLAUNER und GLOCKER [23])

Durch geometrische Gründe ist eine Anisotropie bedingt, die
sich in Unterschieden in der Eindringtiefe von eingeschlos-
senen Korpuskeln äußert. Abb. 11 zeigt dies für Krypton-Ionen
mit verschiedener Einschußrichtung in Al-Kristallen. Eine
deutliche Bevorzugung der ⟨110⟩-Richtung ist zu erkennen.

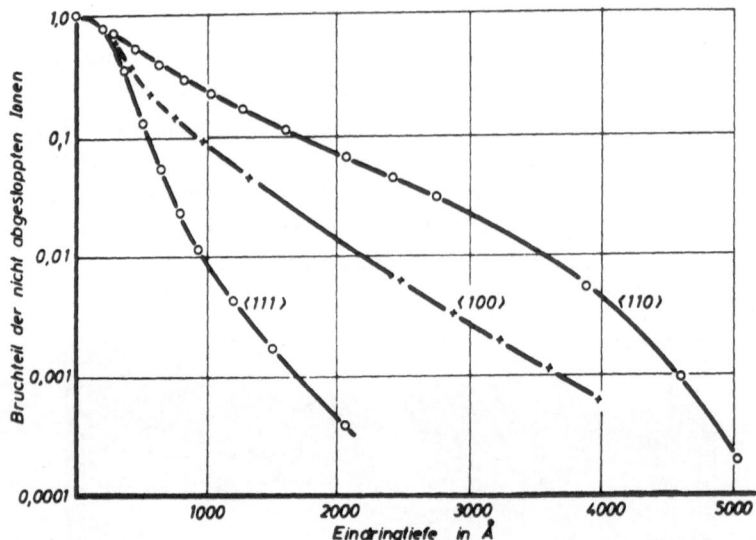

Abb.11. Eindringtiefe von Kr^{85}-Jonen (40 keV) in Alumini-
umkristallen bei verschiedener Einschußrichtung (nach
PERCY u. a. [24])

Parallel zu dieser Richtung verlaufen im kub.flz.Gitter Kan
mit geringsten Hindernissen. Im Einklang damit stehen Rück-
streuversuche mit Protonen an Au- und W-Kristallen (Abb.12)
Die Richtungen bevorzugter Rückstrahlung treten klar zutage
Interessant ist, daß eine solche Kanalleitung bereits 1912
von BRAGG diskutiert wurde, der damit die Ergebnisse des Ve
suchs von FRIEDRICH, KNIPPING und LAUE erklären wollte. Er
neigte ja sehr zur Auffassung, daß die Röntgenstrahlen kor-
puskulare Natur hätten und versuchte, die erhaltenen Laue-
Bilder durch bevorzugte Ausbreitung der Korpuskeln entlang
kristallographischer Schichten zu erklären. Übrigens hat auc
STARK eine ähnliche Deutung versucht.

5.1.3 Plastizität

Daß eine kritische Schubspannung für den Beginn deutlicher
Translation in Kristallen verantwortlich ist, hat sich immer
wieder bestätigt. Die Zugspannung, die zum Beginn der Verformung führt, muß daher von der Kristallorientierung abhän-

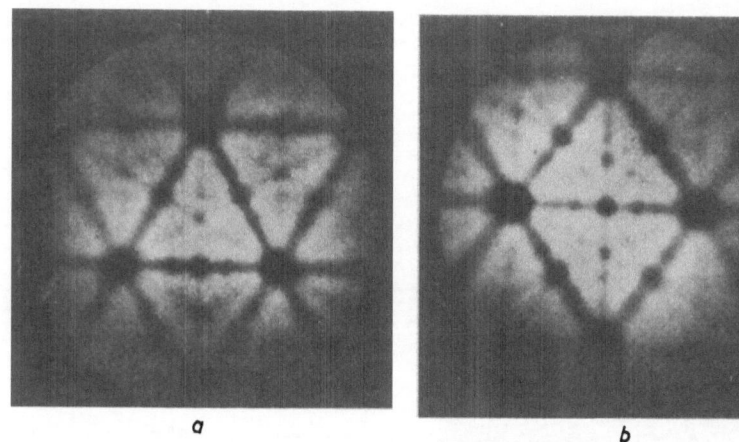

Abb.12. Zur Kanalleitung. Protonenrückstreuung
(nach NELSON [25])
a) Goldeinkristall,
b) Wolframeinkristall

gen. Sie wird einen umso größeren Spannungsbereich umfassen,
je weniger gleichwertige Translationssysteme zur Verfügung
stehen. In Abb. 13 ist die Orientierungsabhängigkeit der
"Streckgrenze" für Oktaedertranslation für kub.flz. Kristalle (sie gilt ebenso für Dodekaedertranslation in kub.rz.) und
für hexagonale Kristalle mit Basistranslation dargestellt.
Während im ersten Fall das Verhältnis der Streckgrenze des
festesten zu der des schwächsten Kristalls nur 1,84 beträgt,
sind bei den hexagonalen Kristallen die durch die Orientierung gegebenen Unterschiede ungleich größer, wie man aus den
Schnitten des Festigkeitskörpers ohne weiteres entnimmt.
Liegt die singuläre Translationsfläche fast parallel oder fast
senkrecht zur Zugrichtung, so treten hier andere Verformungsmechanismen (Zwillingsbildung, Spaltung) in Kraft.

Die Temperaturabhängigkeit der kritischen Schubspannung ist
relativ gering. Sie fällt mit steigender Temperatur. Aber
selbst bei $1^{o}K$ weisen Zn- und Cd-Kristalle noch außerordent-
liche Plastizität durch Basistranslation auf. Neuerdings wurden jedoch auch Abnahmen der Schubfestigkeit eines Gleitsy-

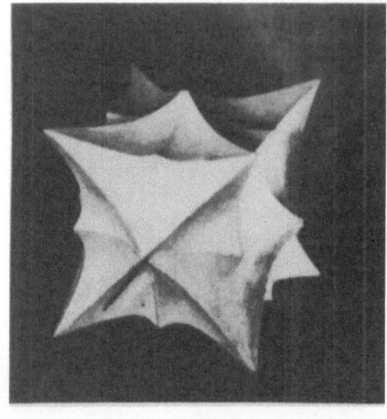
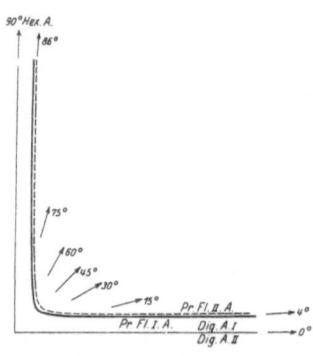

a   b

Abb.13. Fließgefahrkörper von Metallkristallen (nach [2])

a) Oktaedertranslation in kub.flz.Kristallen (Dodekaedertranslation in kub.rz.Kristallen)
b) Basistranslation in hexagon. Kristallen

stems mit fallender Temperatur beobachtet. Abb. 14 zeigt dies
für die Prismentranslation von Be. Während die Basisschubfestigkeit den gewohnten leichten Anstieg mit fallender Temperatur zeigt, ergibt sich bei der Schubfestigkeit des Pris-
mengleitsystems zunächst ein erheblich steilerer Anstieg mit
fallender Temperatur, der aber bei etwa 50°C von einem Ab-
fall gefolgt wird, bis bei weiterer Temperaturabnahme ein
neuerlicher Anstieg folgt. Ähnliche Ergebnisse wurden auch
bei Pb beobachtet, dessen Oktaedertranslationssystem bei
Temperatur-Erniedrigung unter einen von der Kristallorien-
tierung abhängigen Wert einen Schubfestigkeitsabfall aufweist [27].

Die den Plastizitätsbeginn von Kristallen kennzeichnende kritische Schubspannung von Translationssystemen ist eine ausgesprochen fehlstellenabhängige Eigenschaft. Nur 0,6 At% Cd
erhöhen beispielsweise die Basisschubfestigkeit von Reinzink
auf das über 12-fache. In der lückenlosen Mischkristallreihe

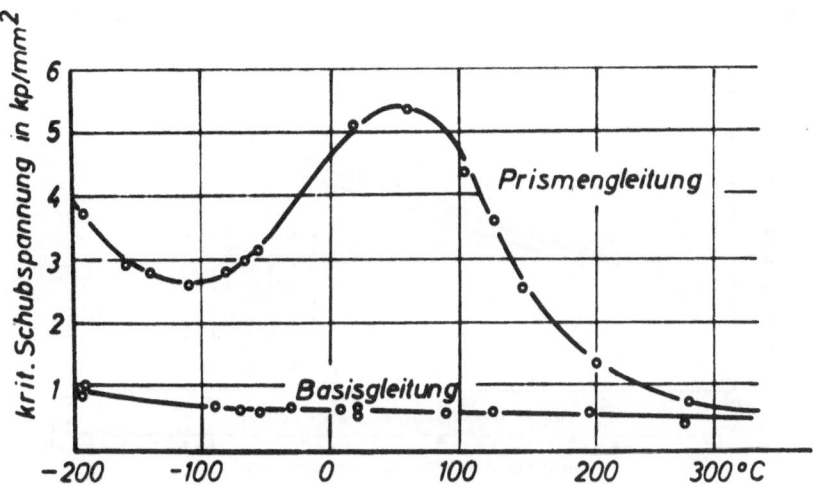

Abb.14. Beryllium-Kristall. Temperaturabhängigkeit der kritischen Schubspannung für Basis- und Prismengleitung (nach REGNIER und DUPOUY [26])

Ag-Au ist bei mittleren Konzentrationen die kritische Schubspannung des Oktaedergleitsystems auf das über 8-fache der Werte für die reinen Metalle gestiegen. Übergang von ungeordneter Atomverteilung in Mischkristallen Cu 25At% Au in die geordnete Verteilung Cu_3Au führt zu einem Absinken der Schubfestigkeit auf etwa die Hälfte (vgl. [2]).

Die Fehlstellenabhängigkeit des Translationsbeginns (sie betrifft auch den weiteren Verlauf der Translation) prägt sich deutlich auch im Verhalten von Kristallen aus, die vor dem Deformationsversuch einer Bestrahlung bei tiefen Temperaturen ausgesetzt waren. Abb. 15 gibt hierfür ein Beispiel für Cu- und Au-Kristalle. Durch eine Steigerung der Dosis von 10^{14} auf 10^{17} n/cm^2 erhöht sich die Schubfestigkeit des Oktaedergleitsystems um eine Größenordnung.

Ungeklärt ist heute noch die Frage der Initialbedingung für Zwillingsbildung. Ist auch hier eine kritische Schubspannung im Gleitsystem für die Bildung der Deformationszwillinge maßgebend, oder gilt eine ganz andere, etwa eine Energiebedingung? Jahrelange Bemühungen haben vor allem wegen der der

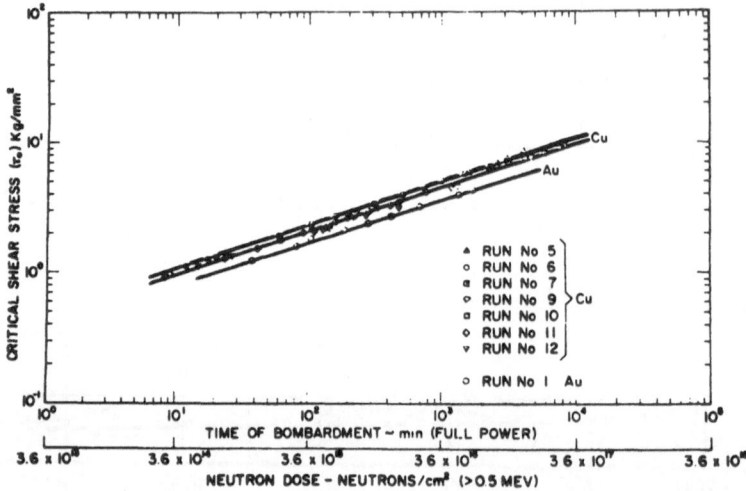

Abb.15. Einfluß von n–Bestrahlung bei 4,2°K auf die kritische Schubspannung von Cu (Au)–Kristallen (nach BLEWITT und
ARENBERG [28])

Zwillingsbildung zumeist vorangehenden Translation keine Lösung gebracht. Es scheint nun, daß es durch Verwendung besonders geeigneter Versuchskristalle gelungen ist, einen
nützlichen Beitrag zu liefern. Im System FeBe schließt ein
Gehalt von etwa 0,4 Gew.% Be, das sind etwa 2 At%, das γ–
Feld ab. Eine Legierung mit 25 At% Be liegt also als kub.rz.
Mischkristall vor. Abschreckung von 1125°C führt auf regellose Verteilung, anschließendes Anlassen durch mehrere Stunden bei 350°C liefert die geordnete Verteilung Fe_3Be. Bei
Stauchversuchen mit derartigen Kristallplättchen tritt die
Translation weitgehend zurück. Die Verformung erfolgt vorwiegend durch Zwillingsbildung mit den für raumzentrierte
Metallkristalle bekannten Zwillingselementen (112) als Gleit-
ebene K_1 und [111] als Gleitrichtung η_1. Abb. 16 zeigt die
Ergebnisse einer großen Zahl von Versuchen, in denen Kristallorientierung, Ordnungsgrad und Temperatur variiert wurden. Man erkennt erstens, daß der mögliche Orientierungsbereich durch die Vielzahl der verwendeten Kristalle weitgehend
überdeckt ist, man erkennt weiterhin, daß für eine bestimmte

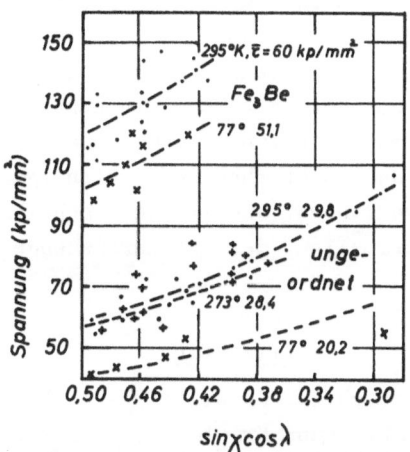

Abb.16. Mechanische Zwillingsbildung von FeBe (25at%)
(nach BOLLING und RICHMAN [29])

Temperatur die beobachteten Spannungwerte einigermaßen Hy-
perbelästen zugeordnet werden können, die jeweils für konstante Schubspannung berechnet wurden. Mit steigender Temperatur steigt, im Gegensatz zum Verhalten bei Translation, die
Schubfestigkeit des Zwillingsgleitsystems an. Ebenfalls im
Gegensatz zur Translation steht der Befund, daß im geordne-
ten Zustand die Schubfestigkeit mehr als doppelt so groß ist,
wie im Zustand mit regelloser Atomverteilung. Gegenüber der
von den Verfassern vertretenen Auffassung, daß auch für Zwillingsbildung ein Schubspannungsgesetz gültig sein kann, wird
von anderen Seiten betont, daß als Keim der Zwillingsbildung,
für die ein Versetzungsmechanismus vorausgesetzt wird, Span-
nungskonzentrationen wirken, die nicht durch die angelegte
Spannung auszudrücken sind, sodaß es kein Schubspannungsgesetz für Zwillingsbildung geben kann (vgl. z.B. [30]). Weitere experimentelle Beiträge sind daher abzuwarten.

Ich habe versucht, durch einige Beispiele – die Zahl hätte
erheblich vergrößert werden können – das anisotrope Verhalten
der Metallkristalle zu illustrieren. Abgesehen von den nicht
zur Diskussion stehenden magnetischen Eigenschaften sind es
vor allem die elastischen Eigenschaften technischer Werk-

stücke, die durch Schaffung geeigneter Texturen in gewoll-
ter Weise verbessert werden können. Von überragender Bedeutung für die Herstellung und die technologischen Eigenschaften technischer Werkstücke ist das plastische Verhalten
des Einzelkorns. Für dieses gelten die beschriebenen außerordentlich ausgeprägten Anisotropien. Außerhalb des Rahmens
meines Berichtes und daher ohne Berücksichtigung blieben die
Wirkungen der Korngrenzen, die ein sehr schwieriges und komplexes Problem darstellen, dessen ausreichende Lösung heute
noch fern liegt.

Literatur

1. W. Voigt: Lehrbuch der Kristallphysik, B.G. Teubner 1910.

2. E. Schmid und W. Boas: Kristallplastizität, Springer,
   Berlin 1935 – Plasticity of Crystals, Chapman und Hall,
   London 1968.

3. E. Schmid: Festschrift Reinhard Straumann, J.F. Stein-
   kopf, Stuttgart 1952.

4. K. Salmutter und F. Stangler: Z. Metallkde. 51(1960)544.

5. W.L. Bond, W.P. Mason, H.J. Mc.Skimin, K.M. Olsen und
   G.K. Teal: Phys. Rev. 80(1950)176.

6. Nach Ch. Kittel: Introduction to Solid State Physics,
   Wiley a. Sons, New York 1960.

7. E.S. Fischer und H.J. Mc.Skimin: J.Appl.Phys. 29(1958)1473.

8. K. Lintner und E. Schmid: Werkstoffe des Reaktorbaues,
   Springer, Berlin 1962.

9. R. Bacon und Ch.S. Smith: Acta Met. 4(1956)337.

10. K. Fuchs: Proc.Roy.Soc. A 153, S. 622 u. A 157, S. 444
    (1936).

11. D.O. Thompson und D.K. Holmes: J.Appl.Phys. 17(1956)713.

12. F. Stangler: Z. Metallkde. 49(1958)489.

13. F.G. Foote: Proc.Int.Conf. Peaceful Uses of Atomic
    Energie, Genf 9(1955)33 – Nuclear Metallurgy IMD Spec.
    Report AIME 1(1955)65.

14. H.H. Klepfer und P. Chiotti: US.AEC Rep., I.SC – 893
    (1957).

15. Vgl. W. Hartmann: Sitz.-Ber.Abt. II Österr.Akad.d.Wis-
    sensch. 176(1967)121.

16. L.W. Aukermann und K. Lark-Horovitz: BAPS 1(1956)332;
    zitiert n. J.J. Harwood, H.H. Hausner, J.G. Moore und
    W.G. Rauch: Effects of Radiation on Materials, Reinhold,
    New York 1958.

17. K. Honda, H. Masumoto und S. Kaya: Sci.Rep.Tohoku Imp.
    Univ. 17(1928)111.

405

18. S. Kaya: Sci.Rep.Tohoku Imp.Univ. 17(1928)639 u.1157.

19. H.P. Karnthaler und F. Stangler: Acta Met. 15(1967)921.

20. O. Knacke und J.N. Stranski: Ergeb.exakt.Naturw. 26(1952) 383.

21. D. Hauck und F. Stangler: Acta Phys.Austriaca 18(1964)180.

22. J.L. Walter und C.G. Dunn: Acta Met. 8(1960)497.

23. R. Glauner und R. Glocker: Z. Kristallogr. 80(1931)377.

24. G.Q. Piercy, M.Mc.Cargo, F. Brown und J.A. Davies: Canad. J.Phys. 42(1964)116.

25. R.S. Nelson, zitiert n.: Phys.Blätter 23(1967)530.

26. Regnier und J.M. Dupouy: Vortrag Intern.Conf.Strength of Metals and Alloys, Tokyo, 1967.

27. V.J. Startsev, Postovalvo und Femenke: Vortrag Intern. Conf.Strength of Metals and Alloys, Tokyo 1967.

28. T.H. Blewitt und C.A. Arenberg: Vortrag Intern.Conf. Strength of Metals and Alloys, Tokyo 1967.

29. G.F. Bolling und R.H. Richman: Acta Met. 13(1965) 709 u.723.

30. D.G. Westlake: Acta Met. 9(1961)327.

5.2 Properties of Polycrystalline Materials with and without Texture
Eigenschaften vielkristalliner Aggregate mit und ohne Textur

by E. Kröner[+]

Zusammenfassung

Bei der Berechnung der Eigenschaften von zusammengesetzten
Materialien, insbesondere auch von vielkristallinen Aggrega-
ten, stellt die rationelle Beschreibung der Anordnung der
Konstituenten (z.B. Körner) ein zentrales Problem dar.

Obwohl sich die Kugelflächenfunktionen und deren Verallge-
meinerungen für die Textur-Charakterisierung sehr bewährt
haben, sind sie für diejenigen Fälle, bei denen die Beziehungen zwischen den Kornformen und den Orientierungen eine Rolle spielen, nicht sehr brauchbar. Hier haben die statistischen Funktionen, insbesondere die sogenannten Korrelationsfunktionen, offensichtlich einen größeren Wert.

Die Bedeutung der n-Punkt Korrelationsfunktion wird erklärt.
Es wird ferner beschrieben, wie Korrelationsfunktionen für
die Berechnung solcher Material-Parameter, wie Elastizitäts-
moduln, Dielektrizitätskonstanten, Permeabilitäten etc. herangezogen werden können. Exakte Ergebnisse erhält man z. B.
wenn die Verteilung der Kristalliten in einem vielkristallinen Aggregat in 2 oder 3 Dimensionen regellos ist.

5.2.1 Preliminaries

The problem of calculating the properties of composite
materials, in particular those of polycrystalline materials,
from the properties of the constituents has been investigated
by numerous workers during the last 100 years. Though the
methods to be discussed later can be applied to many sorts
of composite materials, we shall here speak mainly of polycrystalline aggregates with and without texture.

It is not self-evident that the problem just stated makes

[+] Institut für Theoretische Physik, Technische Universität
Clausthal, Clausthal, Bundesrepublik Deutschland

sense at all. In fact, if the grain boundaries have a
particular effect on the properties of the aggregate, then
one cannot expect to derive the polycrystal properties from
those of the single grains alone. This is the reason why
all attempts to explain the polycrystal plasticity on the
grounds of the single crystal plasticity have led to more
or less rough approximations only.

There exist properties, however, on which the grain boundaries
have only a little effect as long as the "volume" of the
grain boundaries is small compared to the total volume of the
aggregate. Therefore, although, for instance, the elastic
moduli might be quite different in the grain boundary regions
and in the interior of the grains, this difference will have
very little effect - well below 0,1 percent under normal
conditions - on the macroscopic moduli of the aggregate. In
all such cases it will be a very good approximation to treat
the grains as continuous media with sharp boundaries between
them along which the well-known boundary conditions of the
theory in question apply. By the last remark the problem is
restated in a more mathematical way. However, even if one
considers only linear theories, as we shall do, the problem
is still so complex that so far only in very special cases
could rigourous solutions be obtained. As an example of a
rather trivial case I mention the aggregate the grains of
which are parallel thin disks.

Of particular interest are random or partly random aggregates,
the randomness referring to orientation and grain shapes as
well. As an example of a partly random aggregate, I mention
the polycrystal with ideal fibre texture. In this case the
specimen is ordered in one direction, and random in the two
other directions.

A crucial point in our problem is the rational description
of the arrangement of the grains in the aggregate. The use
for specifying textures of spherical harmonics and their
generalizations is now so well-known that I need not to
discuss this here. The most interesting new development in
our field is rather the introduction of the powerful methods
of statistical mechanics, in particular the use of the so-

called correlation functions which will be explained in the next section. Historically, this development started with preliminary work of LIFSHITZ and ROSENZWEIG [1] and of BROWN [2] and was put forward with increasing intensity after 1963, among others by BERAN and MOLYNEUX [3,4], VOLKOV and KLINSKIKH [5] and LOMAKIN [6].

5.2.2 Quantitative Description of Polycrystalline Aggregates by means of Correlation Functions

We shall explain the general idea by referring to the property of elasticity. It can easily be seen that the idea applies just as well to other properties such as those described by dielectric constants, permeabilities, conductivities etc.

In general, two specimens which behave the same way in all possible macroscopic elasticity experiments will nevertheless be distinguishable in their microscopic state, i.e. in the details of the arrangements of the grains. According to the methods of statistical mechanics, we consider an imagined ensemble of all those specimens which are in the same macrostate but differ in their microstate. All specimens of the ensemble have the same gross elastic moduli.

If two ensembles are different in their macrostate (gross elastic moduli), it follows that the microstates of the first ensemble are so much different from those of the second ensemble that this difference can even be observed in macroscopic experiments. Now, for the theory it is only important to describe the m a c r o s c o p i c features of the microstates. Such a description which works in terms of macroscopic functions will be distinguished only between specimens which belong to different ensembles but not between the specimens of the same ensemble.

The problem of the macroscopic description of the microstates is solved by the infinite set of correlation functions which we introduce in the following.

Let us perform experiments at a point $\underline{r}$ of a specimen of the particular ensemble. Since we do not know the details of the constitution of this specimen, only expectation values can be predicted for experiments on the microscopic scale. These expectation values are the average values which would be found if we extended the same measurements, always at the same points r, over all specimens in the ensemble. If the expectation values are independent of point r, then the ensemble is called statistically homogeneous. In this case one can replace the ensemble averages by volume averages in a single specimen of the set (ergodic theorem).

Now assume statistical homogeneity and let the experiment be the measurement of the elastic moduli at point r of all specimens in the ensemble. So we find the expectation value which obviously provides a certain information about the aggregate. It has been proved by HILL in 1952 [7] that, having just this information, one can establish two bounds between which the gross elastic moduli of all specimens possessing this expectation value must lie. These two bounds are known as the values of Voigt and Reuss respectively.

To improve the information let us perform this experiment: Measure an elastic modulus at 2 points r_1 and r_2 and find the expectation of the product of the 2 values by either going over all specimens in the ensemble or by taking the corresponding volume average in a single specimen[+]. In general, the result will be a function of the relative position of the 2 points in question. This function is called the 2-point correlation function of the modulus distribution because it tells us to what extent the elastic moduli at the points r_1 and r_2 are correlated. Clearly the knowledge of the 2-point correlation function provides additional information about the aggregate. It has been proved by Beran and Molyneux that this additional information allows us to shift the two bounds mentioned above closer together, a very sensible result.

[+] The latter measurement requires the shifting of the 2 points over the specimen keeping the relative position fixed. Observe that certain difficulties arise if the specimen is not infinitely extended.

Further information can be gathered if we measure the elastic
moduli at 3 points r_1, r_2, r_3 in the specimens and obtain
the expectation value for the 3-fold product as before. The
function of the relative position of the 3 points is called
the 3-point correlation function of the moduli distribution.
Again the bounds can be shifted closer due to this information.

It is clear now how the procedure goes on. By increasing
information in terms of higher order correlations the bounds
come closer and closer. They finally coincide if all
correlations up to infinite order are included. This result
shows that our initial underlying idea is correct according
to which the macroscopic elastic moduli of an aggregate are
uniquely determined by the complete set of correlation
functions. In this manner of speaking the expectation values
of the elastic moduli are classified as 1-point correlation
functions.

5.2.3 Results

It cannot be the purpose of this brief lecture to develop
the theory which allows expression of the macroscopic moduli
of the aggregate in terms of the single crystal moduli and
the correlation functions. This expression has been given in
the form of a sum of multiple integrals in ref. [8]. The
satisfaction of the basic equations of elasticity is a crucial
part of the theory.

The result means that given the correlation functions, the
rest is pure mathematics. Unfortunately the integrals are
extremely involved except for special forms of the
correlation functions. An important case in which the integrals
could be calculated is that of perfect disorder which by
definition describes an aggregate in which the values of the
elastic moduli at a point r, say, are independent (in the
sense of statistics) of the values at any other point r'.
Aggregates of this sort are macroscopically isotropic. Real
materials do not possess such ideal distributions of the
moduli. In fact, if the distances of the points r and r' are
smaller than the grain diameters there is a larger
probability to find the same values of the moduli at the two

points than to find any other values. Clearly this means a
correlation.

However, if the specimen is very big – which implies that
the number of grains is very large – then the error in
assuming the real specimen to be perfectly disordered goes
to zero with increasing grain number. Hence, the result of
the integration with those correlation functions which
describe perfect disorder is rigourous for infinite specimens
in which the relative orientations and shapes of the grains
are statistically independent.

For perfectly disordered aggregates of cubic crystallites the
integration gives the result

$$G = \bar{\mu} \left[1 - \frac{12}{5} \frac{\alpha(\frac{\mu-\nu}{5\bar{\mu}})^2}{1 - \frac{2}{5}\alpha\frac{\mu-\nu}{5\bar{\mu}} - \frac{24}{25}\alpha^2(\frac{\mu-\nu}{5\bar{\mu}})^2} \right] \quad (1)$$

where G is the macroscopic shear modulus, $\bar{\mu} = (3\mu + 2\nu)/5$
its Voigt limit, and

$$\alpha = \frac{3\bar{k} + 6\bar{\mu}}{3\bar{k} + 4\bar{\mu}}, \quad \bar{k} = (c_{11} + 2c_{12})/3, \quad \mu = c_{44},$$

$$\nu = (c_{11} - c_{12})/2. \quad (2)$$

For the shear modulus of polycrystalline copper the above
formula gives, when the single crystal values measured by
BRADFIELD [10] are inserted, G = 4.77 (in units of 10^{-12}
dyn/cm^2), whereas Bradfields experimental polycrystal shear
modulus is 4.83. This agrees within about 1%. Considering
how difficult it is to account for the everpresent deviations
from random arrangement of the grains, the result is certainly
satisfactory. This applies all the more since copper is one
of the more anisotropic metals. The agreement in aggregates
of less anisotropic crystallites is even better.

I cannot yet report results of our present calculations on
polycrystalline aggregates with fibre texture. There is
reasonable hope that this problem will also permit a
rigourous solution. The same is not so clear with more

complicated textures although I personally feel that in those cases there also exists a real chance for exact solutions under idealized conditions.

The method has been applied to other properties than elasticity, namely to those which are described by 2nd rank tensors. REICHSTEIN [11] obtained the following result for the dielectric constant of a perfectly disordered aggregate:

$$\epsilon = \bar{\epsilon} - \frac{1}{3} \sum_{\nu=1}^{3} \frac{(\epsilon_\nu - \bar{\epsilon})^2}{\epsilon_\nu + 2\bar{\epsilon}} \tag{3}$$

where ϵ is the macroscopic dielectric constant, $\bar{\epsilon} = \frac{1}{3} \sum_{\nu=1}^{3} \epsilon_\nu$ its Voigt limit and ϵ_ν are the components of the single crystal dielectric tensor in the diagonal representation.

5.2.4 Conclusion

Modern experimental techniques allows us to obtain elastic moduli or other material parameters with an accuracy of 0,1 percent and better. This accuracy does not help much in the registering of polycrystal values because the conditions of the aggregate (such as texture and dislocations) are not defined within such an accuracy. For this reason most of the single crystal values reported are much more reliable than those reported for random polycrystalline aggregates. The above formula gives the possibility of registering the latter quantities with an accuracy which is about the same as that of the corresponding single crystal values. The values so obtained together with the Voigt and Reuss bounds would be the values to be registered in tables of the material parameters of polycrystalline aggregates. Similarly one can plan for later to register the corresponding values for the most important ideal textures.

Of course, experimentally determined deviations from the ideal values provide information about deviations from the considered idealized states. This effect can be used to sort out "good" from "bad" specimens (for instance when deviations from the ideal texture are considered bad) by measurements of elastic moduli or other material parameters. Since such measurements can easily be performed in series, the method might have some relevance to techniques. Another application could be the reinforcement of a weak body by inclusions of a harder material.

References

1. I.M. Lifshitz and L.N. Rosenzweig: JETP (russ.) 11(1946) 967.

2. W.F. Brown: J.Chem.Physics 23(1955)1514.

3. M.J. Beran and J. Molyneux: Nuovo Cim. 30(1963)1403.

4. M.J. Beran and J. Molyneux: Quart.Appl.Math. 24(1966)107.

5. S.D. Volkov and N.A. Klinskikh: Physics Metals, Metallogr. (Transl. from Russian) 19(1965)24.

6. V.A. Lomakin: J.Appl.Math.Mech. (Transl. from Russian) 29(1965)1048.

7. R. Hill: Proc.Phys.Soc. (Lond.) A 65(1952)349.

8. E. Kröner: J.Mech.Phys. Solids 15(1967)319.

9. G. Kneer: phys.stat.sol. 9(1965)825.

10. See G. Kneer in ref. [9].

11. K.-H. Reichstein: Ph.D.Dissertation, Clausthal 1968.

5.3 Texture Hardening
Texturverfestigung

by W. F. Hosford[+)]

Zusammenfassung

Unter Texturverfestigung versteht man die Erhöhung des Fließ-
widerstandes unter verschiedenen Belastungsbedingungen durch
bestimmte Texturen. Die Texturverfestigung ist sowohl anwendbar auf die Umformung von Metallen, z. B. durch Tiefziehen,
Walzen oder Biegen, wobei Fließen erforderlich ist als auch
auf Konstruktionen im Maschinenbau, bei denen Fließen unter
einachsiger oder zweiachsiger Zugbeanspruchung, Torions etc.
nicht eintreten darf.

Die Kontinuums-Theorien des Fließens haben einen bedeutenden
praktischen Wert für die Beurteilung der allgemeinen Form
der plastischen Anisotropie und ihre Anwendung auf praktische
Probleme. Grundlegend aufgestellte Fließkriterien müssen jedoch mit Vorsicht betrachtet werden. Experimentell bestimmte
Fließgrenzendarstellungen stimmen häufig nicht mit der Theorie überein.

Ein grundlegendes Verständnis der Abhängigkeit der Anisotropie von der Textur verlangt die Berücksichtigung von Gleitung und Zwillingsbildung in den Körnern eines Metalls. Für
kubische Metalle wurde eine quantitative Theorie entwickelt,
die in recht guter Übereinstimmung mit Experimenten über die
erzwungene Formänderung von Kristallen ist. Bei hexagonalen
Metallen ist wegen der großen Anzahl von Verformungsmechanismen nur eine qualitative Deutung der Anisotropie möglich.

Während die ideale Textur für viele Anwendungszwecke festgestellt werden kann, sind noch keine Verfahren zu ihrer Erzeugung entwickelt worden.

5.3.1 Introduction

In recent years there has been a growing interest in the

[+)] University of Michigan, Ann Arbor, Michigan, U.S.A.

concept of "texture hardening" or the strengthening of metals by control of crystallographic texture [1-4]. The basis of this stengthening is simply that the plastic properties of textured metals are anisotropic because plastic deformation is restricted to slip and twinning on specific crystallo- graphic systems. Useful strengthening results when the orientation of the deformation systems makes yielding difficult under the applied loading.

The importance of plastic anisotropy in service and during plastic working is easy to overlook because frequently these involve combined stress loading. To assess the role of anisotropic behavior in yielding under combined stresses, consideration should be given to the form of the yield locus.

5.3.2 Yield Loci

5.3.2.1 Theory

In dealing with anisotropic materials, it is convenient to express the stress and strain components relative to the axes of material symetry, (or principal axes of anisotropy). With rolled sheet or plate, these would be the rolling direction -x, the transverse direction -y, and the rolling plane normal -z, while with rod or tube x, y, and z would refer to the rod or tube axis, the circumferential and radial directions respectively.

The level of hydrostatic stress, $1/3(\sigma_x+\sigma_y+\sigma_z)$ has very little effect on yielding, which is sensitive only to shear stresses. Therefore the yield criterion should be of the form

$$f(\sigma_x-\sigma_y, \ -\sigma_z, \ \sigma_z-\sigma_x, \ \sigma_{yz}, \ \sigma_{zx}, \ \sigma_{xy}) = C \qquad (1)$$

Yielding occurs when the value of $f(\sigma_{ij})$ reaches the constant C. There are five independent stress terms (only two of the first three terms may be chosen independently). Consequently a graphical representation of a general yield criterion would require five-dimensional space!

However simplification can be made with some loss of generality. With plates, rods and tubes, for example, real situations do not usually involve the stress strains σ_{yz} or

σ_{zx}. If in addition, attention is restricted to cases involving only normal stresses σ_x, σ_y, and σ_z (ie $\sigma_{xy} = 0$), the yield function involves only two independent parameters and may be represented graphically as a cylinder whose axis is equally inclined to the σ_x, σ_y, and σ_z axes (Fig. 1a).

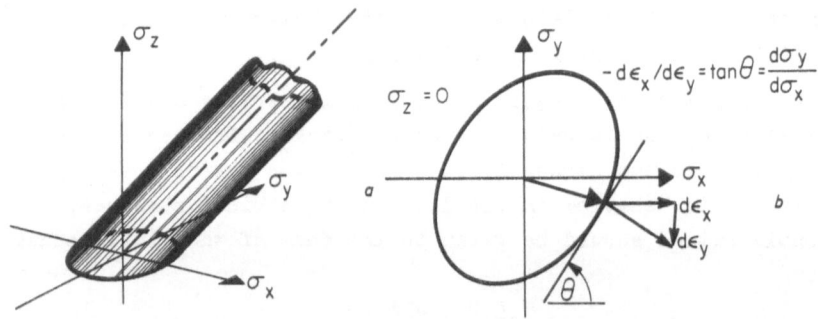

Fig. 1a and b. a) The yield surface of a material in which yielding is not sensitive to hydrostatic stress. The locus is a cylinder with an axis equally inclined to the three normal stress directions.

b) The $\sigma_z=0$ section of the locus above. The vector representing the plastic strains on yielding are normal to the yield locus at the point of yielding (eqs. 2 and 3)

No loss of generality is incurred in a two dimensional section on the $\sigma_z = 0$ plane (Fig. 1b). Any stress state for which $\sigma_z = 0$ is represented on such a plot by an e q u i v a l e n t stress state,

$$\sigma_x' = \sigma_x - \sigma_z$$

$$\sigma_y' = \sigma_y - \sigma_z \qquad (2)$$

$$\sigma_z' = \sigma_z - \sigma_z = 0$$

which differs from the σ_x, σ_y, σ_z state only by a hydrostatic component.

On very general theoretical grounds two significant statements can be made about the yield locus [5]. First it can not be concave outward at any point. Second the strains which occur during yielding must obey the flow rule

$$d\epsilon_{ij} = d\lambda\frac{\delta f(\sigma_{ij})}{\delta\sigma_{ij}} \qquad (3)$$

This rule may be interpreted as a statement that the strain vector must be normal to the yield locus (Fig. 1b). In other words,

$$\tan \Theta = d\sigma_y/d\sigma_x = - d\epsilon_x/d\epsilon_y \qquad (4)$$

Besides these two statements, little can be said regarding the shape of the yield locus without assuming something about the nature of the material or resorting to experiment.

5.3.2.2 Experimental

Combined stress tests are difficult to make, and often the geometric form of a material precludes their use. One method, however, which has been successfully employed on thin-walled tubes is the simultaneous loading by internal or external pressure and axial load [6.7]. Fig. 2 is a partial yield locus for Zircaloy-2 tubing determined in this manner.

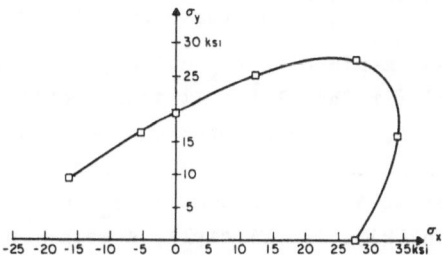

Fig.2. Partial plane-stress ($\sigma_z=0$) yield locus for an extended Zircaloy-2 tube at 580°F. determined by simultaneous stressing under internal pressure and axial stress. The values of σ_x and σ_y (hoop and axial stresses respectively) correspond to an plastic strain of 0,5%

Alternatively a series of uniaxial and plane-strain tests may be employed [8-10]. This is illustrated in Fig. 3. Points A, B, Ā, and B, can be determined directly from uniaxial tension and compression tests in the x and y direction.

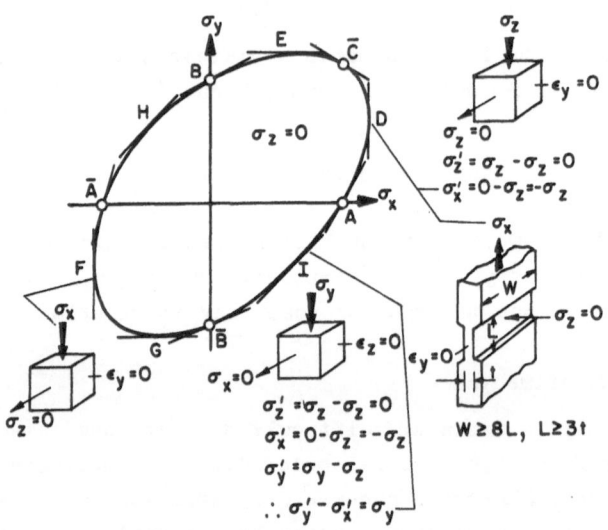

Fig.3. Yield locus determination by uniaxial and
plane strain testing. Points A and B, and $\bar{A}$, $\bar{B}$,
and $\bar{C}$ and the slopes at these points can be found
by uniaxial tension and compression testing. Other
tangents can be determined from plane-strain
compression or tension tests as illustrated

Furthermore measurements of ϵ_y/ϵ_x during these tests together
with equation (4) fix the slopes at these points. Similarly,
point $\bar{C}$ and the slope at this point can be determined by a
z direction compression test, because uniaxial compression
σ_z, is equivalent to a biaxial tension $\sigma_x' = \sigma_y' = -\sigma_z$ and
$\sigma_z' = 0$ (equation (2)). In principle, a point in the third
quadrant could be established by a through-thickness tension
test, but this is impractical at least in the case of sheets.

Plane strain compression and tension tests can be used to
establish tangents to the locus on the various quadrants. The
basis for this is the flow rule (equations (3) and (4)). For
example tangent D can be established by a σ_z compression,
maintaining $\sigma_x = 0$, $\epsilon_y = 0$, this being equivalent to $\sigma_x' =
-\sigma_z$, $\sigma_z' = 0$. Alternatively this tangent could be determined
by a plane-strain tension test. Other orientations of test
can be used to establish the other tangents as indicated in

419

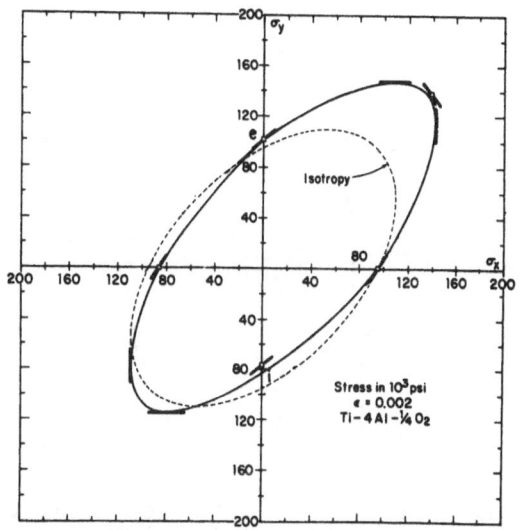

Fig.4. The plane-stress ($\sigma_z = 0$) yield locus for a
textured sheet of Ti-4Al-2-1/20 (adapted from ref.
[9]). Points and tangents were determined from
uniaxial and plane-strain tests. The von Mises locus
for isotropy is shown for comparison

Fig.3. Yield loci established in this manner for titanium [9]
and magnesium [10] sheets are shown in Fig. 4 and 5. The yield
locus of an isotropic material, obeying the von Mises
criterion is also shown for comparison.

It has been suggested that the approximate shape of the yield
locus might be established by appropriate plotting of the
results of six orientations of Knoop hardness tests [11], as
indicated on Fig. 6. While this method may have some value,
it is difficult to justify on an theoretical ground. Furthermore, tests on textured magnesium indicate that this method
significantly underestimates any differences between yield
stresses in the first and third quadrants and somewhat
overestimates the strength on the second and fourth quadrants
[12].

5.3.2.3 Hill's Analysis
For most engineering purposes the anisotropic yield criterion

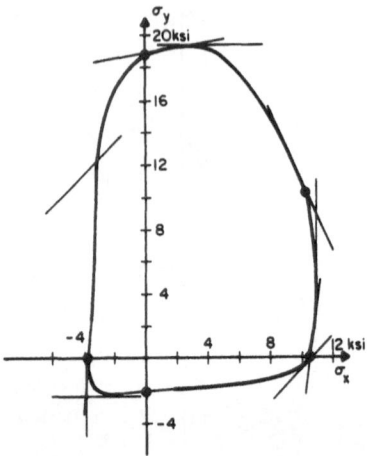

Fig.5. The plane-stress ($\sigma_z=0$) yield locus for a textured magnesium sheet, (adapted from ref. [10]). Stress levels correspond to plastic strains of 1%. Points and tangents were determined by uniaxial and plane-strain tests

postulated by HILL [13] provides a simple means of evaluating anisotropic yielding behaviour. If loading is restricted to normal stresses along x, y, and z, Hill's yielding criterion reduces to:

$$R(\sigma_y-\sigma_z)^2 + P(\sigma_z-\sigma_x)^2 + RP(\sigma_x-\sigma_y)^2 = P(1+R)X^2$$
$$= R(1+P)Y^2 \qquad (5)$$

Here X and Y are the yield stresses in uniaxial tension (and compression) in the x and y directions and R and P are the ratios of lateral strain in the uniaxial tests:

$$R = \epsilon_y/\epsilon_z \text{ in an x direction test}$$
$$P = \epsilon_x/\epsilon_z \text{ in a y direction test.}$$

Thus two tests are sufficient to determine the parameters X, Y, R, and P, only three of which are needed to establish the yield locus. Sometimes it may be advantageous to establish these parameters or equivalent parameters by other

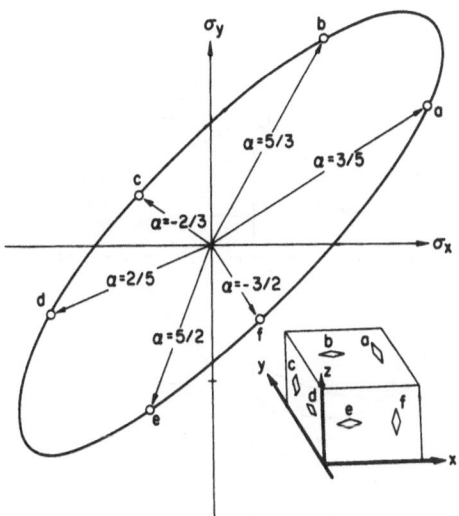

Fig.6. Proposed method of determining the shape of
the plane-stress ($\sigma_z=0$) yield locus from Knoop
hardness data, (ref. [11]). Points are plotted at
$\sigma_x=KHN/\sqrt{1-\alpha+\alpha^2}$, $\sigma_y=\alpha\sigma_x$ where KHN is the Knoop
hardness number. The appropriate values of α for
each test are indicated. The data are for a Zirc-
aloy-2 plate (adapted from ref. [3])

types of tests.

The corresponding flow rules may be expressed as

$$d\epsilon_x : d\epsilon_y : d\epsilon_z = [RP(\sigma_x-\sigma_y)+P(\sigma_x-\sigma_z)]:$$

$$[R(\sigma_y-\sigma_z+RP(\sigma_y-\sigma_x))]:[P(\sigma_z-\sigma_x)+R(\sigma_z-\sigma_y)] \qquad (6)$$

The $\sigma_z = 0$ yield loci corresponding to equation 5 plot as
ellipses centered on the origin (Fig. 7). With increasing
values of R and P (corresponding to increased resistance to
sheet thinning) the loci extend further into the first
quadrant.

Hill's general theory is capable of dealing with forms of
loading in which the shear terms relative to x, y, and z do
not vanish. While this is particularly important in

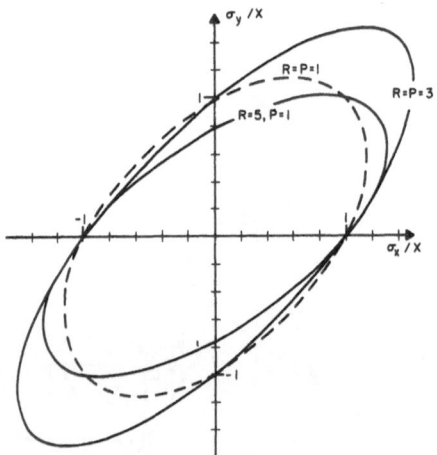

Fig.7. Plane-stress ($\sigma_z=0$) yield loci for anisotro-
pic materials according to the Hill criterion. The
R=P=1 locus corresponds to isotropy (von Mises).
Increasing values of R and P lead to texture
       hardening in the first and third quadrants

considerations of the directionality of properties on the
plane of a sheet material, it will not be discussed here.

It should be noted that the Hill criterion is merely
postulated and may not adequately describe the form of the
entire yield locus. The predictions may be quite good if
loading paths used to establish its parameters are not widely
divergent from those of primary interest. On the other hand,
it would be foolish to evaluate the behavior of material
under biaxial ($\sigma_x = \sigma_y$) compression from uniaxial tension
tests in the x and y directions. Because the Hill theory
predicts yield loci centered on the origin, it cannot
adequately describe the types of loci in Fig. 4 and 5.

5.3.2.4 Practical Implications

A number of practical aspects of texture hardening can be
illustrated by superimposing various loading paths on a yield
locus of the type common to rolled sheets of α-titanium alloys

(Fig.8). In this material a high resistance to sheet thinning extends the locus into the first quadrant loading to an obvious strengthening. The loading path $\alpha = \sigma_y/\sigma_x = 1$ corresponds to balanced biaxial tension as occurs in the walls of a spherical pressure vessel or during a bulge test. The $\alpha = 1/2$ loading path occurs in the walls of a cylindrical pressure vessel.

The feasibility of using α-titanium alloys in such applications has been investigated by pressure testing of welded cylindrical and spherical vessels of several Ti alloys [14]. Fig. 9 shows the results for a commercial Ti-5Al-2.5Sn sheet material. In the spherical vessels yielding occurred at stress levels which are about 40% higher than those predicted from isotropic theory, and the burst strengths were about 75% higher than the strengths in uniaxial tension.

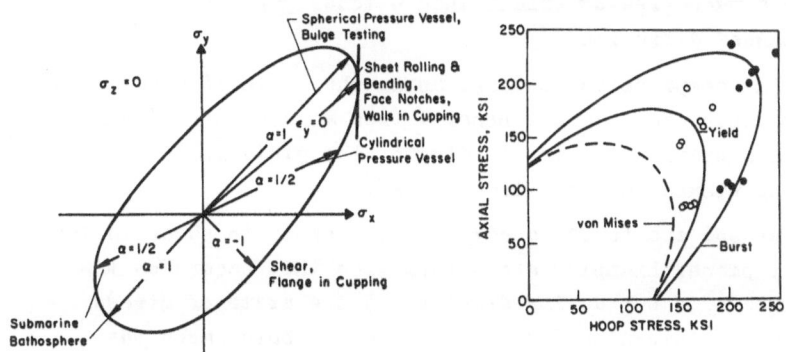

Fig.8 (left). Schematic ($\sigma_z = 0$) yield locus indicating various loading paths of practical interest

Fig.9 (right). Texture hardening of a Ti-5Al-2.5Sn sheet material under biaxial tension. Data were determined by pressure testing of welded vessels and the curves drawn with the Hill criterion assuming R=P. Isotropic behavior is shown for comparison (adapted from ref. [14])

Plane-strain loading is not uncommon in structures and plastic working operations. The loading path corresponding to $\epsilon_y = 0$ intersects the yield locus where it is vertical and consequently represents the maximum texture hardening with respect to σ_x. This is the loading path that corresponds to sheet rolling. The roll pressure, $-\sigma_z$, is equivalent to the value of σ_x on the yield locus. (In rolling $\sigma_x = 0$ and $\sigma_z \neq 0$, but on the yield locus $\sigma_z' = 0$ so $\sigma_x' = -\sigma_z$). The high mill forces experienced in rolling of α-titanium alloys are to a large extent a manifestation of this strengthening.

The stress states in sheet bending also correspond to plane-strain. In the bending of highly textured beryllium sheet, the stress, σ_x, may rise sufficiently to cause fracture. Tubes of textured titanium have a strong tendency to flatten during bending rather than deform in plane-strain $\epsilon_y = 0$. Ball mandrels used to prevent flattening may be entrapped. If flattening is successfully prevented by the mandrels, the high axial stresses may result.

Material at the base of a notch deforms in plane-strain; the component of strain along the notch being zero. In materials with high R values, notches parallel to the rolling plane habe a much greater effect than notches parallel to the thickness direction.

Biaxial compressive loading, as in the walls of a submarine or bathosphere, may not benefit from as much strengthening as does biaxial tension loading of the yield locus is foreshortened in the third quadrant.

The second and fourth quadrants correspond to shear in the x - y plane. Examples are deformation that occurs on the flanges of a cup during drawing and the state of stress beneath a thickness direction notch. In both cases the texture softening that accompanies high R values is desirable.

5.3.3 Structural Bases of Anisotropic Plasticity

The metallurgist should be interested in the structural origin of the anisotropy as well as its potential application. While crystallographic texture is the primary source of anisotropic behavior, it is not the only source possible.

One obvious microstructural feature of wrought metals is the mechanical fibering or alignment of second phases. The possible role of directional microstructures can be illustrated by considering two extreme geometric distributions of two isotropic phases of different strength. In the case of the

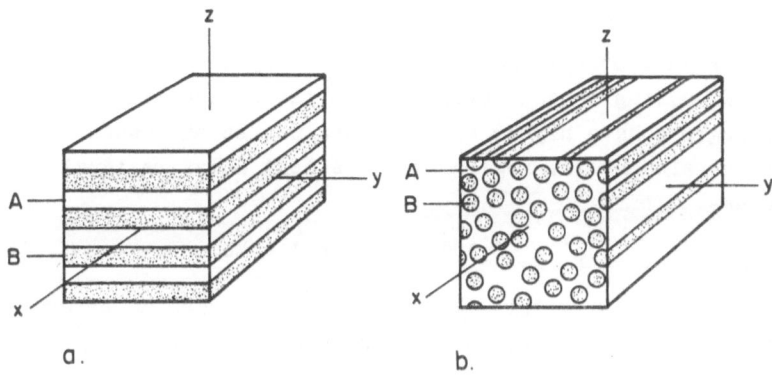

Fig.10. Two extreme geometric distributions of phases
which could cause plastic anisotropy if the phase B
is much harder than phase A

layered structure, (Fig. 10a), a simple plastic theory predicts
that the uniaxial yield stress in the z direction is the same
as that in the x and y directions: $Y = f_A Y_A + f_B Y_B$ where f_i
is the volume fraction of the i phase and Y_i is its yield
stress.

However, under shear stresses σ_{yz} and σ_{zx} yielding should
occur only in the softer phase, A, at a shear stress of
$Y_A/\sqrt{3}$ whereas under shears σ_{xy} both phases must yield so the
corresponding shear stress would be $(f_A Y_A + f_B Y_B)/\sqrt{3}$.

In the case of the fibered structure, (Fig. 10b), the yield
stress in the x direction would be considerably greater than
that in the y and z directions. The strength under a shear
σ_{yz} would probably be higher than those under shears σ_{xz}
and σ_{xy}. Thus some anisotropy in materials with elongated
inclusions, second phases or even grain boundaries may be
attributed to directional microstructures. In most practical
cases, however, the effects are probably small.

In cold worked materials, the Bauschinger effect, arising
from local residual stresses may be considered as a form of
anisotropy. After tensile elongation in the x direction, not

only is the yield stress in x direction compression lower
than that in tension, but the yield stresses in other
directions are affected as well.

Consider an isotropic material, which contains hard and soft
regions. During tensile straining in the x direction, the
hard regions have higher elastic strains, so that when the
load is released the regions will be left under residual
stresses

$$\sigma_x' = -\Delta \quad \text{and} \quad \sigma_y' = \sigma_z' = k\Delta \tag{7}$$

where k is approximately 1/2. On reloading under an external
stress σ_x, σ_y, σ_z, yielding should be observed when

$$(\sigma_y'' - \sigma_z'')^2 + (\sigma_z'' - \sigma_x'')^2 + (\sigma_x'' - \sigma_y'')^2 = 2Y^2 \tag{8}$$

where the actual stress in the soft regions $\sigma_i'' = \sigma_i' + \sigma_i$.
In terms of the applied stresses, then new $\sigma_z = 0$ yield
locus will be

$$\sigma_x^2 - \sigma_x\sigma_y + \sigma_y^2 + \Delta(1+k) \ (-2\sigma_x + \sigma_y) = Y^2 - \frac{\Delta^2(1+k)^2}{2} \tag{9}$$

which corresponds to a displacement of the yield locus to the
right, (Fig. 11). Prior straining under other loadings cause
similar shifts in other directions.

SVENSSON [15] has measured 0.05% yield stresses in both
tension and compression in various directions on cold rolled
aluminum and steel sheets. Although comparison of rolling
and transverse direction properties is complicated by
crystallographic texture, it is noteworthy that in rolling
direction tests, the yield stress in compression was lower
than in tension, while the reverse was true for transverse
direction tests. Furthermore his observation that low
temperature annealing markedly reduced the differences without a general softening is consistent with residual stresses
as the origin of the effect.

It should be noted that the extent of anisotropy (or
displacement of the yield locus) here is sensitive to the
definition of yielding. If yielding is defined in terms of
very small plastic strains, the anisotropy can be significant.

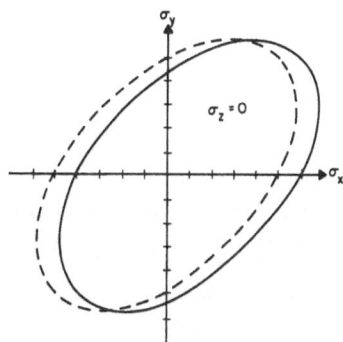

Fig.11. The plane-stress ($\sigma_z = 0$) yield locus resulting
from local residual stresses after tensile prestraining
in the x direction (solid curve) predicted by equation
9. The effect is one of shifting the isotropic locus
(dashed) to the right

On the other hand residual stresses should have a much lower
effect on the stresses necessary to cause plastic strains
which are large relative to elastic strains.

5.3.4 Texture-induced Anisotropy

5.3.4.1 HCP Metals

It is widely accepted that crystallographic texture is the
major cause of plastic anisotropy. The role of texture is
most easily visualized in the hcp metals where easy slip is
generally confined to close-packed $\langle 11\bar{2}0 \rangle$ directions. Slip
in these directions, whether primarily on the basal plane
as in beryllium and magnesium, or whether also on the prism
and pyramidal planes as on titanium and zirconium, produces
no tensile or compressive strain along the c-axis. Twinning
must be activated to produce c-axis strains. Generally the
most easily activated twinning mode is $\{10\bar{1}2\}\langle 10\bar{1}\bar{1} \rangle$ which
causes an extension along the c-axis (except in zinc and
cadmium for which $c/a\sqrt{3}$). Compression along the c-axis
requires twinning on other systems which are usually harder
to activate.

Rolled sheets of hcp metals are usually characterized by a

strong basal pole alignment with the thickness direction.
Such sheets have high R values and strong thinning resistances
which account for the texture hardening under biaxial tension
(first quadrant) [1]. Indeed, for balanced biaxial tension,
$\sigma_x = \sigma_y$, yielding must occur by either slip on the unfavorably
oriented basal planes or by twinning on one of the harder
systems. Strengthening under biaxial compression may be
considerably lower because of the possibility of $\{10\bar{1}2\}$
twinning.

Titanium and zirconium on the other hand, slip with relative
ease on prism planes. The strengthening under biaxial tension,
where yielding occurs primarily by $\{11\bar{2}1\}$ twinning, can be
quite high relative to the uniaxial strength. The
foreshortening of the yield locus in the third quadrant is
considerably less than that of magnesium as can be seen by
comparing Fig. 4 and 5. This is probably due to a smaller
difference in the critical stresses for operation of the
different twinning modes. Fig. 12 is a yield locus calculated
for a hcp metal of this type.

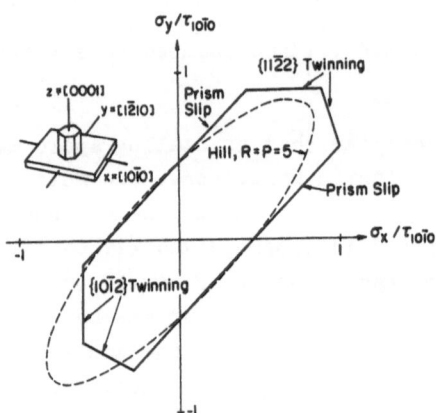

Fig. 12. The plane-stress ($\sigma_z=0$) yield locus for an
ideally textured sheet material, calculated on the
assumption that the critical stresses for prism slip,
$\{10\bar{1}2\}$ twinning, and $\{11\bar{2}2\}$ twinning are in the ratio
of 1:1.5:2. The locus is not centered on the origin
because of the different twinning modes operating in
the first and third quadrants. The Hill locus for R=
P=5 is shown for comparison (ref. [3])

Except in the case of very simple orientations, it would be
extremely difficult to calculate the anisotropic behavior
of hcp metals without making gross oversimplifications. The
underlying reason for this is that in general four different
types of deformation modes would have to be considered in
each grain, each mode having a different critical stress for
operation. That the anisotropic behavior of single crystals
is closely related to that of single crystals is, however,
suggested strongly by the similar shapes of the stress-strain
curves for magnesium single crystals and polycrystals in plane
strain [10].

5.3.4.2 Cubic Metals

The cubic metals by comparison are less anisotropic, even
when strongly textured. The problem of how anisotropy is
related to crystallographic orientation can, however, be
treated quantitatively by an extension of the analyses of
TAYLOR [16] or of BISHOP and HILL [17,18]. Basically these
analyses involve assumption that: (1) each grain in a
polycrystal undergoes the same strains relative to the
polycrystal axes as the polycrystal itself, and deformation
within the grain is homogeneous; (2) every equivalent slip
system operates at the same resolved shear stress regardless
of what combination of slip systems is active; and (3) the
deformation occurs in such a way as to minimize the energy
expended or equivalently to maximize the virtual work. Because
of the first assumption these analyses should be regarded as
establishing upperbounds to strength. Except in unusual cases,
however, they are probably close to reality and yield the
best analytic solution available.

The Taylor method of calculating the strength of an individual
grain of known orientation involves finding which combination
of slip systems can produce the required shape change with a
minimum amount of slip. If the work/volume dw needed for an
increment of strain is supplied by a single external stress
σ_x, we can write

$$dw = \sigma_x d\epsilon_x = \tau d\gamma \qquad (10)$$

where $d\epsilon_x$ is the net strain resulting from the amount of slip $d\gamma$ on all of the active slip systems and τ is the shear stress to activate those systems. Hence the strength relative to the basic shear stress,

$$M = \sigma_x/\tau = d\gamma/d\epsilon_x \qquad (11)$$

The calculations thus involve finding which combination of slip systems give the minimum value of $d\gamma/d\epsilon_x$.

The Bishop and Hill method, on the other hand, is based on the principal of maximum virtual work

$$dw = \Sigma\sigma_{ij}d\epsilon_{ij} \qquad (12)$$

where σ_{ij} and $d\epsilon_{ij}$ are expressed relative to the cubic axes. Since only a finite number of stress states σ_{ij} are capable of activating enough slip systems for an arbitrary shape change, and since $d\epsilon_{ij}$ may be calculated relative to $d\epsilon_x$ from the imposed shape change, the values of

$$M = \frac{dw}{\tau d\epsilon_x} = \frac{\Sigma\sigma_{ij}(d\epsilon_{ij}/d\epsilon_x)}{\tau} \qquad (13)$$

corresponding to each permissible stress state are calculated and the maximum selected as appropriate.

Fig. 13 represents the calculated orientation dependence on M for fcc crystals undergoing axisymmetric flow. These values are relavent to the strength of materials with a fiber texture when tested in uniaxial stress along the fiber axis. The apply equally well to sheets with a rotationally symetric texture under equal biaxial tension in the plane of the sheet. The greatest strengthening here occurs for [111] and [110] textures for which $M = 1.5\sqrt{6}$, which is about 20% higher than the $M = 3.06 = 1.25\sqrt{6}$ for randomly oriented material and about 60% greater than the $M = 0.93\sqrt{6}$ for the weakest orientation.

Under plane-strain flow the range of anisotropic behavior

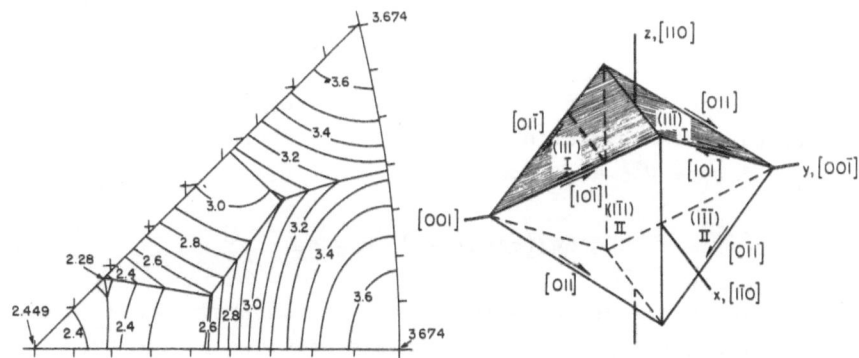

Fig.13 (left). Orientation dependence of $M = \sigma/\tau$ for fcc crystals deforming with axisymmetry about [hkl]. The M values give the relative strengths of various fiber textures under uniaxial stress ‖ [hkl] and under balanced biaxial stress ⊥ [hkl]. (Adapted from ref. [2] with corrections near the [100] corner.)

Fig.14 (right). An octahedron representing the {111} slip planes and ⟨110⟩ slip directions in a fcc crystal. With this orientation, slip would occur only on the shaded planes for plane-strain $\varepsilon_x=0$, but both the shaded and unshaded planes would have to operate for plane-strain $\varepsilon_y = 0$

is considerably greater. The yielding of a fcc crystal under normal stresses along x = [1$\bar{1}$0], y = [00$\bar{1}$] and z = [110] (Fig. 14) illustrates this point [19]. The normal strains resulting from slip $d\gamma_I$ on the shaded planes and $d\gamma_{II}$ on the unshaded planes are:

$$d\varepsilon_x = d\gamma_{II}/\sqrt{6}$$
$$d\varepsilon_y = d\gamma_I/\sqrt{6} - d\gamma_{II}/\sqrt{6} \qquad (14)$$
$$d\varepsilon_z = -d\gamma_I/\sqrt{6}$$

The work, per unit volume, dw, is simply

$$dw = \tau(d\gamma_I - d\gamma_{II}) = \sigma_x d\varepsilon_x + \sigma_y d\varepsilon_y + \sigma_z d\varepsilon_z \qquad (15)$$

When plane strain $d\varepsilon_x = 0$, $d\varepsilon_z = -d\varepsilon_y$ is imposed, $d\gamma_{II} = 0$

and $d\gamma_I = \sqrt{6}\ d\epsilon_y$. Therefore

$$M = dw/(\tau d\epsilon_y) = \sqrt{6} = \sigma_y - \sigma_z \qquad (16)$$

On the other hand if plane strain $d\epsilon_y = 0$, $d\epsilon_z = -d\epsilon_x$ is imposed, both sets of slip systems must operate, so that $d\gamma_I = d\gamma_{II} = \sqrt{6}\ d\epsilon_x$. Now

$$M = dw/(\tau d\epsilon_x) = 2\sqrt{6} = \sigma_x - \sigma_z \qquad (17)$$

The results of this calculation are represented on a $\sigma_z = 0$ yield locus (Fig. 15). Thus a simple rotation of the axis of zero strain by 90° causes a factor of two difference in the plane-strain strengths. (A practical implication of this calculation is that internally pressurized cylinders would benefit strongly from a texture with $\langle 110 \rangle$ direction aligned with the tangential and radial directions and $\langle 001 \rangle$ in the axial direction).

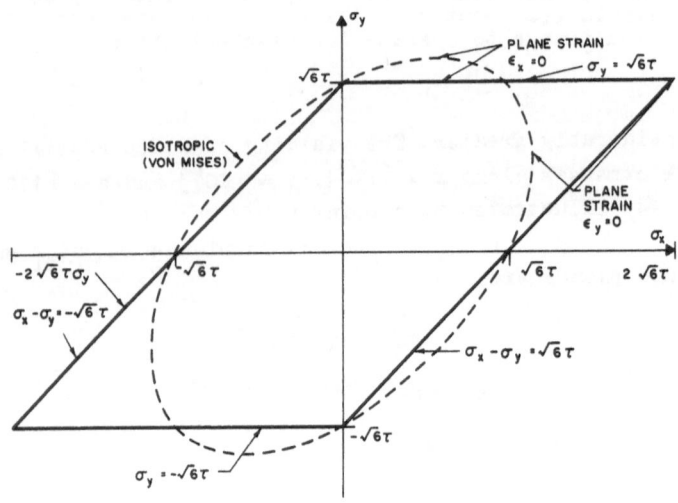

Fig.15. The plane-stress ($\sigma_z=0$) yield locus for the orientation illustrated in Fig.14. Note the extreme departure from isotropy (dashed curve). The plane-strain strength for $\epsilon_y = 0$ is twice that for $\epsilon_x = 0$

For comparison, the $M = 2\sqrt{6}$ calculated for the $(110)[1\bar{1}0]$ orientation is 48% greater than the value of $M = 1.35\sqrt{6}$ for randomly oriented grain and 2.45 times as great as the minimum value of $M = 2$ for plane-strain loading.

Strengths under other shape changes have been calculated and it is possible, at least in principle, to calculate in a rational manner R values for textured sheets [19].

While the arguments above are specifically for fcc metals, the methods can be extended to other cubic crystals. The anisotropy of bcc crystals is very similar to that for fcc crystals. Indeed if slip is restricted to ⟨111⟩{110} systems it is identical. Even if ⟨111⟩ pencil glide occurs, the M factors must be between 0.866 and 1.0 times the corresponding values for the fcc case. Calculations have been made for ⟨111⟩ {112}, ⟨111⟩{123} and mixed ⟨111⟩{110}, ⟨111⟩{112} and ⟨111⟩ {123} slip in axisymmetric flow [20].

More interesting are the extremes in anisotropy that can result from ⟨111⟩{112} or ⟨112⟩{111} twinning alone [21,22] or in combination with ⟨110⟩{111} slip [21,23]. Table 1 gives the M values for a few orientations with different operative deformation modes.

5.3.5 Conclusions

It is clear that an understanding of the anisotropy of plastic properties can be employed in such a way as to achieve a substantial degree of texture hardening. We know that the principal cause of anisotropy is crystallographic texturing. In most cases we can identify what ideal texture would be for a given application, and at least for the cubic metals estimate the degree of texture strengthening for the ideal orientations. What is lacking for full exploitation of the texture hardening concept is the ability to produce materials with taylored textures. We must learn how we can control the variables plastic working and annealing processes to do this in an economic manner.

Acknowledgements

The support of The U. S. Army Research Office - Durham in

Table 1. M factors for simple orientations of cubic metals
with various deformation modes

Orientation	Deformation Mode				
	⟨110⟩ {111} slip[+]	⟨111⟩ {112} slip[+]	mixed ⟨111⟩{110} ⟨111⟩{112} ⟨111⟩{123} slip[+]	{112}⟨111⟩ twinning[++] tension parallel to x	compression parallel to x
Axisymmetric flow about – x					
x=[111]	3.674	3.182	3.182	3.182	6.364
x=[110]	3.674	3.182	3.182	4.243	5.303
x=[100]	2.449	2.121	2.121	4.243	2.121
minimum	2.28	2.06	2.04	2.94	2.121
random (ave)	3.067	2.954	2.754	——	——
Plane-strain $\epsilon_y=0$, $\epsilon_z=-\epsilon_x$, $(\epsilon_{xy}=\epsilon_{yz}=\epsilon_{zx}=0)$[+++]					
x=[100],z=[001]	2.449	2.828	2.449	4.243	4.243
x=[110],z=[001]	2.449	2.121	2.121	2.121	3.243
x=[110],z=[1T0]	4.899	4.243	4.243	6.364	6.364
x=[110],z=[1T1]	4.082	4.007	——	6.364	5.657
x=[110],z=[T12]	3.266	4.007	——	3.657	4.950
x=[1T1],z=[T12]	3.674	4.007	——	5.657	6.364

+ slip plane and slip directions can be interchanged without
affecting M-values.

++ for {112}⟨111⟩ twinning tension and compression values are
interchanged.

+++ in each case the minimum M value for plane strain is 2.0.

preparing this paper is gratefully acknowledged. The interest
of the author in this subject has been stimulated by Professor
Backofen and more recently, by Dr. G.Y. Chin.

References

1. W.A. Backofen, W.F. Hosford, and J.J. Burke: Trans.Amer.
Soc.Metals 55(1962)264.

2. W.F. Hosford and W.A. Backofen in: Fundamentals of Deformation Processing, Syracuse Univ. Press 1964, p.259.

3. W.F. Hosford: Met.Eng.Quart. 6(1966)13.

4. D.V. Wilson: J.Inst.Metals 94(1966)84.

5. D.C. Drucker: Proc. 1st U.S.Nat.Cong. of Appl.Mech. (1951) 487.

6. R.L. Mehan: Trans.ASME ser.D, 83(1961)499.

7. S.H. Miller and J. Swota, Jr.: Knolls Atomic Power Lab. Report KAPL-2296 (1963).

8. D. Lee and W.A. Backofen: Trans.Met.Soc. AIME 236(1966) 1077.

9. D. Lee and W.A. Backofen: Trans.Met.Soc. AIME 236(1966) 1696.

10. E.W. Kelley and W.F. Hosford: Trans.Met.Soc. AIME 242 (1968)654.

11. R.G. Wheeler and D.R. Ireland: Electrochem.Technology 4 (1966)313.

12. W.W. Wilkening and B.C. Wonsiewicz: Submitted for publication. ("Comparison of Knoop Hardness Yield Locus with Conventional Yield Locus").

13. R. Hill: Mathematical Theory of Plasticity, Oxford Univ. Press 1950, Chapter XII.

14. A.W. Babel, D.A. Eitman, and R.W. McIver: Trans.ASME ser.D, 89(1967)13.

15. N.L. Svensson: J.Inst.Metals 94(1966)284.

16. G.I. Taylor: J.Inst.Metals 62(1938)307.

17. J.F.W. Bishop and R.Hill: Phil.Mag. 42(1951)1298.

18. J.F.W. Bishop: Phil.Mag. 44(1953)51.

19. W.A. Backofen and W.F. Hosford: Discussion to I.L. Dillamore, Trans.Amer.Soc. Metals 58(1965)769.

20. G.Y. Chin and W.L. Mammel: Trans.Met.Soc. AIME 239(1967) 1400.

21. G.Y. Chin, W.L. Mammel, and M.T. Dolen: To be published in Trans.Met.Soc. AIME ("Effect of Mechanical Twinning on the Strength of Crystals under Conditions of Axisymetric Flow").

22. W.F. Hosford and G.Y. Chin: Submitted for publication ("Stress States for {111}⟨112⟩ Multiple Slip and Twinning").

23. G.Y. Chin, W.F. Hosford and D.R. Mendorf: To be published in Proc.Roy.Soc. ("Accomodation of Constrained Deformation in FCC Metals by Slip and Twinning").

5.4 The Prediction of Anisotropic Yield for Textured Sheets
Die Vorhersage von anisotropen Fließgrenzen bei Blechen mit
Textur

by H.R. Piehler[+] and W.A. Backofen[++]

Zusammenfassung

Fließen und plastische Verformung bei isotropen Blechen werden üblicherweise durch nur eine zweidimensionale Fließgrenzendarstellung beschrieben, deren Koordinaten zwei in der
Blechebene wirkende Hauptspannungen sind. Anisotrope Bleche
können jedoch häufig durch eine einzelne Fließgrenzendarstellung nicht beschrieben werden, da die Hauptachsen für Spannung und Formänderung häufig nicht zusammen fallen. Um diese Schwierigkeit zu überwinden, werden zwei getrennte zweidimensionale Fließgrenzen vorgeschlagen, die als "ebene Span-
nungsgrenze" und "ebene Formänderungsgrenze" bezeichnet werden.

Es wird die Bestimmung von Fließgrenzendarstellungen sowohl
für Einkristalle als auch für vielkristallines Material mit
mehreren, verschiedenen Orientierungen besprochen und zwar
für Gleiten auf {111}⟨110⟩ und {110}⟨111⟩. Weiterhin wird
die Fließgrenzendarstellung für kub.rz. Metalle erörtert, die
pencil-glide zeigen.

5.4.1 Introduction

Thus far, theoretical attempts to describe plastic anisotropy
have frequently made use of the ratio of width to thickness
strains (R) as measured in a conventional tension test. BACK-
OFEN, HOSFORD, and BURKE [1], in first introducing "texture
hardening", used R as a parameter to describe the yielding
and plastic flow of a metal with varying degrees of "normal"
anisotropy. This was done by making use of a modified form of
the anisotropic continuum locus suggested by HILL [2]. To
arrive at this result, it was also necessary to assume that

[+] Carnegie-Mellon University, Department of Metallurgy and
Materials Science, Pittsburgh, Pa., U.S.A.
[++] Massachusetts Institute of Technology, Department of
Metallurgy, Cambridge, Mass., U.S.A.

the uniaxial yield strength did not change as R changes and
that the principal axes of stress and strain coincided, both
conditions which are often absent in actual textured metals.

A more direct theoretical description of plastic anisotropy
is possible which follows from a knowledge of the constituent
textural components and a description of the crystallographic
slipping process. Yielding and plastic flow in anisotropic
sheets can be described by crystallographically based
anisotropic yield loci, much as the familiar v. Mises and
Tresca loci are used for isotropic sheets.

5.4.2 Planar-stress and Planar-strain Yield Loci

A difficulty arises in describing yielding and plastic flow
in anisotropic metals because the principal axes of stress
and strain frequently do not coincide. This lack of coincidence
is apparent in a single crystal which is pulled in tension
and slips only on a single system. The state of stress
referred to the specimen axes is a principal stress state,
while the state of strain referred to these same axes
contains both normal and shear components. This same lack of
coincidence also often occurs under conditions of polyslip.
Principal stresses may cause both normal and shear strains,
while the enforcement of a state of principal strain may
require the application of both normal and shear stresses. As
a result of this lack of coincidence of the principal axes of
stress and strain, yielding and plastic flow in anisotropic
metals can not be simply described by a single two-dimensional
yield locus taken normal to the hydrostatic line in principal-
stress space. To overcome this difficulty, two separate
mathematically and physically meaningful two-dimensional loci
are suggested.

Of the two-dimensional yield loci chosen to describe
anisotropic plastic behavior, one will be called the "planar-
stress" or "principal-stress" and the other the "planar-
strain" or "principal-strain" yield locus. Both will have as
variables two normal stresses acting in the plane of a sheet
and will be chosen such that the third normal stress, which
acts perpendicular to the plane of the sheet, is zero. The

distinction between the two types of loci is that, in the
planar-stress locus, the normal stresses are chosen to be
principal stresses and the resulting strain state may contain
both normal and shear strains, while, in the planar-strain
locus, the strain state is chosen to contain only principal
strains and the requisite stress state may contain both
normal and shear components. Hence, the planar-strain locus
cannot always show the complete state of stress necessary to
cause the material to experience only normal strains along
the locus coordinate directions. The planar-stress locus, on
the other hand, will often have associated with it additional
shear-strain components which cannot be represented along
its coordinate axes. When the principal directions of stress
and strain coincide, the planar-stress and planar-strain loci
are identical.

5.4.3 Yield Loci for Single Crystals Deforming by {111}⟨110⟩ or {110}⟨111⟩ Slip

Planar-stress and planar-strain yield loci for single crystals
deforming by {111}⟨110⟩, or, equivalently, {110}⟨111⟩, can
be calculated using existing analytical procedures. Planar-
stress loci can be found directly from the generalized
SCHMID's law [3]. Using the notation of BISHOP and HILL [4],
the generalized Schmid's law can be written as

$$
\begin{aligned}
A \pm G \pm H &= \pm \sqrt{6\,k} \\
B \pm H \pm F &= \pm \sqrt{6\,k} \\
C \pm F \pm G &= \pm \sqrt{6\,k},
\end{aligned}
\tag{1}
$$

where $A = \sigma_{22} - \sigma_{33}$, $B = \sigma_{33} - \sigma_{11}$, $C = \sigma_{11} - \sigma_{22}$, $F = \sigma_{23}$,
$G = \sigma_{31}$, $H = \sigma_{12}$; σ_{ij} is the stress state referred to the
cubic axes; and k is the critical resolved shear stress. If
σ_{xx} and σ_{yy} are the principal stresses acting along the
chosen x and y axes of the single crystal, the stresses σ_{ij}
are found from transformation

$$
\sigma_{ij} = \ell_{ix}\ell_{jx}\sigma_{xx} + \ell_{iy}\ell_{jy}\sigma_{yy}.
\tag{2}
$$

Substituting equation (2) into the generalized Schmid's law

of equation (1) gives the planar-stress yield locus for the single crystal.

Planar-strain yield loci for single crystals deforming by {111}⟨110⟩ slip can be found using the procedure of Bishop and Hill. Now the first step is to express the strains $d\epsilon_{ij}$ referred to the cubic axes in terms of the imposed principal strains $d\epsilon_{xx}$, $d\epsilon_{yy}$, and $d\epsilon_{zz}$ acting in the x, y, and z directions chosen for the particular single crystal. This is done by using the transformation equation

$$d\epsilon_{ij} = \ell_{ix}\ell_{jx}d\epsilon_{xx} + \ell_{iy}\ell_{jy}d\epsilon_{yy} + \ell_{iz}\ell_{jz}d\epsilon_{zz}. \quad (3)$$

The values of $d\epsilon_{ij}$ are then substituted into the work expression

$$dw = -Bd\epsilon_{11} + Ad\epsilon_{22} + 2Fd\epsilon_{23} + 2Gd\epsilon_{31} + 2Hd\epsilon_{12}. \quad (4)$$

The work expression is then evaluated for the 56 permissible Bishop-Hill polyslip stress states, and the operative stress state (or states) which maximizes the work is selected. The normal stresses referred to the textural axes can be found from the Bishop-Hill stress state σ_{ij} referred to the cubic axes by using the transformations

$$
\begin{aligned}
\sigma_{xx} &= \ell_{xi}\,\ell_{xj}\,\sigma_{ij} \\
\sigma_{yy} &= \ell_{yi}\,\ell_{yj}\,\sigma_{ij} \\
\sigma_{zz} &= \ell_{zi}\,\ell_{zj}\,\sigma_{ij}
\end{aligned}
\quad (5)
$$

The additional requirement that σ_{zz} vanish amounts to adjusting the hydrostatic pressure and allows σ_{xx} and σ_{yy} to be determined uniquely. By repeating this procedure for various strain states, a complete planar-strain yield locus can be generated for a particular single crystal orientation or, equivalently, a particular textural component.

Representative planar-stress and planar-strain yield loci are shown in Fig.1. Loci are calculated for three symmetric orientations each for crystals whose z axes lie along the [100], [110], and [111] axes. These components are designated as cube-on-face ((001)[110]), cube-on-edge ((110)[001]), and cube-on-corner ((111)[11$\bar{2}$]) textures because of the orientation of the unit cube relative to the plane of the sheet.

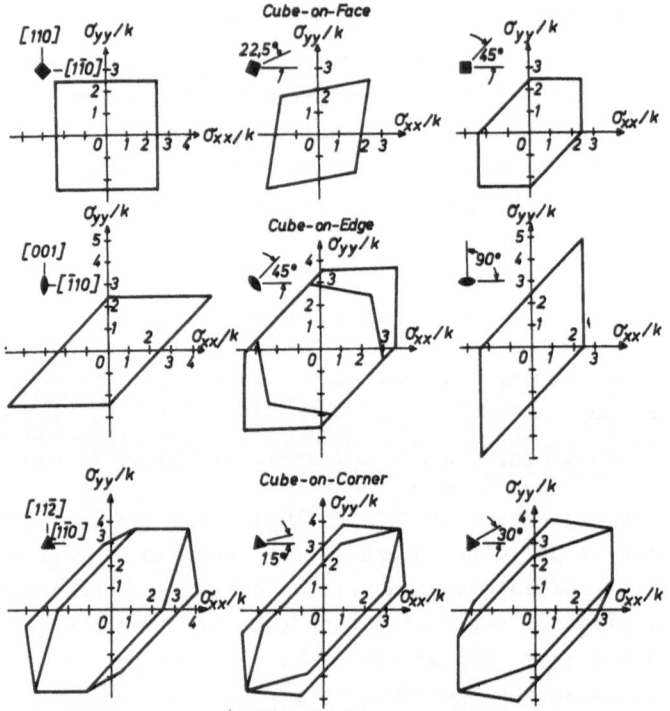

Fig.1. Representative planar-stress (inner) and planar-strain (outer) component yield loci for the cube-on-face, cube-on-edge, and cube-on-corner textures deforming by {111}⟨110⟩ or {110}⟨111⟩ slip. The orientations of the single-crystal components are specified by the rotations indicated

5.4.4 Composite Yield Loci for Metals Deforming by {111}⟨110⟩ or {110}⟨111⟩ Slip

In order to find the yield locus for a textured metal containing more than one textural component, it is necessary to average in some manner the contributions from the various components. The choice of averaging procedure is apparent in the case of planar-strain loci. The values of the average normal stresses acting in the plane of the sheet are found by averaging the stresses necessary to impose the same strain state in all constituent textural components. The average values of the two normal stresses for a given imposed strain state determine the coordinates of a point on the average planar-strain yield locus, and the orientation of the strain

vector, which must be normal to the yield locus, determines
the slope at that point. The normality of the strain vector
to the yield locus can be derived thermodynamically [5] and
can also be shown to follow from the critical resolved shear
stress law for single crystals [4]. Again, additional shear
stresses are often required to enforce the imposed principal-
strain state, and these shear stresses may vary from component
to component subjected to the same strain state, but this
information is not contained in planar-strain yield loci.

An analogous procedure can be used to calculate composite
planar-stress yield loci. The values of the required average
normal stresses, which are principal stresses in this case,
can be found by averaging at the same stress ratio σ_{xx}/σ_{yy}
the principal stresses which are operative in each constituent
textural component. The resulting strain state will not, in
general, be a principal-strain state. Additional shear strains
may be present and the magnitudes of these shear strains may
differ in the various textural components that are being
subjected to the same principal-stress ratio.

The averaging procedures decribed for the composite planar-
strain and a planar-stress yield loci will give upper and
lower bounds to the actual yield locus, much as the TAYLOR
[6] and SACHS [7] average give upper and lower bounds to the
uniaxial yield stress. The upper-bound character of the
composite planar-strain locus is readily apparent, since
enforcing the same strain state in each textural component
immediately satisfies in each component the displacement
boundary conditions imposed on the aggregate. The composite
planar-stress locus, however, must be considered a lower
bound in an average sense. Different crystals, though
subjected to the same stress ratio σ_{xx}/σ_{yy}, may yield at
numerically different levels of σ_{xx} and σ_{yy}. Hence, the
composite planar-stress averages calculated for various stress
ratios will form a lower bound to the actual yield locus
only in an average sense.

5.4.5 Yield Loci for BCC Metals Deforming by ⟨111⟩ Pencil
Glide

Crystallographically based anisotropic yield loci can also

be calculated for BCC metals deforming by pencil glide, i.e.,
slip with equal ease on any plane containing the $\langle 111 \rangle$ slip
directions. In this case, one must make use of the pencil-
glide equivalents of the generalized Schmid's law and the
Bishop-Hill procedure [8]. The pencil-glide equivalent of
the generalized Schmid's law describes yielding on the most
highly stressed plane in each of the four $\langle 111 \rangle$ slip
directions. The pencil-glide equivalent of the Bishop-Hill
procedure allows one to enforce an arbitrary imposed strain
by the simultaneous operation of three or four $\langle 111 \rangle$ slip
directions.

It is interesting to note that the shapes of the crystallographically based yield loci calculated for either $\{110\}$
$\langle 111 \rangle$ slip or $\langle 111 \rangle$ pencil glide do not differ significantly.
This can be conveniently demonstrated from single-crystal
yield loci for orientations in which the principal axes of
stress and strain coincide for both $\{110\}\langle 111 \rangle$ slip and $\langle 111 \rangle$
pencil glide. Four such orientations will be considered:
the 0° and 45° orientations for the cube-on-face texture and
the 0° and 90° orientations for the cube-on-edge as
designated in Fig.1.

The pencil-glide equivalent of the generalized Schmid's law
can be expressed as

$$(A \pm G \pm H)^2 + (B \pm H \pm F)^2 + (C \pm F \pm G)^2 = 9k^2 . \qquad (6)$$

The pencil-glide planar-stress (and planar-strain) yield
loci for the four chosen orientations can be found by
inserting the appropriate stress states found from equation
(1) into equation (8). The shapes of three of the four yield
loci in question will not change. The pencil-glide yield
locus for the 0° orientation of the cube-on-face texture will
remain a sqare, while the yield loci for the 0° and 90°
orientations of the cube-on-edge texture will remain parallelo-
grams. The only change is that the pencil-glide loci are
shrunk as compared to the $\{110\}\langle 111 \rangle$ loci, since, in these
orientations, pencil-glide deformation occurs on $\{112\}$ rather
than $\{110\}$ planes. The pencil-glide yield locus for the 45°
orientation of the cube-on-face texture will be a v. Mises

ellipse inscribed in the Tresca locus for {110}⟨111⟩ slip.
In none of these special cases is the shape of the yield
locus significantly different for either {110}⟨111⟩ slip or
⟨111⟩ pencil glide, a result which is quite general for all
single-crystal and composite yield loci.

References

1. W.A. Backofen, W.F. Hosford, Jr., and J.J. Burke: Trans.
   Amer.Soc.Metals 55(1962)264.

2. R. Hill: Proc.Roy.Soc. A193(1948)281.

3. E. Schmid: Z. Elektrochem. 37(1931)447.

4. J.F.W. Bishop and R. Hill: Phil.Mag. 42(1951) 414 and
   1298.

5. D.C. Drucker: U.S.Nat.Cong.App.Mech. (1951) p. 487.

6. G.I. Taylor: J.Inst.Metals 62(1938)307 - Stephen Timo-
   shenko 60th Anniversary Volume, Macmillan Company, New
   York 1958, p. 218.

7. G. Sachs: Z.VDI 72(1928)734.

8. H.R. Piehler: Sc.D.Thesis, Department of Metallurgy,
   Massachusetts Institute of Technology, Cambridge,
   Massachusetts (1967).

5.5 Relation between Texture and r-Value in Steel Sheets
Zusammenhang zwischen Textur und r-Wert bei Stahlblechen

by S. Nagashima, H. Takechi and H. Kato[+]

Zusammenfassung

Die Verformung von vielkristallinen Aggregaten wurde für den
Fall der Mehrfachgleitung betrachtet. Mit Hilfe des hierauf
aufbauenden Modells konnte die Verformung bei Zugbeanspruchung berechnet und die Beziehungen zwischen dem Parameter
der plastischen Anisotropie $\bar{D}$ und der Textur eingehend erörtert werden.

Für geglühte Proben mit mittleren Verformungsgraden zeigen
die aus der Textur abgeleiteten $\bar{D}$-Werte mit den im Zugversuch gewonnenen recht gute Übereinstimmung. Die Textur von
Blechen mit Kaltwalzgraden zwischen 65 und 70% und einer
Glühtemperatur von 700°C ist durch hohe Belegungsdichte nahe
und bei ⟨111⟩ parallel der Blechnormalen gekennzeichnet.

5.5.1 Introduction

Many studies have been made on the texture of low carbon
steel sheets, since it was shown that preferred orientations
are closely related to deep drawability of steel sheets. The
studies on the relation between deep drawability and preferred
orientation have been developed passing through the following
stages; in the early stage it was confirmed that the plastic
anisotropy had an important role on a press formability of
steel sheets, in the next stage it was clarified that deep
drawability or the r-value of low carbon steel sheet can be
estimated from the preferred orientation, and now we are in
the new stage and are studying to control textures in order
to obtain better deep drawability.

It was a common concept that the isotropic sheet material was
profitable for press forming, until LANKFORD, SNYDER and
BAUSHER [1] had clarified the fact that the steel sheet having

---

[+] Technical Research Institute, Yawata Iron and Steel Co.,
Kitakyushu, Japan

certain preferred orientation showed good press formability.
As the characteristic parameter relating to plastic anisotropy, Lankford et al. proposed the r-value, which is the
ratio of width strain to thickness strain obtained by a
simple tensile test.

BURNS and HEYER [2] also studied on the correlation among
deep drawability, mechanical properties, magnetic and
crystallographic anisotropy, and clarified that a sheet which
had an appropriate orientation showed good drawability and
high r-value. Further, they tried to estimate r-value taking
simple slip process into consideration on the three major
orientations. The result was that cube on corner orientation
is favorable for deep drawing since r-values of this
orientation are in the range between 1.4 to 2.4.

After BURNS et al. [2], HOSFORD [3], the present authors [4],
and VIETH and WHITELEY [5] studied the calculation of r-
value for various orientations of single crystal in
consideration of slip system, r-values along any direction
of single crystalline sheets were calculated.

Further, OKAMOTO, SHIRAIWA and FUKUDA [6,7] tried to
calculate r-valued of polycrystalline steel sheet by adding
up the strains of main components of preferred orientation.

ELIAS, HEYER and SMITH [8] have studied on obtaining r-value
of textured steel sheets by a unique idea to use the (111)
pole figure and iso-r or iso-D nets. In the present paper,
the authors have tried to calculate r-value of a textured
sheet specimen by the following concept: (1) Set up a model
to describe the relation between the resolved shear stress
and the amount of slip on each active slip system, (2)
represent the texture by a distribution of axis density which
is determined directly from the inverse pole figure, (3)
introduce a plastic anisotropy parameter to estimate a
deformation of textured polycrystalline aggregate.

5.5.2 Deformation by Slip in Textured Polycrystalline Aggregate

5.5.2.1 A Model of Deformation in a Polycrystalline Sheet

In order to discuss the deformation in polycrystalline

aggregate on the basis of slip behavior, the following factors must be taken into account: orientation of individual crystallite, influence of neighbours on a crystallite under deformation, and influence of grain boundary.

5.5.2.1.1 Model of deformation of an aggregate. An assumption is made on the correlation between the deformation in an individual crystallite and in macroscopic deformation of the aggregate as a whole.

(a) Strains along the tensile axis ϵ_1 in each crystallite are all identical. While, strains along width and thickness directions ϵ_w and ϵ_t, respectively, are not equal but the average strain $\bar{\epsilon}_w$ and $\bar{\epsilon}_t$ are equal to macroscopic strains; that is,

$$\bar{\epsilon}_1 = \epsilon_1 = E_1, \quad \bar{\epsilon}_t = E_t, \quad \bar{\epsilon}_w = E_w \tag{1}$$

where E_1, E_w, E_t are macroscopic strains.

(b) Change in textures during tensile elongation is neglected. In other words, the rotation of crystallite during deformation is not considered.

(c) It is assumed that the slip behaviour of a crystallite with any orientation may occur by following way. The actual slip may occur when the Schmid factor of any slip system becomes larger than 0.25, and the amount of slip is proportional to the Schmid factor. In other words, if the order of slip system in a crystallite is expressed by suffix n, the amount of slip in each slip system is represented as $d\gamma n$, and the Schmid factor of the system as M_{xn}, we obtain

$$
\begin{aligned}
d\gamma n &= K(M_{xn}-0.25) \quad M_{xn} > 0.25 \\
&= 0 \quad\quad\quad\quad\quad M_{xn} < 0.25
\end{aligned}
\tag{2}
$$

(d) Shear strains γ_{YX}, γ_{ZX}, γ_{ZY} are not taken into consideration. In fact they are not zero in each orientation, but in a polycrystalline aggregate the total sum of them over whole crystallite of all orientations becomes zero because of the symmetry of sheet specimen.

5.5.2.1.2 Introduction of plastic anisotropy parameter D.
Let us define the parameter D as

$$
\left.\begin{aligned}
E_t &= -E_1 \cdot \frac{1-D}{2} \\
E_w &= -E_1 \cdot \frac{1+D}{2}
\end{aligned}\right\}
\tag{3}
$$

where E_1, E_w, E_t are the macroscopic strains along longitudinal, thickness and width directions, respectively. The relation between D-and R-values is

$$
D = \frac{E_t - E_w}{E_1} = \frac{r-1}{r+1}
\tag{4}
$$

$$
r = \frac{1+D}{1-D}
\tag{5}
$$

About a crystallite of any orientation, we can define the same parameter D (α, $\beta:\theta$) as

$$
D(\alpha, \beta : \theta) = \frac{\epsilon_t - \epsilon_w}{\epsilon_1}
\tag{6}
$$

where α, β, θ are variables determining the orientation of a crystallite in the specimen and shown in Fig. 1.

The D(α, $\beta:\theta$) value of a crystallite determined by a set of angles α, β, θ may be derived using above equations. When slip occurs on a n-th slip system, the Schmid factors along X, Y, Z axes are represented by M_{xn}, M_{yn}, M_{zn}, respectively, and we obtain

$$
\epsilon_x = \Sigma M_{xn} \, d\gamma n
$$

$$
\epsilon_y = \Sigma M_{yn} \, d\gamma n
$$

$$
\epsilon_z = \Sigma M_{zn} \, d\gamma n
$$

Summation is made about the whole possible slip systems. From the assumption (c), we get

$$
D(\alpha, \beta:\theta) = \frac{\epsilon_y - \epsilon_z}{\epsilon_x} =
$$

$$
\frac{\sum_n{}'(D_{2n} E_{2n} - D_{3n} E_{3n})(D_{1n} E_{1n} - 0.25)}{\sum_n{}' D_{1n} E_{1n}(D_{1n} E_{1n} - 0.25)}
\tag{8}
$$

Where D, E are direction cosines as shown in Table 1. Subscript n represents n-th slip system. The value of $D_{1n} E_{1n}$ is always taken as positive, and the summation Σ' is made on the values of $D_{1n} E_{1n}$ larger than 0.25 .

Table 1. The direction cosines between specimen axis
coordinates X-Y-Z and slip system

	slip direction	slip plane normal
X	D_1	E_1
Y	D_2	E_2
Z	D_3	E_3

5.5.2.1.3 D-value of textured polycrystalline aggregate. Let us consider the method of representation of D-value when the specimen has preferred orientation. At first, if we consider that the texture of polycrystalline aggregate is expressed by a distribution function $P(\alpha, \beta : \theta)$, and the specimen is deformed by an external force along an axis at an angle ρ from the X axis, as shown in Fig. 2, we obtain D-value from

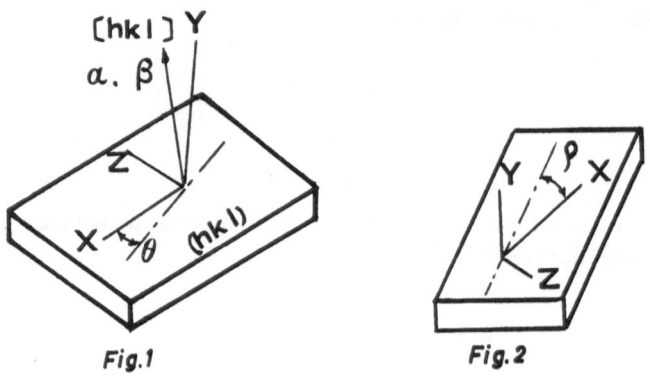

Fig.1

Fig.2

Fig.1. Determination of α, β and θ
Fig.2. External shape and tensile direction

the Eq.(1) based upon the assumption (a),

$$D(\rho) = \frac{E_y - E_z}{E_x} = \left(\frac{\epsilon_y - \epsilon_z}{\epsilon_x}\right) \qquad (9)$$

which leads to the equation

$$D(\rho) = \iiint D(\alpha,\beta:\theta)P(\alpha,\beta:\theta-\rho)d\alpha d\beta d\theta \qquad (10)$$

Next, average value of D with angle ρ is expressed as

$$\bar{D} = \frac{\int D(\rho)d\rho}{\int d\rho} = \frac{1}{\int d\rho} \iiint D(\alpha,\beta:\theta)[\int P(\alpha,\beta:\theta-\rho)d\rho]d\alpha d\beta d\theta \qquad (11)$$

If we define $\bar{P}$ (α,β) and $\bar{D}$ (α,β) as

$$\int P(\alpha,\beta:\theta)d\theta = \bar{P}(\alpha,\beta)$$

$$\frac{\int P(\alpha,\beta:\theta)d\theta}{\int d\theta} = \bar{D}(\alpha,\beta) \qquad (12)$$

We obtain

$$\bar{D} = \iint \bar{D}(\alpha,\beta)\bar{P}(\alpha,\beta)d\alpha d\theta \qquad (13)$$

This means that the individual D-values of a certain crystallographic plane along some orientation cannot be obtained by these equations but only the average value over the values of θ from $0°$ to $360°$ is obtained. Where $\bar{P}(\alpha,\beta)$ is a probability when the crystallographic plane determined by α, β is parallel to the plane of sheet surface, that is, the axis density shown in ND inverse pole figure. In other words, $\bar{D}$ means the integrated value of $\bar{D}(\alpha,\beta)$ applying weight of density about the orientation determined by α, β in inverse pole figure.

In the present work, axis density was obtained by the intensity measurement on about twenty crystallographic planes parallel to sheet surface, since it is rather complicated to compute a precise inverse pole figure. Then Eq. (13) becomes

$$\bar{D} = \Sigma \bar{D}_{hkl} \cdot \bar{P}_{hkl} \qquad (14)$$

5.5.2.2 Calculation of D-value in single crystals of bcc iron [9,10].

If we assume that 48 slip systems which are a combination of

450

⟨111⟩ slip direction and one of {110}, {112} and {123} slip
planes may operate under the same critical shear stress, and
apply the assumption (c) described above, D-value may be
calculated by the use of a computer.

In the present study, the crystallographic plane was defined
by a set of angles α, β as shown in Fig. 3. The D-value of
specimen was calculated and denoted as D (α, β : θ), when the
specimen was stretched in the plane determined by α, β and
along the direction with an angle θ from the reference
direction OX shown in Fig. 3. Where values of α, β and θ were

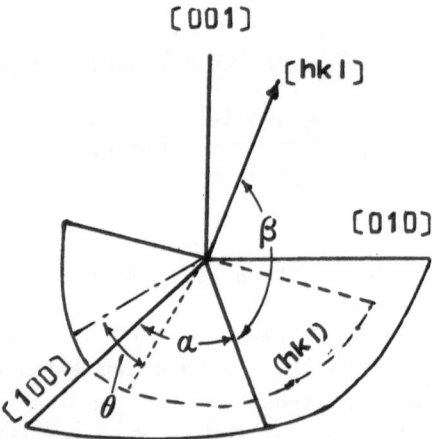

Fig.3. Angles α, β determine the (hkl) plane in the
[100] – [010] – [001] coordinate. θ is an angle from
a standard direction on (hkl) plane

taken as follows, as shown in Fig. 4:

$$\alpha = 0^{\circ}, 9^{\circ}, 18^{\circ}, \ldots\ldots, 45^{\circ}$$
$$\beta = 0^{\circ}, 7^{\circ}, 14^{\circ}, \ldots\ldots, 35^{\circ}$$
$$\theta = 0^{\circ} - 180^{\circ}$$

From the result shown in Fig. 5, change in D (α, β : θ) are
roughly devided into 4 types:

(a) D-values of (111) and the neighbouring planes are
always positive and have maxima at 30°, 90° and 150°

451

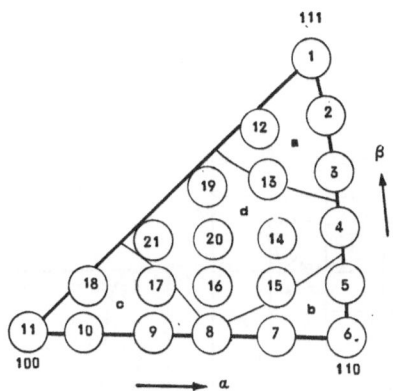

Fig.4. (α,β) representing the position of the pole of crystallographic plane. $\alpha = 0^{\circ}$, 9°, $18^{\circ}.....,45^{\circ}$; $\beta = 0^{\circ}$, 7°, $14^{\circ},.....,35^{\circ}$; $\theta = 0^{\circ} - 180^{\circ}$

of θ, and minima at 60° and 120°.

(b) Planes (110) and the neighbours having maxima at 0° and 180°, submaximum at 90° minima at 45° and 135°.

(c) (100) and the neighbours show zero of D-value even at maximum points where θ is 0°, 90° or 180°, and minima at 45° and 135°.

(d) Planes of intermediate position between three main directions show the change in D $(\alpha, \beta : \theta)$ of averaged value of them.

Since we have got $\bar{D}$ (α, β) as the mean value of D $(\alpha, \beta : \theta)$, we can obtain $\bar{D}$-value from Eq.(8), if we measure the pole density F (α, β) of textured polycrystalline specimen. Values of $\bar{D}$ (α, β) are shown in Fig. 6.

5.5.3 Texture of Low Carbon Steel Sheets

5.5.3.1 Measurement of Inverse Pole Figure

The method of determining inverse pole figure with high precision [11-13] requires laborious numerical calculations. However, the axis densies of the locations corresponding to low Miller indices can be obtained from the measurement of

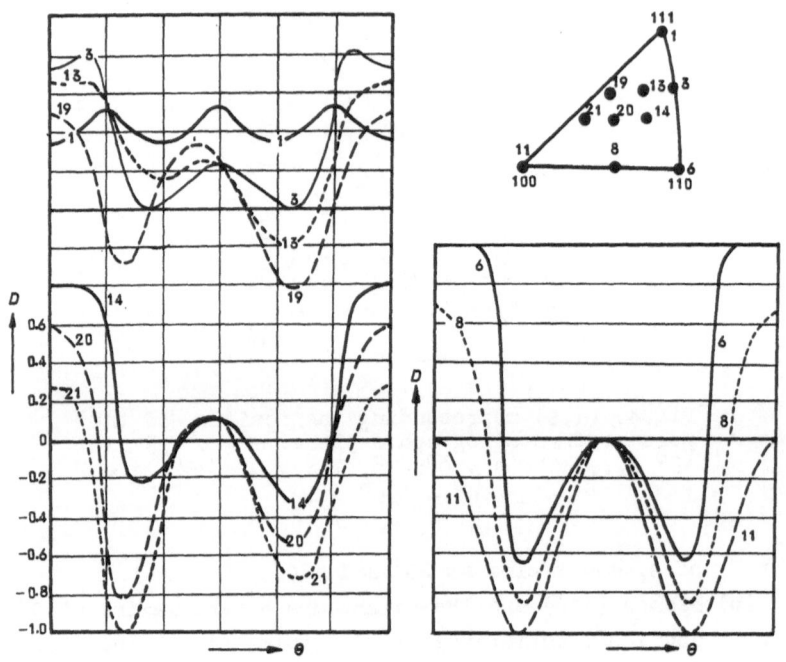

Fig.5. Calculated $D(\alpha,\beta:\theta)$ of b.c.c. iron

diffraction peak intensities. That is, the densities of
particular points in a stereographic triangle can be obtained
from simple method by the use of an X-ray diffractometer.
Therefore, it is possible to estimate an inverse pole figure
with tolerable precision by measuring as many points as
possible.

In order to obtain high index diffraction peaks for inverse
pole figure determination, Zirconium-filtered Mo Kα radiation
was used in conjunction with a scintillation counter and im-
pulse-height analyzer. The integrated intensity of the
diffraction peak was measured and the intensity ratio of the
specimen to random sample was obtained.

Inverse pole figures being determined by the precision method

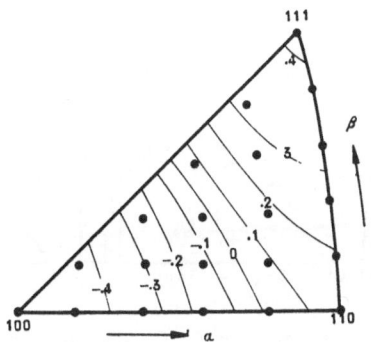

Fig.6. $\bar{D}(\alpha,\beta)$ in stereographic triangle

of JETTER et al., [11] and by the above simple method were
shown in Fig.7 [14]. It is clear that the inverse pole figure
determined by the simple method shows fairly good agreement
with that of precise method.

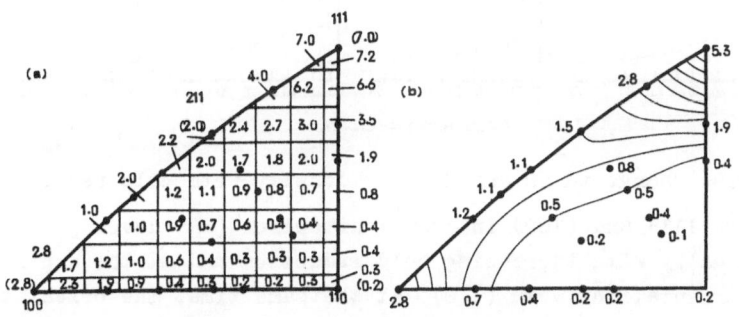

Fig.7a and b. ND inverse pole figures of 70% cold rolled sheet.
a) By precision method,
b) by convinient method

5.5.3.2 Texture of Steel Sheets

5.5.3.2.1 Cold rolled texture. The typical inverse pole figure
of cold rolled sheet with 70 pct reduction is shown in Fig. 7
and 8 [10]. From Fig. 9, it is apparent that the orientations

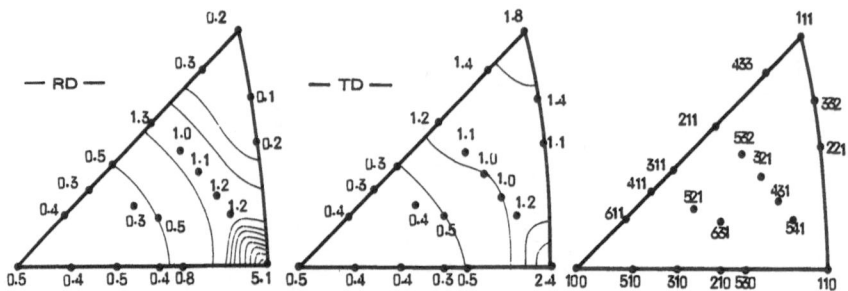

Fig.8. Inverse pole figures of 70% cold rolled sheet

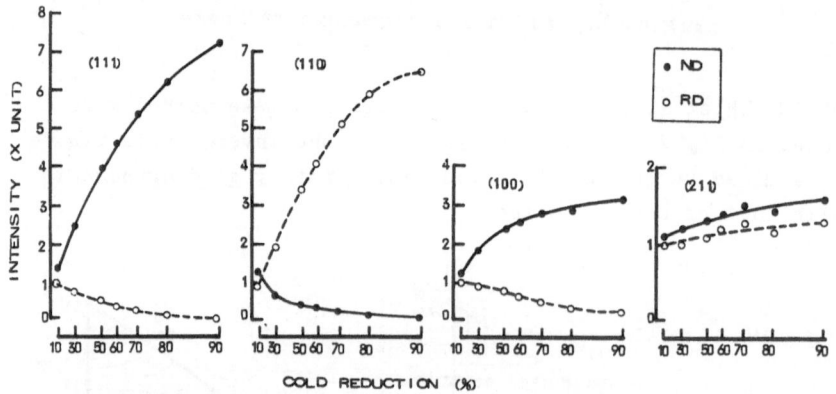

Fig.9. Change in ND and RD axis density with cold reduction

near (111) and (100) in normal direction (ND) develop
gradually with increasing cold reduction and at the expence
of orientations near (110). At the same time, the orientations
near (110) in the rolling direction (RD) develop at the
expence of near (111) and (100). Fig. 10 is a schematic
diagram of the changes in the ND and RD axes density with
increase of cold reduction.

5.5.3.2.2 Annealing texture with medium cold reduction. The
hot rolled sheet was cold rolled 70 pct and annealed at 700°C
for 4hr. In the inverse pole figure shown in Fig. 11, density
of ND pole is high near (111), with a peak at (111), similar
to the cold rolled state. However, the density near (100)

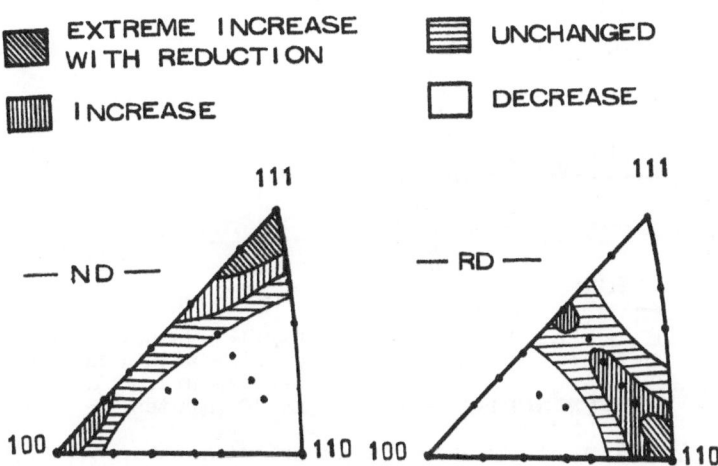

Fig.10. Schematic diagram of change in ND and RD axis
densities with cold reduction

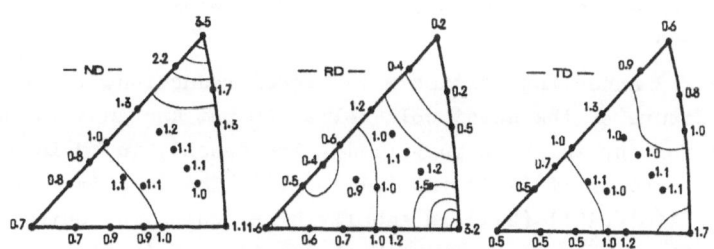

Fig.11. Inverse pole figures of 70% cold rolled sheet annealed
at 700° for 4 hrs

decreases and near (110) increases, contrary to the cold
rolled state. In the RD and TD inverse pole figures, the
density along (100)-(321)-(211) is high, which indicates that
the major component (111) parallel to the sheet surface is
not greatly changed. The decrease of (100) density in ND and
the increase of (100) in RD indicate that the crystals with

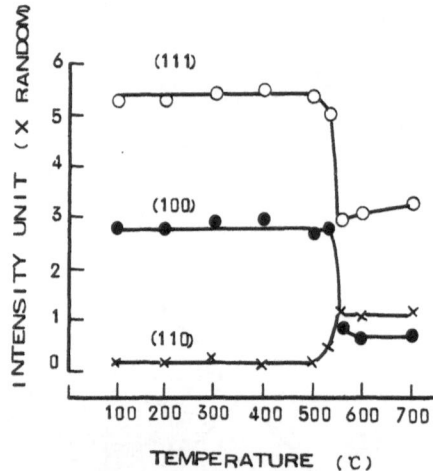

Fig.12. Change in ND axis density with annealing temperature. Holding time is 10 minutes

(100) in the sheet plane change in orientation from (100)[001] to (110)[001].

The change in ND density of principal axes is as follows when the sheet with 70 pct reduction was annealed for 10 min at various temperatures between 100° and 700°C. As shown in Fig. 12, at the recrystallization temperature, densities of (111) and (100) decrease sharply, while the density of (110) increases.

5.5.3.2.3 Annealing texture with higher reduction. The inverse pole figure of the sheet cold rolled 90 pct and annealed at 800°C for 4hr is shown in Fig. 13. The density in ND is high in the regions of (111) and (411)-(311). The density in RD along (110)-(321)-(211) is equally high, which is contrary to the cold rolled state in which the density of (110) is much higher.

Annealing leads to an increase in the density of ND pole figure near (411)-(311) at the expense of (111) and (100).

5.5.4 Estimation of D-value of Steel Sheets

5.5.4.1 D-value of normally Processed Sheets

In order to obtain steel sheets having wide range of r-value, 8 ingots of Al-killed steel were hot rolled and cold rolled with 65 pct cold reduction. Sheets were then annealed with simulated box annealing; heated up to 700°C with various kinds of heating rate, annealed for 4 hrs and cooled.

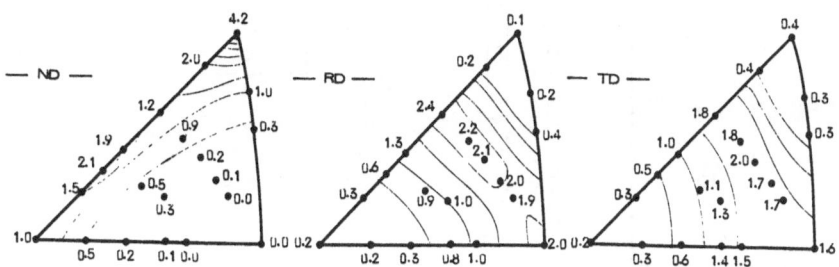

Fig.13. Inverse pole figures of 90%
cold rolled sheet annealed at 800°C
for 4 hrs

The content and heating rate were shown in Table 2. r-values of these specimens were measured by tensile test with 15 pct elongation, and $\bar{D}$-values were calculated by

$$\bar{D} = \frac{1}{12} \{D_0 + D_{90} + 2(D_{15} + D_{30} + D_{45} + D_{60} + D_{75})\} \qquad (15)$$

Where suffices 0, 15, ..., 90 mean the angles between the rolling direction and the tensile axes (Table 3).

There is no clear relation between the Al content and $\bar{D}$, but some tendency is observed; when heating rate is small, $\bar{D}$ decreases with Al content, while by larger heating rate $\bar{D}$ increases with Al as shown in Fig. 14.

As described in above section, $\bar{D}$-values are calculated by Eq. (9). $\bar{D}_{hkl}$ is already known as shown in Table 3. P_{hkl} is just an axis density of (hkl) plane parallel to the sheet surface and is measured easily by an X-ray deffractometer. The texture of these specimens belongs to nearly the same type although $\bar{D}$-values scatter in a wide range. The inverse pole figure of some specimens is shown in Fig. 15. It is clear from Fig. 16 that $\bar{D}_{exp}$ determined by tensile test and $\bar{D}_{cal}$ estimated from axis density have fairly good linear relation.

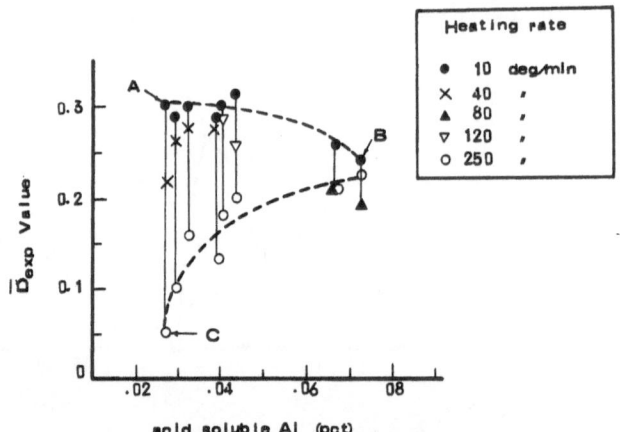

Fig. 14. Relation between $\bar{D}_{exp}$ and content
of acid soluble Al

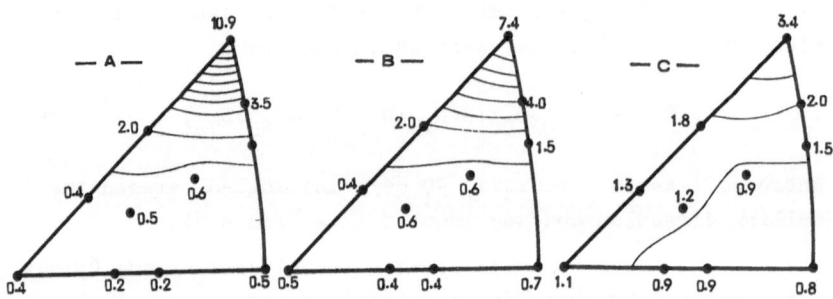

Fig.15. ND inverse pole figures of specimens A,B,C shown in
Fig.14

different, i.e., r_{exp}-values are

$$r_0 = 0.24, \quad r_{45} = 0.61, \quad r_{90} = 0.29,$$
$$\bar{r}_{exp} = 0.43$$

while $\bar{r}_{cal}$ is

$$\bar{D}_{cal} = -0.22, \quad \bar{r}_{cal} = 0.64$$

Table 2. Al (acid soluble) content and heat treatment of specimens

Speci-men	Al %	Heating rate				
		10°C/hr	40°C/hr	80°C/hr	120°C/hr	250°C/hr
31	0.029	31–1	31–4			31–3
33	0.027	33–1	33–4			33–3
34	0.039	34–1	34–4			34–3
47	0.032	47–1	47–4			47–3
48	0.066	48–1		48–5		48–3
49	0.072	49–1		49–5		49–3
54	0.040	54–1			54–2	54–3
57	0.043	57–1			57–2	57–3

Table 3. $\bar{D}_{hkl}$ for main crystallographic plane

hkl	100	110	111	210	211	221	310	321	332	411	521
$\bar{D}_{hkl}$	−0.50	0.16	0.41	−0.17	0.13	0.32	−0.34	0.16	0.37	−0.27	−0.21

5.5.4.2 D-values of Sheet with Special Orientation

As described above, a sheet specimen cold rolled with 90 pct
reduction has a peak at (411)–(311) as high as at (111) (Fig.
13). A specimen which has such a type of texture changes the
preferred orientation during tensile deformation. Fig. 17
shows the change in the inverse pole figure after elongation.
From the pole figure shown in Fig. 18, it is also clear that
crystallites having near (411)–(311) orientation rotate toward
(100). This rotation is described as follows:

$$\text{ND} : (311) \rightarrow (411) \rightarrow (611) \rightarrow (100)$$
$$\text{RD} : [13\bar{6}] \simeq [01\bar{1}] \pm 25° \rightarrow [01\bar{1}]$$

Since $\bar{D}$-value of (100) is low, it is supposed that $\bar{D}_{exp}$
obtained by tensile test becomes smaller than $\bar{D}_{cal}$ estimated
from texture. In fact, $\bar{D}_{exp}$ and $\bar{D}_{cal}$ of above specimen are

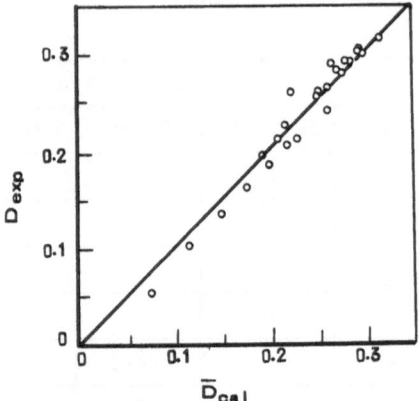

Fig.16. Relation between calculated $\bar{D}$-value ($\bar{D}_{cal}$) and one obtained from tensile test ($\bar{D}_{exp}$)

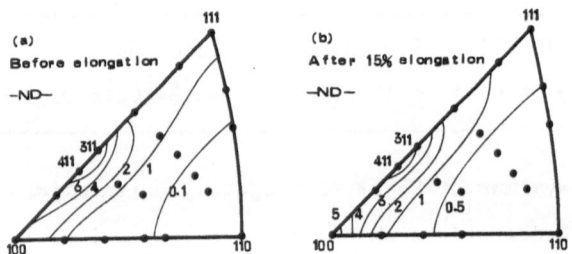

Fig.17 a and b. Change in ND inverse pole figure during tensile deformation. Cold rolled 97%, annealed at 850°C for 4 hrs

5.5.5 Discussion

5.5.5.1 Relation between $\bar{D}_{exp}$ obtained by Tensile Test and $\bar{D}_{cal}$ estimated from Texture

Experimental data on the Al-killed steel showed a fairly good agreement between $\bar{D}$-values obtained from two kinds of method, about the range of $\bar{D}$-value 0.05 and 0.3.

Further, the comparison of $\bar{D}$-values obtained by tensile test,

461

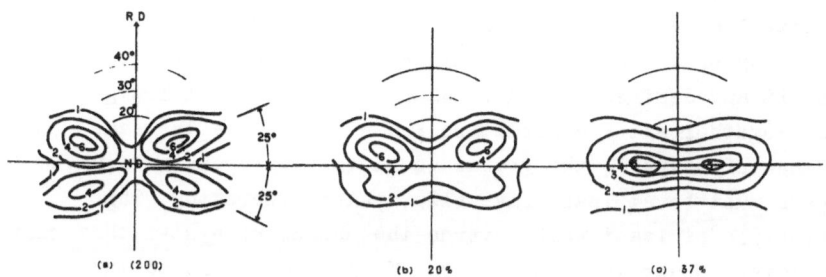

Fig. 18 a - c. Change in (200)-pole figure during tensile deformation. The same specimen as in Fig. 17

a) Before elongation,
b) and c) after elongation

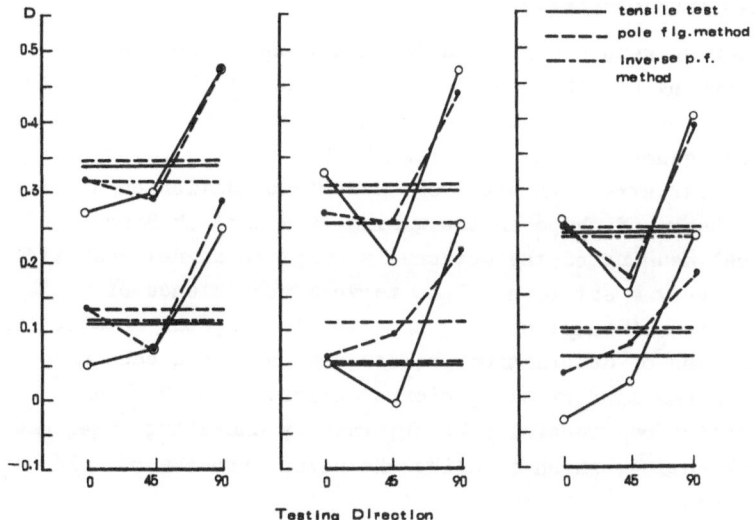

Fig. 19. Comparison of D-values obtained by conventional pole figure method and inverse pole figure method with those of tensile test

conventional pole figure method [9], and the present method [10,11] was made by AOKI et al [16]. Fig. 19 shows the result, from which it is concluded that D-values obtained by each method agree fairly well.

5.5.5.2 Discussion of the Assumption on Deformation by Slip

TAYLOR [17] has studied on the plastic deformation of polycrystalline metals having fcc lattice by a more rigorous way
than any previous one. The main point of his theory is: "To
permit any desired change of shape - such as will let grains
fit together after deformation and will produce the same
change of shape in the grains as in the aggregate as a whole -
there must be at least five slip systems operating. The
principle of least work governs the choice of system that must
operate." Under this assumption, Taylor has obtained the
crystal rotation during tensile and compressive deformation,
and also obtained the mean Schmid factor.

In the present study, bcc metal which has four times as many
slip systems as in fcc metal is under consideration, and
further anisotropic parameter D must be taken into account.
Therefore, the calculation making use of Taylor's theory
becomes much complex.

In order to keep clear the difficulty, assumption (c)
described by Eq. (2) was applied in stead of the "principle
of least work" which has a definite physical meaning, and also
applied assumption (a) described by Eq. (1) and (d) which
leave shear strain in stead of Taylor's condition on continuity
"same change of shape in the grains as a whole." Since the
physical meaning of the authors assumption is not definite,
this is only a set of model to serve a convenience of
calculation. Assumption (b), ignoring the crystal rotation in
the process of deformation, holds good only in a case when the
specimen has appropriate preferred orientation. Such an
orientation may generally be obtained by annealing after medium
cold reduction, which is alike the normal process of cold strip
production.

Acknowledgements

The authors are indebted to Mr. N. Takahashi in making sheet
with variety of r-values by his excellent idea, to Mr. Shimizu
and Drs. Aoki, Hayami and their colleagues, and also to
authors' colleagues for helpful discussions relating the
estimation of D-value. The authors express their deep thanks

to Messrs. R.H. Heyer and J.A. Elias and their colleagues
in Armco Steel Corp. for the many suggestions and discussions
relating to this work.

References

1. W.J. Lankford, S.C. Snyder and T.A. Bausher: Trans.Amer.
   Soc.Metals 42(1950)1197.

2. R.S. Burns and R.H. Heyer: Sheet Metal Ind. 35(1958)261.

3. W.F.Hosford: Trans.Met.Soc. AIME 227(1963)272.

4. S. Nagashima, S. Sekino and H. Kato: J.Jap.Inst. Metals
   27(1963) - Trans.JIM. 5(1964)244.

5. R.W. Vieth and R.L. Whiteley: 3rd International Colloquium
   of IDDRG, London 1964.

6. T. Okamoto, T. Shiraiwa and M. Fukuda: Sumitomo Metals
   14(1962)211.

7. T. Okamoto, T. Shiraiwa and M. Fukuda: 3rd International
   Colloquium of IDDRG, London (1964).

8. J.A. Elias, R.H. Heyer and J.H. Smith: Trans.Met.Soc. AIME
   224(1962)679.

9. S. Nagashima, H. Takechi and H. Kato: Trans.JIM. 5(1964)
   274 - J.Jap.Inst. Metals 29(1965)393.

10. H. Kato, H. Takechi and S. Nagashima: J. Japan Soc.
    Technology Plasticity 7(1966)13.

11. L.K. Jetter, C.J. McHargue and R.O. Williams: J.Appl.Phys.
    27(1956)368.

12. H.J. Bunge: Z.Metallkde. 51(1960)535.

13. R.J. Roe: J.Appl.Phys. 36(1965)2024.

14. H.Takechi: J.Japan Soc. Technology Plasticity: To be
    published.

15. H. Takechi, H. Kato and S. Nagashima: Trans.Met.Soc. AIME
    242(1968)56.

16. K. Aoki, S. Hayami and M. Matsuo: Advances in X-Ray
    Analysis, Vol.10(1967) Plenum Press, p. 342.

17. G.I. Taylor: J.Inst.Metals 62(1938)307.

5.6 Die Bedeutung des r-Wertes für das Tiefziehverhalten von Blechen
Significance of r-Value for the Behaviour of Sheets during Deep Drawing

von W. Panknin[+)]

Summary

The investigations have demonstrated that an influence of the
material on the maximum deep drawing ratio is only possible
about the normal anisotropy (r-value) when a certain lowest
elongation exists which is necessary for the formation of
the bottom. The r-value affects mainly the fraction stress
in the bottom region. On the contrary there is only a little
influence on the required drawing stress. With increasing r-
value the fraction stress in the bottom and therewith the
maximum transferable load will increase, for that very reason
because a later possibly appearing fracture will be displaced
outwards on account of the stress composition during the
formation of the bottom. In the case of an optimum bottom
tearer, a greatest possible load transmission, and therewith
a greatest possible maximum deep drawing ratio, is reached.
In this region an influence of the r-value seams not to exist.
For materials with low r-value, however, it is not possible
to get an optimum bottom tearer.

The external displacement of the fraction with increasing
r-value is superimposed with the strain hardening and friction.
With growing strain-hardening exponent the fracture will
displace inwards, if the difference in strength existing in
the bottom cannot be received by friction contact.

The assumption having plane strain taken as a basis for the
mathematical determination of the fraction stress in the
bottom seems to be inadmissible, because the fraction set in
normally on another point.

The criteria for plastic flow for anisotropic materials by

[+)]Institut für Verformungskunde, Technische Universität Berlin

Hill seems to reply correctly the material condition in
biaxial tension, while a different condition was found out
for combined tensile and compression stresses. Appearing
dispersion in the results for the relation between the
maximum drawing-ratio and the r-value are attributed partly
thereon, that the point of fraction and therewith the load
carrying capacity is not only influenced by the r-value but
also by the strain hardening and friction coefficient.

Das Tiefziehen ist ein in der Blechverarbeitung häufig ange-
wendetes Umformverfahren. Es werden unter diesem Begriff verschiedene Verfahren verstanden, die sich in der Werkstoffbe-
anspruchung unterscheiden. Das Umformen eines ebenen Bleches
zu einem Hohlkörper kann entweder durch Streckziehen oder
durch Tiefziehen erfolgen. Beim Streckziehen liegt eine vorwiegend ein- oder zweiachsige Zugbeanspruchung in der Blechebene vor. Der Werkstoff fließt aus der Dicke in tangentialer
und radialer Richtung. Auf diese Weise wird das Blech z. B.
über einen Stempel gereckt und nimmt so dessen räumlich ge-
krümmte Form an. Die Grenze für die maximal mögliche Dehnung
ist neben den geometrischen Einflüssen und der Reibung vom
Werkstoff abhängig. Die Werkstoffkenngröße, die die mögliche
Dehnung am meisten beeinflußt, ist die Spannungsverfestigungsfunktion, sofern es sich um schmeidige Werkstoffe handelt.
Die Spannungsverfestigungsfunktion wird in der sogenannten
Fließkurve (k_f = f (φ)) dargestellt und kann durch den Verfestigungsexponenten n erfaßt werden, wenn die Fließkurve der
Gleichung k_f = a . φ^n genügt.

Beim Tiefziehen (Abb. 1) wird die Ziehkraft vom Stempel über
den Boden und die Zarge in die Formgebungszone übertragen.
Die Umformung findet zwischen Ziehring und Blechhalter unter
vorwiegend radialen Zug- und tangentialen Druckspannungen
statt. Mit im Verhältnis zum Stempeldurchmesser d_o zunehmendem Zuschnittdurchmesser D_o (Ziehverhältnis β_o = D_o/d_o) steigt
die erforderliche Ziehkraft an. Das größtmögliche Ziehverhältnis $\beta_{o\ max}$ (Grenzziehverhältnis) ist erreicht, wenn die
erforderliche Ziehkraft etwas kleiner ist als die Bruchfestigkeit des Bodens, da ein etwas größerer Zuschnittdurchmesser zum Bodenreißer führen würde.

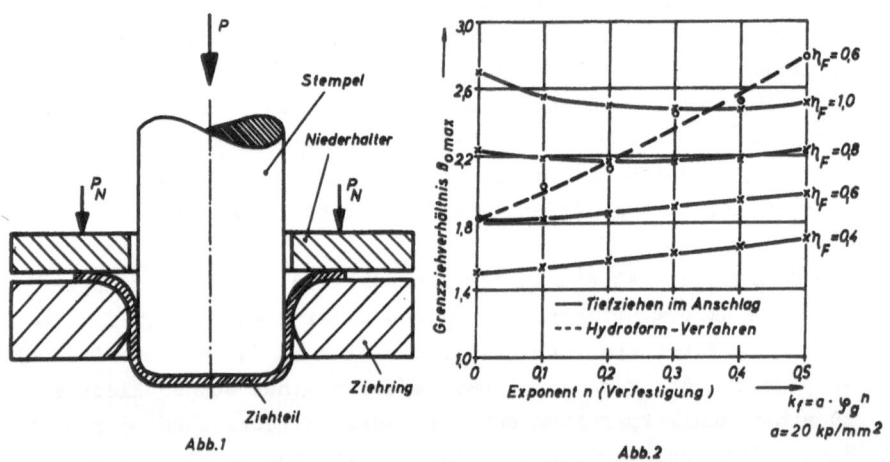

Abb.1   Abb.2

Abb.1. Tiefziehen im Anschlag

Abb.2. Einfluß des Verfestigungsexponenten n auf das Grenzziehverhältnis ß$_{o\ max}$ beim Tiefziehen im Anschlag bei verschiedenen Wirkungsgraden η$_F$

Die folgenden Ausführungen sollen sich auf die Bedeutung des
Werkstoffs auf das maximal mögliche Ziehverhältnis beim Tiefziehen zylindrischer Teile beschränken. Dabei soll lediglich
der Fall des Bodenreißers betrachtet werden, nicht dagegen
der Einfluß von Versprödung usw., der zu Trennungsbrüchen in
Form von Umfangs- oder Längsreißern in den am stärksten verformten Zonen führen kann. Die Einflüsse des Streckziehens
sollen soweit behandelt werden, wie sie für die Ausbildung
des Bodens und die Beanspruchungsverhältnisse im Boden wichtig sind. Auf Werkzeuggeometrie und Schmierung soll nicht
eingegangen werden, soweit sie nicht für die Deutung von
Werkstoffeinflüssen wichtig sind.

Das Grenzziehverhältnis eines Werkstoffs wird um so größer
sein, je mehr die Bodenfestigkeit gegenüber der erforderlichen Ziehkraft angehoben wird. Es ist daher zu untersuchen,
wie sich diese Kräfte bzw. Spannungen durch Werkstoffeinflüs-
se verändern. Es soll zunächst betrachtet werden, welchen
Einfluß die Fließkurve bzw. der Verfestigungsexponent aus-

übt. Die rechnerischen Ergebnisse sind in Abb. 2 wiedergegeben [1]. Es zeigt sich, daß der Einfluß des Verfestigungsexponenten klein ist. Jedoch muß ein bestimmter Mindestwert
vorhanden sein, damit der Boden ohne vorzeitigen Reißer geformt werden kann. (Dieser Fall wurde im dargestellten Bild
nicht berücksichtigt.) Bei relativ dickwandigen Teilen, wie
z.B. bei der Näpfchenprüfung ($\eta_F = 0,8$), ist sogar kein Einfluß des Verfestigungsexponenten zu erkennen.

Der Verfestigungsexponent geht demnach bei isotropem Werkstoff etwa gleichstark in die erforderliche Ziehkraft und
die Bodenabreißkraft ein und wirkt sich deshalb nicht im
Grenzziehverhältnis aus. Experimentelle Ergebnisse bestäti-
gen die Rechnung, sofern bestimmte Mindestwerte für den Verfestigungsexponenten vorhanden sind [1,2].

Bei anisotropem Werkstoff wird die Vergleichsspannung, bei
der Fließen einsetzt, von der Art der Beanspruchung abhängig. Da Boden und Flansch eines Ziehteils unterschiedlich
beansprucht werden, ist es möglich, daß infolge Anisotropie
verschiedene Fließkurven wirksam werden und so das Grenzziehverhältnis beeinflußt werden kann. Falls z. B. die Bodenfestigkeit gegenüber der erforderlichen Ziehkraft angehoben wird, müssen sich größere Ziehverhältnisse erzielen
lassen. Auf diese Gesetzmäßigkeiten soll im folgenden Teil
eingegangen werden.

Wenn isotrope Bleche im Zugversuch verformt werden, ist die
Breitenformänderung gleich der Dickenformänderung. Bei anisotropen Werkstoffen dagegen ergeben sich Unterschiede, wenn
Anisotropie zwischen Blechebene und Blechebenennormale vorhanden ist. Diese Anisotropie wird vielfach durch den r-Wert
angegeben, der das Verhältnis von logarithmischer Breitenformänderung φ_b zur logarithmischen Dickenformänderung φ_s im
einachsigen Zugversuch angibt.

$$r = \frac{\varphi_b}{\varphi_s} = \frac{\varphi_b}{\varphi_l - \varphi_b} \tag{1}$$

Falls die Bestimmung der Dickenformänderung infolge der auftretenden Meßfehler zu ungenau wird, kann sie über die loga-
rithmische Längenformänderung φ_l bestimmt werden.

Experimentelle Untersuchungen haben ergeben, daß das Tief-
ziehverhalten vom r-Wert abhängig ist [3]. Abb. 3 gibt den
Zusammenhang zwischen r-Wert und Grenzziehverhältnis vorwiegend von beruhigten und unberuhigten Stahlblechen wieder
[4]. In Abb. 4 ist diese Abhängigkeit vorwiegend für Nicht-
eisenmetalle aufgetragen [5]. In beiden Fällen wurde das
Grenzziehverhältnis mit verhältnismäßig kleinen Stempeldurch-
messern bestimmt (d_o = 50 bzw. 33 mm), so daß der Reibungs-
einfluß nicht sehr groß ist und die Grenzziehverhältnisse
vergleichbar sind.

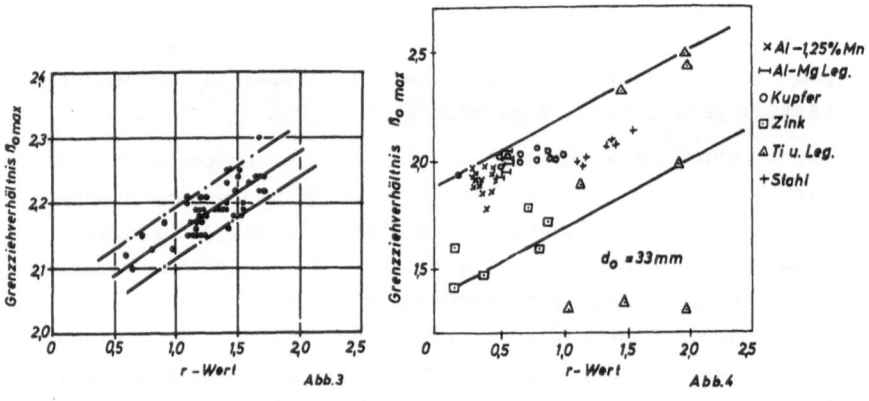

Abb.3. Einfluß des r-Wertes auf das Grenzziehverhältnis beim
　　　　Tiefziehen zylindrischer Teile
　　　　(Stempeldurchmesser $d_o \cong$ 51 mm)

Abb.4. Einfluß des r-Wertes auf das Grenzziehverhältnis beim
　　　　Tiefziehen zylindrischer Teile
　　　　(Stempeldurchmesser d_o = 33 mm)

Den Bildern kann entnommen werden, daß das Grenzziehverhältnis zwar vom r-Wert abhängig ist, daß aber ein großer Streubereich vorhanden ist, selbst wenn die in Abb. 4 herausfal-
lenden drei Werkstoffe nicht berücksichtigt werden. Bei diesen drei Werkstoffen liegt wegen ungünstiger Verfestigungsexponenten vermutlich ein anderer Versagensfall vor (vorzeitiger Reißer).

Es erhebt sich nun die Frage, wie sich der Einfluß des r-
Wertes auf das Grenzziehverhältnis erklären läßt. Wenn man
gemäß WHITELEY [6] und HOLCOMB und BACKOFEN [7] annimmt, daß

der Flansch des Ziehteils gemäß Abb. 5 unter Zug-Druck-Beanspruchung steht und die Blechdicke etwa konstant bleibt (ebene Formänderung), erhält man unter Berücksichtigung der Fließbedingung für anisotropes Material [6,10] (Abb. 6) für einen ideal-plastischen Werkstoff [7,8]:

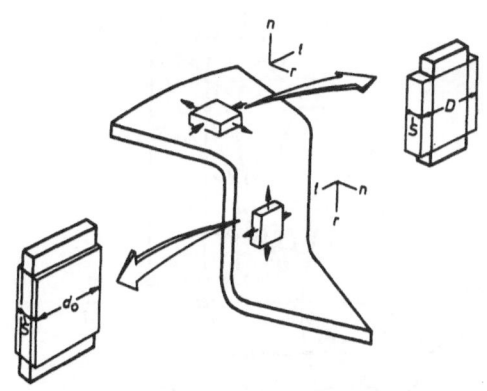

Abb.5. Schematische Darstellung des Spannungs- und Formänderungszustandes für die Berechnung des Grenzziehverhältnisses gemäß Gleichung (2), (3) und (5)

$$\sigma_{r\ max} = (1 + \eta) \sqrt{\frac{2\ (1+r)}{1 + 2r}} \cdot k_f \cdot \ln \frac{D_o}{d_o} \qquad (2)$$

Darin bedeuten $\sigma_{r\ max}$ = größte erforderliche Ziehspannung, η = Reibungsanteil, bei Näpfchenversuch η = 0,2 bis 0,3, r = r-Wert (Mittelwert), k_f = Formänderungsfestigkeit im einachsigen Zugversuch, D_o = Zuschnittdurchmesser, d_o = Stempeldurchmesser. Für die Berechnung der Bruchspannung des Bodens wird ebenfalls ideal-plastischer Werkstoff vorausgesetzt, so daß der Fließbeginn der Bruchspannung σ_{Br} gleichgesetzt werden kann. Unter der Voraussetzung eines zweiachsigen Zugspannungszustandes (Radial- und Tangentialspannung) und ebener Formänderung gemäß Abb. 5 (tangentiale Formänderung = 0) erhält man

$$\sigma_{Br} = \sqrt{\frac{(1 + r)^2}{1 + 2r}} \cdot k_f \qquad (3)$$

Für den Fall, daß das Ziehverhältnis gerade über dem Grenz-

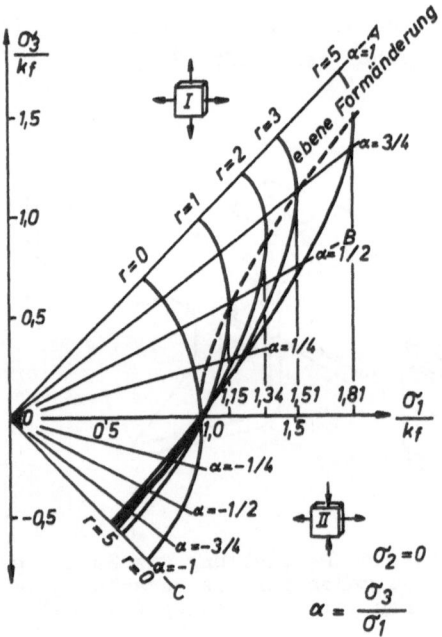

Abb.6. Fließbedingungen nach Hill für Anisotropie
zwischen Blechebene und Blechebenennormale

ziehverhältnis liegt, erhält man die Bodenabreißbedingung:

$$\sigma_{r\ max} = \sigma_{Br} \tag{4}$$

Dann wird

$$\ln \left(\frac{D_o}{d_o}\right)_{max} = \frac{1}{1 + \eta} \sqrt{\frac{r + 1}{2}} \tag{5}$$

Dabei soll $\beta_{o\ max} = \left(\frac{D_o}{d_o}\right)_{max}$ noch als Grenzziehverhältnis
bezeichnet werden, obwohl es bereits den beginnenden Boden-
reißer angibt.

In Abb. 7 sind die Grenzziehverhältnisse gemäß Gleichung (5)
für verschiedene r-Werte dargestellt (ausgezogene Linie). Die
Kurve für den Reibwert $\eta = 0$ entspricht dem reibungsfreien

Fall, während $\eta = 0,25$ etwa der Näpfchenprüfung und damit den Versuchsbedingungen der Abb. 3 und 4 entspricht. Die eingezeichneten Mittelwertkurven von Abb. 3 und 4 sind punktiert bzw. strichpunktiert dargestellt. Es zeigt sich, daß die Versuchswerte flacher verlaufen, als die Rechnung gemäß Gleichung (5) ergibt.

Es soll darum untersucht werden, ob die Berücksichtigung der Verfestigung eine andere Tendenz ergibt. Man erhält dann für die erforderliche Ziehspannung beim Ziehkraftmaximum für übliche Werkstoffe [1]

$$\sigma_{r\,max} = \frac{k_{f_m}}{\eta_F} \ln \frac{D^x}{d_o} \sqrt{\frac{2(1+r)}{1+2r}} = \frac{k_{f_m}}{\eta_F} \ln \sqrt{0,6\,\beta_o^2 + 0,4} \cdot$$

$$\sqrt{\frac{2(1+r)}{1+2r}} \qquad (6)$$

Für die Abreißspannung kann man näherungsweise setzen:

$$\sigma_{Br} = 1,15 \cdot \sigma_B \sqrt{\frac{(1+r)^2}{1+2r}} \qquad (7)$$

Damit ergibt sich gemäß Gleichung (4) für das Grenzziehverhältnis:

$$\ln \sqrt{0,6\,\beta_{omax}^2 + 0,4} = \eta_F \cdot \frac{1,15 \cdot \sigma_B}{k_{f_m}} \sqrt{\frac{r+1}{2}} \qquad (8)$$

In diesen Gleichungen bedeuten: D^x = Flanschdurchmesser beim Ziehkraftmaximum, k_{f_m} = mittlere Formänderungsfestigkeit im Ziehteilflansch im Augenblick des Kraftmaximums, η_F = Formänderungswirkungsgrad. Das Verhältnis σ_B/k_{f_m} ist vom Werkstoff und dem Ziehverhältnis abhängig. Bei üblichen Fließkurvencharakteristiken ist vorwiegend das Ziehverhältnis von Einfluß. In Abb. 7 ist das Grenzziehverhältnis gemäß Gleichung (8) gestrichelt dargestellt für den reibungsfreien Fall ($\eta_F = 1,0$) und den Näpfchenversuch ($\eta_F = 0,8$). Es zeigt sich, daß sich kein wesentlicher Unterschied gegenüber dem ideal-plastischen Werkstoff ergibt.

Um die Unterschiede zwischen Rechnung und Versuch näher zu untersuchen, soll zunächst die Gültigkeit von Gleichung (6)

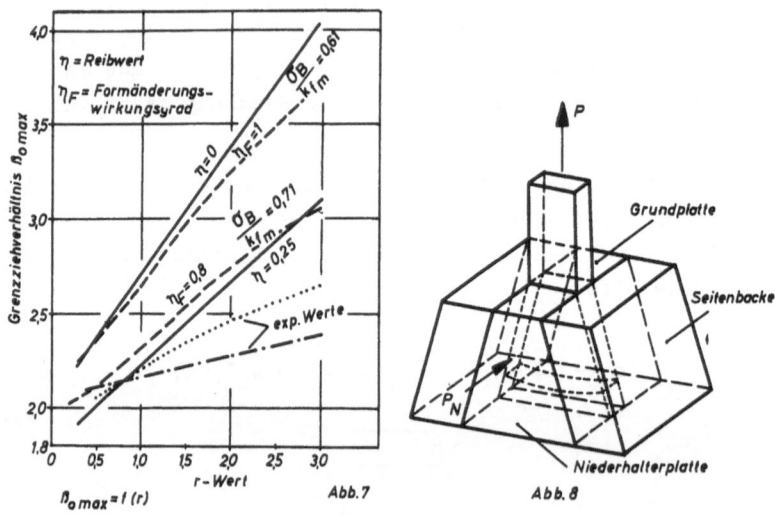

Abb.7. Einfluß des r-Wertes auf das Grenzziehverhältnis, Vergleich von rechnerischen und experimentellen Werten

Abb.8. Keilziehversuch nach Sachs

überprüft werden. Da jedoch der Tiefziehversuch wegen der
auftretenden Reibungs- und Biegeverluste schwer auswertbar
ist, wurden Keilziehversuche nach SACHS [9] durchgeführt
(Abb. 8). Die Meßwerte wurden durch eine geeignete Meßschal-
tung von der Reibung kaum beeinflußt. Auf diese Weise konnten die Ziehspannungen direkt ermittelt werden. Diese Werte
wurden mit den rechnerischen nach Gleichung (6) für $\eta_F = 1$
verglichen. Für die mittlere Formänderungsfestigkeit k_{f_m}
wurden Fließkurven zugrunde gelegt, die unter verschiedenen
Beanspruchungen ermittelt worden waren [2]. Abb. 9 gibt die
Ergebnisse für Ms 63 wieder. Da für diesen Werkstoff $\bar{r} = 1,0$
war, ergaben sich auch bei den verschiedenen Beanspruchungen
gleiche Fließkurven (Abb. 10). Auch die rechnerisch und experimentell bestimmten maximalen Ziehspannungen fallen zusammen. Bei einem beruhigten Stahlblech dagegen ergaben sich je
nach der Beanspruchung unterschiedliche Fließkurven (Abb.11).
Der zweiachsige Zugversuch und der Abwalzversuch ergaben gegenüber dem einachsigen Zugversuch eine Hochlage. Der mitt-

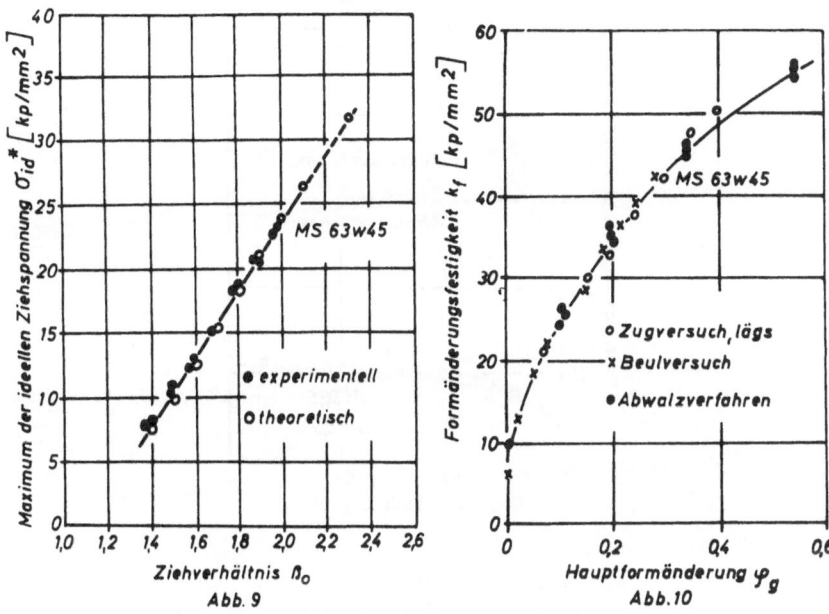

Abb.9. Das Maximum der ideellen Ziehspannung in Abhängigkeit
vom Ziehverhältnis beim Keilziehversuch an Messingblech

Abb.10. Fließkurve von Ms 63 nach verschiedenen Ermittlungs-
verfahren

lere r-Wert dieses Werkstoffs war $\bar{r} = 2,1$. Beim Vergleich
der gemessenen und gerechneten maximalen Ziehspannung (Abb.
12) fällt auf, daß die gemessene Ziehspannung höher liegt,
als die, die sich aus der Fließkurve bei einachsiger Zugbeanspruchung ergibt. Wenn man den r-Wert gemäß Gleichung (6)
berücksichtigt, müßte die Ziehspannung sogar unter der liegen, die aus dem Zugversuch berechnet worden ist (gestrichelte Linie). Es bleibt späteren Untersuchungen vorbehalten zu
klären, worauf diese Unterschiede zurückzuführen sind. Insbesondere soll untersucht werden, ob die Fließbedingungen
für anisotropes Material die Verhältnisse bei verschiedenen
Beanspruchungsarten richtig wiedergeben.

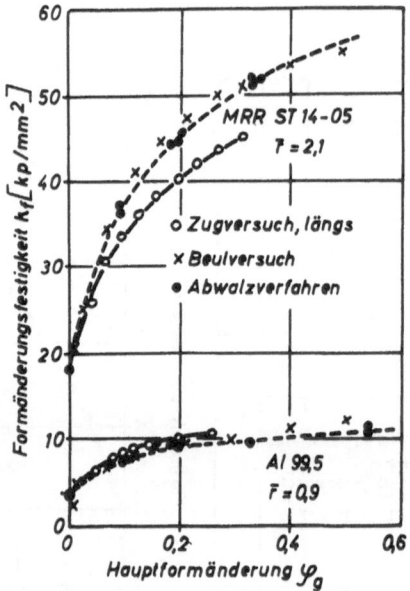

Abb.11. Fließkurven von Stahl und
Aluminium aus verschiedenen Er-
mittlungsverfahren

In Abb. 13 sind die rechnerischen und experimentellen Werte
für Al 99,5 dargestellt. Die Übereinstimmung der Werte ist
gut, wenn man die Zugfließkurve zugrunde legt. Da $\bar{r}$ = 0,88
war, ergibt sich noch eine geringe Verschiebung der rechnerischen Werte zu den gemessenen hin, wenn der R-Wert berücksichtigt wird.

Wenn man die in den Keilziehversuchen für ein Ziehverhältnis
ermittelten maximalen Ziehspannungen verschiedener Werkstoffe
auf die aus der Zugfließkurve ermittelten ideellen Ziehspannungen bezieht und über dem r-Wert aufträgt (Abb. 14), ist
zu ersehen, daß die Meßwerte vom r-Wert in dem untersuchten
Bereich nicht abhängig zu sein scheinen. Daher dürfte der
Einfluß des r-Wertes auf das Grenzziehverhältnis von der
Ziehspannung her gesehen nicht groß sein. Jedoch ist anzunehmen, daß bei kleinen r-Werten mit einem Anstieg der Ziehspannung gerechnet werden muß.

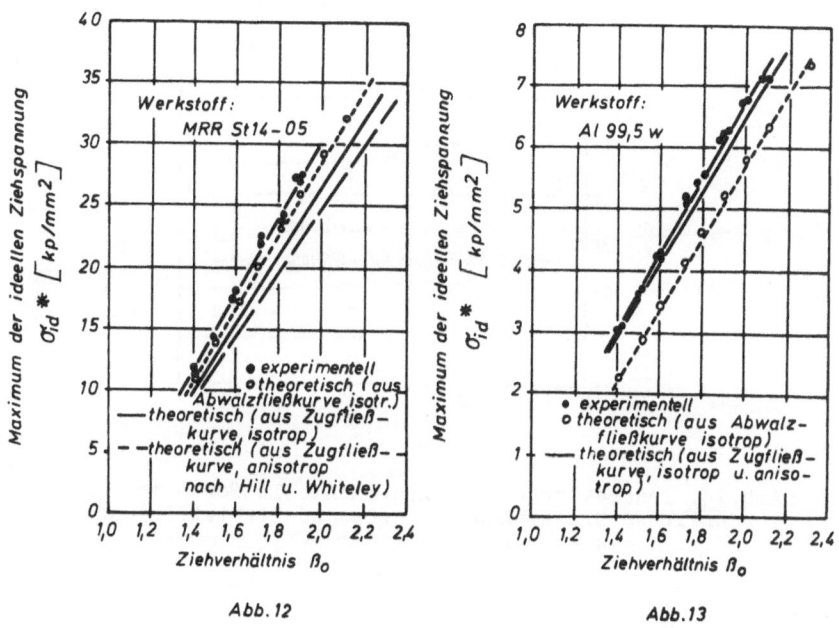

Abb. 12

Abb. 13

Abb.12. Das Maximum der ideellen Ziehspannung in Abhängigkeit vom Ziehverhältnis beim Keilziehversuch an Stahl

Abb.13. Das Maximum der ideellen Ziehspannung in Abhängigkeit vom Ziehverhältnis beim Keilziehversuch an Aluminium

Im folgenden Teil soll der Einfluß des r-Wertes auf die Bodenabreißkraft untersucht werden. Hierbei sind zunächst drei
verschiedene Bodenreißer zu unterscheiden (Abb. 15). Rechts
im Bild ist ein vorzeitiger Reißer dargestellt. Er tritt ein,
bevor sich der Boden voll ausgebildet hat. Die Ursache ist in
zu kleiner Dehnfähigkeit und damit zu kleinem Verfestigungsexponenten zu suchen. Da die übertragbaren Kräfte bei dieser Versagensart klein sind, ergeben sich auch nur kleine
Grenzziehverhältnisse. Das Ziehverhältnis ist in diesem Bereich vom Verfestigungsexponenten abhängig. Wenn der Werkstoff jedoch eine von der Stempelgeometrie und der Schmierung
abhängige Mindestdehnung erreicht, treten die beiden links
im Bild dargestellten Bruchformen auf. Der optimale Reißer

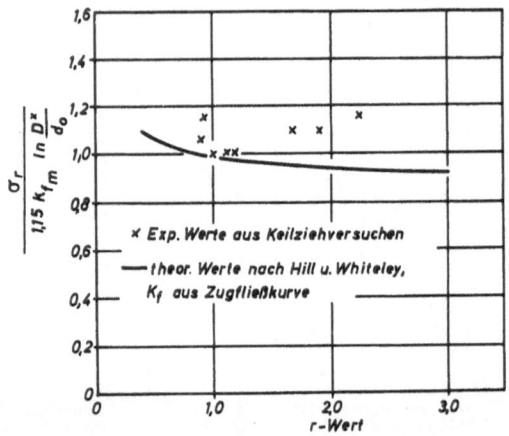

Abb.14. Theoretische und experimentelle Werte der erforderlichen Ziehspannung σ_r für den Ziehteilflansch in Abhängigkeit vom r-Wert

optimaler Reißer   Bodenreißer   vorzeitiger Reißer

Abb.15. Die möglichen Versagensarten durch Reißer im Ziehteilboden

läßt die größtmögliche Kraftübertragung zu und ergibt daher die günstigsten Grenzziehverhältnisse. Der normale Bodenreisser ist entweder durch den Werkstoff, zu geringe Haftreibung oder die Stempelgeometrie bedingt. Diese Bruchform tritt an sich am häufigsten auf.

In Abb. 16 sind die Formänderungen im Ziehteilboden eines zylindrischen Werkstücks für zwei verschiedene Stempelradien über der Abwicklung des Bodens aufgetragen. In der Mitte des Ziehteilbodens ergibt sich sowohl Radial- als Tangentialdeh-

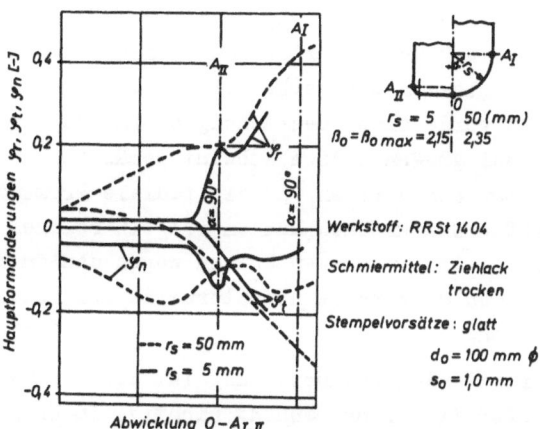

Abb. 16. Hauptformänderungen im Ziehteilboden bei verschiedenen Stempelradien

nung (φ_r und φ_t), während die Blechdicke abnimmt. Falls ein
vorzeitiger Reißer eintritt, liegt er in diesem Bereich. Während die radiale Formänderung φ_r von der Bodenmitte nach außen
hin ständig zunimmt, wird die tangentiale Formänderung φ_t
kleiner, durchläuft den Wert 0 und geht dann in Stauchung
über. Die Blechdickenformänderung φ_n ist im gesamten Boden-
bereich negativ. Für den optimalen Reißer ergibt sich daraus,
daß er in einem Bereich liegt, in dem radiale Dehnung und
tangentiale Stauchung stattgefunden hat. Der übliche Bodenreißer kann dagegen in drei verschiedenen Bereichen liegen:

1. radiale und tangentiale Dehnung, Blechdickenabnahme,
2. radiale Dehnung, Blechdickenabnahme, tangentiale Formänderung Null (ebene Formänderung),
3. radiale Dehnung, tangentiale Stauchung, Blechdickenabnahme.

Zunächst soll untersucht werden, mit welchen Abreißspannungen
beim optimalen Reißer gerechnet werden kann. Da der Bruch in
diesem Falle im zylindrischen Teil des Werkstücks liegt, ist
die Beanspruchung durch die Ziehspannung vorwiegend durch
einachsigen radialen Zug gekennzeichnet. Tangential- und Nor-

malspannungen können sich durch Querkontraktion ausbilden. Infolge des Ziehspalts zwischen Ziehring und Stempel liegt das
Ziehteil gewöhnlich nicht so eng am Stempel an, daß dieser
Einfluß groß sein kann. Außerdem kann während des Kraftan-
stiegs ein Nachgleiten des Werkstoffs über die Stempelrundung
eintreten. Das würde einen mehrachsigen Spannungszustand zur
Folge haben. Bei großen Radien scheint gemäß Abb.16 jedoch
kenn Nachgleiten einzutreten, da die radiale Formänderung
stetig verläuft. Bei kleinen Radien ist ein solcher Einfluß
denkbar. Jedoch wird in diesen Fällen wegen des in der Rundung auftretenden Querdrucks normalerweise kein optimaler
Reißer eintreten.

Neben der auftretenden Beanspruchung ist die vorangegangene
Formänderung für die Bruchspannung wichtig. Im Übergang vom
Stempelradius zum zylindrischen Teil liegt sowohl bei kleinen
als auch bei großen Stempelrundungen radiale Dehnung und tan-
gentiale Stauchung vor, wie aus Abb. 16 zu entnehmen ist. Die
Ausbildung des Bodens hat an dieser Stelle also unter ähnli-
chem Spannungszustand stattgefunden wie die Umformung im
Flansch des Ziehteils.

Der Betrag der Formänderung liegt nur bei kleinen Rundungen
unter der Gleichmaßdehnung des Werkstoffs (Abb.16). Wenn aber
die Gleichmaßdehnung in der Vorverformung überschritten wurde, führt eine nachfolgende einachsige Zugbeanspruchung ohne
wesentliche weitere Dehnung zum Bruch. Aus diesem Grunde fallen Zugfestigkeit und Formänderungsfestigkeit in diesem Bereich etwa zusammen (Abb. 17). Um die Bruchspannung für anisotrope Werkstoffe zu erhalten, ist lediglich noch zu klären,
welche Fließkurve von Bedeutung ist. Dabei hat sich gezeigt,
daß die Zugfestigkeit die Fließkurve wiedergibt, die für die
vorangegangene Formänderung maßgeblich gewesen ist. So z. B.
ergab die Zugfestigkeit an durch Kaltwalzen vorverformten
Proben nicht die Fließkurve des einachsigen Zugversuchs, sondern etwa die des höher liegenden mehrachsigen Zugversuchs
(Abb. 11). Daraus folgt, daß die Abreißspannung beim optimalen Reißer etwa den gleichen Gesetzmäßigkeiten bezüglich des
Einflusses der Anisotropie folgen sollte wie die erforderliche Ziehspannung. Gemäß den Abb. 9, 12 und 13 ist in die

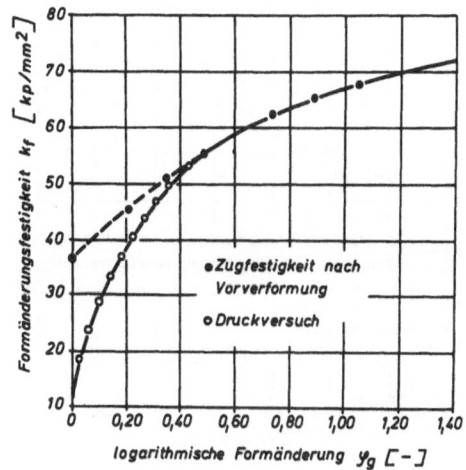

Abb. 17. Fließkurve und Zugfestigkeit nach verschiedener Vorverformung von Ms 63

Ziehspannung die jeweils höchste Fließkurve eingegangen. Das
entsprach bei Werkstoffen mit r > 1 der Beulfließkurve und
bei Werkstoffen mit r < 1 der Zugfließkurve. Wenn man die
Bodenabreißspannung für den optimalen Bodenreißer jeweils
auf diese Fließkurven bezieht (Abb. 18), erhält man unabhängig vom r-Wert Meßwerte bei $\sigma_{Br}/k_f = 1$. Demnach ist auch der
r-Wert auf den optimalen Bodenreißer ebenso ohne größeren
Einfluß wie auf die Ziehspannung (Abb. 14). Deshalb dürfte
das Grenzziehverhältnis vom r-Wert nicht beeinflußt werden,
wenn ein optimaler Reißer erreicht wird.

Anders dagegen scheinen die Verhältnisse beim normalen Bodenreißer zu liegen. Hier tritt der Bruch in der Stempelrundung ein, und es liegt ein mehrachsiger Spannungszustand vor.
Die tangentiale Formänderung geht in diesem Bereich gemäß
Abb. 16 von positiven zu negativen Werten über. An der Stelle
$\varphi_t = 0$ liegt der in Gleichung (3) zugrunde gelegte ebene
Werkstofffluß vor. Für diese Stelle gilt die in Abb. 18 dar-
gestellte Kurve $\varphi_t = 0$. Die auf die Formänderungsfestigkeit
k_f bezogene Bruchspannung steigt mit zunehmendem r-Wert
stark an. Falls also Bruch an dieser Stelle eintritt, ist

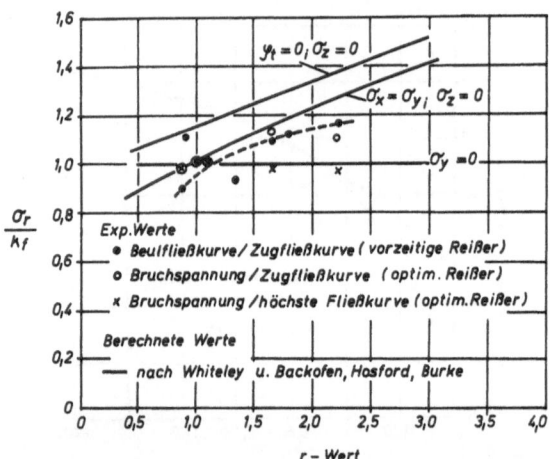

Abb.18. Theoretische und experimentelle
Werte für die Bruchspannung im Ziehteilboden bzw. am Übergang in die Zarge

das Grenzziehverhältnis vom r-Wert abhängig, da die maximal
übertragbare Spannung mit r ansteigt, während die erforderliche vom r-Wert nicht oder kaum abhängig ist.

Weiter zur Mitte des Bodens ergibt sich ein zweiachsiger Zug-
spannungszustand, der zunehmend dem Beulversuch ähnlich wird.
Für diesen Fall ($\sigma_x = \sigma_y$; $\sigma_z = 0$) erhält man die Abreißspannung zu [10]:

$$\sigma_r = \frac{1}{\sqrt{2 - \frac{2r}{r+1}}} \cdot k_f \qquad (9)$$

Diese Funktion ergibt mit zunehmendem r-Wert ebenfalls eine
steigende Tendenz (Abb. 18), so daß beim Eintreten des Ver-
sagensfalls in diesem Bereich das Grenzziehverhältnis vom
r-Wert abhängig ist. Um zu überprüfen, ob Gleichung (9) das
anisotrope Verhalten der Werkstoffe richtig wiedergibt, wurde
das Verhältnis von gemessenen Beulfließkurven zu Zugfließkur-
ven für verschiedene Werkstoffe eingezeichnet. Die gestrichelte Kurve gibt die Tendenz wieder, die sich näherungsweise mit der nach Gleichung (9) deckt. In diesem Bereich
scheint also die Fließbedingung für anisotrope Werkstoffe

das Werkstoffverhalten etwa richtig zu beschreiben.

Wenn man die Formänderungen im Boden des Ziehteils von der Stelle der ebenen Formänderung aus ($\varphi_t = 0$) nach außen verfolgt (Abb. 16), läßt sich in der Rundung eine Stelle ermitteln, wo bei Vernachlässigung der Normalspannung nur eine radiale Zugspannung wirksam gewesen ist (einachsiger Spannungszustand). Bei isotropem Werkstoff ist an dieser Stelle die tangentiale Formänderung φ_t gleich der Dickenformänderung φ_n. Für diesen Bereich gilt, daß die Formänderungsfestigkeit und damit auch in erster Näherung die Bruchspannung der Zugfließkurve entspricht.

Die Radialspannung σ_r, die gleich der örtlichen Bruchfestigkeit σ_{Br} gesetzt werden soll, läßt sich von der Bodenmitte bis zur Stelle der einachsigen Zugspannung berechnen zu [10]:

$$\frac{\sigma_r}{k_f} = \frac{1}{\sqrt{1 + (\frac{\sigma_t}{\sigma_r})^2 - \frac{\sigma_t}{\sigma_r} \cdot \frac{2r}{r+1}}} \qquad (10)$$

In Bodenmitte ist $\sigma_t = \sigma_r$ bzw. $\varphi_t = \varphi_r$. Damit geht Gleichung (10) in Gleichung (9) über. Für $\sigma_t = 0$ wird $\sigma_r/k_f = 1$. Gemäß Abb. 16 nimmt die tangentiale Stauchung von diesem Punkt aus in Richtung zum Übergang in die Zarge zu. Im Bereich, wo $|\varphi_t| > |\varphi_n|$ ist, treten bei der Ausformung des Bodens tangentiale Druckspannungen auf. σ_t wird also negativ. Für den Versagensfall in diesem Bereich soll, wie bereits beim optimalen Reißer ausgeführt wurde, angenommen werden, daß der Werkstoff dann versagt, wenn die örtliche Radialspannung den Wert der Formänderungsfestigkeit erreicht, die für das Ausformen des Bodens maßgeblich gewesen ist (Zug-Druck-Spannungszustand). Dann erhält man für die örtliche Abreißspannung σ_{Br}:

$$\frac{\sigma_{Br}}{k_f} = \frac{1}{\sqrt{1 + (\frac{\sigma_t}{\sigma_r})^2 - \frac{\sigma_t}{\sigma_r} \frac{2r}{r+1}}} \; (1 - \frac{\sigma_t}{\sigma_r}) \qquad (11)$$

Diese Gleichung gilt nur, wenn bei Vernachlässigung der

Normalspannung die Tangentialspannung bei der Ausbildung des
Bodens negativ und die Radialspannung positiv gewesen ist,
die Formung des Bodens bei Bruchbeginn beendet gewesen ist
und die Vorverformung über der Gleichmaßdehnung des Werkstoffs gelegen hat. Diese Annahmen sind normalerweise für
den üblichen Bodenreißer zulässig.

Wenn man die Gleichungen (10) und (11) für verschiedene Span-
nungsverhältnisse graphisch darstellt (Abb. 19), erhält man
für positive Werte von σ_t/σ_r den Bereich im Ziehteilboden von
der Stelle der einachsigen Beanspruchung zur Bodenmitte hin
und für negative Werte von σ_t/σ_r den Bereich von dort zur
Zarge hin. An der Stelle der einachsigen Beanspruchung ist
die relative Festigkeit für die verschiedenen r-Werte am
kleinsten. Sie steigt zur Bodenmitte mit zunehmendem r-Wert
stärker an und nach außen hin geringer. Wenn man für die örtliche Formänderung, die nach außen hin zunimmt, die zugehörigen Formänderungsfestigkeiten und den Einfluß der Reibung
berücksichtigt, kann man feststellen, daß Werkstoffe mit
großem r-Wert die Tendenz zu optimalen Reißern haben müssen,
weil der höchste Anstrengungsgrad außen auftritt. Mit abneh-
mendem r-Wert steigt die relative Festigkeit nach außen hin
stärker an und der höchste Anstrengungsgrad wird damit nach
innen verlagert, weil unter Berücksichtigung der örtlichen
Formänderungsfestigkeit ein größeres Festigkeitsgefälle auftritt. Infolge der Verschiebung der Bruchlage nach innen ergeben sich kleinere übertragbare Kräfte und damit auch kleinere Grenzziehverhältnisse. Bei kleinen r-Werten fällt der
höchste Anstrengungsgrad oftmals in den Bereich der zweiachsigen Zugbeanspruchung. Das liegt darin begründet, daß der
Absolutwert der Festigkeit an dem Punkt der einachsigen Beanspruchung wegen der Verfestigung höher liegen kann als weiter innen, trotz der größeren relativen Festigkeit. Bei Werkstoffen mit kleinen r-Werten ist es selbst bei sehr großem
Reibwert am Stempel (gerändelte Stempel) aus den genannten
Gründen nicht möglich, einen optimalen Reißer zu erzeugen.

Wenn man in einem Ziehteilboden an verschiedenen Stellen Ab-
reißversuche macht und die radial wirkende Abreißspannung
auf die Ausgangszugfestigkeit des Werkstoffs bezieht und über

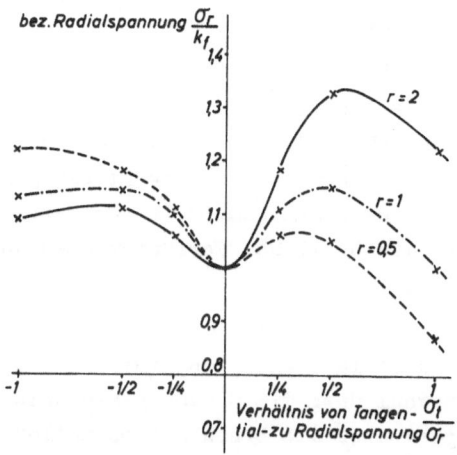

Abb.19. Rechnerische Festigkeitsverteilung im Boden eines Ziehteils aufgrund der bei der vorangegangenen Formung des Bodens vorhanden gewesenen Spannungsverteilung

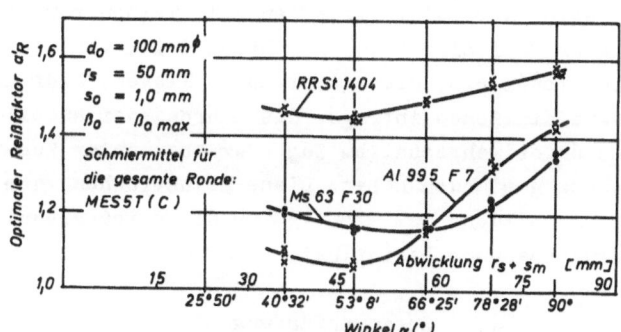

Abb.20. Gemessene Festigkeitsverteilung im Ziehteilboden bei verschiedenen Werkstoffen

der Abwicklung des Stempelradius aufträgt, erhält man Abb. 20 [11]. Diese Abb. gibt die in Abb. 19 gezeigte Tendenz wieder. Bei Werkstoffen mit kleinem r-Wert ist zum Auslauf aus der Rundung hin die relativ größte Festigkeitszunahme festzustellen (Al 99,5), während ein Werkstoff mit großem r-Wert nur eine verhältnismäßig kleine Steigerung zeigt. Ms 63 mit einem r-Wert um eins liegt zwischen den anderen beiden Werkstoffen. Auch der Festigkeitsanstieg zur Mitte hin als Folge der ebenen Formänderung kündet sich in dem Bild an. Der Punkt der ebenen Formänderung selbst ist nicht mehr erfaßt worden.

Die Unterschiede der Absolutwerte in Abb. 20 zu den gerechneten relativen Werten sind zum Teil darauf zurückzuführen, daß Abb. 20 auf die Ausgangszugfestigkeit bezogen ist im Gegensatz zu Abb. 19. Um einen unmittelbaren Vergleich zu ermöglichen, müßten die örtlichen Formänderungen und die Fließkurve berücksichtigt werden.

Zusammenfassend über den Einfluß des r-Wertes im Bereich des normalen Bodenreißers läßt sich feststellen, daß er das Grenzziehverhältnis sowohl über die Spannungsmechanik als auch damit zusammenhängend über die Bruchlage beeinflußt. Mit zunehmendem r-Wert verschiebt sich die Bruchlage bei sonst gleichbleibenden Bedingungen nach außen zu optimalen Reißern hin und ergibt damit eine Zunahme des Grenzziehverhältnisses. Die oftmals in Rechnungen gemachte Annahme, daß der Bruch im Bereich der ebenen Formänderung erfolgt, erscheint nicht zulässig zu sein. Nur bei Werkstoffen mit kleinen r-Werten könnte dieser Fall eintreten. Der Bruch tritt vielmehr in solchen Bereichen ein, die weiter außen liegen. Hier kommen insbesondere die Zonen infrage, die während der Ausbildung des Bodens unter mehrachsigem Zug oder aber unter Zug-Druck-Beanspruchung gestanden haben. Diese Beanspruchung des Bodens bei seiner Ausbildung ist wichtig für die Fließkurvenlage, die die Bruchspannung beeinflußt.

Zusammenfassung

Die Untersuchungen haben gezeigt, daß ein Werkstoffeinfluß auf das Grenzziehverhältnis nur über die normale Anisotropie (r-Wert) möglich ist, wenn eine bestimmte Mindestdehnung, die für die Ausbildung des Bodens benötigt wird, vorhanden ist. Der r-Wert beeinflußt vorwiegend die Bodenabreißspannung. Dagegen geht er wenig in die erforderliche Ziehspannung ein. Mit zunehmendem r-Wert steigt die Bodenabreißkraft und damit die maximal übertragbare Kraft vorwiegend deshalb an, weil sich ein später eventuell auftretender Bruch auf Grund der

Spannungszusammensetzung während der Ausbildung des Bodens nach außen verlagert. Falls ein optimaler Bodenreißer vorliegt, ist eine größtmögliche Kraftübertragung und damit ein größtmögliches Grenzziehverhältnis erreicht. In diesem Bereich scheint ein Einfluß des r-Wertes nicht vorhanden zu sein. Es ist jedoch bei Werkstoffen mit kleinen r-Werten nicht möglich, einen optimalen Reißer zu erzeugen.

Der Verlagerung des Bruchs nach außen mit zunehmendem r-Wert sind überlagert die Verfestigung und die Reibung. Mit zunehmendem Verfestigungsexponenten wandert der Bruch nach innen, wenn das im Boden vorhandene Festigkeitsgefälle nicht über Reibungsschluß aufgenommen werden kann.

Die für die rechnerische Ermittlung der Bodenabreißspannung zugrunde gelegte Annahme des ebenen Werkstoffflusses erscheint nicht zulässig zu sein, da der Bruch normalerweise an anderer Stelle eintritt. Die Fließbedingungen für anisotrope Werkstoffe nach Hill scheinen das Werkstoffverhalten im Zug-Zug-Gebiet richtig wiederzugeben, während im Zug-Druck-Gebiet ein abweichendes Verhalten ermittelt wurde. Die bei Versuchen auftretenden Streuungen zwischen Grenzziehverhältnis und r-Wert dürften zu einem Teil darauf zurückzuführen sein, daß die Bruchlage und damit die maximal übertragbare Ziehkraft nicht nur vom r-Wert, sondern auch vom Verfestigungsexponenten und dem Reibungskoeffizienten beeinflußt wird.

Literatur

1. W. Panknin: Bänder-Bleche-Rohre (1961), S.133, 201 und 264.

2. W. Ziegler: Dissertation TU Berlin, 1966.

3. W.T. Lankford, S.C. Snyder and J.A. Bauscher: Trans.Amer. Soc.Metals 42(1950)1197.

4. R.L. Whiteley, D.E. Wise and D.J. Blickwede: Sheet Met. Ind. 38(1961) No. 409, S. 349.

5. J.L. Wright: Sheet Metal Ind. 39(1962)887.

6. R. Hill: The Mathematical Theory of Plasticity, Oxford Engineering Science Series, At the Clarendon Press 1950.

7. R.L. Whiteley: Trans.Amer.Soc. Metals 52(1960)154.

8. R.T. Holcomb and W.A. Backofen: Sheet Met. Ind. 43(1966)479.

9. G. Sachs: Metallw. 9(1930)213.

10. W.A. Backofen, W.F. Hosford and J.J. Burke: Trans.Amer. Soc.Metals 55(1962)264.

11. E. Doege: Dissertation TU Berlin, 1963.

5.7 Mechanische Anisotropie von aushärtbaren Aluminiumlegierungen durch Preßeffekt

Mechanical Anisotropy of Extruded and Age-Hardened Rods of
Aluminium-Alloys

von P. Wincierz[+)]

Abstract

Extruded and age-hardened Al alloys show an anisotropy in
mechanical properties, in Germany known as "extrusion effect"
(Preßeffekt). This effect is manifested by a higher strength
of the specimens parallel than perpendicular to the rod axis.
The extrusion texture of the rods is retained during solution
annealing. Extruded rods, which are cold-drawn before solution
annealing, show no extrusion effect as a cause of random
recrystallization in the age-hardened state. The changes of
mechanical anisotropy during age-hardening can be expressed
by anisotropy ratios which compare the ultimate tensile
strength, yield strength and total elongation of the extruded
rods with the corresponding values of the cold-drawn rods.

An evaluation of this anisotropy ratio for commercial alloys
of the AlCuMg-, AlZnMg- and AlMgSi-type including published
data has shown that the anisotropy ratio of the yield strength
has its maximum in the solution-annealed state. It decreases
during ageing at room temperature as well as at higher
temperatures. Higher ageing temperatures and longer times
lead to values of the mechanical properties similar to those
of texture-free samples with the same treatment. The reason
for this behaviour is discussed on the basis of a disturbance
of the texture strengthening by precipitates formed during
the age-hardening process.

An stranggepreßten Stangen aus Aluminiumlegierungen vom Typ
AlCuMg und AlMgSi beobachtete DREYER [1] im ausgehärteten
Zustand anomal hohe Werte für die in Preßrichtung bestimmte
Zugfestigkeit und Streckgrenze bei verminderter Dehnung. In

[+)] Metall-Laboratorium der Metallgesellschaft A.-G., Frankfurt (Main), Bundesrepublik Deutschland

radialer Richtung entnommene Proben besaßen demgegenüber eine
geringe Festigkeit [2]. Diese als Preßeffekt [3] bezeichnete
mechanische Anisotropie zeigten vor dem Lösungsglühen noch
zusätzlich durch Ziehen kaltverformte Preßstangen nicht. Da
in manganfreien AlCuMg-Legierungen der Preßeffekt gleichfalls
nicht auftrat und das Material im Gegensatz zu den manganhal-
tigen Legierungen vollständig rekristallisiert war, sahen
Dreyer und HANSEN [4] sowie DREYER und SEEMANN [5] das Unter-
bleiben der Rekristallisation beim Lösungsglühen als Voraussetzung für den Preßeffekt an. UNCKEL [2] und WASSERMANN [6]
wiesen nach, daß der Preßeffekt an die Erhaltung der Preßtex-
tur beim Lösungsglühen [7] gebunden ist. Eine Kaltverformung
der Preßstangen von z. B. 20 % führt beim Lösungsglühen zu.
regelloser Rekristallisation.

Zur Charakterisierung der mechanischen Anisotropie von ausge-
härtetem Material mit Preßeffekt ist einerseits die Überhöhung
(Differenz) der in Längsrichtung bestimmten Festigkeitswerte
[vergl. z. B. 8,9] einer derartigen Stange gegenüber den Werten einer nach dem Pressen noch kaltgezogenen (nachgezogenen)
und beim Lösungsglühen regellos rekristallisierten Stange verwendet worden. Andererseits diente hierzu das Verhältnis der
Festigkeitswerte in Längs- und Radialrichtung [10], das mit
Mikrozerreißproben aus einer Stange mit Preßeffekt bestimmt
wurde.

Eine weitere Möglichkeit zur Kennzeichnung des anisotropen
Charakters des Preßeffektes stellen Verhältniszahlen der in
einer beliebigen Richtung, am häufigsten in Längsrichtung
bestimmten Festigkeitswerte von Stangen aus dem gleichen Material mit bzw. ohne Preßeffekt dar, z. B. $\sigma_{0,2}$ gepreßt/
$\sigma_{0,2}$ nachgezogen. Der Wert 1 dieser Verhältniszahlen entspricht fehlender mechanischer Anisotropie. Für diese hier
gewählte Art der Auswertung sprechen vor allem zwei Gesichtspunkte: Der eine ist die Vergleichsmöglichkeit mit theoretischen Abschätzungen der mechanischen Eigenschaften von textur-
behaftetem vielkristallinem Material. Ausgehend von Einkri-
stalldaten [11] gelangt man dabei zu auf den regellos orientierten Vielkristall [12] bezogenen Verhältniszahlen [2,13],
in denen als wichtigster Beitrag zur Texturanisotropie der-

jenige der Aluminiummatrix zum Ausdruck kommt. Neben der
Texturanisotropie sind die geometrischen Anisotropiefaktoren
von Bedeutung [14]. Beide Einflüsse führen gemeinsam zur
mechanischen Anisotropie. Der zweite der obengenannten Gesichtspunkte besteht in der prinzipiell gegebenen Abtrenn-
barkeit derjenigen geometrischen Anisotropiefaktoren, die
durch das Lösungsglühen und die im Fall der nachgezogenen Preßstangen dabei ablaufende Rekristallisation nicht verändert
werden und daher bei den Stangen mit und ohne Preßeffekt übereinstimmen. Hierher gehören bei den handelsüblichen aushärtbaren Aluminiumlegierungen die meist zeilenförmig angeordneten
Teilchen intermetallischer Phasen, die insbesondere Mangan,
Eisen und Silizium neben Aluminium enthalten. Weiter sind
Oxideinschlüsse und Poren sowie in einigen Fällen die Gestalt
der Aluminiummatrixkörner (Verhältnis Länge/Durchmesser) jedoch z. B. nicht ihre Größe zu nennen.

Natürlich reichen Zugversuche, die zudem noch auf in Preß-
richtung entnommene Proben beschränkt sind, nicht aus, um
ein vollständiges Bild der als Preßeffekt bezeichneten mechanischen Anisotropie zu gewinnen. Hierzu wäre die experimentelle Ermittlung von Fließspannungsellipsen (yield loci) erforderlich. Doch kann man in relativ einfach gelagerten Fällen
auf der Basis üblicher Zugversuche durchaus zu weiterführen-
den Einblicken in die Natur des Preßeffektes gelangen. Die
auf diese Weise verfolgte Änderung der mechanischen Anisotropie im Verlauf der Aushärtung von Stangen mit und ohne Preßeffekt ist ein Beispiel hierfür. Sie kann unmittelbar mit
den bei der Auslagerung gebildeten Ausscheidungen in Zusammenhang gebracht werden, da sowohl die obengenannten nach dem
Lösungsglühen vorhandenen geometrischen Anisotropiefaktoren
als auch die Textur während der Auslagerung konstant bleiben.
Inwieweit während der Auslagerung eintretende Änderungen der
mechanischen Anisotropie auf Einflüsse des geometrischen oder
des Texturanisotropieanteiles zurückgeführt werden können, ist
eine bisher wenig [8,9,10] untersuchte Frage. Der insbesondere bei der Bildung von allseitig kohärenten plättchen- oder
nadelförmigen Ausscheidungen (GP-Zonen) gegebene kristall-
graphisch gesetzmäßige Orientierungszusammenhang mit der Aluminiummatrix führte zu der Hypothese einer als anisotrope

Aushärtung [8,9,15] bezeichneten Zunahme der Texturanisotropie. Im folgenden wird daher anhand von in der Literatur vorliegenden und eigenen Versuchsergebnissen an handelsüblichen Aluminiumlegierungen der Typen AlCuMg, AlZnMg und AlMgSi die Änderung der mechanischen Anisotropie während der Aushärtung erneut untersucht. Der Vergleich dieser drei Legierungen, die bei der Aushärtung Ausscheidungspartikel verschiedener Gestalt (Plättchen, Nadeln) mit unterschiedlichem Orientierungszusammenhang mit der Aluminiummatrix bilden [16-20,7], sollte zur Klärung der Frage nach dem vorherrschenden Einfluß bei der Änderung der mechanischen Anisotropie während der Aushärtung beitragen.

DREYER und SEEMANN [21] beobachteten mit Zugversuchen an Längsproben die Änderung der mechanischen Eigenschaften von gepreßten sowie nach dem Pressen um 17 % gezogenen Stangen aus einer Legierung vom Typ AlCuMg während der Auslagerung bei $20^{\circ}C$. Sie fanden, daß Zugfestigkeit und Streckgrenze $\sigma_{0,2}$ der Stangen mit Preßeffekt bereits im lösungsgeglühten Zustand höher lagen, zu Beginn der Aushärtung steiler zunahmen und auch während der übrigen Auslagerungsdauer größere Werte aufwiesen als bei den nachgezogenen Stangen. Die Bruchdehnung verhielt sich umgekehrt. Sie war bei den gepreßten Stangen nach dem Lösungsglühen wesentlich geringer als bei den Stangen ohne Preßeffekt und blieb nahezu konstant während sie bei diesen mit fortschreitender Auslagerungszeit abnahm. In Abb.1 sind anhand dieser Ergebnisse in der oben beschriebenen Weise gebildete Verhältniszahlen $\sigma_{0,2}$ gepreßt $/ \sigma_{0,2}$ nachgezogen; σ_B gepreßt $/ \sigma_B$ nachgezogen und σ_{10} gepreßt $/ \sigma_{10}$ nachgezogen zur Kennzeichnung der Anisotropie in Abhängigkeit von der Auslagerungszeit bei Raumtemperatur dargestellt. Es zeigt sich, daß die Anisotropie-Verhältniszahl der Streckgrenze mit ca. 1,5 höher ist als für die Zugfestigkeit mit ca. 1,2. Die Änderungen während der Auslagerung sind gering. Im Fall der Bruchdehnung ist mit der Auslagerung eine deutliche Zunahme der nach dem Lösungsglühen 0,5 betragenden Verhältniszahl verbunden, d.h. eine Verringerung der mechanischen Anisotropie festzustellen. In analoger Weise sind ebenfalls in Abb. 1 nach Werten von RICHTER und WASSERMANN [22] (Abb. 5 in [22]) ermittelte Verhältniszahlen für eine AlCuMg-Legierung nach Aus-

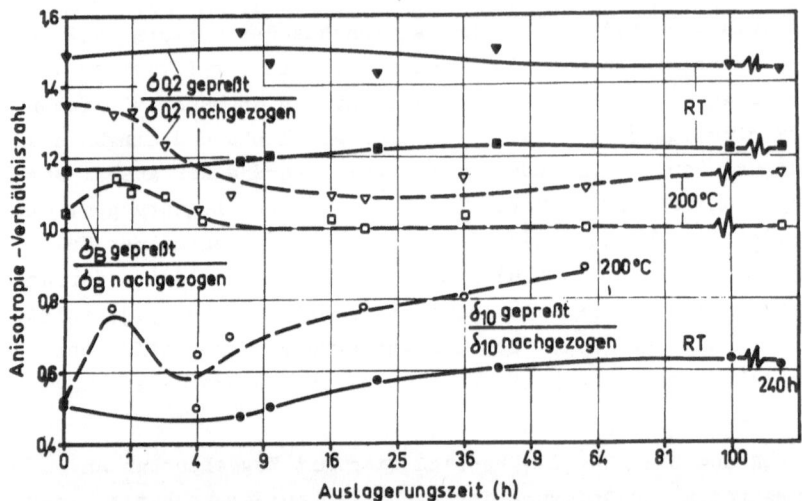

Abb.1. Anisotropie-Verhältniszahlen von AlCuMg-Legierungen in Abhängigkeit von Auslagerungstemperatur und -zeit. Für Zugfestigkeit σ_B, Streckgrenze $\sigma_{0,2}$ und Bruchdehnung δ_{10} berechnet aus Meßwerten für stranggepreßte sowie nach dem Strangpressen um 17 bzw. 20 % kaltgezogene Stangen aus Al-4,1 Cu-0,8 Mg-0,3 Mn (Auslagerung bei Raumtemperatur, Abb. 1b in [5]) und Al-4,22 Cu-1,22 Mg-0,77 Mn (Auslagerung bei 200°C, Abb. 5 in [22]

lagerung bei 200°C eingetragen. Wiederum kommt in den $\sigma_{0,2}$-Verhältniszahlen die Anisotropie stärker zum Ausdruck als in den σ_B-Verhältniszahlen. Jene sind für den lösungsgeglühten Zustand am größten und nehmen mit wachsender Auslagerungsdauer ab, während diese nach Durchlaufen eines flachen Maximums den Wert 1 erreichen. Die Anisotropie-Verhältniszahlen der Bruchdehnung zeigen einen Rückgang der Anisotropie.

In analoger Weise wurde die Aushärtung einer Legierung vom Typ AlZnMg1 an Stangen mit und ohne Preßeffekt anhand von Zugversuchen in Längsrichtung in Abhängigkeit von Auslagerungstemperatur und -zeit untersucht. Aus Platzgründen muß auf eine Wiedergabe der Isothermen der mechanischen Eigenschaften verzichtet werden. Hier sei lediglich vermerkt, daß die Aushärtung bei 220°C über ein Maximum bei ca. 1 Stunde Auslagerungszeit verläuft, während bei den niedrigeren Auslagerungstemperaturen das Festigkeitsmaximum nach 24-stündiger Auslagerung noch nicht erreicht wird. In Abb. 2a sind die Anisotropie-Verhältniszahlen für die Streckgrenze $\sigma_{0,2}$ in Abhängigkeit

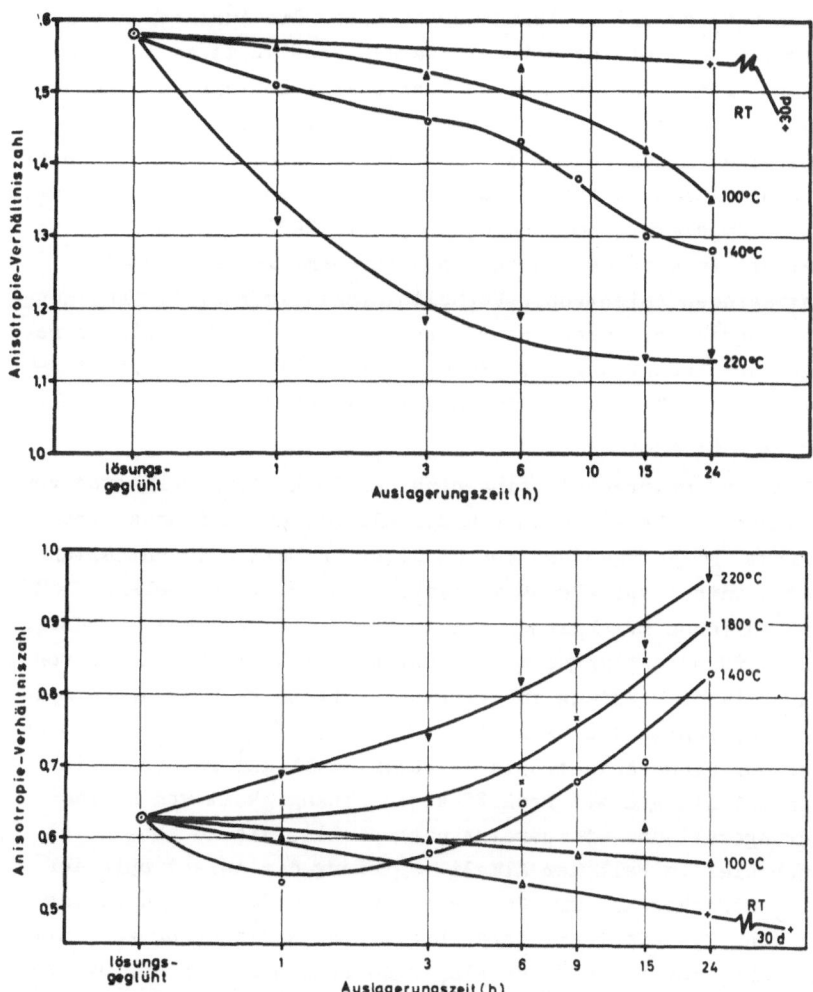

Abb.2 a und b. Anisotropie-Verhältniszahlen einer AlZnMg1-
Legierung in Abhängigkeit von Auslagerungstemperatur und -zeit

a (oben)) Für Streckgrenze $\sigma_{0,2}$ berechnet aus Meßwerten für
stranggepreßte sowie nach dem Strangpressen um 25 % kaltgezogene Stangen aus Al-4,5 Zn-1,15 Mg-0,25 Mn-0,22 Cr

b (unten)) Für Bruchdehnung δ_5, sonst wie a)

von der Auslagerungszeit dargestellt. Der höchste Wert für
diese Größe wird im lösungsgeglühten Zustand gefunden. Bei
allen untersuchten Auslagerungstemperaturen, auch bei Raumtemperatur, wird eine mit der Zeit fortschreitende Abnahme
der mechanischen Anisotropie beobachtet. Der Rückgang erfolgt

umso rascher und vollkommener je höher die Auslagerungstemperatur ist. Die zeitliche Lage des Festigkeitsmaximums markiert sich in diesen Kurven nicht. Abb. 2b zeigt die Auswertung der an denselben Proben aus der AlZnMg1-Legierung ermittelten Bruchdehnung δ_5. Aufgetragen sind die Anisotropie-Verhältniszahlen für δ_5 in Abhängigkeit von Auslagerungszeit und -temperatur. Im Gegensatz zu Abb. 2a ist in Abb. 2b eine weitere Zunahme der Anisotropie (Abnahme der hier stets ≤ 1 betragenden Anisotropie-Verhältniszahl) im Fall der Kaltaushärtungstemperaturen (Raumtemperatur und 100°C) sowie zu Beginn der Aushärtung bei 140°C festzustellen. Im Temperaturbereich der Warmaushärtung nimmt die Anisotropie der Bruchdehnung hingegen ab.

DIES [15] untersuchte eingehend das Aushärtungsverhalten von Stangen aus Legierungen vom Typ AlMgSi1 mit und ohne Preßeffekt. Bildet man aus den in dieser Arbeit [15] enthaltenen Daten Anisotropie-Verhältniszahlen für $\sigma_{0,2}$, so gelangt man zu ähnlichen Ergebnissen wie in Abb. 2 für AlZnMg1. Da dieser Autor für die Stangen mit Preßeffekt auch die Resultate von in radialer Richtung ausgeführten Mikrozerreißversuchen mitteilte, konnte das Verhältnis $\sigma_{0,2 \text{ längs}}/\sigma_{0,2 \text{ radial}}$ bestimmt werden. Diese Anisotropie-Verhältniszahl ist in Abb. 3 für AlMgSi1-Stangen mit Preßeffekt in Abhängigkeit von Auslagerungszeit und -temperatur dargestellt. Wie in Abb. 2 ist auch hier im Fall der AlMgSi-Legierung die Anisotropie der $\sigma_{0,2}$-Streckgrenze im lösungsgeglühten Zustand am ausgeprägtesten. Sie nimmt mit fortschreitender Auslagerungszeit mit Ausnahme der bei 80° ausgelagerten Proben stetig ab. Die in Abb. 3 vorhandenen Überschneidungen bedürfen noch der Aufklärung. Es zeigt sich aber deutlich, daß auch bei Wahl der Streckgrenze in Radialrichtung als Bezugsgröße und der von regelloser Orientierung verschiedenen Textur die Anisotropie der Streckgrenze bereits bei der Raumtemperaturauslagerung deutlich abnimmt.

Auch bei der Prüfung der Dauerschwingfestigkeit macht sich die durch den Preßeffekt bewirkte mechanische Anisotropie besonders augenfällig an der Ausbildung der Bruchflächen von Umlaufbiegeproben bemerkbar. Hierüber berichtete BRENNER [23]

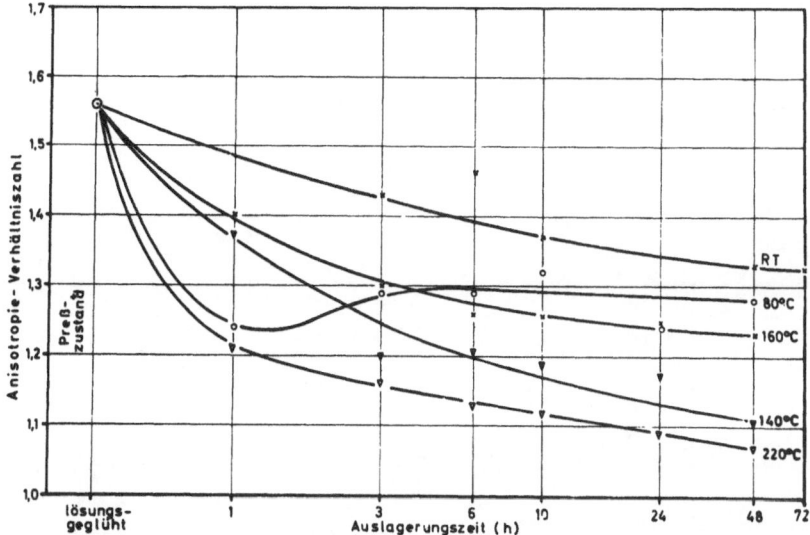

Abb. 3. Anisotropie-Verhältniszahlen einer AlMgSi1-Legierung in Abhängigkeit von Auslagerungstemperatur und -zeit. Für Streckgrenze $\sigma_{0,2}$ berechnet aus Meßwerten (Abb. 3a und 5a in [15]) von Längs- und Radialproben stranggepreßter Stangen aus Al-0,75 Mg-1,00 Si-0,81 Mn

aufgrund von Beobachtungen an einer Legierung vom Typ AlZnMgCu. Eigene Untersuchungen an Legierungen vom Typ AlZnMg1 ergaben, daß durch Nachziehen vor dem Lösungsglühen preßeffektfreie Stangen nach Auslagerungsbehandlungen zwischen Raumtemperatur und 220°C stets Normalbrüche (Bruchfläche unter 90° zur Preß-richtung) ergaben, während das Material mit Preßeffekt zu Schiebebrüchen (Bruchverlauf im allgemeinen unter ca. 45° zur Preßrichtung) Anlaß gab. Jedoch führte eine 15-stündige Aus-lagerung bei 220°C deutlich zu einem dem Normalbruch bereits nahe kommenden Bruchgefüge. Auf Abbildungen muss hier aus Platzgründen verzichtet werden.

Die in den Abb. 1 bis 3 zum Ausdruck kommenden Veränderungen der mechanischen Anisotropie können auf der Basis der die Aushärtung bewirkenden Entmischungs- und Ausscheidungsvorgänge diskutiert werden. Als Beispiel sind hierzu in Abb. 4a bis c die durch steigende Auslagerungstemperaturen ausgelösten Ge-fügeveränderungen anhand eigener elektronenmikroskopischer Untersuchungen wiedergegeben. Die größeren, abgerundeten und stark im Kontrast stehenden Teilchen sind beim Lösungsglühen

494

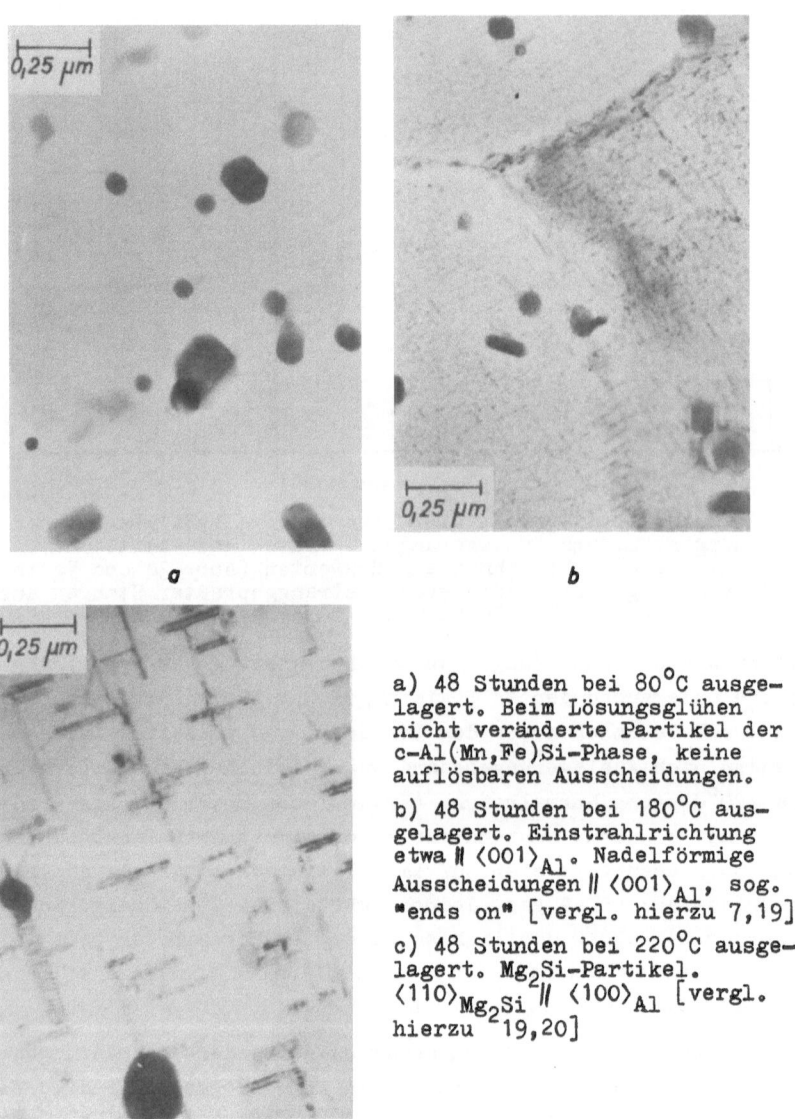

a) 48 Stunden bei 80°C ausgelagert. Beim Lösungsglühen
nicht veränderte Partikel der
c-Al(Mn,Fe)Si-Phase, keine
auflösbaren Ausscheidungen.

b) 48 Stunden bei 180°C ausgelagert. Einstrahlrichtung
etwa $\parallel \langle 001 \rangle_{Al}$. Nadelförmige
Ausscheidungen $\parallel \langle 001 \rangle_{Al}$, sog.
"ends on" [vergl. hierzu 7,19]

c) 48 Stunden bei 220°C ausgelagert. Mg_2Si-Partikel.
$\langle 110 \rangle_{Mg_2Si} \parallel \langle 100 \rangle_{Al}$ [vergl.
hierzu 19,20]

Abb.4a-c. Elektronenmikroskopische Gefügebilder (Direktdurch-
strahlung) einer AlMgSi1-Legierung in Abhängigkeit von der
Auslagerungstemperatur. Legierung Al-0,65 Mg-1,02 Si-0,78 Mn.
Lösungsglühung: 30 min. 535°C/H_2O.

unverändert gebliebene c-Al(Mn,Fe)Si-Phase. Ihre Dichte beträgt ca. $10^{13}/cm^3$. Die 48-stündige Auslagerung bei 80°C (langsamer Anstieg der Streckgrenze) ließ keine mit dem Siemens-Elmiskop 1 auflösbaren Ausscheidungen erkennen (Abb. 4a). Nach 48-stündiger Auslagerung bei 180°C (nach Überschreiten des Aushärtungsmaximums) findet man in den Aluminium-Matrixkörnern zahlreiche punktförmig erscheinende Ausscheidungsteilchen. Hierbei handelt es sich nach Ausweis von Elektronen-Feinbereichbeugungsbildern um nadelförmige parallel $\langle 100 \rangle_{Al}$ ausgeschiedene Teilchen. Ihre Länge beträgt ca. 250-500 Å, ihr Durchmesser ca. 30 bis 50 Å und ihre Dichte ca. 10^{15} bis $10^{16}/cm^3$ (Abb. 4b). Ein wesentlich anderes Bild erhält man nach 48-stündiger Auslagerung bei 220°C (Überalterung). Die langen dünnen Plättchen sind Mg_2Si-Ausscheidungen mit folgendem Orientierungszusammenhang $\langle 110 \rangle_{Mg_2Si} \parallel \langle 100 \rangle_{Al}$. Ihre Dichte beträgt nur noch ca. $3 \cdot 10^{14}/cm^3$ (Abb. 4c).

Im Zusammenhang mit den Ergebnissen in Abb. 3 folgt aus den Gefügeänderungen, daß trotz des Vorliegens kristallographisch gesetzmäßig orientierter Ausscheidungen im gesamten untersuchten Temperaturbereich der Auslagerung keine Erhöhung der mechanischen Anisotropie gegenüber dem lösungsgeglühten Zustand, sondern eine Abnahme eingetreten ist. Da die Aluminiumtextur und die primäre, vom Lösungsglühen herrührende geometrische Anisotropie unverändert blieb, ist dieser Befund im Sinne einer Störung der Texturanisotropie durch die Ausscheidungsteilchen zu deuten. Die Dichte der Ausscheidungen spielt hier anscheinend eine geringere Rolle als die Behinderung der Versetzungsbewegung bei den im allgemeinen nicht schneidfähigen Ausscheidungen der Gleichgewichtsphasen. Diese treten aber erst nach längeren Auslagerungszeiten bei hohen Auslagerungstemperaturen auf. Hiermit sind Ergebnisse von DICKEN-SCHEID und SEEMANN [10] im Einklang, die an einer Legierung vom Typ AlCuMg bei Überalterung eine Abnahme der mechanischen Anisotropie von Zugfestigkeit und Streckgrenze feststellten. Einen Hinweis auf die Existenz einer anisotropen Aushärtung, im Sinne einer Verstärkung der Texturanisotropie der Fließgrenze, gegen die vorstehenden Befunde nicht. In diesem Zusammenahng ist auf die grundlegende Untersuchung von TOR-RALBA und WASSERMANN [24] hinzuweisen, die an ausgehärteten

Aluminium-Kupfer-Einkristallen die Gültigkeit des Schmid'schen Schubspannungsgesetzes bestätigten und schlossen, daß eine die Schubspannungswerte übersteigende Aushärtungsanisotropie nicht existiert. Die bei den untersuchten Legierungen gefundenen Höchstwerte für das Anisotropieverhältnis der Streckgrenze sind wesentlich größer als dieses aufgrund der ⟨111⟩+⟨100⟩-Textur von der Theorie [13] angegeben wird.

Literatur

1. K.L. Dreyer: Metallwirtsch. 17(1938) 185-187.

2. H. Unckel: Metallwirtsch. 19(1940)37-44.

3. Zusammenfassende Darstellung in: G. Wassermann und J. Grewen: Texturen metallischer Werkstoffe, Springer, Berlin 1962, S. 651.

4. K.L. Dreyer und M. Hansen: Z. Metallkde. 33(1941)193.

5. K.L. Dreyer und H.J. Seemann: Aluminium 23(1941)437.

6. G. Wassermann: Jb. Dtsch. Luftf.-Forschg. (1941)610.

7. Zum Kombinierten Einfluß von Preß- und Lösungsglühtemperatur vergl.: K. Dies und P. Wincierz: Z. Metallkde. 57 (1966)141 u. 227.

8. H.J. Seemann: Z. VDI 91(1949)621.

9. H.J. Seemann: Z. Metallkde. 40(1949)441.

10. W. Dickenscheid und H.J. Seemann: Z. Metallkde. 49(1958) 527.

11. E. Schmid: Z. Metallkde. 19(1927)154.

12. G. Sachs: Z. VDI 72(1928)734.

13. W.F. Hosford, jr., und W.A. Backofen in: Fundamentals of Deformation Processing (Proceedings of the Ninth Sagamore Army Materials Research Conference, 1962), Syracuse University Press, Syracuse, N.Y. 1964, S. 259-298.

14. J. Grewen und G. Wassermann: Texturen als Ursache anisotropen Verhaltens bei der Umformung in: Mechanische Umformtechnik, O. Kienzle, Hrsg., Springer, Berlin 1969.

15. K. Dies: Aluminium 41(1965)747.

16. H. Schmalzried und V. Gerold: Z. Metallkde. 49(1958)291.

17. V. Gerold und H. Haberkorn: Z. Metallkde. 50(1959)568.

18. G. Thomas und J. Nutting: J. Inst. Metals 88(1959)81.

19. G. Thomas: J. Inst. Metals 90(1961)57.

20. H. Lambot: Rev. Mét. 47(1950)709.

21. K.L. Dreyer and H.J. Seemann: Z. Metallkde. 36(1944)13.

22. H. Richter und G. Wassermann: Aluminium 34(1958)193.

23. P. Brenner: Aluminium 32(1956)756.

24. M. Torralba und G. Wassermann: Z.Metallkde. 59(1968)467.

5.8 Textureinfluß auf die Gittereigendehnungen von Armcoeisen nach plastischer Deformation
Influence of Texture on Lattice Extensions of Armco Iron after Plastic Deformation

von R. Prümmer[+]

Abstract

Plastic deformation of polycrystalline materials leads to lattice extensions. Based on linear theory of elasticity these are due to residual stresses. There exist, however, in different publications some contradicting results. Thus, it is possible that measurements of different reflections of iron strained in tension show different amounts of residual stresses and even residual stresses with different signs. The investigation shows that textures are responsible for these findings. If appropriate sample preparation is used, the obvious contradicting results are reproducible.

5.8.1 Einleitung

In den letzten Jahren wurden viele Experimente zur Ermittlung der Gittereigendehnungen nach plastischer Deformation vielkristalliner Proben durchgeführt. Insbesondere mit Hilfe des Standardverfahrens [1] der röntgenographischen Spannungsmessung wurde diesen Untersuchungen zu einem neuen Aufschwung verholfen. Es besteht der übereinstimmende Befund, daß nach einachsiger Zugverformung Gittereigendehnungen in vielkristallinen Proben zurückbleiben. Diese werden überwiegend in einer durch die Probenlängsachse und das Oberflächenlot bestimmten Ebene gemessen. Dem Gitterdehnungsanstieg ordnet man formal Eigenspannungen zu. Über die Art der Eigenspannungen bestehen Unklarheiten. Insbesondere wird bisher nicht verstanden, warum an derselben Probe an verschiedenen Netzebenen Gitterdehnungsanstiege mit verschiedenem Vorzeichen auftreten können. Im folgenden wird gezeigt, daß ein Textureinfluß hierfür verantwortlich ist.

[+] Institut für Werkstoffkunde I, Universität Karlsruhe, Karlsruhe, Bundesrepublik Deutschland

5.8.2 Grundlagen

Der Ausgangspunkt für röntgenographische Gitterdehnungsmes-
sungen ist die Braggsche Gleichung. Nach ihr werden Röntgenstrahlen der Wellenlänge λ an bestimmten Netzebenen unter
dem Braggschen Winkel θ reflektiert. Eine Änderung des Netz-
ebenenabstandes D bedingt eine Änderung des Braggwinkels θ
um

$$d\theta = - \, tg\theta \cdot \frac{dD}{D} = - \, tg\theta \, \frac{da}{a} \qquad (1)$$

wobei a die Gitterkonstante ist. Das größte Auflösungsvermögen wird bei großen θ-Winkeln erzielt, weshalb fast ausschließlich im Rückstrahlgebiet entweder mit Film- oder
Zählrohrregistrierung der Debye-Scherrer-Interferenzen gearbeitet wird. Zur Bestimmung der Spannungskomponente σ_1
eines ebenen Spannungszustandes werden in der durch ihre
Richtung und das Oberflächenlot L aufgespannten Ebene in
mehreren Richtungen ψ gegenüber dem Lot Gitterdehnungsmes-
sungen durchgeführt (Abb.1). Auf Grund der linearen Elastizitätstheorie gilt dann die in Abb. 1 angegebene Beziehung

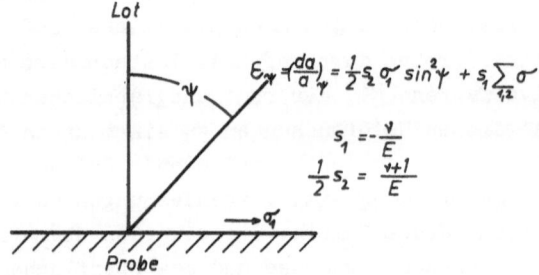

Abb.1. Dehnungs-Spannungszusammenhang in der durch
die Spannung σ_1 und das Lot aufgespannten Ebene bei
elastischer Beanspruchung eines isotropen Körpers

zwischen ε_ψ und σ_1 und der Hauptspannungssumme $\Sigma\sigma$ des ebenen
Spannungszustandes. Trägt man die ermittelten ε_ψ-Werte gegenüber $\sin^2\psi$ auf, ergibt sich ein linearer Zusammenhang mit
einer zur wirksamen Spannung σ_1 proportionalen Steigung

$$\frac{\delta\varepsilon_\psi}{\delta\sin\psi} = \frac{1}{2} \, s_2 \cdot \sigma_1 \qquad (2)$$

Ein positiver Dehnungsanstieg entspricht einer Zugspannung,
ein negativer einer Druckspannung. Bei der Umrechnung der
Gitterdehnungen in Spannungen ist aber zu berücksichtigen,
daß das röntgenographische Verfahren selektiv arbeitet. Da
unter verschiedenem ψ Kristallite immer nur in ganz bestimmten kristallographischen Richtungen vermessen werden, kommt
die elastische Anisotropie der den Vielkristall aufbauenden
Einzelkristallite zur Auswirkung. Dieser Einfluß findet seine
Berücksichtigung, indem anstelle der Voigtschen Konstanten
$1/2$ s_2 und s_1 die röntgenographischen Elastizitätskonstanten
benutzt werden. Sie können entweder nach den Theorien von
VOIGT und REUSS [2] oder KRÖNER [3] mit Hilfe der Einkri-
stallkoeffizienten berechnet werden oder sind durch einen
Zugversuch experimentell zugänglich [4].

5.8.3 Bisherige Befunde

Zur Schilderung der in den bisher vorliegenden Messungen auftretenden Diskrepanzen sind in Abb. 2 Gittereigendehnungsbe-
stimmungen an Eisen wiedergegeben. Abb. 2a bezieht sich auf
Messungen [5] an 25% verformtem Eisen mit 0,018 % C. An {310}-
Ebenen wird eine Dehnungsverteilung mit negativer Steigung
ermittelt. Hingegen wird ein positiver Anstieg der Gitterdehnungsverteilungen an {211}- und {220}-Ebenen gefunden.
Bei formaler Anwendung der elastizitätstheoretischen Beziehungen müßte demnach in gleichen Probenbereichen im einen
Fall von Druckeigenspannungen, in den beiden anderen Fällen
von Zugeigenspannungen gesprochen werden. Abb. 2b gibt ähnliche Messungen an Armcoeisen bei untereinander vergleichbaren Deformationsgraden wieder [4]. Übereinstimmend wird an
allen Netzebenen eine Gitterdehnungsverteilung mit negativer
Steigung ermittelt. Werden die an diesem Eisen experimentell
ermittelten röntgenographischen Elastizitätskonstanten $1/2$
$s_2{}^{rö} = 4,9 \times 10^{-5}$ mm^2/kg für {211}- und {220}-Ebenen sowie
$1/2$ $s_2{}^{rö} = 5,9 \times 10^{-5}$ mm^2/kg für {310}-Ebenen zur Umrechnung der Gitterdehnungsanstiege in Eigenspannungen benutzt,
ergeben sich innerhalb der Meßgenauigkeit gleiche Eigenspannungsbeträge unabhängig vom Netzebenentyp. Diese beiden ge-
gensätzlichen Befunde sind charakterisitsch für die Unterschiede, die in den letzten Jahren an verschiedenen Werk-

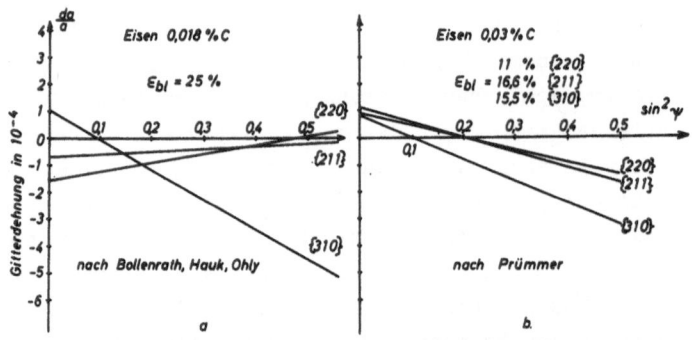

Abb.2. An zugverformtem Eisen an verschiedenen Netzebenen des Ferrits ermittelte Gittereigendehnungsverteilungen

  a) Eisen mit 0,018 %C (nach [5])
  b) Eisen mit 0,03 %C (nach [4])

stoffen mehrfach beobachtet wurden. Ein großer Teil der bestehenden Untersuchungen über die Eigenspannungsausbildung
nach plastischer Deformation von Eisen zeigt Unterschiede
in den Vorzeichen der aus den Gitterdehnungsanstiegen bestimmten Eigenspannungswerte [5-10], wenn verschiedene Netzebenen zur Dehnungsmessung herangezogen werden. Ähnliche Befunde ergeben sich auch bei Messung am Ferrit eines verformten unlegierten Stahles [11]. Gleiche Vorzeichen in den Eigenspannungen bei Messung an verschiedenen Netzebenen ergeben sich hingegen bei einer Reihe weiterer Untersuchungen
[12-15]. An allen Netzebenen identische Eigenspannungen werden jedoch nur erhalten [4], wenn der Einfluß der elastischen
Anisotropie berücksichtigt wird. Die folgenden gezielten Untersuchungen haben den Zweck, die Ursache der geschilderten
Diskrepanzen zu beleuchten.

5.8.4 Versuchsdurchführung, Ergebnisse und Diskussion

Für die Experimente diente Armcoeisen mit 0,03 %C, 0,1 %Si,
0,15 %Mn und je 0,01 %P und S. Das Rohmaterial wurde geglüht,
dann ein Teil davon bei 75 % Querschnittsabnahme gehämmert.
Anschließend wurden zylindrische Versuchsproben herausgear-

beitet, 2 h bei 700°C im Vakuum mit langsamer Ofenabkühlung geglüht und elektrolytisch poliert.

Der Teil der Proben, die gehämmert und spannungsfrei geglüht vorlagen, wurde 23% bleibend zugverformt. In diesem Zustand erfolgten Gittereigendehnungsmessungen in der durch die Probenlängsachse und das Oberflächenlot aufgespannten Ebene mit Chrom-, Kobalt- und Eisenstrahlung jeweils an {211}-, {310}- und {220}-Ebenen. Nach sukzessivem elektrolytischem Abtragen von Oberflächenschichten wurden dieselben Messungen an mehreren tiefer gelegenen Probenbereichen wiederholt. Abb. 3 zeigt zunächst die an der Oberfläche ermittelten Eigendehnungen als Funktion von $\sin^2\psi$. Innerhalb der Meßgenauigkeit wird ein linearer ε_ψ-$\sin^2\psi$-Zusammenhang erhalten. Die Steigungen der Ausgleichsgeraden weichen für die einzelnen Netzebenen stark voneinander ab. Eine Zuordnung von Eigenspannungen mit Hilfe der Elastizitätskonstanten $1/2\ s_2 = 6{,}1\ \times\ 10^{-5}\ mm^2/kg$ ergibt jeweils eine Eigenspannung von -17, -6 und $-1{,}5\ kg/mm^2$ für die {310}-, {211}- und {220}-Ebenen. Als

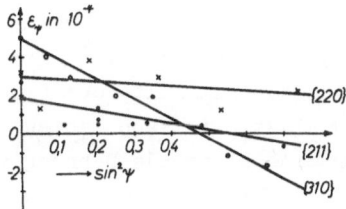

Abb. 3. An Armcoeisen, das vor dem Spannungsfreiglühen bei 70 % Querschnittsabnahme gehämmert wurde, nach Zugverformung an verschiedenen Netzebenen des Ferrits ermittelte Gittereigendehnungsverteilungen und die sich bei formaler Anwendung der elastizitätstheoretischen Beziehungen daraus ergebenden Eigenspannungsbeträge

Funktion des Probenquerschnitts wird der in Abb. 4 gezeigte Verlauf der Eigenspannungen gemessen. Die an den {310}-Ebenen ermittelten Druckeigenspannungen nehmen nach dem Probeninneren schwach ab, die wesentlich kleineren der {211}-Ebenen nähern sich dem Wert Null, ebenso die der {220}-Ebenen mit überwiegendem Zugcharakter.

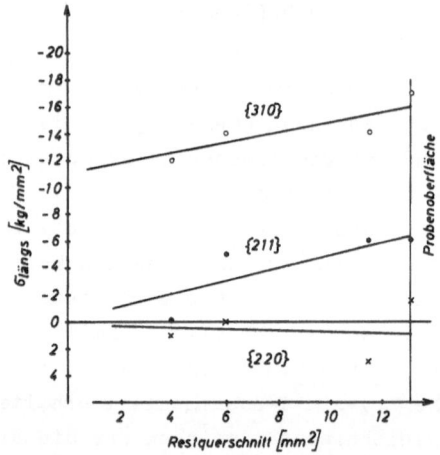

Abb.4. Aus den Gittereigendehnungsverteilungen der
verformten Eisenprobe (Abb.3) berechnete Eigenspannungen in Abhängigkeit vom Probenquerschnitt bei Messung an verschiedenen Netzebenen des Ferrits

Völlig andere Resultate werden an dem ungehämmerten Ausgangsmaterial erhalten. Die Proben wurden 6, 11 und 17 % bleibend
verformt. Es wird stets ein linearer Zusammenhang zwischen
der ermittelten Gitterdehnung und $\sin^2\psi$ mit negativer Steigung erhalten. Die daraus mit Hilfe der früher bestimmten
röntgenographischen Elastizitätskonstanten berechneten Eigenspannungen sind für eine bleibende Dehnung von 6 und 17 % in
Abb. 5 als Funktion des Probenrestquerschnitts für Messungen
an {211}-, {310}- und {220}-Ebenen wiedergegeben [16]. An
allen Netzebenen werden im Vorzeichen und Betrag innerhalb
der Meßgenauigkeit gleiche Eigenspannungswerte erhalten. Bei
6 % plastischer Deformation nehmen die Eigenspannungen nach
dem Probeninneren stark ab, bei 17 % bleibender Dehnung nur
unwesentlich.

Am gleichen Ausgangsmaterial sind also bei verschiedenem
Probenzustand die eingangs geschilderten Unterschiede in den
Gittereigendehnungsverteilungen reproduzierbar. Die beiden
Probenzustände weichen im Texturzustand voneinander ab. Das
Ausgangsmaterial ließ mit Hilfe von Debye-Scherrer-Aufnah-
men keine Textur erkennen. Das gehämmerte Probenmaterial

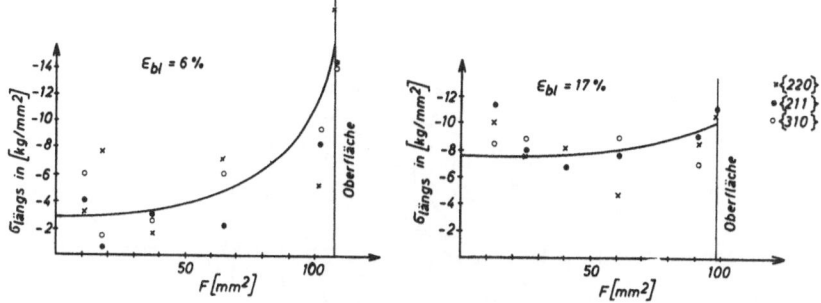

Abb.5. An einer ungehämmerten, texturfreien Eisenprobe nach
verschiedenen Deformationsgraden ermittelte Eigenspannungen
als Funktion des Probenquerschnitts. Innerhalb der Messge-
nauigkeit besteht eine Übereinstimmung zwischen den an {211}-,
{220}- und {310}-Ebenen ermittelten Eigenspannungen

zeigte dagegen eine Textur. Abb. 6 gibt die {110}-Polfigur
des gehämmerten Materials wieder. Sie wurde an einer Fläche,
die den Durchmesser und die Längsachse der Probe enthält,
ermittelt und zeigt, daß die Textur nur schwach ausgeprägt
ist.

Die an diesem Probenmaterial an verschiedenen Netzebenen erhaltenen Gitterdehnungsverteilungen sind zumindest qualitativ
mit dem Teil der Befunde identisch, bei denen an verschiedenen Netzebenen unterschiedliche Vorzeichen in den Eigenspannungen zugverformten Eisens erhalten werden [5-10]. Wenn Unterschiede in den Vorzeichen auftreten, geschieht dies immer so, daß an {310}-Ebenen große Druckeigenspannungen, an
{211}-Ebenen um Null schwankende und an {220}-Ebenen überwiegend kleine Zugspannungen ermittelt werden. Die Versuchsproben der zitierten Untersuchungen lagen jedoch zum Teil
als Rund- und Flachproben mit unterschiedlicher Wärmebe-
handlung und damit verschiedener Textur vor. Ob daraus der
Schluß gezogen werden kann, daß die Art der in den Versuchsproben vorliegenden Textur zumindest qualitativ wenig von
Einfluß auf die Gittereigendehnungsverteilungen zugverformten, texturbehafteten vielkristallinen Eisens ist, muß Ge-
genstand weiterer Untersuchungen sein. Eine Ausnahme bilden
Befunde [17], nach denen die von der Elastizitätstheorie ge-

504

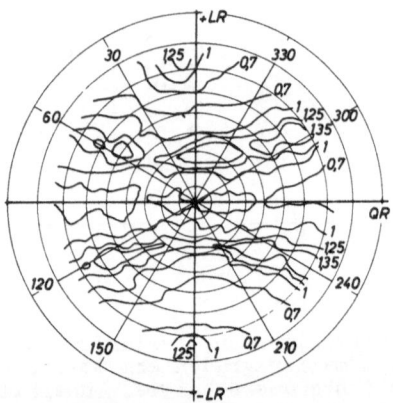

Abb.6. {110}-Polfigur des bei 70 % Querschnittsabnahme gehämmerten und anschließend spannungsfrei geglühten Eisens

forderte Linearität der Gitterdehnungsverteilung mit $\sin^2\psi$
verletzt ist. Innerhalb der hier geschilderten Untersuchungen
trat eine solche Abweichung nicht auf.

Die hier geschilderten Untersuchungen erklären nicht nur die
von verschiedenen Autoren erhaltenen Diskrepanzen in den
Oberflächeneigenspannungen zugverformter Eisenproben. Auch für
die Deutung von Gittereigendehnungen an sukzessiv abgedünn-
ten Proben sind die hier mitgeteilten Befunde von Bedeutung.
Bisher wurde geschlossen, daß eine über den Probenquerschnitt
konstante Gitterdehnungsverteilung ihren Ursprung in Mikro-
spannungen hat. Eine Abnahme der Gitterdehnungsanstiege nach
dem Probeninneren dagegen wurde auf den Einfluß von Makro-
eigenspannungen zurückgeführt. Aus Messungen an texturbehaf-
tetem Eisen kann damit je nach der Wahl der vermessenen Netz-
ebene entweder auf einen Makro- oder einen Mikroeigenspan-
nungszustand geschlossen werden.

An allen Netzebenen identische Befunde werden hingegen an
texturfreiem Eisen erhalten. Dabei treten bei kleinen Deformationsgraden überwiegend Eigenspannungen auf, die sich in
großen Probenbereichen kompensieren. Sie sind eine Folge einer
Fließspannungsinhomogenität zwischen Rand- und Kernbereichen

[18] der Versuchsprobe. Bei größeren Deformationsgraden hin-
gegen sind die über den Probenquerschnitt nahezu unveränder-
lichen Gittereigendehnungsverteilungen eine Folge von Spannungen, die sich in kleinsten Probenbereichen kompensieren.
Hierfür ist die mit plastischer Deformation fortschreitende
Zellbildung ausschlaggebend. Indem eine Kompensation der
Mikroeigenspannungen zwischen Zellinnerem und Zellwand [19]
erfolgt und im wesentlichen nur das relativ versetzungsarme
Zellinnere zur Röntgeninterferenz beiträgt, werden dann vom
Probenort unabhängige Gittereigendehnungsverteilungen ver-
mittelt.

Literatur

1. E. Macherauch und P. Müller: Z.ang.Physik 13(1961)305.

2. H. Möller und G. Martin: Mitt.KWI Eisenforsch. 21(1939)
   261.

3. E. Müller: Dissertation, TH Aachen, 1965.

4. R. Prümmer: Dissertation Universität Karlsruhe, 1967.

5. F. Bollenrath, V. Hauk und W. Ohly: Z.Metallkde. 57(1966)
   464.

6. G.B. Greenough: Nature 160(1947)258.

7. R.I. Garrod und J.H. Auld: Acta Met. 3(1955)190.

8. R.I. Garrod: Nature 165(1950)241.

9. S. Taira und Y. Yoshioka: 7th Japan Congress on Test.Mat.,
   März 1964.

10. T. Tamaru und K. Kojima: J.Soc.Mat.Sci.Japan 16(1967)972.

11. R.E. Ricklefs und W.P. Evans: 15th Ann.Conf., X-Ray
    Analysis, Denver 1966.

12. R.I. Garrod und G.A. Hawkes: Brit.J.appl.Phys. 14(1963)
    422.

13. K. Kolb and E. Macherauch: Arch.Eisenhüttenw. 36(1965)9.

14. G. Faninger: Acta Physica Austriaca XXXII(1966)272.

15. G. Faninger: Z.Metallkde. 58(1967)201.

16. G. Hellwig: Dipl.Arbeit, Universität Karlsruhe, 1968.

17. W. Weidemann: Dissertation TH Aachen, 1966.

18. S. Karashima, R. Prümmer und E. Macherauch: Materialprüf.
    10(1968)262.

19. B.D. Cullity: Trans.Met.Soc.AIME 227(1963)356.